GENETICS, GENOMICS AND BREEDING OF VEGETABLE BRASSICAS

Genetics, Genomics and Breeding of Crop Plants

Series Editor

Chittaranjan Kole
Department of Genetics and Biochemistry
Clemson University
Clemson, SC
USA

Books in this Series:

Published or in Press:

- Jinguo Hu, Gerald Seiler & Chittaranjan Kole: *Sunflower*
- Kristin D. Bilyeu, Milind B. Ratnaparkhe & Chittaranjan Kole: *Soybean*
- Robert Henry & Chittaranjan Kole: *Sugarcane*
- Kevin Folta & Chittaranjan Kole: *Berries*
- Jan Sadowsky & Chittaranjan Kole: *Vegetable Brassicas*
- James M. Bradeen & Chittaranjan Kole: *Potato*
- C.P. Joshi, Stephen DiFazio & Chittaranjan Kole: *Poplar*
- Anne-Françoise Adam-Blondon, José M. Martínez-Zapater & Chittaranjan Kole: *Grapes*
- Christophe Plomion, Jean Bousquet & Chittaranjan Kole: *Conifers*

Books under preparation:

- Dave Edwards, Jacqueline Batley, Isobel Parkin & Chittaranjan Kole: *Oilseed Brassicas*
- Marcelino Pérez de la Vega, Ana María Torres, José Ignacio Cubero & Chittaranjan Kole: *Cool Season Grain Legumes*

GENETICS, GENOMICS AND BREEDING OF VEGETABLE BRASSICAS

Editors

Jan Sadowski

Department of Biotechnology
Institute of Molecular Biology and Biotechnology
Faculty of Biology
Adam Mickiewicz University
Poznań, Poland
and
Laboratory of Functional Genomics
Institute of Plant Genetics
Polish Academy of Sciences
Poznań, Poland

Chittaranjan Kole

Department of Genetics and Biochemistry
Clemson University
Clemson, SC
USA

CRC Press
Taylor & Francis Group
an **informa** business
www.crcpress.com

6000 Broken Sound Parkway, NW
Suite 300, Boca Raton, FL 33487
270 Madison Avenue
New York, NY 10016
2 Park Square, Milton Park
Abingdon, Oxon OX 14 4RN, UK

Science Publishers
Enfield, New Hampshire

Published by Science Publishers, P.O. Box 699, Enfield, NH 03748, USA
An imprint of Edenbridge Ltd., British Channel Islands

E-mail: info@scipub.net Website: www.scipub.net

Marketed and distributed by:

CRC Press
Taylor & Francis Group
an informa business
www.crcpress.com

6000 Broken Sound Parkway, NW
Suite 300, Boca Raton, FL 33487
270 Madison Avenue
New York, NY 10016
2 Park Square, Milton Park
Abingdon, Oxon OX 14 4RN, UK

ISBN 978-1-57808-706-8

Cover Photo: Ginny Booker, Christopher S. Bond, Life Sciences Center, University of Missouri

Library of Congress Cataloging-in-Publication Data

Genetics, genomics and breeding of vegetable Brassicas / editors: Jan
Sadowski, Chittaranjan Kole. -- 1st ed.
p. cm. -- (Genetics, genomics and breeding of crop plants)
Includes bibliographical references and index.
ISBN 978-1-57808-706-8 (hardcover)
1. Brassica--Genetics. 2. Brassica--Breeding. I. Sadowski, Jan. II.
Kole, Chittaranjan. III. Series: Genetics, genomics and breeding of crop
plants.
SB299.B7G46 2011
635'.34--dc22

2010046623

Printed in the United States of America

Dedication

The authors dedicate this monography to Professor César Gómez-Campo, who recently passed away and whose knowledge and remarkable work have been a great inspiration for the *Brassica* community.

Preface to the Series

Genetics, genomics and breeding has emerged as three overlapping and complimentary disciplines for comprehensive and fine-scale analysis of plant genomes and their precise and rapid improvement. While genetics and plant breeding have contributed enormously towards several new concepts and strategies for elucidation of plant genes and genomes as well as development of a huge number of crop varieties with desirable traits, genomics has depicted the chemical nature of genes, gene products and genomes and also provided additional resources for crop improvement.

In today's world, teaching, research, funding, regulation and utilization of plant genetics, genomics and breeding essentially require thorough understanding of their components including classical, biochemical, cytological and molecular genetics; and traditional, molecular, transgenic and genomics-assisted breeding. There are several book volumes and reviews available that cover individually or in combination of a few of these components for the major plants or plant groups; and also on the concepts and strategies for these individual components with examples drawn mainly from the major plants. Therefore, we planned to fill an existing gap with individual book volumes dedicated to the leading crop and model plants with comprehensive deliberations on all the classical, advanced and modern concepts of depiction and improvement of genomes. The success stories and limitations in the different plant species, crop or model, must vary; however, we have tried to include a more or less general outline of the contents of the chapters of the volumes to maintain uniformity as far as possible.

Often genetics, genomics and plant breeding and particularly their complimentary and supplementary disciplines are studied and practiced by people who do not have, and reasonably so, the basic understanding of biology of the plants for which they are contributing. A general description of the plants and their botany would surely instill more interest among them on the plant species they are working for and therefore we presented lucid details on the economic and/or academic importance of the plant(s); historical information on geographical origin and distribution; botanical origin and evolution; available germplasms and gene pools, and genetic and cytogenetic stocks as genetic, genomic and breeding resources; and basic information on taxonomy, habit, habitat, morphology, karyotype, ploidy level and genome size, etc.

Classical genetics and traditional breeding have contributed enormously even by employing the phenotype-to-genotype approach. We included detailed descriptions on these classical efforts such as genetic mapping using morphological, cytological and isozyme markers; and achievements of conventional breeding for desirable and against undesirable traits. Employment of the in vitro culture techniques such as micro- and megaspore culture, and somatic mutation and hybridization, has also been enumerated. In addition, an assessment of the achievements and limitations of the basic genetics and conventional breeding efforts has been presented.

It is a hard truth that in many instances we depend too much on a few advanced technologies, we are trained in, for creating and using novel or alien genes but forget the infinite wealth of desirable genes in the indigenous cultivars and wild allied species besides the available germplasms in national and international institutes or centers. Exploring as broad as possible natural genetic diversity not only provides information on availability of target donor genes but also on genetically divergent genotypes, botanical varieties, subspecies, species and even genera to be used as potential parents in crosses to realize optimum genetic polymorphism required for mapping and breeding. Genetic divergence has been evaluated using the available tools at a particular point of time. We included discussions on phenotype-based strategies employing morphological markers, genotype-based strategies employing molecular markers; the statistical procedures utilized; their utilities for evaluation of genetic divergence among genotypes, local landraces, species and genera; and also on the effects of breeding pedigrees and geographical locations on the degree of genetic diversity.

Association mapping using molecular markers is a recent strategy to utilize the natural genetic variability to detect marker-trait association and to validate the genomic locations of genes, particularly those controlling the quantitative traits. Association mapping has been employed effectively in genetic studies in human and other animal models and those have inspired the plant scientists to take advantage of this tool. We included examples of its use and implication in some of the volumes that devote to the plants for which this technique has been successfully employed for assessment of the degree of linkage disequilibrium related to a particular gene or genome, and for germplasm enhancement.

Genetic linkage mapping using molecular markers have been discussed in many books, reviews and book series. However, in this series, genetic mapping has been discussed at length with more elaborations and examples on diverse markers including the anonymous type 2 markers such as RFLPs, RAPDs, AFLPs, etc. and the gene-specific type 1 markers such as EST-SSRs, SNPs, etc.; various mapping populations including F_2, backcross, recombinant inbred, doubled haploid, near-isogenic and pseudotestcross; computer software including MapMaker, JoinMap, etc. used; and different types of genetic maps including preliminary, high-resolution, high-density, saturated, reference, consensus and integrated developed so far.

Mapping of simply inherited traits and quantitative traits controlled by oligogenes and polygenes, respectively has been deliberated in the earlier literature crop-wise or crop group-wise. However, more detailed information on mapping or tagging oligogenes by linkage mapping or bulked segregant analysis, mapping polygenes by QTL analysis, and different computer software employed such as MapMaker, JoinMap, QTL Cartographer, Map Manager, etc. for these purposes have been discussed at more depth in the present volumes.

The strategies and achievements of marker-assisted or molecular breeding have been discussed in a few books and reviews earlier. However, those mostly deliberated on the general aspects with examples drawn mainly from major plants. In this series, we included comprehensive descriptions on the use of molecular markers for germplasm characterization, detection and maintenance of distinctiveness, uniformity and stability of genotypes, introgression and pyramiding of genes. We have also included elucidations on the strategies and achievements of transgenic breeding for developing genotypes particularly with resistance to herbicide, biotic and abiotic stresses; for biofuel production, biopharming, phytoremediation; and also for producing resources for functional genomics.

A number of desirable genes and QTLs have been cloned in plants since 1992 and 2000, respectively using different strategies, mainly positional cloning and transposon tagging. We included enumeration of these and other strategies for isolation of genes and QTLs, testing of their expression and their effective utilization in the relevant volumes.

Physical maps and integrated physical-genetic maps are now available in most of the leading crop and model plants owing mainly to the BAC, YAC, EST and cDNA libraries. Similar libraries and other required genomic resources have also been developed for the remaining crops. We have devoted a section on the library development and sequencing of these resources; detection, validation and utilization of gene-based molecular markers; and impact of new generation sequencing technologies on structural genomics.

As mentioned earlier, whole genome sequencing has been completed in one model plant (Arabidopsis) and seven economic plants (rice, poplar, peach, papaya, grapes, soybean and sorghum) and is progressing in an array of model and economic plants. Advent of massively parallel DNA sequencing using 454-pyrosequencing, Solexa Genome Analyzer, SOLiD system, Heliscope and SMRT have facilitated whole genome sequencing in many other plants more rapidly, cheaply and precisely. We have included extensive coverage on the level (national or international) of collaboration and the strategies and status of whole genome sequencing in plants for which sequencing efforts have been completed or are progressing currently. We have also included critical assessment of the impact of these genome initiatives in the respective volumes.

Comparative genome mapping based on molecular markers and map positions of genes and QTLs practiced during the last two decades of the last century provided answers to many basic questions related to evolution, origin and phylogenetic relationship of close plant taxa. Enrichment of genomic resources has reinforced the study of genome homology and synteny of genes among plants not only in the same family but also of taxonomically distant families. Comparative genomics is not only delivering answers to the questions of academic interest but also providing many candidate genes for plant genetic improvement.

The 'central dogma' enunciated in 1958 provided a simple picture of gene function—gene to mRNA to transcripts to proteins (enzymes) to metabolites. The enormous amount of information generated on characterization of transcripts, proteins and metabolites now have led to the emergence of individual disciplines including functional genomics, transcriptomics, proteomics and metabolomics. Although all of them ultimately strengthen the analysis and improvement of a genome, they deserve individual deliberations for each plant species. For example, microarrays, SAGE, MPSS for transcriptome analysis; and 2D gel electrophoresis, MALDI, NMR, MS for proteomics and metabolomics studies require elaboration. Besides transcriptome, proteome or metabolome QTL mapping and application of transcriptomics, proteomics and metabolomics in genomics-assisted breeding are frontier fields now. We included discussions on them in the relevant volumes.

The databases for storage, search and utilization on the genomes, genes, gene products and their sequences are growing enormously in each second and they require robust bioinformatics tools plant-wise and purpose-wise. We included a section on databases on the gene and genomes, gene expression, comparative genomes, molecular marker and genetic maps, protein and metabolomes, and their integration.

Notwithstanding the progress made so far, each crop or model plant species requires more pragmatic retrospect. For the model plants we need to answer how much they have been utilized to answer the basic questions of genetics and genomics as compared to other wild and domesticated species. For the economic plants we need to answer as to whether they have been genetically tailored perfectly for expanded geographical regions and current requirements for green fuel, plant-based bioproducts and for improvements of ecology and environment. These futuristic explanations have been addressed finally in the volumes.

We are aware of exclusions of some plants for which we have comprehensive compilations on genetics, genomics and breeding in hard copy or digital format and also some other plants which will have enough achievements to claim for individual book volume only in distant future. However, we feel satisfied that we could present comprehensive deliberations on genetics, genomics and breeding of 30 model and economic plants, and their groups in a few cases, in this series. I personally feel also happy that I could work with many internationally celebrated scientists who edited the book volumes on the leading plants and plant groups and included chapters authored by many scientists reputed globally for their contributions on the concerned plant or plant group.

We paid serious attention to reviewing, revising and updating of the manuscripts of all the chapters of this book series, but some technical and formatting mistakes will remain for sure. As the series editor, I take complete responsibility for all these mistakes and will look forward to the readers for corrections of these mistakes and also for their suggestions for further improvement of the volumes and the series so that future editions can serve better the purposes of the students, scientists, industries, and the society of this and future generations.

Science publishers, Inc. has been serving the requirements of science and society for a long time with publications of books devoted to advanced concepts, strategies, tools, methodologies and achievements of various science disciplines. Myself as the editor and also on behalf of the volume editors, chapter authors and the ultimate beneficiaries of the volumes take this opportunity to acknowledge the publisher for presenting these books that could be useful for teaching, research and extension of genetics, genomics and breeding.

Chittaranjan Kole

Preface to the Volume

Vegetable Brassicas are of economic importance worldwide.They alongwith cereals provide the basis of world supplies. Vegetable Brassicas contain beneficial phytochemicals—anticancerogenic compounds including, minerals, vitamins, and fiber. *Brassica* species offer unique opportunities to widen our knowledge about plant growth, development and rapid phenotypic evolution. Among the *Brassica* species, *Brassica oleracea* and *B. rapa* make up most of the vegetable crops and offer numerous choices of edible forms within each species. On the basis of their origin, the greatest genetic and phenotypic variability of *B. oleracea* is observed in Europe, while Asia is the main area of diversification of vegetable *B. rapa* crops.

In the last century, the Brassicaceae family members acquired new roles as ideal model plants in basic scientific research. Furthermore, interspecific hybridization among *Brassica* spp. is relatively easy compared to other crucifer species and this character has been successfully applied to obtain synthetic amphidiploids, improve Brassica vegetable crops, study intergenomic relationships and introduce genetic variability in the genus. Thus, the genetic diversity of *Brassica* species has helped the development of many lines of research on molecular biology and plant genetics.

The phylogenetic relationship of *Arabidopsis thaliana* to *Brassica* crops offers an exceptional opportunity to benefit from the information and technology derived from the Arabidopsis Genome Project. It is already recognized, that the Arabidopsis genome sequence provides a valuable resource for identification of genes that may account for the genetic control of complex traits in Brassica. Besides, the small size of the *Brassica* genomes and their close relationship to *Arabidopsis*, provide excellent models for whole-genome sequencing. Evidently, there is potential to greatly increase our understanding of the *Brassica* biology by application of emerging genomic technologies.

This volume has been written by specialists with extensive experience in the genetics, molecular breeding and genomics of these very special and important crop plants. Chapter 1 provides basic information on vegetable Brassica crops. Breeding and genetic analysis are presented in Chapter 2. Diversity and molecular taxonomy is described in Chapter 3. Strategies of molecular linkage maps construction for vegetable Brassicas are introduced in Chapter 4, followed by molecular mapping of complex traits in Chapter 5. In Chapter 6, the paleopolyploidy nature and extensive cytogenetic studies of vegetable Brassicas are presented. Strategies and current status of *Brassica rapa* genome sequencing can be found in Chapter 7. Results of exploring the *Brassica* genome architecture are presented in Chapter 8. Transcriptomics and related topics are outlined in Chapter 9 and the growing importance of proteomics and metabolomics in Chapter 10. Finally, the field of bioinformatics, public databases and a corresponding network of tools is described in Chapter 11.

We hope that recent achievements and new technologies presented in this book will further support Brassica research development and improvement of vegetable Brassica crops.

Jan Sadowski
Chittaranjan Kole

Contents

List of Contributors

Danuta Babula-Skowrońska
Institute of Plant Genetics, Polish Academy of Sciences, Strzeszyńska 34, 60-479 Poznań, Poland.
Email: *dbab@igr.poznan.pl*

Jacqueline Batley
School of Land, Crop and Food Sciences and, ARC Centre of Excellence for Integrative Legume Research, University of Queensland, Brisbane, QLD 4072, Australia.
Email: *j.batley@uq.edu.au*

Beom-Seok Park
Genomics Division, Department of Agricultural Biotechnology, National Academy of Agricultural Science, RDA, 150 Suin-ro, Gwonseon-gu, Suwon 441-707, Korea.
Email: *pbeom@korea.go.kr*

Guusje Bonnema
Laboratory of Plant Breeding, Wageningen University, 6708PB, Wageningen, The Netherlands.
Email: *Guusje.Bonnema@wur.nl*

Zoran Boskovic
Australian Centre for Plant Functional Genomics, School of Land, Crop and Food Sciences, Institute of Molecular Biosciences, University of Queensland, Brisbane, QLD 4072, Australia.
Email: *Z.Boskovic@uq.edu.au*

Dunia Pino Del Carpio
Laboratory of Plant Breeding, Wageningen University, 6708PB, Wageningen, The Netherlands.
Email: *Dunia.Pino@wur.nl*

María Elena Cartea
Department of Plant Genetics, Misión Biológica de Gallicia, Spanish Council for Scientific Research (CSIC), P.O. Box 28, E-36080 Pontevedra, Spain.
Email: *cartea@mbg.cesga.es*

Agata Cieśla
Institute of Plant Genetics, Polish Academy of Sciences, Strzeszyńska 34, 60-479 Poznań, Poland.
Email: *agataciesla@gmail.com*

Wallace A. Cowling
School of Plant Biology and Institute of Agriculture, The University of Western Australia, 35 Stirling Highway, Crawley WA 6009, Australia.
Email: *wcowling@cyllene.uwa.edu.au*

Chris Duran
Australian Centre for Plant Functional Genomics and School of Land, Crop and Food Sciences, University of Queenslands, Brisbane, QLD 4072, Australia.
Email: *c.duran@uq.edu.au*

David Edwards
Australian Centre for Plant Functional Genomics and School of Land, Crop and Food Sciences, University of Queenslands, Brisbane, QLD 4072, Australia.
Email: *dave.edwards@uq.edu.au*

Marta Francisco
Department of Plant Genetics, Misión Biológica de Gallicia, Spanish Council for Scientific Research (CSIC), P.O. Box 28, E-36080 Pontevedra, Spain.
Email: *mfrancisco@mbg.cesga es*

Waheeb K. Heneen
Department of Plant Breeding and Biotechnology, Swedish University of Agricultural Sciences, P.O.B. 101; Alnarp 23053 Sweden.
Email: *Waheeb.Heneen@ltj.slu.se*

Izabela Ibragimow
Department of Biotechnology, Institute of Molecular Biology and Biotechnology, Adam Mickiewicz University, Umultowska 89, 61-614 Poznań, Poland.
Email: *Ibiza@amu.edu.pl*

Małgorzata Kaczmarek
Institute of Plant Genetics, Polish Academy of Sciences, Strzeszyńska 34, 60-479 Poznań, Poland.
Email: *mrum@igr.poznan.pl*

Krzysztof Krasowski
Department of Biotechnology, Institute of Molecular Biology and Biotechnology, Adam Mickiewicz University, Umultowska 89, 61-614 Poznań, Poland.
Email: *krasowsk@amu.edu.pl*

Soo-Jin Kwon
Genomics Division, Department of Agricultural Biotechnology, National Academy of Agricultural Science, RDA, 150 Suin-ro, Gwonseon-gu Suwon 441-707, Korea.
Email: *sjkwon@rda.go.kr*

Margarita Lema
Department of Plant Genetics, Misión Biológica de Gallicia, Spanish Council for Scientific Research (CSIC), P.O. Box 28, E-36080 Pontevedra, Spain.
Email: *mlema@mbg.cesga.es*

Yong Pyo Lim
Department of Horticulture, Chungnam National Univeristy, Gung-Dong, Yuseong-Gu, Daejeon 305-764, Korea.
Email: *yplim@cnu.ac.kr*

Agnieszka Ludwików
Department of Biotechnology, Institute of Molecular Biology and Biotechnology, Adam Mickiewicz University, Umultowska 89, 61-614 Poznań, Poland.
Email: *ludwika@amu.edu.pl*

Martin A. Lysak
Department of Functional Genomics and Proteomics, Masaryk University, Kamenice 5, 625 00 Brno Czech Republic.
Email: *lysak@sci.muni.cz*

Lucyna H. Misztal
Department of Biotechnology, Institute of Molecular Biology and Biotechnology, Adam Mickiewicz University, Umultowska 89, 61-614 Poznań, Poland.
Email: *lmisztal@amu.edu.pl*

Jeong-Hwan Mun
Genomics Division, Department of Agricultural Biotechnology, National Academy of Agricultural Science, RDA, 150 Suin-ro, Gwonseon-gu, Suwon 441-707, Korea.
Email: *munjh@rda.go.kr*

Matthew N. Nelson
School of Plant Biology and Institute of Agriculture, The University of Western Australia, 35 Stirling Highway, Crawley WA 6009, Australia.
Email: *mnelson@plants.uwa.edu.au*

Suhyoung Park
Vegetable Division, National Institute of Horticultural and Herbal Science, RDA, Suwon 441-440, Korea.
Email: *psh@korea.kr*

Isobel A. Parkin
Agriculture and Agri-Food Canada Saskatoon, Research Centre 107 Science Place Saskatoon, SK, S7N 0X2, Canada.
Email: *isobel.parkin@agr.gc.ca*

Nirala Ramchiary
Department of Horticulture, Chungnam National Univeristy, Gung-Dong, Yuseong-Gu, Daejeon 305-764, Korea.
Email: *nrudsc@gmail.com*

Stephen J. Robinson
Agriculture and Agri-Food Canada Saskatoon, Research Centre 107 Science Place Saskatoon, SK, S7N 0X2, Canada.
Email: *RobinsonS@agr.gc.ca*

Jan Sadowski
Department of Biotechnology, Institute of Molecular Biology and Biotechnology, Adam Mickiewicz University, Umultowska 89, 61-614 Poznan, Poland and Institute of Plant Genetics, Polish Academy of Sciences, Strzeszyńska 34, 60-479 Poznań, Poland.
Email: *jsad@amu.edu.pl*

Pablo Velasco
Department of Plant Genetics, Misión Biológica de Gallicia, Spanish Council for Scientific Research (CSIC), P.O. Box 28, E-36080 Pontevedra, Spain.
Email: *pvelasco@mbg.cesga.es*

Tae-Jin Yang
Department of Plant Science, Plant Genomics and Breeding Institute, and Research Institute for Agriculture and Life Sciences, College of Agriculture and Life Sciences, Seoul National University, Seoul, 151-921, Korea.
Email: *tjyang@snu.ac.kr*

Leo Zeef
Bioinformatics Core Facility Faculty of Life Sciences, Michael Smith Building B.1083, The University of Manchester, Oxford Rd, Manchester M13 9PT, United Kingdom.
Email: *Leo.Zeef@manchester.ac.uk*

Jianjun Zhao
Laboratory of Plant Breeding, Wageningen University, 6708PB, Wageningen, The Netherlands.
Email: *jianjun.zhao@wur.nl*

Piotr A. Ziolkowski
Department of Biotechnology, Institute of Molecular Biology and Biotechnology, Adam Mickiewicz University, Umultowska 89, 61-614 Poznań, Poland.
Email: *pzio@amu.edu.pl*

Abbreviations

2-DE	Two dimensional electrophoresis
2-D DIGE	Two-dimensional fluorescence difference gel electrophoresis
3-MSP	3-Methylsulfinylpropyl
4-MSB	4-Methylsulfinylbutyl
ABTS	2,2′-Azino-bis(3-ethylbenzthiazoline-6-sulphonic acid
AC	Affinity chromatography
ACK	Ancestral crucifer karyotype
ADH	Alcohol dehydrogenase
AdVaB	Adding Value to the UK Brassica Crop Science Community Consortium
AFLP	Amplified fragment length polymorphism
AgNOR	silver staining of nucleolar organizer region
APS	Acid phosphatase
ASTRA	*Brassica* ASTRA database
ATH1	GeneChip Arabidopsis ATH1 Genome Array
ATP	Adenosine triphosphate
BAC	Bacterial artificial chromosome
BAGI	*Brassica*/Arabidopsis Genomics Initiative
BBSRC	Biotechnology and Biological Sciences Research Council
BC	Backcross
BES	BAC-end sequences
BG21	Biogreen21 Program
BGG	*Brassica* Genome Gateway
BLAST	Basic Local Alignment Search Tool
BNK1	*Brassica napus* kinase 1
Bras-EDB	European *Brassica* Database
BrGP	*Brassica rapa* Genome Project
BrGSP	Multinational *Brassica rapa* Genome Sequencing Project
Ca	Calcium
CAGS	Conserved *Arabidopsis* genome sequence
CAPS	Cleaved amplified polymorphic sequence
CCP	Comparative chromosome painting
cDNA	Complementary DNA

CDS	Coding sequence
CentBr	*Brassica rapa* centromer satellite repeat sequences
CIM	Composite interval mapping
cM	CentiMorgan
CMS	Cytoplasmic male sterility
CO	*CONSTANS*
COT	Clutamateoxaloacetate transarninase
CP	Condensation pattern
CRBs	Centromeric retrotransposons of *Brassica*
CrGC	Crucifer Genetics Cooperative
CSHL	Cold Spring Harbor Laboratory
CYP79B2	Cytochrome P450
DAPI	4′, 6-Diamidino-2-phenylindole
DArT	Diversity arrays technology
dbGaP	Genotype and phenotype database
DBP	Di-N-Butyl phthalate
DFFS	Diversity fixed foundation sets
DH	Doubled haploid
DIM	3,3′-Diindolylmethane
DMD	Digital Micromirror Device
DNA	Deoxyribonucleic acid
EccDNA	Extrachromosomal circular DNA
ECP/GR	European Cooperative Program for Crop Genetic Resources Networks
EMS	Ethylmethane sulfonate
EQTL	Expression QTL
ESI	Electrospray ionization
EST	Expressed sequence tag
ET	Ethylene
FAFLP	Fluorescent amplified fragment length polymorphism
FAO	Food and Agricultural Organization
FISH	Fluorescence *in situ* hybridization
FLC	*FLOWERING LOCUS C*
FT	*FLOWERING LOCUS T*
GBIF	Global Biodiversity Information Facility
gDNA	Genomic DNA
GEO	Gene Expression Omnibus database
GFP	Green fluorescent protein
GISH	Genomic *in situ* hybridisation
GLS	Glucosinolates
GOT	Glutamateoxaloacetate transarninase
GSS	Genome survey sequences
HICF	High-information content fingerprinting

HPLC	High-pressure liquid chromatography
HPLC/DAD	HPLC with diode array detector
HRI	Human Resources International
HTGS	High-throughput genome sequence
I3C	Indole-3-carbinol
IAA	Indole-3-acetic acid
IBPGR	International Board for Plant Genetic Resources
IG	Indole methyl glucosinolate
IMSORB	Integrated Market System for Oilseed Rape Breeding
In/Del	Insertion/deletion
ISSR	Inter simple sequence repeat
ITCs	Isothiocyanates
iTRAQ	Isobaric tag for relative and absolute quantitation
ITS	Internal transcribed spacer
JCVI	J. Craig Venter Institute
JIC	John Innes Center
Kb	kilo base
KBGP	Korea *Brassica* Genomics Project
LAP	Leucine aminopeptidase
LC	Liquid chromatography
LC/MS	Liquid chromatography/mass spectrometry
LD	Linkage disequilibrium
LF	*LEAFY* locus
LFY	*LEAFY* locus
LG	Linkage group
LINE	Long interspersed nuclear element
LS	Least-square
LTR	Long terminal repeats
MALDI	Matrix-assisted laser desorption ionization
MALDI-TOF	Matrix-assisted laser desorption/ionization time of flight mass spectrometry
MAS	Marker-assisted selection
MBGB	Multinational Brassica Genome Project
MeJA/JA	Methyl jasmonate/jasmonic acid
Mg	magnesium
MIAME	Minimum information about a microarray experiment
MIM	Multiple interval mapping
MJ	Methyl jasmonate
ML	Maximum-likelihood method
MLPK	*M* locus protein kinase
MPSS	Massively parallel signature sequencing
MS	Mass-spectrometry
MS/MS	Tandem mass-spectrometry

MYA	Million years ago
NCBI	National Center for Biotechnology Information
NIL	Nearly isogenic line
NIHHS	National Institute of Horticultural & Herbal Science
NMR	Nuclear magnetic resonance
NNK	Nicotine-derived nitrosamine ketone
NNK	Nicotine-derived nitrosamine ketone, 4-(methylnitrosamino)-1-(3-pyridyl)-1-butanone
NOR	Nucleolar organizer region
NovoHMM	Hidden Markov Model for *de novo* peptide sequencing
OREGIN	Defra funded Oilseed Rape Genetic Improvement Network
ORF	Open reading frame
PAC	P1-derived artificial chromosome
PAGE	Polyacrylamide gel electrophoresis
PC1	Pack choi of group 1
PC-3	Prostate cancer
PCK	Proto-calepineae karyotype
PCR	Polymerase chain reaction
PCRBr	Peri-centromeric retrotransposons of *Brassica rapa*
PFGE	Pulsed field gel electrophoresis
PGI	Phosphoglucoisomerase
PGM	Phosphoglucomutase
PGRU	Plant Genetic Resources Unit
PM	Perfect-match probe
PMF	Peptide mass fingerprinting
PR1a	Pathogenesis-related protein 1a
PTM	Posttranslational modification
QPCR	Real-time quantitative PCR
QTL	Quantitative trait loci
QTN	Quantitative trait nucleotide
RAPD	Random(ly) amplified polymorphic DNA
RCBr	Rapid-cycling *Brassica*
rDNA	Ribosomal DNA
RFLP	Restriction fragment length polymorphism
RIL	Recombinant inbred line
ROS	Reactive oxygen species
RP-HPLC	Reverse phase chromatography
RPS2	Resistance protein
SA	Salicylic acid
SAGE	Serial analysis of gene expression
SCAR	Sequence-characterized polymorphic DNA
SCX	Strong cation exchange

SDS-PAGE	Sodium dodecyl sulfate polyacrylamide gel electrophoresis
SFP	Single feature polymorphism
SI	Self-incompatibility
SINE	Short interspersed nuclear element
siRNA	Small interfering RNA
SLG12	S locus glycoprotein
SNF	Sulforaphane
SNP	Single nucleotide polymorphism
SRAP	Sequence-related amplified polymorphism
SSH	Suppression subtractive hybridization
SSR	Simple sequence repeat
STS	Sequence tagged site
STS-PCR	Sequence tagged site-polymerase chain reaction
TAC	Transformation-competent artificial chromosome
TAIR	Arabidopsis Information Resource
TE	Transposable element
TIGR	The Institute for Genomic Research
TOF	Time of flight
TR	Terminal repeat
TRIM	Terminal-repeat retrotransposons in miniature
TuMV	Turnip mosaic virus
UK	United Kingdom
UPGMA	Unweighted pair-group method with arithmetic mean
UPM	Universidad Politécnica de Madrid
US	United States
VIR	Vavilov Research Institute for Plant Industry
VRN2	*VERNALIZATION2*
WFP	Wisconsin fast plant
WGD	Whole-genome duplication
WGS	Whole genome shotgun
XSP30	Xylem sap protein
YAC	Yeast artificial chromosome

1

Basic Information on Vegetable Brassica Crops

María Elena Cartea, Margarita Lema, Marta Francisco,* and *Pablo Velasco*

ABSTRACT

The *Brassica* genus belongs to the Brassicaceae family and is economically the most important genus within the tribe Brassiceae, consisting of 37 different species. This genus includes a group of six interrelated species of worldwide economic importance, three diploid: *Brassica nigra, Brassica oleracea,* and *Brassica rapa* and three amphidiploid species: *Brassica carinata, Brassica juncea,* and *Brassica napus*. The genus is categorized into oilseed, forage, condiment and vegetable crops by using their buds, inflorescences, leaves, roots, seeds and stems. *Brassica oleracea* and *B. rapa* include most of the vegetable Brassica crops and they display many choices of edible forms within each species. The greatest genetic and phenotypic variability of *B. oleracea* is found in Europe, while Asia represents the main area of diversification of vegetable *B. rapa* crops. Vegetable Brassicas are of great economic importance throughout the world. Currently, Brassica crops together with cereals represent the basis of world supplies. Vegetable Brassica crops have the nutritional characteristics of other vegetable crops, as for example a low fat and protein content and a high value in vitamins, fiber and minerals. Besides, they show high quantities of phenolic compounds and have a singular kind of compounds which differentiate these crops from other vegetables; these compounds are called glucosinolates. These compounds gave Brassicas their antioxidant and anticarcinogenic properties. Finally, the remarkable morphological diversity of vegetable *Brassica* species and their relatives offers valuable opportunities to increase our knowledge about plant growth and development and our understanding of rapid phenotypic evolution. Moreover, the genetic diversity displayed by *Brassica* species has allowed the development of

---

Department of Plant Genetics, Misión Biológica de Galicia, Spanish Council for Scientific Research (CSIC), P.O. Box 28, E-36080 Pontevedra, Spain.
*Corresponding author: *ecartea@mbg.cesga.es*

new crop types but also the substantial increase on the knowledge and research on molecular biology and plant genetics. In the 20th century, members of the Brassicaceae family acquired new roles as an ideal model plant for their use in basic scientific research and subsequently as educational tool.

Keywords: Breeding, Genetics, Germplasm, Nutritional value, Plant model, Taxonomy

1.1 Taxonomy

The *Brassica* genus belongs to the Brassicaceae family (previously known as Cruciferae) and economically is the most important genus within the tribe Brassiceae, consisting of 37 different species. The taxonomy of this genus is complex. Gómez-Campo (1999) presented a complete classification of the genus *Brassica* and its allied genera, indicating subgenera, sections, species and subspecies.

In that classification, Gómez-Campo subdivided the *Brassica* genus in two subgenera, i.e. *Brassica* and *Brassicaria*. Subsequently, the same author separated these two subgenera into two genera, i.e. *Brassica* and *Guenthera* (= subgenus *Brassicaria*), as they are morphologically very distinct. This separation is mainly based on the stylar portion of the pistils always without seed primordia and other traits related to the stem and leaves (Gómez-Campo 2003). He suggested including the species belonging to subgenus *Brassicaria* under the generic denomination *Guenthera* Andr. in Bess. Later on, molecular studies (Warwick and Sauder 2005) confirmed that *Guenthera* should be separated from *Brassica*. Still, the main problem would involve the classification into sections within *Brassica* itself. According to the abovementioned findings, an updated classification of the *Brassica* genus is shown in Table 1-1.

The *Brassica* genus includes a group of six interrelated species of worldwide economic importance. U (1935) studied the cytology of the genus and established the relationships among the genomes of the six species. The three diploid *Brassica* species, *Brassica nigra* (L.) Koch ($2n = 16$), *Brassica oleracea* L. ($2n = 18$) and *Brassica rapa* L. ($2n = 20$), form the classic Triangle of U (see Fig. 6.1 in Chapter 6). In nature, these species have hybridized in different combinations to give rise to the three amphidiploid species, namely *Brassica carinata* A. Braun ($2n = 4x = 34$), *Brassica juncea* (L.) Czern. ($2n = 4x = 36$) and *Brassica napus* L. ($2n = 4x = 38$). Genomes of *B. rapa*, *B. nigra* and *B. oleracea* have been named A, B and C, respectively. Therefore, the resulting amphidiploids become AB, AC and BC for *B. juncea*, *B. napus* and *B. carinata*, respectively. These species have complex genomes, highly duplicated with intra- and inter-genomic conservation of Linkage blocks, which permit homoeologous recombination (Prakash and Hinata 1980).

Table 1-1 A taxonomic classification of *Brassica* genus, according to Gómez-Campo (1999, 2003).

Section *Brassica*

- *B. oleracea* L.
- *B. montana* Pourret
- *B. incana* Ten. subsp. *vncana*
  - *B. i.* subsp. *cazzae* (Ginz. and Teyb.) Trinajstic
- *B. villosa* Biv. subsp. *villosa*
  - *B. v.* subsp. *bivoniana* (Mazzola and Raimondo) R. and M.
  - *B. v.* subsp. *drepanensis* (Caruel) Raimondo and Mazzola
  - *B. v.* subsp. *tinei* (Lojac.) Raimondo and Mazzola
- *B. rupestris* Rafin subsp. *rupestris*
  - *B. r.* subsp. *brevisiliqua* Raimondo and Mazzola
  - *B. r.* subsp. *hispida* Raimondo and Mazzola
- *B. macrocarpa* Guss.
- *B. insularis* Moris
- *B. cretica* Lam. subsp. *cretica*
  - *B. c.* subsp. *aegaea* (Heldr. and Hal.) Snogerup *et al.*
  - *B. c.* subsp. *laconica* Gustafsson and Snogerup
- *B. botteri* Vis. subsp. *botteri*
  - *B. b.* subsp. *mollis* (Vis.) Trinajstic
- *B. hilarionis* Post.
- *B. carinata* Braun
- *B. balearica* Pers.

Section *Rapa* (Miller) Salmeen

- *B. rapa* L. subsp. *rapa*
  - *B. r.* subsp. *campestris* (L.) Clapman
  - *B. r.* subsp. *chinensis* (L.) Hanelt
  - *B. r.* subsp. *dichotoma* (Roxb.) Hanelt
  - *B. r.* subsp. *narinosa* (Bailey) Hanelt
  - *B. r.* subsp. *nipposinica* (Bailey) Hanelt
  - *B. r.* subsp. *pekinensis* (Lour.) Hanelt
  - *B. r.* subsp. *trilocularis* (Roxb.) Hanelt
- *B. napus* L.
- *B. juncea* (L.) Czern.

Section *Micropodium* DC.

- *B. fruticulosa* Cyr. subsp. *fruticulosa*
  - *B. f.* subsp. *djafarensis* Blanco and Matarranz
  - *B. f.* subsp. *dolichocarpa* Emberger and Maire
  - *B. f.* subsp. *glaberrima* (Pomel) Maire
  - *B. f.* subsp. *mauritanica* (Cosson) Maire
  - *B. f.* subsp. *numidica* Maire
  - *B. f.* subsp. *pomeliana* Maire
  - *B. f.* subsp. *radicata* (Desf.) Maire
- *B. nigra* (L.) Koch
- *B. cossoniana* Boiss. and Reuter
- *B. spinescens* Pomel
- *B. maurorum* Durieu
- *B. procumbens* (Poiret) O.E. Schulz
- *B. cadmea* O.E. Schulz
- *B. desertii* Danin and Hedge

Table 1-1 contd...

Table 1-1 contd...

Section *Brassicoides* Boiss.
B. deflexa Boiss.
Section *Sinapistrum* Willkomm
B. barrelieri (L.) Janka *B. oxyrrhina* Coss. *B. tournefortii* Gouan

The genus is categorized into oilseed, forage, condiment and vegetable crops by using their buds, inflorescences, leaves, roots, seeds and stems. The same species can be utilized for several uses according to different forms or types. Four species, *B. oleracea, B. rapa, B. napus* and *B. juncea,* include crops, which have a horticultural use. The principal vegetable species is *B. oleracea,* which includes vegetable and forage forms, such as kale, cabbage, Broccoli, brussels sprouts, cauliflower and others; *B. rapa* includes vegetable forms, such as turnip, Chinese cabbage, and pak choi, along with forage and oilseed types; *B. napus* crops are mainly used as oilseed (rapeseed), although forage and vegetable types as leaf rape and "nabicol" are also included; finally, the mustard group is formed by three species, *B. carinata, B. nigra* and *B. juncea,* which are mainly used as a condiment because of their seeds although leaves of *B. juncea* are also consumed as vegetables in Asian countries (Nishi 1980; Table 1-2). *Brassica oleracea* and *B. rapa* include most of the vegetable Brassica crops and they display many choices of edible forms within each species. The greatest genetic and phenotypic variability of *B. oleracea* is found in Europe, while Asia represents the main area of diversification of vegetable *B. rapa* crops. They include the enlarged single apical bud of cabbage, the many axillary buds of Brussels sprouts, the arrested inflorescences and floral meristems of cauliflower and broccoli, the swollen stems of kohlrabi, the enlarged root of turnip and the enlarged and twisted leaves of pak choi and Chinese cabbage (Kalloo and Bergh 1993).

1.2 Economic Importance

FAO Statistics (FAOSTAT 2009) about vegetable Brassicas comprise "cabbage and other Brassicas" on the one hand and "cauliflower and broccoli" on the other hand, except for consumption data. It must be pointed out that FAO must include red, white and savoy cabbage, Chinese cabbage, Brussels sprouts and green kale under the name "cabbage and other Brassicas". All these crops belong to *B. oleracea* species, with the exception of Chinese cabbage which, as mentioned above, is *B. rapa* (see Table 1-2). Probably, other *Brassica* species like *B. napus* (leaf rape) and *B. rapa* (turnip tops and turnip greens) are not included in this report, as well as *B. oleracea* crops that are cultivated and traded in local markets all around the world. With regard to consumption, Brassicas are included in the epigraph "Other vegetables" which refers to vegetable Brassicas along with other minor vegetables. For this reason, discussion referred to consumption is based on "Other vegetables" data.

Table 1-2 Main *Brassica* vegetable species and crops and the plant part used for consumption.

Species	Group	Common name	Organ
Brassica oleracea	*acephala*	Kale, collards	Leaves
	capitata capitata	Cabbage	Terminal leaf buds (heads)
	capitata sabauda	Savoy cabbage	Terminal leaf buds (heads)
	costata	Tronchuda cabbage	Loose heads
	gemmifera	Brussels sprouts	Vegetative buds
	botrytis botrytis	Cauliflower	Inflorescences
	botrytis italica	Broccoli	Inflorescences
	gongylodes	Kohlrabi	Stem
	albogabra	Chinese kale	Leaves
Brassica rapa	*rapa*	Turnip	Roots
	rapa	Turnip greens	Leaves
	rapa	Turnip tops	Shoots
	chinensis	Pak choi, bok choi	Leaves
	pekinensis	Chinese cabbage, pe-tsai	Leaves
	parachinensis	Choy sum	Leaves
	ruvo	Broccoleto	Shoots
	perviridis	Komatsuna, Tendergreen	Leaves
Brassica napus	*pabularia*	Leaf rape, nabicol	Leaves
	napobrassica	Swede	Roots
Brassica juncea	*rugosa*	Mustard greens	Leaves
	capitata	Head mustard	Heads
	crispifolia	Cut leaf mustard	Leaves

1.2.1 Trade

Vegetable Brassicas are of great economic importance throughout the world. Currently, Brassica crops together with cereals represent the basis of world supplies. They provide beneficial phytochemicals as well as vitamins, minerals and fiber (more details further in this section). Probably Brassicas were domesticated in the Neolithic period thousands of years ago. Ancient Greeks, Romans, Indians and Chinese valued and used them extensively as a source of food and medicinal compounds. The evolution of Brassica crops was deeply marked by the particular regional preferences over the centuries. Consequently, human selection was focused on different parts of the plant, so much so that some of them are hardly recognizable as being members of the same species.

Import and export FAO (FAOSTAT 2009) estimated data reflect the economic importance of vegetable Brassicas around the world. In 2006, approximately 2,289,843 t of vegetable Brassicas were exported, therefore representing a value of USD1,478 million and Spain being the main exporter (333,077 t) followed by China (292,754 t) and the United States of America (277,179 t) (FAOSTAT 2009). A mean price of USD 511 per metric ton of cabbage and other Brassicas and USD 822 per metric ton of cauliflower and broccoli were paid. Spain is by far the world's main cauliflower and broccoli exporter, reaching 270,000 t with an elevated mean value of USD 1,135 per metric ton (mostly due to winter season greenhouse production), while the United States was the major exporter of cabbage and other Brassicas selling to other countries with a mean price of USD 860 per ton. On the other hand, the major importer of vegetable Brassicas is Canada (204,166 t) followed by Germany (201,439 t), Honk Kong special administrative region of the People's Republic of China (199,879 t), the United States (183,013 t), the United Kingdom (176,914 t) and the Russian Federation (166,444 t). The United Kingdom was the only country where cauliflower and broccoli imports exceed that of the cabbage and other Brassicas, totalling 124,484 t with a value of almost USD 166 million .

1.2.2 Harvested Area

In 2007, vegetable Brassica were under cultivation in more than 142 countries around the world and they occupied more than 4.1 million ha (FAOSTAT 2009). Asia accounts for more than 75% of the global vegetable Brassica growing area, led by China with a total harvested area of 2.2 million ha, followed by India (511,600 ha) and the Russian Federation (171,700 ha) (Table 1-3). In the last 10 years, China has doubled the vegetable Brassica harvested area, in contrast to other Asian (Indonesia, Japan) and European (France, Italy, Poland) countries where the harvested area has shrunk.

1.2.3 Production

In recent decades, world vegetable Brassica production has been in continual expansion and has risen from 26.8 million t in 1961 to a record 88.3 million t in 2007 (Fig. 1-1). In crops, cauliflower and broccoli production has increased six times whereas that of the cabbage and other Brassicas has multiplied three-fold in the last five decades. Undoubtedly, China is the world's number one producer of vegetable Brassicas, growing more that a half of all vegetable Brassica in 2007 (Table 1-3). India ranked next (10.3 million t), followed by the Russian Federation (4.1 million t), the Republic of Korea (3.0 million t), Japan (2.5 million t) and the United States (2.4 million t). From this production, around 88% of the vegetable Brassicas goes for fresh human consumption, while just 1.2% is processed, almost 4% is used as farm animal feed, and the rest is discarded as waste (FAOSTAT 2009).

Table 1-3 Area harvested, production, and yield in the world and in the main vegetable Brassica-producing countries in 2007.

	Area harvested (ha)		Production (1,000 metric tons)		Yield (kg/ha)	
	Cabbage and others	Cauliflower and broccoli	Cabbage and others	Cauliflower and broccoli	Cabbage and others	Cauliflower and broccoli
World	3,088,328	1,022,770	69,214	19,108	22,411	18,682
Africa	103,057	14,226	1,878	291	18,221	20,440
America	90,784	105,710	2,393	1,713	26,356	16,204
Mexico	6,100	22,000	205	305	33,607	13,864
United States of America	29,950	71,570	1,171	1,241	39,110	17,336
Asia	2,396,899	760,255	52,647	14,769	21,965	19,426
China	1,772,000	403,800	36,335	8,585	20,505	21,260
India	232,800	278,800	5,283	5,014	22,694	17,986
Indonesia	57,000	11,200	1,250	168	21,930	15,000
Japan	54,000	12,000	2,390	132	44,259	11,000
Korea, Democratic People's Republic of	34,000	—	700	—	20,588	—
Korea, Republic of	48,000	—	3,000	0.1	62,500	—
Pakistan	4,800	11,400	75	209	15,625	18,333
Europe	496,603	131,179	12,250	2,160	24,668	16,466
France	7,000	27,000	195	370	27,857	13,704
Italy	17,753	18,038	345	433	19,444	24,019
Poland	36,400	15,500	1,376	277	37,799	17,884
Romania	46,000	1,800	1,120	24	24,348	13,333
Russian Federation	170,000	1,700	4,054	14	23,847	8,529
Spain	8,250	24,500	270	450	32,727	18,371
Ukraine	64,000	1,700	1,300	15	20,312	8,823
United Kingdom	9,800	16,600	253	186	25,867	11,229
Oceania	985	11,400	46	175	47,086	15,351

Source: FAOSTAT (2009).

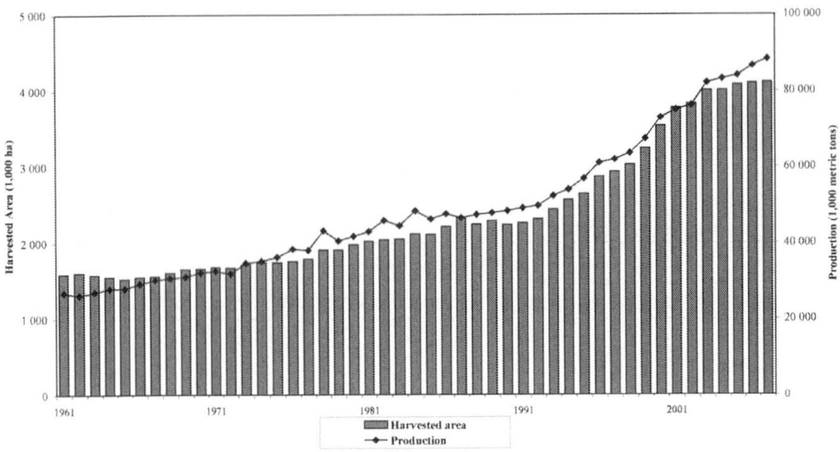

Figure 1-1 World series of harvested area and production in the main Brassica vegetable crops between 1961 and 2007.

1.2.4 Yield

Although Oceania and Africa are minor vegetable Brassica producing regions (accounting for 0.3 and 2.5%, respectively of world production in 2007), they provided the best yields. Oceania was the leader in cabbage yield and other Brassicas as well, with more than 47.1 t ha^{-1}, far more than other regions, and Africa was the leader in cauliflower and broccoli yield with 20.4 t/ha. In cabbage and other Brassicas, some of the world's best yields are recorded from the United Republic of Tanzania—on average, around 136.7 t ha^{-1}, followed by El Salvador (91.2 t ha^{-1}), the Republic of Korea (62.5 t ha^{-1}) and New Zealand (58.7 t ha^{-1}). The best cauliflower and broccoli yield was in Kuwait with 45.6 t/ha, followed by New Zealand (42.9 t ha^{-1}) and the United Arab Emirates (34.8 t ha^{-1}).

1.2.5 Consumption

FAO data related to vegetable consumption around the world showed that in 2003 Asia consumed more that 70% of the world's vegetable supply, which means around 540 million t, followed by Europe (85 million t) and America (67 million t). Led by China with 355 million t, the world's top 10 vegetable consuming countries included India, the United States, Turkey, the Russian Federation, Japan, Egypt, the Islamic Republic of Iran, Italy and the Republic of Korea. However, Greece had the highest consumption per capita (275 kg per person per year), while China was very close (270 kg). Several other Mediterranean countries, viz. Lebanon (243 kg), Turkey (230 kg), Libyan Arab Jamahiriya (228 kg), Israel (221 kg) and Tunisia (199 kg), also figure among the greatest vegetable consumers.

Looking in depth at the vegetable consumption around the world, it can be observed that large differences exist between developed and developing countries (Fig. 1-2). Consumption in the world's developed countries increased steadily in the last 42 years (from 85 million t in 1961 to 153 t in 2003). Meanwhile, the situation was completely different in the world's developing countries. A rapid increase from 109 to 588 million t between 1961 and 2003 was accounted for, especially since the 1980s. Currently, vegetable consumption in the world's developing countries is almost four times that of the developed world. This remarkable increase in consumption is clearly led by China. However, due to the huge population of the developing countries, vegetable consumption per capita is very similar in the world's developed and developing countries, accounting for 115 and 118 kg per person and year, respectively. It is necessary to point out that the consumption per capita in developing countries increased progressively since the 1970s from 44 kg in 1972 to 118 kg in 2003. Therefore, this made the consumption in developing countries more, for the first time, than that of the developed world (Fig. 1-2).

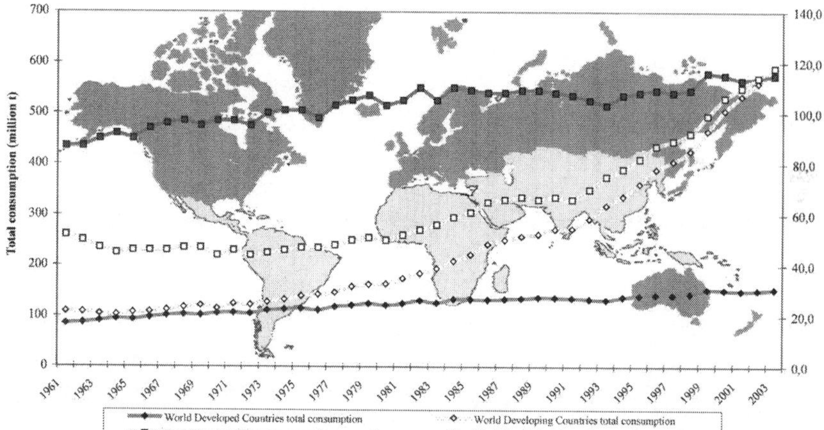

Figure 1-2 World series of vegetable Brassicas consumption between 1961 and 2003 and world series of vegetable Brassicas per capita consumption between 1961 and 2003.

1.3 Nutritional Quality

Nowadays, consumers are aware of the need for a constant supply of nutrients contained in plants to get optimal health benefits and there is a growing tendency to demand quality products with a higher added value. In this aspect, the popularity of Brassica crops is increasing because of their nutritional value and anticancer, antioxidant and anti-inflammatory properties. Vegetable Brassica crops have the general characteristics of other vegetable crops, as for example a low fat and protein content and a high value in vitamins, fiber and minerals. Besides, they show high quantities of phenolic compounds and have singular kinds of compounds which differentiate these crops from other vegetables; these compounds are called glucosinolates. Below, a short review about the most interesting nutritional compounds of vegetable Brassicas is presented. Data about vitamins, minerals as well as the nutritional components of the main vegetable Brassicas discussed below is based on the information provided by the USDA National Nutrient Database for Standard Reference (USDA 2008) (Table 1-4).

1.3.1 Vitamins

In general, vegetable Brassica crops have a high content of vitamins. If they are compared to 30 other vegetable crops (e.g., artichoke, beans, beet, carrot, sweet corn, onion, lettuce or spinach), Brassica crops have the highest level of vitamin C. With regard to vitamin E, only cilantro showed a higher level, while other crops like asparagus, soybean and spinach showed a similar level of this vitamin. Brassicas also have high levels of vitamin B-6 (exceeded by garlic, pepper and spinach), vitamin A and β-carotene (only exceeded by carrot), lutein, zeaxanthin and vitamin K. Vitamin C, E and carotenoids have the potential to prevent and treat malignant and degenerative diseases (Jahangir et al. 2009). Besides, Brassica crops show high levels of folate (similar to many other crops like eruca, bean, thistle, etc.), which is a scarce and important vitamin related to the reduced risk of vascular diseases, cancer and neural tube defects (Jahangir et al. 2009) and choline (exceeded by asparagus and similar to spinach, pea, garlic and eggplant).

1.3.2 Minerals

Brassica crops have been found to be rich in many minerals (Table 1-4). They have high levels of calcium, being higher than most of the compared vegetable crops. According to Lucarini et al. (1999), calcium present in Brassicas shows an excellent bioavailability, because of the low levels of oxalic and phytic acids, which make Brassicas a good source of calcium. They also show high levels of potassium, which is an essential mineral involved in different metabolic processes (i.e. synthesis of proteins, carbohydrate metabolism, etc.) as well as in the maintenance of human health (blood pressure, heart diseases, osteoporosis, etc.) and adequate levels for the rest of the minerals, including selenium, which is an important element from a biological, environmental and health point of view. Selenium deficiency is very common all around the world and nutritional supplements have been recommended to increase daily selenium intake (Pyrzynska 2009).

Table 1-4 Nutritional composition in 100 g of edible portion of different vegetable Brassicas (raw material).

		Brassica oleracea									
		Broccoli	Brussels sprouts	Cabbage	Red Cabbage	Savoy Cabbage	Green Cauliflower	Cauliflower	Collards	Kale	Kohlrabi
Water	g	89	86	92	90	91	90	92	91	84	91
Energy	kcal	34	43	25	31	27	31	25	30	50	27
Protein	g	2.82	3.38	1.28	1.43	2.00	2.95	1.98	2.45	3.30	1.70
Total lipid	g	0.37	0.30	0.10	0.16	0.10	0.30	0.10	0.42	0.70	0.10
Ash	g	0.87	1.37	0.64	0.64	0.80	0.88	0.71	0.89	1.53	1.00
Carbohydrate	g	6.64	8.95	5.80	7.37	6.10	6.09	5.30	5.69	10.01	6.20
Dietary fiber	g	2.6	3.8	2.5	2.1	3.1	3.2	2.5	3.6	2.0	3.6
Sugars	g	1.70	2.20	3.20	3.83	2.27	3.03	2.40	0.46	.	2.60
Minerals											
Calcium	mg	47	42	40	45	35	33	22	145	135	24
Iron	mg	0.73	1.40	0.47	0.80	0.40	0.73	0.44	0.19	1.70	0.40
Magnesium	mg	21	23	12	16	28	20	15	9	34	19
Phosphorus	mg	66	69	26	30	42	62	44	10	56	46
Potassium	mg	316	389	170	243	230	300	303	169	447	350
Sodium	mg	33	25	18	27	28	23	30	20	43	20
Zinc	mg	0.41	0.42	0.18	0.22	0.27	0.64	0.28	0.13	0.44	0.03
Copper	mg	0.049	0.070	0.019	0.017	0.062	0.041	0.042	0.039	0.290	0.129
Manganese	mg	0.210	0.337	0.160	0.243	0.180	0.247	0.156	0.276	0.774	0.139
Selenium	mcg	2.5	1.6	0.3	0.6	0.9	0.6	0.6	1.3	0.9	0.7
Vitamins											
Vitamin C	mg	89.2	85.0	36.6	57.0	31.0	88.1	46.4	35.3	120.0	62.0
Thiamin	mg	0.071	0.139	0.061	0.064	0.070	0.080	0.057	0.054	0.110	0.050
Riboflavin	mg	0.117	0.090	0.040	0.069	0.030	0.102	0.063	0.130	0.130	0.020
Niacin	mg	0.639	0.735	0.234	0.418	0.300	0.734	0.526	0.752	1.000	0.400
Pantothenic acid	mg	0.573	0.309	0.212	0.147	0.187	0.696	0.652	0.267	0.091	0.165
Vitamin B-6	mg	0.175	0.219	0.124	0.209	0.190	0.222	0.222	0.165	0.271	0.150
Folate	mcg	63	61	43	18	80	57	57	166	29	16

Table 1-4 contd...

Table 1-4 contd...

		Brassica oleracea									
		Broccoli	Brussels sprouts	Cabbage	Red Cabbage	Savoy Cabbage	Green Cauliflower	Cauliflower	Collards	Kale	Kohlrabi
Choline	mg	18.7	19.1	10.7	17.1	12.3	57.1	45.2	23.2	.	12.3
β-Carotene	mcg	361	450	42	670	600	93	8	3842	9226	22
Vitamin A	IU	623	754	98	1116	1000	155	13	6668	15376	36
Lutein + zeaxanthin	mcg	1403	1590	30	329	77	42	33	8932	39550	0
Vitamin E	mg	0.78	0.88	0.15	0.11	0.17	0.04	0.08	2.26	.	0.48
Vitamin K	mcg	101.6	177.0	76	38.2	68.8	20.2	16.0	510.8	817.0	0.1

		Brassica rapa						*Brassica napus*		*Brassica juncea*
		Pe-tsai	Pak choi	Turnip greens	Broccoleto	Tendergreen	Turnips	Leaf rape	Swede	Mustard greens
Water	g	94	95	90	93	92	92	87	90	91
Energy	kcal	16	13	32	22	22	28	42	36	26
Protein	g	1.20	1.50	1.50	3.17	2.20	0.90	2.80	1.20	2.70
Total lipid	g	0.20	0.20	0.30	0.49	0.30	0.10	0.60	0.20	0.20
Ash	g	0.98	0.80	1.40	0.93	1.40	0.70	1.28	0.81	1.40
Carbohydrate	g	3.23	2.18	7.13	2.85	3.90	6.43	8.32	8.13	4.90
Dietary fiber	g	1.2	1.0	3.2	2.17	2.8	1.8	1.7	2.5	3.3
Sugars	g	1.41	1.18	0.81	0.38	.	3.80	.	5.60	1.60
Minerals										
Calcium	mg	77	105	190	108	210	30	205	47	103
Iron	mg	0.31	0.80	1.10	2.14	1.50	0.30	3.00	0.52	1.46
Magnesium	mg	13	19	31	22	11	11	88	23	32
Phosphorus	mg	29	37	42	73	28	27	62	58	43
Potassium	mg	238	252	296	196	449	191	450	337	354
Sodium	mg	9	65	40	33	21	67	70	20	25
Zinc	mg	0.23	0.19	0.19	0.77	0.17	0.27	0.37	0.34	0.20
Copper	mg	0.036	0.021	0.350	0.042	0.075	0.085	0.243	0.040	0.147
Manganese	mg	0.190	0.159	0.466	0.395	0.407	0.134	0.649	0.170	0.480

Table 1-4 contd...

Table 1-4 contd...

		Brassica rapa						*Brassica napus*		*Brassica juncea*
		Pe-tsai	**Pak choi**	**Turnip greens**	**Broccoleto**	**Tendergreen**	**Turnips**	**Leaf rape**	**Swede**	**Mustard greens**
Selenium	mcg	0.6	0.5	1.2	1.0	0.8	0.7	0.9	0.7	0.9
Vitamins										
Vitamin C	mg	27.0	45.0	60.0	20.2	130.0	21.0	130.0	25.0	70.0
Thiamin	mg	0.040	0.040	0.070	0.162	0.068	0.040	0.070	0.09	0.080
Riboflavin	mg	0.050	0.070	0.100	0.129	0.093	0.030	0.060	0.04	0.110
Niacin	mg	0.400	0.500	0.600	1.221	0.678	0.400	1.300	0.700	0.800
Pantothenic acid	mg	0.105	0.088	0.380	0.322	0.178	0.200	0.076	0.160	0.210
Vitamin B-6	mg	0.232	0.194	0.263	0.171	0.153	0.090	0.227	0.100	0.180
Folate	mcg	79	66	194	83	159	15	28	21	187
Choline	mg	7.6	6.4	.	18.3	.	11.1	.	14.1	0.4
β-Carotene	mcg	190	2681	6952	1573	.	0	.	1	6300
Vitamin A	IU	318	4468	11587	2622	9900	0	3100	2	10500
Lutein + zeaxanthin	mcg	48	40	12825	1121	.	0	.	0	9900
Vitamin E	mg	0.12	0.09	2.86	1.62	.	0.03	.	0.30	2.01
Vitamin K	mcg	42.9	45.5	251.0	224.0	.	0.1	.	0.3	497.3

Source: USDA National Nutrient Database for Standard Reference (USDA 2008).

1.3.3 Glucosinolates

In the past decade, much interest has been devoted to the positive effects of glucosinolates. Glucosinolates are a class of phytochemicals that are present in the Brassicaceae family, whose breakdown products are reported to possess cancer preventive activity, as well as anti-inflammatory, antibacterial (*Helicobacter pylori*), cardioprotection properties (Rosa 1999; Mithen et al. 2000; Fahey et al. 2001; Smith et al. 2005; Traka and Mithen 2008). Research impetus was initially generated when the breakdown products were found to be partly responsible for the characteristic flavor of Brassica vegetables (Fenwick et al. 1983). More recently, numerous studies have related the decreased risk of cancer to a diet rich in Brassica vegetables.

Glucosinolates are the major class of secondary metabolites found in Brassica crops. The molecule comprises a β-thioglucoside N-hydroxysulfate containing a side chain and a β-D-glucopyranose moiety. Glucosinolates can be grouped into three chemical classes: aliphatic, indolic and aromatic, according to whether their amino acid precursor is methionine, tryptophan or an aromatic amino acid (tyrosine or phenylalanine), respectively (Giamoustaris and Mithen 1996). The most important glucosinolates found in Brassica vegetables are methionine-derived glucosinolates (Mithen et al. 2003).

Although approximately 120 classes of glucosinolates have been identified in plants, each plant species contains up to four different glucosinolates in significant amounts (Fahey et al. 2001). Glucosinolates are not bioactive until they have been enzymatically hydrolyzed to various bioactive breakdown products by the endogenous plant enzyme called myrosinase, including isothiocyanates, nitriles, thiocyanates, epithionitriles, oxazolidine-2-thiones, and epithioalkanes (Grubb and Abel 2006). At physiological pH, isothiocyanates are the major products, whereas nitriles are formed at more acid pH (Halkier and Du 1997).

As components of feed and food, some glucosinolates degradation products have been recognized for long for their distinctive benefits to human nutrition and plant defense. This feature has led to considering *Brassica* foods as possible functional foods. The term "functional foods" describes foods that, if they are normal dietary constituents, can provide sufficient amounts of bioactive components that are valuable for health improvement.

Comparative studies of glucosinolate profiles indicate significant differences among cruciferous crops (VanEtten et al. 1976; Carlson et al. 1987; Kushad et al. 1999; Ciska et al. 2000). A summary of the major glucosinolates found in the main vegetable Brassicas belonging to *B. oleracea*, *B. rapa* and *B. napus* species has been recently reported by Cartea and Velasco (2008) (Table 1-5). Apart from the glucosinolate profile, large differences in the levels of both aliphatic and indole glucosinolates have been observed

Table 1-5 Principal glucosinolates identified in *Brassica* vegetable crops.

Crop	Aliphatic glucosinolates										Indole glucosinolates				Aromatic
	GIB	PRO	SIN	GAL	GRA	GNA	GBN	GIV	GER	GNL	GBS	NGBS	4HGBS	4MGBS	GST
Brassica oleracea															
White cabbage	+*	+	+*	+	+	+	+	+	+	-	+*	+	+	+	+
Savoy cabbage	+*	+	+*	-	+	+	+	+	-	-	+*	+	-	+	+
Red cabbage	+*	+	+*	-	+	+	-	+	-	-	+*	+	-	-	-
Kale	+*	+	+*	-	+	+	-	+	-	-	+*	+	+	+	+
Collard	+	+*	+*	-	-	-	-	+	+	-	+*	-	-	-	-
Tronchuda cabbage	+*	+	+*	+	+	+	+	+	-	-	+*	+	+	+	+
Broccoli	+	+*	+	+	+*	+*	+	-	+	+	+*	+	+	+	+
Brussels sprouts	+	+*	+*	-	+	+	-	+	-	-	+*	+	-	-	-
Cauliflower	+*	+*	+*	-	+	-	-	+	-	-	+*	+	-	-	-
Kohlrabi	+	+	+*	+	+	+	-	+	-	-	+*	+	+	+	-
Brassica rapa															
Turnip	+	+*	-	-	-	+*	+	-	+	+	+	+	+	+	+*
Turnip greens	+	+	-	+	+	+*	+*	+	-	+	+	+	+	-	+
Turnip tops	+	+	-	-	-	+*	+*	+	-	-	+	+	+	-	+
Chinese cabbage	+	+	-	-	-	+*	+*	-	-	+*	+	+	-	+	+
Brassica napus															
Swede	-	+	-	-	+	-	+	-	-	+	+	+*	+	+*	+
Leaf rape	-	+	-	+	-	+	+*	+	-	+	+	+	-	+	+

*Major glucosinolates found in each crop.

GIB: glucoiberin (3-methylsulfinylpropyl); PRO: progoitrin (2-hydroxy-3-butenyl); SIN: sinigrin (2-propenyl); GAL: glucoalyssin (5-methylsulphinylpentyl); GRA: glucoraphanin (4-methylsulphinylbutyl); GNA: gluconapin (3-butenyl); GBN: glucobrassicanapin (4-pentenyl); GIV: Glucoiberverin (3-methylthiopropyl); GER: glucoerucin (4-methylthiobutyl); GNL: gluconapoleiferin (2-hydroxy-4-pentenyl); GBS: glucobrassicin (3-indolylmethyl); NGBS: neoglucobrassicin (1-methoxy-3-indolylmethyl); 4HGBS: 4-hydroxyglucobrassicin (4-hydroxy-3-indolylmethyl); 4MGBS: 4-methoxyglucobrassicin (4-methoxy-3-indolylmethyl), GST: Gluconasturtiin (2-phenylethyl).

Adapted from Cartea and Velasco (2008).

in Brassica plants, due to the stage of development, tissue type and environmental conditions (Velasco et al. 2007).

Different epidemiological studies have indicated that diet and cancer are closely interlinked. Over the past 30 years, various studies have shown that fruits and vegetables contain natural phytochemicals such as glucosinolates that have anticarcinogenic properties (Block et al. 1992; Talalay and Zhang 1996; Talalay and Fahey 2001; Hecht 2000; Anilakumar et al. 2006). Consequently, cruciferous vegetables have been attracting interest in the last few years for their potential use in cancer chemoprevention (Rosa et al. 1997; Farnham et al. 2004; Smith et al. 2005). Results have consistently shown that the chemoprotective agents derived from this class of vegetables of the Cruciferae family have an influence on carcinogenesis during the initiation and promotion phases of cancer development. Similarly, reports from epidemiological studies and clinical trials also support this theory (Traka and Mithen 2008).

Isothiocyanates and indoles are two major groups of autolytic breakdown products of glucosinolates. Both of them exhibit protective activities against many types of cancer. In vitro and in vivo studies have reported that these compounds affect many stages of cancer development, including the induction of detoxification enzymes (Phase II enzymes) and the inhibition of activation enzymes (Phase I enzymes) (Zhang and Talalay 1994; Hecht 2000; Fahey et al. 2002; Anilakumar et al. 2006). Cartea and Velasco (2008) and Traka and Mithen (2008), have reviewed the most studied glucosinolates and glucosinolate degradation products and the anticarcinogenic mechanisms of isothiocyanates.

Apart from the medicinal value of these sulfur compounds, glucosinolates are responsible for the characteristic flavor and odor of Brassica vegetables. Glucosinolate-derived isothiocyanates produce a pungent and bitter flavor and sulfurous aroma playing a significant organoleptic role in Brassica products (Fenwick et al. 1983; Rosa et al. 1997; Padilla et al. 2007b). However, the direct relation between glucosinolate content and sensory properties is complex.

1.3.4 Phenolic Compounds

Besides glucosinolates, the contribution of Brassica vegetables to health improvement has generally been associated with their antioxidant property, which is mainly due to phenolic compounds (Jahangir et al. 2009). "Phenolic compounds" is a generic term that refers to a large number of compounds (more than 8,000) widely dispersed throughout the plant kingdom. Phenolics range from simple, low molecular-weight, single aromatic-ringed compounds to large and complex tannins and derived polyphenols (Crozier et al. 2006). They can be classified based on the number and arrangement of their carbon atoms in flavonoids (flavonols, flavones, flavan-3-ols, anthocyanidins, flavanones, isoflavones and others) and non-flavonoids (phenolic acids, hydroxycinnamates, stilbenes and others) (Crozier et al. 2006), and are commonly found conjugated to sugars and organic acids. The most widespread and diverse group of the polyphenols are the flavonoids, which are built upon the C6–C3–C6 flavone skeleton. Phenolic compounds, especially flavonoids, possess different biological activities, but the most important ones are the antioxidant activity, the capillary protective effect and the inhibitory effects elicited in the various stages of a tumor. In many in vitro studies, phenolic compounds demonstrated a higher antioxidant activity than antioxidant vitamins and carotenoids (Podsedek 2007).

Podsedek (2007) has an extensive review on phenolic profiles in different *Brassica* species. Studies have mainly focused on broccoli florets and other *B. oleracea* crops such as cabbage and cauliflower (Nielsen et al. 1993; Price et al. 1997; Llorach et al. 2003; Vallejo et al. 2004) but, in the last years other species like *B. rapa* have been widely studied (Fernandes et al. 2007; Francisco et al. 2009). Broccoli is a source of flavonol and hydroxycinnamoyl derivatives. The main flavonol glycosides present in broccoli florets are quercetin and kaempferol 3-O-sophoroside. Three minor glucosides of these Aglycones were also detected, viz. isoquercitrin, kaempferol 3-O-glucoside and kaempferol diglucoside. The predominant hydroxycinnamoyl acids were identified as 1-sinapoyl-2-feruloylgentiobiose, 1,2-diferuloylgentiobiose, 1,2,2′-trisinapoylgentiobiose and neochlorogenic acid (Vallejo et al. 2003).

Nielsen et al. (1993) showed that cabbage contains a mixture of more than 20 compounds, seven of which have been identified as 3-O-sophoroside-7-O-glucosides of kaempferol and quercetin with and without further acylation with hydroxycinnamic acids. Llorach et al. (2003) found 22 compounds that were identified as being several derivatives of kaempferol, ferulic and sinapic acids. Similar compounds were found in *B. rapa* leaves with the presence of isoramnethin (Fernandes et al. 2007; Francisco et al. 2009).

The content of polyphenols in vegetables, like the levels of other phytochemicals, may be influenced by various factors such as varieties, climatic conditions and cultural practices, maturity at harvest and storage conditions. In the case of phenolic compounds, sample preparation method is also very important. Phenolics in vegetables exist in both free and conjugated forms. Generally, only conjugated flavonoids are present in fresh vegetables, but aglycones may be found as a result of food processing. After hydrolysis, HPLC analysis showed that quercetin was the predominant flavonol aglycone in Brassica vegetables.

Some Brassicas provide a variety of anthocyanins (Jahangir et al. 2009). Red pigmentation of red cabbage and purple cauliflower is caused by anthocyanins. Cauliflower and red cabbage showed differences in their anthocyanin profiles: cyanidin-3,5-diglucoside was absent in cauliflower, while it was well represented in red cabbage, together with the characteristic anthocyanin of *Brassica* genus (cyanidin-3-sophoroside-5-glucoside). The *p*-coumaryl and feruloyl esterified forms of cyanidin-3-sophoroside-5-glucoside were predominant in cauliflower, while the sinapyl ester was mostly present in red cabbage (Scalzo et al. 2008).

1.4 Brassica as a Model Plant in Genetics and Breeding

The remarkable morphological diversity of vegetable *Brassica* species and their relatives offers valuable opportunities to increase our knowledge about plant growth and development and our understanding of rapid phenotypic evolution. Plants of this genus are distinguished from many other plant taxa by their tendency to develop new morphological variants rapidly, as it becomes evident from the interfertility among common vegetables such as cabbage, cauliflower, broccoli, Brussels sprouts and kohlrabi. The genetic diversity displayed by *Brassica* species has allowed the development of new crop types and also a substantial increase in the knowledge and research on molecular biology and plant genetics. In the 20th century, members of the Brassicaceae family acquired new roles as an ideal model plant for their use in basic scientific research and subsequently as educational tools. The Brassicaceae family includes two excellent models, the thale cress *Arabidopsis thaliana* (L.), Heynh, and the rapid-cycling populations of the major *Brassica* species, also known as "Wisconsin Fast Plants". On the other hand, interspecific hybridization among *Brassica* spp. is easy compared to other crucifer species and this fact has been successfully applied to develop synthetic amphidiploids, improve Brassica vegetable crops, clarify intergenomic relationships and create genetic variability in the genus.

1.4.1 Rapid-cycling Brassica Populations

These populations are commonly known as Wisconsin Fast Plants (WFP) because they were developed at the University of Wisconsin-Madison by Paul Williams in 1970 (*http://www.fastplants.org*). Williams and coworkers selected accessions of the six *Brassica* species of the U-triangle for a short time to flower and speed up seed maturation. Over multiple generations of recurrent selection and breeding, rapid-cycling populations of each of the six species were developed (Williams and Hill 1986). These populations could produce up to 10 generations of seed per year and their life cycle is completed in 35–45 days (*B. rapa*) to 60 days (*B. oleracea*). Over the past 30 years, rapid-cycling *Brassica* (RCBr) populations especially those of *B. rapa* and *B. oleracea* have been developed. The Wisconsin Fast Plants (WFP) Program utilizes rapid-cycling *B. rapa*, which has a life cycle as short as that of *A. thaliana,* therefore making them easier to study. The short life cycle along with their small size and easiness to grow make them an ideal organism for teaching and for applied and basic plant biology research as genetics, molecular and cell biology, plant biochemistry and plant breeding.

RCBr populations were originally developed as a model for testing the plant/pathogen interactions. Musgrave (2000) reviewed the different applications where RCBr are excellent models, including research on plant disease, genetics, plant physiology and development, self-incompatibility and in vitro culture. Regarding this last feature, different techniques using rapid-cycling plants have been successful as protoplast, anthers, embryo culture techniques, and transformation procedures by using *Agrobacterium tumefaciens* and *A. rhizogenes.*

In addition, RCBr populations have been reported to be valuable laboratory tools for students to research into inheritance and selection along with evolution. A central idea of the WFP Program is that students learning biological concepts should have the opportunity to design and carry out their own experiments that involve living organisms. Populations of WFP are successfully used for genetic analysis of both Mendelian and complex traits and to study the genotype × environment interaction. Their use has also been reported successfully for studies on plant tissue culture in the educational situation. A review detailing the use of RCBr as a model system in education has been recently published by Himelblau et al. (2004). A particular application of RCBr has been the study of their use for developing crops to be grown in space. For instance, astronauts aboard the Mir space station have used RC *B. rapa* in microgravity experiments. Such studies have revealed a difference in shoot carbohydrate content and root zone hypoxia between plants growing in space and control plants growing on Earth (Stout et al. 2001).

1.4.2 *Arabidopsis*

Arabidopsis thaliana is a well-studied plant model worldwide. It was originally identified as an experimental tool by Laibach in Germany (cited by Dixon 2007) in the middle of the 20th century. Its short life cycle (average 60 days), the huge seed production, the estimated genome size of 125–130 Mb, which is the smallest among the higher plants known, and the low content of repetitive sequences are the characteristics which make this small plant an ideal model organism. This species has thus become the universal research tool for plant biology, including different aspects of metabolism, development, biochemistry, plant environment interactions, genetics and genomic analysis. *Arabidopsis* was the first plant for which the whole genome was sequenced (Arabidopsis Genome Initiative 2000); this provides genes for many different types of genomic research and currently, many *Arabidopsis* genes have been cloned and characterized.

Species of the *Brassica* genus are closely related to the *A. thaliana*, which also belongs to the Brassicaceae family. The close relationship between the two genera, *Arabidopsis* and *Brassica*, is reflected by an average identity of exon sequences at the nucleotide level, which is estimated to be 87% (Cavell et al. 1998). The phylogenetic relationship of *Arabidopsis* to Brassica crops offers an exceptional opportunity to test if the information derived from the *Arabidopsis* Genome Project can be transferred to crop plants. The *Arabidopsis* genome sequence provides a valuable resource for identifying genes that may account for the genetic control of complex traits in Brassica. As an example, the results obtained from *Arabidopsis* are being applied to enhance a new insight into how important genes involved in the agronomy work and the nutritional aspects of *Brassica* species, including the genes responsible for head formation in cauliflower and broccoli (Lan and Paterson 2000) and the genes involved in glucosinolate biosynthesis (Gigolashvili et al. 2009). For details of comparative genomics between *Arabidopsis thaliana* and *Brassica* ssp. readers are suggested to see Chapter 9.

In addition, an increasing amount of information is currently available. This information provides links between *Brassica* genetic maps and the *Arabidopsis* genome. Recent studies have demonstrated the potential for comparisons among genetic and cytogenetic maps between *Arabidopsis* and different Brassica crops. The development of tools that facilitate the transfer of knowledge from *Arabidopsis* to vegetable Brassicas would help the breeding of these crops having desirable characteristics. For instance, expressed sequenced tags (ESTs) from *Arabidopsis* have been used as restriction fragment length polymorphism (RFLP) markers in *B. oleracea*, for comparison of the genomes of both species (Babula et al. 2003). Significant progress will be achieved in the coming years through the integration of candidate gene approaches in Brassica crops by using the detailed information now available for the *Arabidopsis* genome. Comparative mapping between the *Brassica* genomes and *Arabidopsis* is becoming a common procedure to identify candidate genes and markers for mapping studies and to expedite positional gene cloning (Qiu et al. 2009). Genome sequencing projects for *B. rapa* and *B. oleracea* are in process. This will provide an excellent opportunity to study the genome changes associated with the origin and evolution of these species in relation to *A. thaliana*.

1.4.3 Chromosome Evolution (Polyploidy Studies)

The Brassicaceae family includes one of the most flexible plants in terms of interspecific and intergeneric crosses. The use of the three diploid species of the U-triangle and their corresponding amphidiploids has been reported in scientific literature as an excellent model system for studying the mechanisms and the control of polyploidization and homoeologous recombination from a breeding perspective. Embryo rescue, protoplast fusion and other methods have been used to intercross the six *Brassica* species of the U-triangle. Due to the distribution of a series of closely related genomes and combined amphidiploids, the *Brassica* genus provides the opportunity to study rapid genome changes associated with polyploidy as well as chromosome evolution.

Each one of the amphidiploid species (*B. napus*, *B. juncea* and *B. carinata*) can be resynthesized by hybridizing the diploid species and then doubling the chromosomes (Song et al. 1993). This results in homozygous polyploid lines that should not segregate, but variation in DNA restriction fragments and phenotype has been observed among progeny derived from self-pollination of single resynthesized polyploids (Song et al. 1995). In addition, the close relationship with *Arabidopsis* allows comparisons from the levels of the chromosome segment up to the sequence level. It may be possible to survey large numbers of *Brassica* genes for changes in expression by using DNA microarrays based on *Arabidopsis* gene sequences. Comparisons of newly derived polyploids with their exact diploid progenitors should provide an insight into the consequences of polyploidy on plant evolution. Genome duplication or polyploidy is a widespread phenomenon throughout the plant kingdom, and provides a mechanism and context for genetic and evolutionary adaptations. *Brassica* species provide excellent experimental systems to understand the consequences of genome duplication. There are well studied examples of whole-genome duplication that result in allopolyploidy (combining two or more distinct genomes) or autopolyploidy (duplication of a diploid genome). New allopolyploid species such as *B. napus* often combine parental phenotypes with novel traits and adaptations and the fixation of hybrid vigor.

1.4.4 Tissue Culture

Vegetable *Brassica* species had pioneered the development of in vitro culture techniques to produce interspecific hybrids by means of protoplast fusion as well doubled haploid lines (DH) either via anther culture or microspore culture. Microspore derived embryogenesis was first successfully completed in broccoli from isolated anthers by Keller et al. (1975). Doubled haploid technology is widely applied in vegetable Brassica crop improvement and research programs. The use of DH lines has allowed the development of tools and resources to improve quantitative traits in a more effective way. Successful anther and microspore cultures have been reported for several *B. oleracea* crops including cabbages (Kuginuki et al. 1999) and kales (Arnison and Keller 1990). Now both methods are routine in most long term breeding programs, therefore saving the necessity of working through several generations of inbreeding. Due to the strong self-incompatibility system, most vegetable Brassica crops are outbreeding species with a high degree of heterozygosity in natural populations and open-pollinated crops. For this reason, the value of microspore culture technology in these types of crops is evident. Doubled haploids have been valuable to detect quantitative trait loci (QTLs), map genes and determine their linkage relationship to other genes important in plant breeding.

1.4.5 Transgenic Technology

Vegetable Brassica crops were amongst the first to be targeted for transgenic genetic modification, for finding traits such as herbicide resistance and modification of male sterility. Genetic transformation technology has allowed the development of novel Brassica varieties by introducing transgenes from unrelated sources, something that cannot be achieved by conventional breeding methods. Conventional Brassica breeding is an intensive and time-consuming endeavor; it takes eight to ten generations to develop a new variety. In contrast, genetic transformation provides direct means to introduce a specific gene or trait into a selected genotype without adversely affecting their desirable genetic background. Moreover, most traits introduced by the gene-transfer method are dominant. Various gene transfer methods have been highly effective. The most general approach for genetic transformation is to use *A. tumefaciens* and *A. rizhogenes*. Some transgenic plants have been recovered from major Brassica vegetables and there are abundant reports of vegetable Brassica field trials by using transgenic broccoli, cabbage, cauliflower, Chinese cabbage, kale, pak choi, swede and turnip cultivars with enhanced resistance or horticultural breeding (see for review Christey and Braun 2008). While transgenic oilseed Brassica crops (such as canola) are widely grown, as far as we know, currently there are no commercial genetically-modified vegetable Brassica crops, although according to the US National Sustainable Agriculture Information Service (*www.attra.ncat.org*) in 2006 cabbage, cauliflower, and mustard had been added to the list under development for commercial release. Notable advances in the improvement of Brassica crops should be achieved by using genetic marker technologies. However, most of the new cultivars produced until now still result from traditional procedures.

1.4.6 Breeding Systems and Technologies

1.4.6.1 Male Sterility

Cytoplasmic male sterility (CMS) occurs although infrequently in natural populations of *Brassica* species but it provides an excellent tool to study genetic interactions between mitochondria and nucleus during flower development. CMS plants can be selected either by following sexual crosses between different species of the same family or by somatic hybridization between unrelated species. Several systems and sources of CMS have been characterized and used in *Brassica* and CMS has increasingly replaced self-incompatibility in hybrid *Brassica* production. A single restorer locus appears to be capable of restoring different forms of *Brassica* CMS.

1.4.6.2 Self-incompatibility

As vegetable *Brassica* species are largely outbreeding with a well described self-incompatibility system, they can provide valuable information concerning the distribution of genetic variation in outbreeding crops and in investigating and evaluating regeneration methodology. *Brassica* species have been established as the model sporophytic pollen self-incompatibility system for research, with a well-characterized and diverse allelic series of "*S*-" alleles in both *B. oleracea* and *B. rapa* having strong to weak incompatibility reactions. The *S*-locus has been well characterized at the genomic and sequence level. Alleles of genes at the *S*-locus display considerable variability, with a very high level of non-synonymous substitution mutation. Comparative studies indicate that this variation appears to have occurred prior to the speciation of *B. rapa* and *B. oleracea*. Before the adoption of CMS systems, self-incompatibility was widely used in the development of F_1 hybrid Brassica crops.

1.4.7 Molecular Breeding

The ability to intercross morphotypes within *B. oleracea* and *B. rapa* has made them suitable organisms to conduct detailed genetic studies, including comparative mapping and QTL analysis. Besides, the size of the *Brassica* genomes (the genomes of diploid *Brassica* species are 3–5 times the size of the *Arabidopsis* genome) (see Table 1-6) and their close relationship to *Arabidopsis*, make them excellent for a whole-genome sequencing. Molecular markers have been widely used to assist breeding and selection procedures in numerous crops. During the last decades, great progress has been made in Brassica breeding due to a better understanding of the plant genome and the application of techniques for its manipulation at a molecular level. The development of DNA-based genetic markers and genetic mapping has made possible the study of structure, origin and evolution of *Brassica* genomes (Sadowski et al. 1996, Quiros et al. 2001). Marker development in Brassica crops has been an active subject since the early 1980s with the development of the first RFLP linkage maps for *B. oleracea*, *B. rapa*, and *B. napus* (reviewed by Quiros 2001). Genetic maps have been produced for all *Brassica* species although more effort was focused on *B. napus*, *B. oleracea* and *B. rapa*. In *B. oleracea*, several maps have been developed independently

Table 1-6 Size of the *Brassica* genomes and genomic relationships among the six *Brassica* species and *Arabidopsis*.

Species	Genome	*n*	Genome size (Mbp)
B. rapa	A	10	500–550
B. nigra	B	8	470
B. oleracea	C	9	600–650
B. juncea	AB	18	1100–1500
B. napus	AC	19	1130–1240
B. carinata	BC	17	1540
Arabidopsis thaliana		5	125–130

involving crosses between different crops. The chromosome maps are also being used for location of useful genes in vegetable Brassica crops. Among the genes being mapped by various laboratories, disease resistance, CMS restorers and oil seed quality are the most prominent ones. Eventually, markers linked to these genes will be used for marker-assisted selection, gene isolation, molecular characterization and manipulation. Until recently, there has been relatively little widespread integration of maps with public-domain sequence-tagged common markers. There is a constant effort to generate such integrated information and to standardize nomenclature and orientation of linkage groups. Different mapping populations of DH lines derived from specific crosses have been used to develop several published linkage maps (Quiros and Paterson 2004). For instance, *B. oleracea* mapping populations and the associated linkage map are publicly available at *http://www.grc.warwick.ac.uk/Brassica/Brassica_mapping.html*.

QTL have been identified for a wide range of morphological, physiological and crop traits in the different Brassica crop types. The ability to resolve QTL is dependent upon the access to a larger number of recombinant individuals (segregating populations, substitution or near-isogenic lines), as well as to high-density genetic and physical maps. There is a continuing requirement to provide the tools that allow researchers to resolve QTL to the level of physical maps, bacterial artificial chromosome (BAC) contigs, and candidate genes. This can be achieved through an integration of physical BAC contigs anchored with genetic markers onto common reference genetic linkage maps, as well as access to comparative genomic data between *Brassica* and *Arabidopsis* (Qiu et al. 2009). New methodologies, as single nucleotide polymorphism (SNP) markers, studies of allele-trait association and candidate gene approaches, have been reviewed by Quiros and Paterson (2004) and developed mainly in rapeseed. These advances will be applied in the next years to the breeding of other Brassica crops including vegetable types.

1.5 Germplasm and Gene Pools

1.5.1 Germplasm, Conservation and GeneBanks

The genetic resources available in Brassica crops are directly related to the limits of their primary, secondary and tertiary gene pools. According to Branca (2008), in *Brassica* genus, the primary gene pool of each crop is represented by its corresponding species of the U-triangle. The secondary gene pool is represented by the other species of the U-triangle, since different phylogenetic studies confirm the cytogenetic relationship among A, B, and C genomes. Finally, the tertiary gene pool gathers species and genera related to Brassica crops in 36 cytodemes that are able to change genetic materials by using special methodologies. The main cytodemes are represented by the genera *Diplotaxis, Eruca, Erucastrum, Sinapis,* among others. These gene pools can confer favorable alleles and useful characteristics in some cases, but they are considered to be non-adaptive in others.

Brassica oleracea and *B. rapa* have much more phenotypic and genetic diversity than *B. napus* and *B. juncea*. The main reason for such diversity is that all these crops have been selected over many centuries by producing divergent plant forms that are markedly different in their use and morphological, agronomical and nutritional traits. Morphological and molecular characterizations carried out in the last years have demonstrated a great genetic diversity in several *B. oleracea* crops such as cauliflower and broccoli (Hu and Quiros 1991; Branca 2008), kales (Cartea et al. 2003; Vilar et al. 2008), cabbages (Cassian and Echeverrigaray 2000; Padilla et al. 2007a) and *B. rapa* crops (Ren et al. 1995; Padilla et al. 2005; Zhao et al. 2005). In contrast, such diversity has not been found in vegetable *B. napus* (Cartea et al. 2005; Rodríguez et al. 2005; Soengas et al. 2006).

In Brassica crops, as it also happens in most crops, the pressure for more intensive agriculture has led to a progressive reduction of the natural genetic diversity within the gene pool of each crop, which is a process called "genetic erosion". Besides, breeding programs have led to a reduction of the allelic variation in modem elite varieties. Selection within each species for adaptation to different uses and growing areas has thus resulted in different pools of genetic variation within the cultivated germplasm (Gómez-Campo and Prakash 1999). Efforts have been made to collect and conserve the Brassica germplasm. Around 70 years ago, N.I. Vavilov proposed the use of wild forms as sources to improve crop plants and seed banks were created to maintain genetic diversity for many crops to be used used by researchers and breeders.

From a historical point of view, the N.I. Vavilov Research Institute for Plant Industry (VIR) at St. Petersburg (Russia) stands out. Its Brassica collection in 1996 comprised 6,879 accessions. The basic working collections are stored in St. Petersburg (Shashilova 1997).

Under the aegis of the European Cooperative Program for Crop Genetic Resources Networks (ECP/GR), a working group on Brassicas was established in 1991. The Centre for Genetic Resources, the Netherlands (CGN) developed the European Brassica Database (Bras-EDB) in the framework of the Brassica Working Group of the European Cooperative Program for Crop Genetic Resources Networks. The Bras-EDB contains passport data of most of the European *Brassica* collections (Boukema et al. 1997). At present, it contains passport data of about 19,600 accessions from 36 collections coming from 22 countries. This database can be accessed at (*http://documents.plant.wur.nl/cgn/pgr/brasedb*). The database has shown its value as a tool in developing core collections, in the identification of duplicates in the European collections and in defining a sound policy of safety duplication (Van Soest et al. 2004). Major updates of Bras-EDB were done in 2001 and 2005, supported financially by the European Commission by the project RESGEN CT99 109–112: "*Brassica* collections for broadening agricultural use, including characterizing and utilizing genetic variation in *Brassica carinata* for its exploitation as an oilseed crop". This project was an important attempt to unify efforts on Brassica resources within the EU. The participant countries were France, Germany, Greece, Italy, the Netherlands, Portugal, Spain, Sweden and the United Kingdom. Besides documenting Brassica collections, other objectives of the project were to create a core collection of the four species included in the project (*B. oleracea, B. rapa, B. napus* and *B. carinata*); characterization and regeneration of collections with priority on accessions included in the core collection; evaluation of different properties with priority also in material included in the cores; rationalization and safety duplication of collections and recommendations for further collection (van Soest et al. 2004).

Other European countries that maintain collections of vegetable Brassica crops are Bulgaria, Croatia, the Czech Republic, Hungary, Poland, Russia, Switzerland and Turkey. There are also collections of cole crops in Israel, Ethiopia, South Africa, India, the Philippines and Taiwan (Swarup and Brahmi 2005).

In the US, Prof. P. H. Williams, at the University of Wisconsin, established in 1982 the Crucifer Genetics Cooperative (CrGC) to acquire, maintain and distribute seed stocks and pollen of Brassicaceae. Presently, the program provides seed stocks and information of rapid-cycling Brassicas, concretely the six species of the U-triangle. Most *Brassica* species are maintained at the Plant Genetic Resources Unit (PGRU) located at the Geneva campus (New York) of Cornell University and formed in 1986. The PGRU maintains some accessions from wild Brassicas and 1,480 accessions of *B. oleracea*.

The conservation of plant genetic resources in genebanks involves the selection of accessions to be conserved and the maintenance of these accessions for current and future users. Choices concerning both these questions need knowledge about the distribution of genetic diversity within and between accessions sampled from the gene pool, but also about the changes in variation of these samples as a result of regenerations. In outbreeding species, such as *B. oleracea* and *B. rapa*, the accessions within the core collection retain a high degree of heterozygosity and heterogeneity, which make the evaluation of quantitative traits difficult. Different research groups are currently developing "Diversity Fixed Foundation Sets" (DFFS) based on core collections to facilitate exploitation of the genetic variation in *B. napus*, *B. oleracea* and wild C genome *Brassica* ssp. (Pink et al. 2008). Diversity fixed foundation sets have been defined as: "an informative set of genetically fixed lines representing a structured sampling of diversity across a gene pool" (*http://www.brassica.info/resource/plants/diversity_sets.php*). They are proposed to represent the majority of allelic variation within respective gene pools in a form appropriate for long term exploitation. They are similar to diversity core collections, but genetically fixed lines are being generated for each of the founder accessions. Genetically fixed lines are currently being produced at HRI by microspore culture and/or single seed descendent (where an accession is not responsive to double haploid production).

1.5.2 Wild Allied Species as Resource of Donor Genes

Wild allied species have been widely used in plant breeding programs to broaden the germplasm base in most crop species and as a source of donor elite genes (elite alleles). Brassica crops are derived from wild species, which still exist in nature and which are able to be intercrossed. The wild relatives of vegetable Brassicas offer a largely genetic resource for beneficial alleles controlling economically important quantitative traits.

The UPM (Universidad Politécnica de Madrid) created in 1966 a Plant Germplasm Bank, which has one of the most complete collections of wild crucifers in the world (1,027 taxa, 4,863 accessions), especially the Brassiceae tribe, which has several species of economic interest including relatives of crops belonging to the genera *Brassica, Sinapis, Raphanus, Eruca,* and others. Most of the accessions were collected by Prof. Gómez-Campo and his collaborators. In 1982 the UPM Plant Germplasm Bank was designated the base-bank for Brassicaceae by IBPGR (International Board for Plant Genetic Resources, these days known as Biodiversity International). The Plant Germplasm Bank–UPM is partly included in the GBIF (Global Biodiversity Information Facility) database (*www.gbif.es*).

There are many wild relatives that have potential as sources for oil, condiments, genes for resistance to different diseases and pests. They also serve as sources for cytoplasmic male sterility for hybrid seed production. Gómez-Campo (1999) introduced the term "coenospecies" to refer to a cytogenetic concept of species and subspecies of the genus *Brassica* and closely related genera (e.g., *Raphanus, Sinapis, Eruca*). For instance, *Sinapis alba* is a source of genes for resistance against the blackleg pathogen (Sjödin and Glimelius 1988), *Moricandia arvensis* is a potential source of genes for improvement of photosynthetic ability and resistance to clubroot (Rawsthorne et al. 1998), and *Raphanus sativus* is a source of resistance genes against nematodes (Voss et al. 2000). *Orychophragmus violaceus* (L.) O.E. Schulz (genome OO, $2n = 24$) is a wild crucifer belonging to the tribe Brassiceae. This species possesses agronomically valuable traits such as seed oil with a high content of palmitic acid, oleic acid and linoleic acid, a low percentage of linolenic acid and erucic acid (0.9%), and a high seed yield (Luo et al. 1994). Intergeneric F_1 plants of the cross *B. napus* (genome AACC, $2n = 38$) × *O. violaceus* have been produced by embryo rescue (Luo et al. 1994). In some cases, the opposite has also occurred, where desirable genes have been transferred from cultivated Brassica forms into non-*Brassica* species. Figure 1-3 shows in a schematic way the fertile somatic hybrids obtained within the Brassicaceae family between the six species from the U-triangle and other species belonging to other genera and tribes. A detailed summary of the intrageneric, intergeneric and intertribal somatic hybrids along with the traits of interest incorporated has been published by Glimelius (1999). The level of success in transferring useful genetic variation from wild sources through wide crosses depends on many factors: the extent of diversity that can be accessed to introduce useful variation, the risk to introduce deleterious traits, the possibility to use a particular valuable allele in different genetic backgrounds and the efficiency with which useful alleles can be transferred. However, most of the wild species are difficult to exploit in research programs mainly because of their extended vegetative phase or due to the difficulty in obtaining homozygous

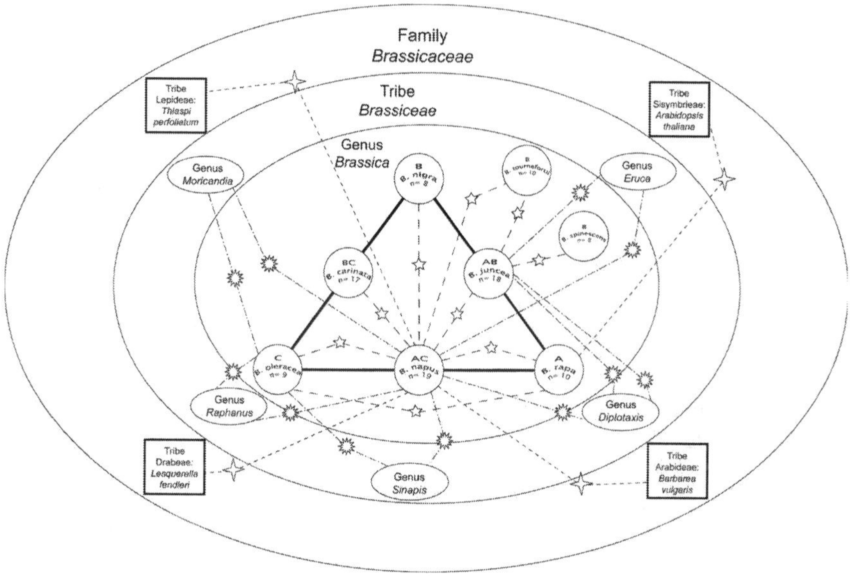

Figure 1-3 Graphic representation of the inter-specific (✫), inter-generic (✪), and inter-tribe (✧) hybrids from the species of the U-triangle within the Brassicaceae family (adopted from Glimelius 1999).

material (e.g., doubled haploid lines) by in vitro culture.

Acknowlegdements

The authors would like to thank and dedicate this chapter to Prof. César Gómez-Campo, who recently passed away and whose remarkable work has been a great inspiration and an enormous help for all of us.

References

Anilakumar KR, Khanum F, Bawa AS (2006) Dietary role of glucosinolate derivatives: a review. J Food Sci Technol (Mysore, India) 43: 8–17.

Arabidopsis Genome Initiative (Multiple Authorship)(2000) Analysis of the genome sequence of the flowering plant *Arabidopsis thaliana*. Nature 408: 796–815.

Arnison PG, Keller WA (1990) A survey of anther culture response of *Brassica oleracea* L. cultivars grown under filed condition. Plant Breed 104: 125–133.

Babula D, Kaczmarek M, Barakat A, Delseny M, Quiros CF, Sadowski J (2003) Chromosomal mapping of *Brassica oleracea* based on ESTs from *Arabidopsis thaliana*: complexity of the comparative map. Mol Genet Genom 268: 656–665.

Block G, Patterson B, Subar A (1992) Fruit, vegetables, and cancer prevention: A review of the epidemiological evidence. Nutr Cancer 18: 1–29.

Boukema IW, van Hintum THJL, Astley D (1997) Creation and composition of the *Brassica oleracea* core collection. Plant Genet Resour News 111: 29–32.

Branca F (2008) Cauliflower and broccoli. In: J Prohens, F Nuez (eds) Vegetables I: Asteraceae, Brassicaceae, Chenopodicaceae, and Cucurbitaceae. Handbook of Crop Breeding. Springer, Verlag, Berlin, Heidelberg, New York, USA, pp 151–186.

Carlson DG, Daxenbichler ME, Tookey HL (1987) Glucosinolates in turnip tops and roots: cultivars grown for greens and/or roots. J Am Soc Hort Sci 112: 179–183.

Cartea ME, Velasco P (2008) Glucosinolates in *Brassica* foods: bioavailability in food and significance for human health. Phytochem Rev 7: 213–229.

Cartea ME, Picoaga A, Soengas P, Ordás A (2003) Morphological characterization of kale populations from northwestern Spain. Euphytica 129: 25–32.

Cartea ME, Soengas P, Picoaga A, Ordás A (2005) Relationships among *Brassica napus* germplasm from Spain and Great Britain as determined by RAPD Markers. Genet Resour Crop Evol 52: 655–662.

Cassian RL, Echeverrigaray S (2000) Discrimination among cultivars of cabbage using random amplified polymorphic DNA markers. HortScience 35: 1155–1158.

Cavell AC, Lydiate DJ, Parkin IAP, Dean C, Trick M (1998) Collinearity between a 30-centimorgan segment of *Arabidopsis thaliana* chromosome 4 and duplicated regions within the *Brassica napus* genome. Genome 41: 62–69.

Christey MC, Braun RH (2004) Production of transgenic vegetable Brassicas. In: EC Pua , CJ Douglas (eds) Biotechnology in Agriculture and Forestry, vol 54: *Brassica*. Springer-Verlag, Berlin, Heidelberg, New York, USA, pp 169–194.

Ciska E, Martyniak-Przybyszewska B, Kozlowska H (2000) Content of glucosinolates in cruciferous vegetables grown at the same site for two years under different climatic conditions. J Agri Food Chem 48: 2862–2867.

Crozier A, Jaganath IB, Clifford MN (2006) Phenols, polyphenols and tannins: An overview. In: A Crozier , MN Clifford ,H Ashihara (eds) Plant Secondary Metabolites: Occurrence, Structure and Role in the Human Diet. Blackwell, Oxford, UK, pp 1–24.

Dixon GR (2007) Vegetable Brassicas and Related Crucifers. CAB Int, London, UK.

Fahey JW, Zalcmann AM, Talalay P (2001) The chemical diversity and distribution of glucosinolates and isothiocyanates among plants. Phytochemistry 56: 5–61.

Fahey JW, Haristoy X, Dolan PM, Kensler TW, Scholtus I (2002) Sulforaphane inhibits extracellular, intracellular, and antibiotic-resistant strains of *Helicobacter pylori* and prevents benzo[a]pyrene-induced stomach tumors. Proc Natl Acad Sci USA 99: 7610–7615.

FAOSTAT data (2009): *http://faostat.fao.org/site/567/default.aspx#ancor*

Farnham MW, Wilson PE, Stephenson KK, Fahey JW (2004) Genetic and environmental effects on glucosinolate content and chemoprotective potency of broccoli. Plant Breed 123: 60–65.

Fenwick GR, Griffiths NM, Heaney RK (1983) Bitterness in Brussels sprouts (*Brassica oleracea* L. var. *gemmifera*): the role of glucosinolates and their breakdown products. J Sci Food Agri 34: 73–80.

Fernandes F, Valentão P, Sousa C, Pereira CA, Seabra RM, Andrade PB (2007) Chemical and antioxidative assessment of dietary turnip (*Brassica rapa* var. *rapa* L.). Food Chem 105: 1003–1010.

Francisco M, Moreno D, Cartea ME, Ferreres F, García-Viguera C, Velasco P (2009) Simultaneous identification of glucosinolates and phenolic compounds in a representative collection of vegetable *Brassica rapa*. J Chromat A 1216: 6611–6619.

Giamoustaris A, Mithen R (1996) Genetics of aliphatic glucosinolates. IV. Side-chain modification in *Brassica oleracea*. Theor Appl Genet 93: 1006–1010.

Gigolashvili T, Berger B, Flugge UI (2009) Specific and coordinated control of indolic and aliphatic glucosinolate biosynthesis by R2R3-MYB transcription factors in *Arabidopsis thaliana*. Phytochem Rev 8: 3–13.

Glimelius K (1999) Somatic hybridization. In: C Gómez-Campo (ed) Biology of Brassica Coenospecies. Elsevier, Amsterdam, The Netherlands, pp 107–148.

Gómez-Campo C (1999) Taxonomy. In: C Gómez-Campo (ed) Biology of Brassica Coenospecies. Elsevier, Amsterdam, The Netherlands, pp 3–32

Gómez- Campo C (2003) The genus *Guenthera* Andr. In Bess. (Brassicaceae, Brassiceae). Ana Jard Bot (Madrid) 60: 301–307.

Gómez-Campo C, Prakash S (1999) Origin and domestication. In: C Gómez-Campo (ed) Biology of *Brassica* Coenospecies. Elsevier, Amsterdam, The Netherlands, pp 33–58.

Grubb CD, Abel S (2006) Glucosinolate metabolism and its control. Trends Plant Sci 11: 89–100.

Halkier BA, Du L (1997) The biosynthesis of glucosinolates. Trends Plant Sci 2: 425–431.

Hecht SS (2000) Inhibition of carcinogenesis by isothiocyanates. Drug Metab Rev 32: 395–411.

Himelblau E, Lauffer D, Teutonico R, Pires JC, Osborn TC (2004) Rapid-cycling *Brassica* in research and education. In: EC Pua ,CJ Douglas (eds) Biotechnology in Agriculture and Forestry, vol 54. *Brassica*. Springer-Verlag, Berlin Heidelberg, New York, pp 13–27.

Hu J, Quiros CF (1991) Identification of broccoli and cauliflower cultivars with RAPD markers. Plant Cell Rep 10: 505–511.

Jahangir M, Kim HK, Choi YH, Verpoorte R (2009) Health-affecting compounds in Brassicaceae. Comp Rev Food Sci Food Safety 8: 31–43.

Kalloo G, Bergh BO (1993) Genetic improvement of vegetable crops. Oxford, Pergamon Press, Oxford, UK.

Keller WA, Rajhathy T, Lacapra J (1975) *In vitro* production of plants from pollen in *Brassica campestris*. Can J Gen Cytol 17: 655–666.

Kuginuki Y, Miyajima T, Masuda H, Hida K, Hirai M (1999) Highly regenerative cultivars in microspore culture in *B. oleracea* L. var. *capitata*. Breed Sci 49: 251–256.

Kushad MM, Brown AF, Kurilich AC, Juvik JA, Klein B, Wallig MA, Jeffery EH (1999) Variation of glucosinolates in vegetable subspecies of *Brassica oleracea*. J Agri Food Chem 47: 1541–1548.

Lan T, Paterson A (2000) Comparative mapping of quantitative trait loci sculpting the curd of *Brassica oleracea*. Genetics 155: 1927–1954.

Llorach R, Gil-Izquierdo A, Ferreres F, Tomás-Barberán FA (2003) HPLC-DAD-MS/MS ESI characterization of unusual highly glycosylated acylated flavonoids from cauliflower (*Brassica oleracea* L. var. *botrytis*) agroindustrial byproducts. J Agri Food Chem 51: 3895–3899.

Lucarini M, Canali R, Cappelloni M, Di Lullo G, Lombardi-Bocia G (1999) *In vitro* calcium availability from Brassica vegetables (*Brassica oleracea* L.) and as consumed in composite dishes. Food Chem 64: 519–523.

Luo P, Lan ZQ, Li ZY (1994) *Orychophragmus-violaceus*, a potential edible-oil crop. Plant Breed 113: 83–85.

Mithen RF, Dekker M, Verkerk R, Rabot S, Johnson IT (2000) The nutritional significance, biosynthesis and bioavailability of glucosinolates in human foods. J Sci Food Agri 80: 967–984.

Mithen R, Faulkner K, Magrath R, Rose P, Williamson G, Marquez J (2003) Development of isothiocyanate-enriched broccoli and its enhanced ability to induce phase 2 detoxification enzymes in mammalian cells. Theor Appl Genet 106: 727–734.

Musgrave ME (2000) Realizing the potential of rapid-cycling *Brassica* as a model system for use in plant biology research. J Plant Growth Regul 19: 314–325.

Nielsen JK, Olsen CE, Petersen MK (1993) Acylated flavonol glycosides from cabbage leaves. Phytochemistry 34: 539–544.

Nishi S (1980) Differentiation of Brassica crops in Asia and the breeding of 'Hakuran', a newly synthesized leafy vegetable. In: C Gómez-Campo (ed) Brassica Crops and Wild Allies. Japan Scient Soc Press, Tokyo, Japan, pp 133–150.

Padilla G, Cartea ME, Rodríguez VM, Ordás A (2005) Genetic diversity in a germplasm collection of *Brassica rapa* subsp. *rapa* L. from northwestern Spain. Euphytica 145: 171–180.

Padilla G, Cartea ME, Soengas P, Ordás A (2007a) Characterization of fall and spring plantings of Galician cabbage germplasm for agronomic, nutritional, and sensory traits. Euphytica 154: 63–74.

Padilla G, Cartea ME, Velasco P, de Haro A, Ordás A (2007b) Variation of glucosinolates in vegetable crops of *Brassica rapa*. Phytochemistry 68: 536–545.

Pink D, Bailey L, McClement S, Hand P, Mathas E, Buchanan-Wollaston V, Astley D, King G, Teakle G (2008) Doubled haploids, markers and QTL analysis in vegetable Brassicas. Euphytica 164: 509–514.

Podsedek A (2007) Natural antioxidants and antioxidant capacity of *Brassica* vegetables: A review. LWT Food Sci Technol 40: 1–11

Prakash S, Hinata K (1980) Taxonomy, cytogenetics, and origin of crop *Brassica*, a review. Opera Bot 55: 1–57.

Price KR, Casuscelli F, Colquhoun IJ, Rhodes MJC (1997) Hydroxycinnamic acid esters from broccoli florets. Phytochemistry 45: 1683–1687.

Pyrzynska K (2009) Selenium speciation in enriched vegetables Food Chem 114: 1183–1191.

Quiros CF (2001) DNA-based marker *Brassica* maps (2001) In: RL Phillips , IK Vasil (eds) Advances in Cellular and Molecular Biology of Plants, vol I. DNA Based Marker in Plants Kluwer Academic Publ, Dordrecht, The Netherlands, pp 201–238.

Quiros CF, Paterson AH (2004) Genome mapping and analysis. In: EC Pua , CJ Douglas (eds) Biotechnology in Agriculture and Forestry. Vol. 54. *Brassica*. Springer, Berlin, Heidelberg, New York, pp 31–42.

Quiros CF, Grellet F, Sadowski J, Suzuki T, Li G, Wroblewski T (2001) *Arabidopsis* and *Brassica* comparative genomics: sequence, structure and gene content in the *ABI1-Rps2-Ck1* chromosomal segment and related regions. Genetics 157: 1321–1330.

Qiu D, Muqiang G, Genyi L, Quiros CF (2009) Comparative sequence analysis for *Brassica oleracea* with similar sequences in *B. rapa* and *Arabidopsis thaliana*. Plant Cell Rep 28: 649–661.

Rawsthorne S, Morgan CL, O'Neill CM, Hylton CM, Jones DA, Frean ML (1998) Cellular expression pattern of the glycine decarboxylase P protein in leaves of an intergeneric hybrid between the C-3–C-4 intermediate species *Moricandia nitens* and the C-3 species *Brassica napus*. Theor Appl Genet 96: 922–927.

Ren J, McFerson JR, Kresovich S, Lamboy WF (1995) Identities and relationships among Chinese vegetable Brassicas as determined by random amplified polymorphic DNA markers. J Am Soc Hort Sci 120: 548–555.

Rodríguez VM, Cartea ME, Padilla P, Velasco P, Ordás A (2005) The nabicol: a horticultural crop in northwestern Spain. Euphytica 142: 237–246.

Rosa EAS (1999) Chemical composition In: C Gómez-Campo (ed) Biology of *Brassica* Coenospecies. Elsevier Science BV, Amsterdam, The Netherlands, pp 315–357.

Rosa EAS, Heaney RK, Fenwick GR, Portas CAM (1997) Glucosinolates in crop plants. HortRev 19: 99–215.

Sadowski J, Gaubier P, Delseny M, Quiros CF (1996) Genetic and physical mapping in *Brassica* diploid species of a gene cluster defined in *Arabidopsis thaliana*. Mol Gen Genet 251: 298–306.

Scalzo RL, Genna A, Branca F, Chedin M, Chassaigne H (2008) Anthocyanin composition of cauliflower (*Brassica oleracea* L. var. botrytis) and cabbage (*B. oleracea* L. var. capitata) and its stability in relation to thermal treatments. Food Chem 107: 136–144.

Shashilova LI (1997) The Russian *Brassica* collection. In: Report of a Working Group on Brassica, Third Meeting, 27–29 Nov 1996, Rome, Italy. International Plant Genetic Resources Institute, Rome, Italy, pp 56–57.

Sjödin C, Glimelius K (1988) Screening for resistance to blackleg *Phoma lingam* within Brassicaceae. J Phytopathol 123: 322–332.

Smith TK, Lund EK, Clarke RG, Bennet RN, Johnson IT (2005) Effects of Brussels sprout juice on the cell cycle and adhesion of human colorectal carcinoma cells (HT29) *in vitro*. J Agri Food Chem 53: 3895–3901.

Soengas P, Velasco P, Padilla G, Ordás A, Cartea ME (2006) Genetic relationships among *Brassica napus* crops based on SSRs markers. HortScience 41: 1195–1199.

Song K, Tang K, Osborn TC (1993) Development of synthetic *Brassica* amphidiploids by reciprocal hybridization and comparison to natural amphidiploids. Theor Appl Genet 86: 811–821.

Song K, Lu P, Tang K, Osborn T (1995) Rapid genome change in synthetic polyploids of *Brassica* and its implications for polyploid evolution. Proc Natl Acad Sci USA 92: 7719–7723.

Stout SC, Porterfield DM, Briarty LG, Kuang A, Musgrave ME (2001) Evidence of root zone hypoxia in *Brassica rapa* L. grown in microgravity. Int J Plant Sci 162: 249–255.

Swarup V, Brahmi P (2005) Cole crops. In: BS Dhillon, RK Tyagi, S Saxena, GJ Randhawa (eds) Plant Genetic Resources: Horticultural Crops. Narosa Publ House, New Delhi, India, pp 75–88.

Talalay P, Zhang Y (1996) Chemoprotection against cancer by isothiocyanates and glucosinolates. Biochem Soc Trans 24: 806–810.

Talalay P, Fahey JW (2001) Phytochemicals from cruciferous plants protect against cancer by modulating carcinogen metabolism. J Nutr 131: 3027–3033.

Traka M, Mithen R (2008) Glucosinolates, isothiocyanates and human health. Phytochem Rev 8: 269–282.

U N (1935) Genome analysis in *Brassica* with special reference to the experimental formation of *B. napus* and peculiar mode of fertilization. Jpn J Bot 7: 389–452.

U.S. Department of Agriculture, Agricultural Research Service (2008) USDA National Nutrient Database for Standard Reference, Release 21. Nutrient Data Laboratory Home Page: *http://www.ars.usda.gov/ba/bhnrc/ndl*

Vallejo F, Tomás-Barberán FA, García-Viguera C (2003) Health promoting compounds in broccoli as influenced by refrigerated transport and retail sale period. J Agri Food Chem 51: 3029–3034.

Vallejo F, Tomás-Barberán FA, Ferreres F (2004) Characterisation of flavonols in broccoli (*Brassica oleracea* L. var. *italica*) by liquid chromatography-UV diode-array detection-electrospray ionisation mass spectrometry. J Chromatogr A 1054: 181–193.

VanEtten CH, Daxenbichler ME, Williams PH, Kwolek F (1976) Glucosinolates and derived products in cruciferous vegetables. Analysis of the edible part from twenty-two varieties of cabbage. J Agri Food Chem 24: 452–455.

Van Soest LJM, Boukema IW, Bas N (2004) The achievements of the EU genres CT99 109–112 project Brassica, including *B. carinata*. Crucifer Newsl 25: 117–119.

Velasco P, Cartea ME, González C, Vilar M, Ordás A (2007) Factors affecting the glucosinolate content of kale (*Brassica oleracea acephala* group). J Agri Food Chem 55: 955–962.

Vilar M, Cartea ME, Padilla G, Soengas P, Velasco P (2008) The potential of kales as a promising vegetable crop. Euphytica 159: 153–165.

Voss A, Snowdon RJ, Luhs W, Friedt W (2000) Intergeneric transfer of nematode resistance from *Raphanus sativus* into the *Brassica napus* genome. In: GJ King (ed) Proc 3th Int Symp on Brassicas and 12th Crucifer Genet Workshop, Wellesbourne, UK, pp 129–134.

Warwick SI, Sauder CA (2005) Phylogeny of tribe Brassiceae (Brassicaceae) based on chloroplast restriction site polymorphisms and nuclear ribosomal internal transcribed spacer and chloroplast *trn*L intron sequences. Can J Bot 83: 467–483.

Williams PH, Hill CB (1986) Rapid cycling populations of *Brassica*. Science 232: 1385–1389.

Zhao J, Wang X, Deng B, Lou P, Wu J, Sun R, Xu Z, Vromans J, Koornneef M, Bonnema G (2005) Genetic relationship between *Brassica rapa* as inferred from AFLP fingerprints. Theor Appl Genet 110: 1301–1314.

Zhang Y, Talalay P (1994) Anticarcinogenic activities of organic isothiocyanates: chemistry and mechanisms. Cancer Res 54: 1976–1981.

2

Classical Breeding and Genetic Analysis of Vegetable Brassicas

Nirala Ramchiary,[1] *Suhyoung Park,*[2] and *Yong Pyo Lim*[1*]

ABSTRACT

Brassicas provide diverse groups of vegetables to mankind in the form of leaves, flowers, stems, and roots. Apart from providing many nutrients including vitamins, minerals, and antioxidants, they are also a potential source of large amounts of dietary fiber. Among the cultivated Brassicas, *Brassica oleracea* and *Brassica rapa* have a maximum diversity of forms which are grown worldwide. Realizing the potential importance of vegetable Brassicas, breeders, since the early part of the 20th century were successful in improving quantity and quality of vegetable Brassicas using diverse morphological and genetic resources. Genetic analysis of economically important traits such as leaf, head size, plant architecture and resistance to insect pests and diseases could identify many genetic resources which could be incorporated into the breeding program. As a result, improved cultivars and hybrid varieties that give high yield, good quality, desired plant architecture and resistance/tolerance to biotic and abiotic stresses of cabbage, cauliflower, Brussels sprout, broccoli, Chinese cabbage, pakchoi etc were developed by breeders. In this chapter, we briefly described the early works by many researchers that involved classical genetic analysis and breeding which led to successful breeding of vegetable Brassicas.

Keywords: vegetable Brassicas, *Brassica rapa*, *Brassica oleracea*, breeding, genetic analysis

[1]Department of Horticulture, Chungnam National University, Gung-Dong, Yuseong-Gu, Daejeon 305-764, Korea.
[2]Vegetable Division, National Institute of Horticultural and Herbal Science, RDA, Suwon 441-440, Korea.
*Corresponding author: *yplim@cnu.ac.kr*

2.1 Introduction

Brassicas are widely known for its diverse groups providing vegetables, fodder and vegetable oils to mankind. *Brassica* vegetables provide leaves, flowers, stems, and roots to man, which are eaten raw or cooked and as a preserved food. Today perhaps vegetable brassicas are the most important vegetable crops grown worldwide as they are a highly diversified group of crops belonging mainly to the species *Brassica oleracea* and *B. campestris*, and to a lesser extent *B. juncea* and *B. napus*. They are rich in many nutrients including vitamins, minerals, and antioxidants and largely supply dietary fiber to mankind. Among the four species of Brassicas grown for vegetables, *B. oleracea* and *B.rapa* have maximum diversity of forms, which are grown worldwide. However, predominance of vegetable brassica cultivation in the western hemisphere including Europe goes to *B. oleracea*, while *B.rapa* is mostly cultivated in Asian countries especially in China, Korea and Japan. The vegetable form of *B. juncea* cultivation and its evolution to different forms are mostly limited to China. The cultivated vegetable *B. oleracea* are found in various types, which are largely divided into six groups i.e. kales comprising green kale, marrow stem kale, collards mainly used for edible forage (var. *acephala*); cabbages (var. *capitata*, var. *sabauda*, var. *bullata*) including headed cabbages, brussel sprouts, savoy cabbage, and others; kohlrabi (var. *gongylodes*); inflorescence kales (var. *botrytis*, var. *italica*) including cauliflower, broccoli, sprouting broccoli, and others; branching bush kales (var. *fruticosa*); and Chinese kale (*B. alboglabra*), used as a leaf vegetable (Snogerup et al. 1990; Rakow 2004). Seven groups of vegetable *B. rapa* types are known, and these are: var. *campestris*, var. *pekinensis*, var. *chinensis*, var. *parachinensis*, var. *narinosa*, var. *japonica* and var. *rapa*. Chinese cabbage and bok choi is cultivated in China, Korea and in Japan. In Korea Chinese cabbage is eaten in the form of preserved food "Kimchi" and is regarded as a national food. Turnip and turnip greens, which also belong to *B. rapa*, are cultivated worldwide. Different groups within *B. oleracea* and *B. rapa* are intercrossable, and the genetic study and transfer of traits are easy. In this chapter we discuss the early genetic studies and breeding of vegetable brassicas especially in different subgroups of *B. oleracea* and *B. rapa*. However, due to limited literature, only a few major crops belonging to these two species are covered. For details of the origin and diversity see Chapter 1.

2.2 *Brassica oleracea*

2.2.1 *Early Development and Breeding History of Brassica Vegetables*

Different *Brassica* vegetables have different origins and breeding history. It is believed that most of the present day cultivated forms of *B. oleracea* have originated from native wild forms, which were found growing originally in the Atlantic coasts of Europe especially northern Spain, western France, southern and southwestern Britain (Snogerup 1980; Snogerup et al. 1990; Hodgkin 1995; Lázaro and Aguinaglade 1998). Hodgkin (1995) believed that this wild species had been brought to the Mediterranean region from the western Atlantic European coast where selection for different morphological types created the present diverse cultivated forms of *B. oleracea*. Further, it is believed that all cole crops originated from the Mediterranean regions through mutation and introgression from wild species during evolution or human selection (Swarup and Brahmi 2005; for details of origin see Chapter 1). *B. oleracea* evolved into a wide range of various subgroups or cultivars that are presently cultivated worldwide for edible roots, leaves, flower stalks or swollen stem. It is shown that plants from different subgroups belonging to this species are closer genetically than morphologically as they hybridize easily among each other. Breeding for different types, forms, quality, duration and other morphological traits by people from different regions created lots of diversity within each subgroup. In the following section the early breeding history of a few major *B. oleracea* crops are described. The authors adapted most of the contents from the review of Shinohara (1984) for the breeding history due to limited literature from other sources/forms.

2.2.1.1 Cabbage

Cabbage is grown for the head formed by several wrapper leaves surrounding the terminal bud. Cabbage or commonly known as heading cabbage developed into a wide range of basic groups, types and forms in West Europe during the 12th to 16th centuries (Shinohara 1984). It is from there that different types of cabbage developed such as common cabbage, red cabbage and Savoy cabbage. The common cabbage group is the widest in diversity, with green leaves, and consists of different head forms such as pointed, drum, oval or round heads of different duration for bolting and flowering and leaf thickness. The red cabbage group is mostly round shaped, thick leaved and susceptibile to hot weather, deep purplish color leaves with different duration or maturity. Savoy cabbages are with crinkled or curling green leaves and deep veins, mostly drumhead shaped.

Cabbage was first introduced to America during the 16th century from Europe and supply of seed materials from Europe continued till World War I. The breeding of brassica vegetables crops in the United States started with the basic material such as cultivars or varieties developed in Germany, Denmark or the Low countries (Shinohara 1984; Dickson and Wallace 1986). The first local variety bred was Charleston Wakefield or Large Wakefield from unfixed population of early Jersey Wakefield in the 1880s (Shinohara 1984). After World War I due to the blockade of seed supply from Europe, the US Government research institutes and people in the seeds business collaborated to develop superior seed stocks for every vegetable crop and established suitable locations for breeding vegetables and started various activities of local vegetable breeding. As a result, the Pacific side of North-West America like Oregon and California were designated as cabbage seed production areas during that time. Breeding for cabbage yellows (Fusarium wilt) resistance started in the United States due to the outbreak of this disease during the beginning of the 20th century. Jones and Gilman in 1911 selected two plants (from Wisconsin grown Hollander and Danish Ballhead cultivars) which survived from highly infected soils at the Agricultural Experimental Station of Kensha County, and made a cross between them. F_2 plants showing 2 to 23% (as against parents 89%) infection were selected for further research by growing the plants in infested soil in 1914. Further selection was done from this basic material up to 1919, by Jones, Walker and Tisdale of Wisconsin Agricultural Experiment Station (Shinohara 1984). This first work of successful resistance breeding encouraged systematic and sophisticated resistance breeding of vegetables in the United States. The Wisconsin Hollander is recorded since 1920. According to Shinohara (1984) the nine most popular varieties cultivated in America before World War II were : Early Jersey Wakefield (early pointed head variety listed), Charleston Wakefield (medium, early, pointed head), Copenhagen market (early, round head), early Winnistadt (medium, pointed but broad based conic head), Glory of Enkhuizen (medium early, round head), All Seasons Vandergaw (medium late, deep drum head), Late Flat Dutch (derived from Brunswick (late, typical drumhead), Danish Ballhead (a strain of Amager —late, round head) and Wisconsin Hollander (late, flatted round head). It is reported that during World War II, the United States became not only self-sufficient to produce sufficient seeds for domestic markets but also became the world's cabbage seed supplier to the allied countries replacing the West European countries.

Until World War II introduction of European, American and Japanese cabbage cultivars was reported in China and by the 1970s cultivation of local varieties were recorded (Shinohara 1984). In Japan, heading cabbage was reported to be introduced from Europe and the US in 1868. Shinohara (1984) reported that due to the climatic similarities with those of Europe and the US, Hokkaido and the north part of Tohoku were the first to grow heading cabbage and develop acclimatized local varieties such as Sapporo Daikyu (large drumhead) and Nambu (medium drumhead) from Vandergaw and Late Flat Dutch at the beginning of the 20th century. Early types of cabbage were developed from many imported cultivars such as Copenhagen Market, Early Crops, etc. However, premature bolting was a big problem in other parts of Japan due to unfavorable climatic conditions. To overcome this problem extensive effort was paid to breed non-bolting heading cabbage. Kurato Nakano near Tokyo and Jiro Ishi from Shizuoka bred non-bolting early types. Kurato Nakano bred two varieties of NAKANO-WASE, the earliest and the early cabbage from the progenies derived from a cross between Early Spring and Sutton`s Earliest (in the year 1916) and progenies of Henderson`s All Head Early and some lines of the former (in the year 1921). Jiro Ishi bred TOYODA-WASE from progenies of hybrid between Henderson`s Early Spring and Sutton`s Flower of Spring in 1917. These two Types, NAKANO-WASE and TOYODA-WASE were the most popularly cultivated non-bolting cultivars developed indigenously in Japan for the autumn season and extended worldwide for its cultivation. These success stories encouraged others to breed more local non bolting cabbage (Shinohara 1984). The improved varieties developed by them were dark green types because these types were tolerant to diseases and easier to handle. Looking at the success of development of autumn cultivars, efforts were made to develop summer growing cultivars using Early Summer and the progenies of different crosses. As a result Nozaki-Natsumahi was developed by T. Nozaki in 1916 in Chita peninsula, Takashi at Atsumi peninsula in around 1930, and Senshu in Senshu in 1920. However, the systematic seed production and breeding of local varieties of vegetables including cabbage came into force after World War II in 1940 due to the blockade of seed supplies from other countries. The Japanese Government consolidated all Japanese seed companies of all of Japan into 18 central seed management companies and forced them to set up systematic breeding and seed testing farms with the help of central and local government research institutes. This project involved collection, analysis and systematic classification of all the vegetables grown in Japan along with their ecological history to identify superior seed resources and set ecologically suited seed production areas specific to crops. This project resulted in the systematic development of seed production in Japan, which not only made it self-sufficient but also the world`s best exporter of cabbage seeds. Development of varieties for year-round growing season was achieved between the 1950s and 1960s by selection of traits for adaptablity to a particular environment or combining the traits from different genotypes and made suitable for growing in different seasons and environments. The first commercial F_1 hybrids were produced by Takii seed company in the 1940s as O-S cross. The more improved F_1 hybrid with heterosis, uniformity and wide adaptability for every season named SHIKIDOORI was developed by Takii Seed Company in 1955 that became widely adapted in the whole of Japan. After that several F_1 hybrids were developed using the local and imported seed parents from foreign countries during 1955 to 1965. This hybrid breeding became one of the best choices for breeders as in heterotic F_1 hybrids they could combine traits from both the parents, besides being able to produce easily using self-incompatibility. Later these hybrids were exported to foreign countries as well.

2.2.1.2 Cauliflower

Cauliflower is believed to be originated from broccoli (Crisp 1982; Gray 1982). The economic part of the cauliflower is curd. The first cauliflower cultivation known as "Originals" was reported from Mediterranean Europe especially in Italy during the 15th century (Swarup and Chatterjee 1972). The spread of this crop to France and the United Kingdom took place during the 16th century with the name of "Cyprus kale". During the 17th and 18th century cauliflower spread all over Europe including Germany and the Netherlands and many local varieties were developed from them. The notable examples are "Northerns" in Yorkshire and Derbyshire, the "Cornish" in Cornwall, the "Angers" and "Roscoff" in Brittany, and "Erfurt" or it's allied "Snowball" in Germany and in the Netherlands (Swarup and Chatterjee 1972). The chief characteristics of the different types of cauliflowers developed in different regions are provided in detail by Swarup and Chatterjee (1972) and by Crisp (1982) and are presented in Table 2-1 and Table 2-2. Cornish were the main favorite in the United Kingdom until 1920 when it was replaced by white curded "Roscoff" from France. In the US, commercial cultivation of cauliflower started in 1920 and both the open-pollinated cultivars and F_1s were selected from snowball cultivars derived from the Erfurt type that was characterized by small compact plants and white compact curds (Branca 2008). Mid season cultivars with larger plants were derived from the French type and late autumn season cultivars with larger head were derived from the Italian types. In India, cauliflower was introduced in 1822 from the United Kingdom and in 1824 from South Africa while in Japan, cauliflower was introduced during the late 19th century (Swarup and Chatterjee 1972). In Italy, cauliflower is believed to have been first cultivated/domesticated in the form of "Originals" that consisted of all types of variations including flower stalk masses and curd colors.

Only early to medium early types of cauliflower are grown in western European countries due to the cold climatic conditions. These are grown mainly in spring and harvested in late summer upto autumn, while around the English channels late varieties of broccoli are grown in early autumn and harvested in spring after over-wintering (Shinohara 1984). The development of medium and medium-late types of cauliflower were developed by the French in the south of France and in its colonized land of Algeria where the winter is mild enough to harvest cauliflower from late autumn through winter. Cauliflower is highly sensitive to frost during harvesting time and for seed production. So its cultivation has been popularized in subtropical Asia and the Indian subcontinent in South East Asia due to the long moderate

Table 2-1 Characteristics of different cauliflower groups given by Swarup and Chatterjee (1972).

Cauliflower types	Origin	Probable date of cultivation	Characters
Originals or Italians	Mediterranean	16th Century	Plants short; leaves erect, broad with round tips, bluish green; curds good, not protected.
Erfurt & Snowball	Germany & the Netherlands	18th Century	Plantsdwarf; leaves short, erect glaucous green; curds solid, well protected.
Cornish	England	Early 19th Century	Plants vigorous, long stalked; leaves loosely arranged, broadly wavy; curds flat, ir regular, loose, not protected, yellow, highly, flavored.
Northerns	England	19th Century	Leaves petiolate, broad, very wavy, serrated; curds good, well protected.
Roscoff	France	19th Century	Plants short; leaves long, erect, slightly wavy with pointed tip, midrib prominent, bluish green; curds white or creamy, hemispherical, well protected.
Angers	France	19th Century	Leaves very wavy, serrated, grayish green; curds solid, white, well protected.
Indian cauliflower	India	Late 19th Century	Plants short, long stalked; leaves loosely arranged, broadly wavy; curds flat, somewhat loose, yellow to creamy, not protected, highly flavored.

*Adapted from Sharma et al. (2005) originally given by Swarup and Chatterjee (1972).

monsoon and a humid climate (neither too cold nor too hot) which is best suited for cauliflower cultivation. Lots of varieties such early to medium to late have been developed in these areas especially in India, China and Taiwan (Shinohara 1984; Sharma et al. 2005). In India, cauliflower is a major vegetable crop and all types of varieties ranging from early to late have been developed of which the earliest variety is "Patna Early" also known as a tropical type and it is believed that Taiwan`s earliest variety and Japan`s extra early varieties have been developed from Patna Early. T. NOZAKI bred the NOZAKI cauliflower variety from Early Snow Ball in 1916 and many varieties were available after the 1920s. The breeding of many local varieties of cauliflower was achieved by several people in the 1940`s. The notable examples are Masuda early, medium and late by T. Masuda. The development of F_1 hybrid cauliflower was started in Japan by many seed companies by using introduced European Early Snow Ball cultivar (maturing in October) and tropical early cultivars from Taiwan as parents to get extra early F_1 hybrids (which are harvested in mid-September) in the 1950s. Presently Japan has all types of cauliflower varieties ranging from early to late. Most of the cauliflowers are white curded but green or violet curded ones are also found among Italian landraces. At present most of the open pollinated cultivars and landraces are replaced by more uniform and high curd quality F1 hybrids.

Table 2-2 Characteristics of different cauliflower groups given by Crisp (1982).

Group	Chief characteristics	Common types
Italian	Very diverse, include both annuals and biennials and curds with peculiar conformations and colors	Jezi, Naples (Autumn Giant), Romanesco, Flora Blanca
North-West European biennials	Derived within the last 300 years from Italian material	Old English, Walcheran, Roscoff, Angers, St. Malo
North European annuals	Developed in Northern Europe for at least 400 years. Origin unknown, perhaps Italian or Eastern Mediterranean	Lecerf, Alpha, Mechelse, Erfurt, Danish
Asian	Recombinants of European annuals and biennials developed within 250 years, adapted to tropics	Four maturity groups are recognized by Swarup and Chatterjee (1972)
Australian	Recombinants of European annuals and biennials and perhaps Italian stock, developed during the last 200 years	Not yet been categorized

*Adapted from Sharma et al. (2005) originally given by Crisp (1982).

2.2.1.3 Broccoli

Broccoli is grown for its economic part known as the curd. Viani (1929) reported that broccoli resulted from wild cabbage through successive improvement. The cultivation of the first form of broccoli took place along the Italian Peninsula and selection resulted in several types of sprouting and a wide range of head size (Branca 2008). Usually the curd of broccoli is green due to the the presence of chlorophyll but landraces of light red color (due to presence to anthocyanin pigment) are also found in the southernmost part of Italy. The widespread presence of landraces in this region are distinguished by harvesting time, cold requirement for flower induction, sprouting habit, leaf color and shape, head size, grain and angle of leaf curvature (Branca 2008). Cultivars with small heads with good aroma content are also found there. The presence of widespread diversity is also reported in Sicily. It is believed that from Italy and Sicily, broccoli was introduced to the United Kingdom and the northern European countries during the 17th and 18th century in Spain (Branca 2008). In North America, broccoli crops were widespread during the 19th century, which was introduced by Italian emigrants. Most of the variety development (single terminal head type) and its cultivation have been extensively done in the US and Japan during the 1950s. Recently, its development and cultivation has been extended to South East Asia including South Taiwan, Thailand and Japan but it has been found that it has a low adaptability in subtropics as compared to cauliflower. NAKAZATO-WASE broccoli was the first local variety bred in Japan by T. NAKAZATO in 1962 (Shinohara 1984) and in the 1970s major seed companies in Japan bred many local varieties of broccoli including F_1 hybrids. Presently green head/curd F_1 hybrids with single small compact to big head having one or two secondary heads and early type are mostly cultivated. At present more uniform and high quality F_1 hybrids are quickly replacing old landraces and open-pollinated varieties developed earlier.

2.2.1.4 *Brussels Sprouts*

Brussels sprouts were believed to have been cultivated in Italy during the Roman times, and possibly as early as the 1200s in Belgium. The modern Brussels sprouts that we are familiar with were first cultivated in large quantities in Belgium, the capital—Brussels as early as 1587 and became prominent in the 18th century. From there, it spread to northern Europe particularly to the Netherlands and parts of UK and become established as an important vegetable crop (Dixon 2006). The development of local open-pollinated landraces took place in these areas and breeding for specific traits according to the needs were developed. Development of early maturing type and F_1 hybrids started in Japan in the 1930s, which later formed the base material for breeding cultivars suited for the freezing industry. Production of Brussels sprouts in the US began around 1800, when French settlers brought them to Louisiana. The first plantings in California's Central Coast began in the 1920s, with significant production beginning in the 1940s. Breeding for different local types started there. With the development of the frozen food industry in the 1940s, the production of Brussels sprouts in California increased to its highest levels over the next 20 years. Presently, F_1 hybrids dominate the worldwide Brussels sprouts cultivation developed from the Netherlands, America and Japan (Dixon 2006). Current objectives of breeding is development of cultivars with regularly spaced, tight, high density sprouts , erect stem and good leaf protection.

2.2.2 Conventional Genetic Analysis in Brassica oleracea Vegetables

Although genetics was not much developed at the beginning of the 20th century, genetic analysis was done for various traits in vegetable brassicas to determine the number and the nature of gene action. The most important traits in vegetable brassicas are shape, size, formation of head and curd, leaf traits, plant and leaf color and other plant morphologies besides diseases and pest resistances. Brassica vegetables having inter-crossable subgroups with wide morphological diversity is very suitable for genetic study by crossing them and studying the trait segregation. Taking these advantages, several genetic analyses has been done by previous researchers in *B. oleracea,* which are extensively reviewed by Ahluwalia et al. (1977), Chatterjee (1993), King (1990) and Yarnell (1956). Detailed gene lists and gene symbols of *B. oleracea* were given by Ahluwalia et al. (1977) and Yarnell (1956) while compilation of gene lists and their linkage information from morphological data was done by Wills (1977). The followings are some of the studies which showed various reports of genetic analysis in vegetable brassicas especially in *B. oleracea.* We adapted most of the references reviewed by Yarnell (1956) for genetic analysis and early linkage analysis and partly from King (1990) as many old and original research papers could not be retrieved. Genetic analysis for different traits along with the number of genes, gene action and references are also provided in Table 2-3 and Table 2-4.

2.2.2.1 Genetics of Head-related Traits

B. oleracea vegetables can be broadly classified into two groups, heading type and non-heading type. The head is an important economic part of cabbage that distinguishes it from other *B. oleracea* groups. Studies involving different genotypes for heading had different opinions about the number of genes involved in the head or heart formation in cabbage by several wrapper leaves surrounding the terminal bud. Genetic analyses by several workers have given a conclusive general agreement that heading is recessive to non-heading and intermediate condition exists in crosses between cabbage and other *B. oleracea* cultivar groups. Pease (1926) in a study involving cabbage and kale reported that heading is governed by two recessive factors/genes $n_1n_1n_2n_2$. He observed F_1 of the cross with a rosette incurved leaves showing paint head and recovery of both the parental types in F_2 besides the observation of continuous variation between the two extremes. $N_1N_1n_2n_2$ and $n_1n_1N_2N_2$ are considered to give the same effect as $N_1n_1N_2n_2$. He obtained similar results while studying cabbage and kohlrabi crosses. This was confirmed by obtaining ratios of progenies 1 $N_1n_1N_2n_2$ (very slight heading):1 $N_1n_1n_2n_2$: 1 $n_1n_1N_2n_2$ (both slight heading): 1 $n_1n_1n_2n_2$ (heading) while backcrossing to cabbage. It is also reported that the loose head of the Savoy cabbage is recessive to the hard head of smooth-leaf cabbage. Allgayer (1928), in a study involving cabbage and kitchen kale cross, proposed one major dominant and three recessive minor factors for heading *KKk1k1k2k2k3k3*. In the second generation of five such crosses he got 562 heading and 168 non-heading plants indicating major dominant gene action.

Table 2-3 Classical genetic analysis of few important traits in *B. oleracea*.

Trait	Crop/subspecies	Identified gene, if any	Nature of gene action	References
Morphology				
Plant height	*B. oleracea* ssp *capitata*	1 major gene many	Tall is dominant over short Unknown	Pease 1926 Magruder 1937 Kwan, 1934
Head shape	*B. oleracea* ssp *capitata*	many	Pointed dominant over round	Tschermak 1916
Heading	*B.oleracea* ssp *capitata*	2 major gene 4	Heading is recessive to non-heading (studied in progenies derived from a cross between cabbage and kale) 1 major dominant, 3 minor recessive (studied in progenies derived from a cross between cabbage and kitchen kale)	Pease 1926 Allgayer 1928
Leaf width	*B. oleracea* ssp *capitata*	1 major gene	Single major dominant gene causes differences between cabbage and kohlrabi	Malinowski 1929
Entire and petiolate leaf	*B. oleracea* ssp *capitata* *kohlrabi*	2 closely linked genes	Closely linked dominant genes Cabbage genotype *petpetEnEN* Kohlrabi genotype *PetPetenen* Brussels sprouts genotype *PetPetEnEn*	Pease 1926
Crinkled Vs Smooth	*B. oleracea* ssp *capitata*	2 genes	Complementary gene action	Kwan 1934
Curly and non curly	*B. oleracea* ssp *capitata* and ssp *acephala*	polygenic	Polygenic/unknown	Magruder 1929
Ascorbic acid	*B. oleracea* ssp *capitata*	several	Polygenic/unknown	Walker and Foster 1946
Head maturity	*B. oleracea* ssp *capitata*	many	Early maturing dominant over late	Detjen and McCue1933

Annual and Biennial	*B. oleracea* ssp *capitata*	many	Annual is dominant over biennial	Detjen and McCue 1933; Walkof 1963
Plant colors	*B. oleracea* ssp *capitata*	4 1	Genotype of red cabbage is *DAbce* and green Cabbage is *daBCE* Dominant homozygous colored blade, heterozygous colored veins	Kristofferson, 1924; Pease 1926; Malinowsk1 1929. Allgayer 1928
Total node number	*B. oleracea* ssp *gemmifera*		Dominant	Hodgkin 1981
Veget. tissue color(anthocyanin)	All *B. oleracea*	1	Dominant	Sampson 1967
Curd color White Orange green	*B. oleracea* ssp *botrytis*		Dominant Recessive Co-dominant	Crisp et al. 1975 Crisp & Angell 1985
Clustered flowers	*B. oleracea* ssp *capitata*	1	recessive	Chiang 1983
Rosetting	*B. oleracea* ssp *capitata*	>1		
Self-incompatibility	all	1 with more than 50 alleles	Dominant recessive	De Nettancourt 1977
Stem thickness	*B. oleracea* ssp *acephara*	1	Dominant gene action	Allgayer 1928
Flower color	*B. oleracea* ssp *capitata*	1	White is simple dominant	Pearson , 1929
Clubroot	*B. oleracea* ssp *capitata*	> 1	Recessive and dominant	Weisaeth 1977 ; Walker & Larsen 1980
Alternaria brassicae	*B. oleracea* ssp *gemmifera*	polygenic	Polygenic/ unknown	Natti et al. 1967

Table 2-3 contd...

Table 2-3 contd...

Trait	Crop/subspecies	Identified gene, if any	Nature of gene action	References
Peronospora parasitica	*B. oleracea* ssp *italica*	1	Recessive and dominance	Hoser-Krauze et al. 1987
Turnip Mosaic Virus	*B. oleracea* ssp *capitata*	> 1	Recessive and dominance	Pink & Walkey 1986
	B. oleracea ssp *gemmifera*	≥ 4		
Cauliflower Mosaic Virus	*B. oleracea* ssp *gemmifera*	≥ 2		
	B. oleracea ssp *gemmifera*	1		
Pest resistance				
Delia radicum	*B. oleracea* ssp *capitata*	> 1		Ellis & Hardman 1975
Aphids	*B. oleracea* ssp *capitata*			Ellis & Hardman 1988
	B. oleracea ssp *italica*			Ellis & Hardman 1988
Lepidoptera	*B. oleracea* ssp *capitata*			Ellis & Hardman 1988

*Prepared from Yarnell (1956) and King (1990).

Table 2-4 Genetics analysis of some cauliflower traits.

Character	Nature of gene action	References
Curd weight	Dominance and epistasis Pronounced over-dominance and epistasis Additive and dominance	Swarup and Pal 1966 Singh et al. 1975 Singh et al. 1976; Jyoti and Vashistha 1986; Gangopadhayay et al. 1997; Sharma et al. 1988
Curd to plant	Partial dominance	Kale et al. 1979
Curd diameter	Predominance of dominance	Lal et al. 1979
Curd size index	Pronounced over-dominance and epistasis Dominance and epistasis Additive dominant gene action Partial dominance	Singh et al. 1975 Swarup and Pal 1966; Lal et al. 1979 Singh et al. 1976; Sharma et al.1988 Kale et al. 1979
Curd angle	Pronounced additive Additive and dominance	Lal et al. 1979 Dadlani 1977; Chand 1980
Curd compactness	Polygenic Dominance and additive Additive	Nieuwhof and Garretson 1961 Lal et al. 1979 Vashistha et al. 1985
Early maturity	Partially dominant gene action. Dominance and epistasis. Predominance of additive Additive gene action Additive and dominant gene action	Watts 1964 Swarup and Pal 1966 Singh et al. 1975, 1976; Lal et al. 1979; Mahajan et al. 1996; Gangopadhayay et al., 1997 Kale et al. 1979 Sandhu and Singh 1977; Sharma et al. 1988
Late maturity	Recessive polygenes	Watts 1963

*Adapted from Sharma et al. (2005) originally given by Chatterjee (1993).

He suggested that $KkK_1K_1K_2K_2K_3K_3$ gives a very loose head and the recessive factors have cumulative effect for head tightness. Three genes governing the head formation was suggested by Malinowski (1929) and Samson (1930) while Kristofferson (1924) suggested many factors governing the heading trait. The variations in the proposed number of factors for heading might be due to the selection of parental genotypes in their study.

The shape of the cabbage head varies from pointed, flat, round or oblate to intermediate shapes. The genetic analysis showed that many factors are involved in governing the head shape (Detjen and McCue 1933; Magruder 1937). Tschermak (1916) reported that the pointed shape is dominant over the round one. Pearson (1934) reported that the number of non-wrapper leaves below the proper head varies from a few to as many as 25. The number of head leaves is also confounded by whether the cabbage is early or late type, the early type having less head leaves. Crosses involving parents with a few and many non-wrapper leaves showed few wrapper leaves to be dominant in F_1 and many modifying factors governing this trait have also been reported. Head size is inherited quantitatively and crosses involving diverse or unrelated lines have been reported to increase the head size (Kotowski 1926; Pearson 1934). The size of the cabbage head is also reported to depend on day-length, long day-length giving a large head while a short day producing a small head.

The genetics of the cabbage core wide and length have also been studied. It has been found that a wide core is dominant over a narrow core. A narrow core is desirable. Core length is controlled by two independent dominant genes for a short core, a short core having < 25% of the head diameter is desirable (Dixon 2006). Another problem in cabbage is that of the head splitting at maturity, this has been reported to be governed by three genes acting additively with partial dominance for early splitting. The narrow sense heritability for splitting was found to be 47% (Chiang 1972). Observation of head splitting revealed that the long core head split at the top while the short core split at the base (Dixon 2006).

Several studies reported that annual habit is dominant over biennial (Detjen and McCue 1933; Walkof 1963). Dominance series of genes/alleles has been suggested to account for the genetic variability in vernalization requirement and earliness (Dickson and Wallace 1986). Vegetative maturity of cabbage head showed that the F_1 derived from a cross involving parents with early maturity and late maturity was early (Rasmusson 1932).

2.2.2.2 Genetics of Curd-related Traits

Curd and its related traits are important in cauliflower. Many studies on curd and its related traits were reviewed by King (1990). The curd formation trait in cauliflower may be either dominant or recessive and is reported to be governed by major genes (Crisp 1982). Polygenic inheritance of curding has also been opined from observation of non-distinct Mendelian segregation ratios in F_2s derived from crosses involving cauliflower and broccoli. Furthermore, poor curd formation in F_1s and F_2s derived from wide crosses within cauliflower lines were also observed (Gray and Crisp 1979). Involvement of some modifier genes in curd formation were also reported (Crisp 1982). Many genes governing curd color have been identified (King 1990). A dominant gene for white color over yellow has been identified. Another gene with partial dominance for white curds has also been reported that causes dwarf orange curd in double recessive condition (Crisp et al. 1975). Crisp and Angell (1985) reported an independent dominant gene for greenless curd. Singh et al. (1978) identified two genes for leaf cover of the curd affecting discoloration by sunlight, while Dickson and Lee (1980) reported two/three dominant genes controlling persistence of whiteness in curd exposed to sunlight. The degree of "sprouting" in broccoli is under the control of many genes affecting several axillary branching (King et al. 1990).

2.2.2.3 Genetics of Plant Morphology

Influence of single major gene to several genes has been reported for plant height. Kwan (1934) reported that several genes govern plant height, while Pease (1926) reported a major dominant gene *T* for height. Most cultivars are recessive for this trait although some cultivars may have the dominant gene. A long stem is undesirable, but for machine harvesting an upright head and not too short a height is required (Dixon 2006). Total node number of Brussels sprouts is the result of the dominant gene action and closely correlated to the rate of node production, and the number of harvested sprouts (Hodgkin 1981).

2.2.2.4 Genetics of Leaf Morphology

A single major dominant gene that is different between cabbage and kohlrabi has been reported for leaf width (Pease 1926; Malinowski 1929). Cabbage has the dominant gene *W* for wide leaf. The presence of modifying factors has also been reported for this trait. The sessile entire leaf of cabbage and petiolate lyrate leaf of mature kohlrabi has been reported to be governed by two genes: *Pet/P* producing petiolate and *En/E* producing entire leaf. The F_1 (*Enen/Petpet*) derived from a cross between cabbage (*EnEn/petpet*) and Kohlrabi (*enen/petPet*) produces entire petiolate leaves like Brussels sprouts (*EnEn/PetPet*). Tschermak (1916) reported that crinkled or rugose (*W*) leaf of Savoy cabbage is dominant over a smooth (*S*) leaf. Kwan (1934) suggested complementary interaction between the genes *W* and *S* in addition to the presence of modifying factors. The phenotype of double recessive for *W* and *S* is smooth. Malinowski (1929) observed continuous variation of leaf from curliness to smoothness in the second generation and suggested that the non-curly leaf of cabbage and the curly leaf of kale are polygenically controlled. Pease (1926) also found similar results. He further suggested that the presence of bizarre protuberances (also known as Asparagodes) was due to a dominant or partially dominant gene, *A/As*.

2.2.2.5 Genetics of Plant and Flower Traits

Genetic studies of anthocyanin pigmentation by several workers in red cabbage revealed inheritance of this trait to be governed by one gene (Allgayer 1928) to many genes (Kristofferson 1924; Pease 1926; Malinowski

1929; Sampson 1967). Allgayer (1928) obtained second generation segregation of 235 colored blade; 358 colored midrib; 166 green, is the ratio approximately equal to the monogenic segregation of 1:2:1. Two factors, *A* and *D*, produce red cabbage, while A is the basic color factor with no effect but producing color with *D* (Kristofferson 1924; Pease 1926; Malinowski 1929). The gene *B* alone produces red midrib, and with *A* produces a dark violet midrib. The factor *C* alone is colorless but with *A* gives a dark violet color. Gene *E* extends the area of dark violet color. *B* and *C* have the same effect as *B* alone (Kristofferson 1924). Based on that, the genotypes of red cabbage and green cabbage are deduced to be *DAbce* and *DaBCE*, respectively. An F_1 plant of green and dark red produces a pink color. The gene *M* produces magenta and *S* produces sun color (Magruder 1937). Inheritance of sun color was shown to be governed by two duplicate genes R_1 and R_2 (Kwan 1934). In another cross involving deep purple and green, Kwan (1934) observed 9 purple; 3 sun color; 4 green in the second generation. He proposed the genotypes *GH, Gh*, and *gH* and *gh* for deep purple, sun, and green colors, respectively. Inheritance of duplicate genes for color was further confirmed by Kristofferson (1921) when he observed 9 dark reddish violets; 3 clear red, 4 white from the cross involving parents of white cabbage and kale. White flower in Cole crops is governed by a single dominant gene *Wh* (Pearson 1929). Segregation of 1 white: 1 yellow was observed in the first generation when a normal plant with a yellow flower was crossed with a mutant with a white flower (Kakizaki 1930).

Male sterility in Cole crops has been found to result from recessive mutation (*ms*) of male fertility gene *Ms* (Neiuwhof 1961, 1968; Borchers 1966; Ahluwalia et al. 1977). Male sterility under the control of duplicate dominant genes with cumulative effect has also been reported (Van der Meer 1985). Ruffio-Chable et al. (1993) reported dominant male sterility in cauliflower.

Self-incompatibility is a mechanism where pollen cannot fertilize an ovary in the same plant. Self-incompatibility may be gametophytic and sporophytic. In gametophytic self-incompatibility, the incompatibility reaction is determined by the genotype of the gamete while in sporophytic, incompatibility reaction is determined by the genotype of the gamete-producing plants. Genetic studies in different *B. oleracea* vegetables showed that self-incompatibility mechanism is a complex system, which is governed by a single major gene *S* with 50–70 alleles (Yarnell 1956; Thompson and Taylor 1966; De Nattancourt 1977; Dickson and Wallace 1986; Ordas and Cartea 2008). An incompatible reaction occurs when identical alleles are inherited by both the pollen and stigma. Recently molecular evidences showed that recognition of self is mediated by *S*-locus specific glycoproteins (*SLGs*). At present many genes involved in self-incompatibility mechanism have been cloned and characterized (Nasrallah et al. 1987, 1988; Trick and Flavell 1989). Sporophytic self-incompatibity is widely used for F_1 hybrid vegetables production in Brassicas (Shinohara 1984; Dickson and Wallace 1986).

2.2.2.6 Genetics of Insect-Pest and Disease Resistance

Insect pests and diseases are the major factors causing crop losses not only in vegetables but also in all other crops. Breeding resistance to diseases and insect-pests involves identification of resistance sources/genotypes and transfer of resistance genes from resistant genotypes to susceptibles. Some important classical genetic analysis for identification of resistant sources for various diseases are listed below. Downy mildew caused by *Peronospora parasitica* is a soil-borne disease and this disease develops first on lower leaves as a sparse to densely packed white mat forming chlorotic to partially chlorotic lesions when the soil is transferred by wind and rain. Genetics of resistance for this disease has been found to be complicated due to the existence of several pathogenic races even though its resistance has been identified in cabbage and broccoli (Natti et al. 1967). The cabbage genotype, PI 245015, has two independent dominant genes for resistance to the races 1 and 2. Williams et al. (1972) identified the resistance source for black rot, a seed-borne disease caused by *Xanthomonas campestris*. They reported that a major gene *f* and two modifier genes, one dominant and the other recessive, govern the resistance to this disease. Powdery mildew caused by *Erysiphe polygoni* D.C. is another disease characterized by a fine necrotic flecking on the exposed head and lower leaves of cabbage. Resistance to this disease has been identified in the cabbage cultivar Globbelle, which is governed by a single dominant gene, although influence of some modifying factors was also assumed (Walker and Williams 1965).

Cabbage yellows caused by soil-borne fungus *Fusarioum oxysorum* f. sp. *conglutinens* is characterized by progressive yellowing followed by brown necrosis of the lower leaves, stunted growth and premature leaf drop. Walker (1930) designated two types of resistance, Type A and Type B. Type A resistance is controlled by a single dominant gene and carried by most resistance cultivars, and type B resistance is controlled by several genes. Type A is not influenced by temperature, whereas type B resistance breaks down at 22°C. Major gene résistance was observed in All Head Early, Copenhagen Market and Glory of Enkhuizen. However, polygenic resistance was observed in Wisconsin Hollander. Resistance to stalk rot caused by *Sclerotinia sclerotiorum* in Indian cauliflower was studied in populations from six generations of six crosses and was found to be polygenic and under the control of recessive genes primarily with additive gene action (Baswana et al. 1991).

Turnip mosaic virus (TuMV) and cauliflower mosaic virus are the two major viral diseases of vegetable brassicas including cabbage, which are characterized by mottling and necrotic spots or ring spots. Complete resistance has been identified for all four races of turnip mosaic virus (identified so far) in Chinese cabbage (Provvidenti 1980). Cabbage cultivar Globelle exhibit quantitative resistance to the races 1 and 2. Non-strain-specific recessive resistance controlled by at least one gene has been reported in F_1 hybrids and open-pollinated cultivars of cabbage with high heritability (Pink and Walkey 1986). The resistance to TuMV is found to be controlled by at least four genes in Brussels sprouts with high heritability (Pink and Walkey 1986). In Brussels sprouts, resistance to cauliflower mosaic virus is generally dominant and is controlled by at least four genes (Pink and Walkey 1986). However, presence of some recessive genes has also been reported.

Clubroot disease, caused by *Plasmodiophora brassicae*, is a major disease causing severe problems in both quality and quantity of *B. oleracea* crops including cauliflower, broccoli, kale and others (for detailed review of genetics of clubroot resistance in *Brassica* species see Piao et al. 2009). Different research groups identified various sources of clubroot resistance through screening of germplasm (Crute et al. 1980; Crisp et al. 1989; Dias et al. 1993; Voorrips and Kanne 1996; Manzanares-Dauleux et al. 2000; Carlsson et al. 2004). In contrast to *B. rapa*, completely resistant accessions have rarely been identified in *B. oleracea* although large numbers of accessions were screened. Crisp et al. (1989) evaluated about 1,000 *B. oleracea* accessions and confirmed resistant sources in some kales including the curly, marrow-stem and 1000-head types and cabbage including Böhmerwaldkohl, Bindsachsener cabbage, Badger Shipper and Ladoszkaya cabbage. Also some open-pollinated Brussels sprouts including Cambridge, Continuity, Rubine and Catskill and forms of South European cabbage, cauliflower and broccoli exhibited lower levels of susceptibility. Among kale, cabbage and winter cauliflower accessions evaluated, only some of the kale accessions, mainly leafy and short leafy kale group, exhibited high levels of resistance to clubroot, while all cabbage and cauliflower accessions were susceptible (Manzanares-Dauleux et al. 2000). Some of these resistant sources are widely used in the breeding program of *B. oleracea*. A detailed genetics of clubroot resistance was studied in *B. oleracea* by several groups and most of them concluded that the inheritance of this trait in *B. oleracea* is polygenic. Yoshikawa (1993), in a study involving progenies of a cross involving resistant cultivar Bohmerwaldkohl and a susceptible cabbage line, observed inheritance of resistance to be govern by one to four genes. Voorrips and Visser (1993), based on the genetic analysis of the three F_1 populations derived from crosses between resistant lines Bohmerwaldkohl, Wisconsin, curly kale and susceptible cabbage line, reported recessive inheritance of clubroot resistance. Laurens and Thomas (1993) observed that resistance to clubroot in kale is controlled by many alleles with additive effects. Based on qualitative and quantitative analysis of the F_1, F_2 and backcross progenies of four crosses derived from four different resistant sources and one common susceptible doubled haploid line, Voorrips and Kanne (1996) observed different mode of inheritance of resistance to clubroot. Of the four resistances studied, one was controlled by two complementary genes. Chiang and Crête (1976) agreed that the resistance was controlled by two loci by using the same resistant source.

Several insect pests damage *B. oleracea* crops, the most important of which are cabbage looper (*Trichoplusia ni*), cabbage worm (*Pieris rapae* L), cabbage aphid (*Brevicorne barssicae*), and diamond back moth (*Plutella xylostella*). Dickson and Eckenrode (1980) found that PI 234599 cauliflower having glossy leaves is immune to cabbage looper, cabbage worm and to diamond back moth. Genetic analysis showed that the resistance is recessive and quantitative and is linked to the glossy trait.

Tip burn, a major physiological disease in cabbage results due to non-translocation of calcium to the rapidly growing leaf tips, is characterized by browning and blackening of leaf tissues due to death of cells. Genetic analyses of resistance to tip burns by different study groups have contrasting views. Walker et al. (1965) reported that resistance was governed by two to three recessive genes, while Dickson (1977) gave evidence for dominance resistance controlled by two to three genes.

2.2.3 Early Linkage Analysis

Linkage analyses for several traits have been done in different studies involving crosses between different cultivar groups of *B. oleracea.* Linkage analyses or association of traits are important for designing breeding strategies in any crop. Linkages of genes for desirable traits are important while linkages of genes for favorable and unfavorable traits are not desirable for crop breeding. Economically important traits are often found to be associated with non-desirable traits in most of the agricultural crops, which hamper the breeding process. The most striking example is the non-occurrence of heading trait of cabbage with curled leaves of kale in F_2 or subsequent generations derived from crosses between heading cabbage having the entire leaf and non-heading kale with curly leaves. It is also observed that in progenies of such crosses head formation decreases when curliness increases and vice versa (Pease 1926; Allgayer 1928; Malinowski 1929; Samson 1930). When cabbage is crossed with kohlrabi and cauliflower, the heading trait of cabbage is not combined with the enlarged stem of kohlrabi and flower clusters of cauliflower in segregating progenies (Pease 1926). From this observation it is clear that genes from a particular botanical variety or cultivar groups of *B. oleracea* tend to remain together and do not combine with the traits of other botanical variety. Malinowski (1929), Pease (1926) and Allgayer (1928) proposed linkage of different gene groups which are: genes for heading and sessile, entire and wide leaves; genes for heading and short stems and non-curly leaves; genes for heading, color (*P*), petiole blade, leaf wide, frilled leaves, and broad vs. narrow stem; and genes for heading, stem length, additional genes for leaf frilling, and the second factor for petiole blade. Observation of recombination between enlarged axil buds within cabbage wrapper leaves was also reported (Pease 1926; Allgayer 1928; Malinowski 1929; Detjen and McCue 1933). Independence of the gene for color has been reported from heading genes (Pease 1926). Recombination of high ascorbic acid content with that of resistance to cabbage yellows has been identified (Walker 1944; Foster and Walker 1946; Walker and Pound 1947).

However, certain percentages of trait recombination in progenies have been observed for some traits while crossing between different *B. oleracea* groups. Notable examples are observation of 20% crossing over between heading factor *K* with purple blade *P* gene by Allgayer (1928) , 10% crossing over between the gene for petiolate *Pet* with N_1 for non-heading, and 30% crossing over between *T* (tall plant) and N_2 (heading) by Pease (1926). He also observed that the gene for purple leaf, *D* is independent of recessive genes for heading, n_1n_2. Kwan (1934) observed 10% crossing over between *R1* and *W* or *R2* and *Sm*. Other studies by Kristofferson (1927) and Malinowski (1929) showed a linkage between color genes *B* with either *A* or *C*, and genes for enlargement of stem in kohlrabi and color gene *D*, respectively. Recessive genes for bolting have been found to be independent of genes for heading and red color (Samson 1930; Detjen and McCue 1933).

Recently, Arus and Orton (1983) studied the inheritance and linkage relationships of variation in isozymes banding patterns for phosphoglucomutase (PGM), phosphoglucoisomerase (PGI), leucine aminopeptidase (LAP), alcohol dehydrogenase (ADH), acid phosphatase (APS) and glutamateoxaloacetate transarninase (GOT) in *B. oleracea*. However, no linkage relationship was observed between the six loci (*Adh-2, Aps- 1L, Pgm-1, Pgm-2, Lap-1,* and *Aps-1*). Recent development in molecular markers particularly DNA markers have been successful in the development of molecular linkage map and linkage analysis of traits of economic importance and molecular markers, and identification, manipulation and cloning of many important traits in *Brassica* vegetables (Quiros 2001; Snowdon and Friedt 2004). The following chapters will be dealing with linkage analysis with molecular markers and traits QTL mapping and their manipulation in *Brassica* vegetable breeding.

2.2.4 Classical Breeding Achievements

Classical breeding and genetic analysis have been instrumental in the present day success of breeding *Brassica* vegetables for various traits. Today with the help of conventional breeding, desired quality and quantities has been achieved in *Brassica* vegetables. During the last few decades, extensive breeding efforts in vegetable brassicas had focused on improvement of agronomic traits, increasing yield, breeding for biotic and abiotic stresses and improvement in quality which are discussed below.

2.2.4.1 Yield and Agronomic Traits

Improvement of uniformity, shape, size, shelf-life and nutritional quality are part of vegetable *Brassica* breeding. In cabbage, breeding of the round shape with compact and solid frames has been achieved compared to earlier traditional varieties with a flat or pointed shape with a loose head and large frames. Because of this improved new type, a large number of plants can be accommodated in a unit area and can be harvested with machines (Dickson and Wallace 1986). Most of the present day cabbage cultivars are non-splitting and non-bolting compared to earlier premature bolting and splitting type. The most important quantitative traits for which success have been achieved in breeding are weight of the head, short growing period, early maturity, frost hardiness, long storage ability and morphological traits related to head development (de Moel and Everaats 1990). Head firmness and color, number of outer leaves, attractive green color of the wrapper leaves, and core dimensions have drawn major attention in breeding cabbage cultivars for commercial cultivation. Breeding for high harvest index by increasing the proportion of reproductive parts to the vegetative parts of the plants has been the major objective in cauliflower and broccoli breeding. Curd and head firmness has been achieved by reducing the floret stem lengths and grain size. This increases the post-harvest quality of the produce.

Another major achievement in vegetable *Brassica* breeding is development of uniform, heterotic and highly productive F_1 hybrids using self-incompatibility system. Of late, use of cytoplasmic male sterility for hybrid vegetable production has achieved major importance by transferring the male sterility system from *Raphanus sativus* into *B. oleracea*. This is more stable compared to self-incompatibility having multiple alleles that makes breeding more complex. Major seed companies these days are more interested in breeding F_1 hybrids because of the highly promising optimum market quality of hybrid vegetable brassicas. The curds/heads uniformity achieved by F_1 cauliflower and broccoli are in high demand for the freezing industry (Acciari et al. 1997).

2.2.4.2 Disease Resistance

Classical genetic analysis and screening of *B. oleracea* germplasms have identified many sources of disease resistance with varying degrees, some which have been exploited in resistance breeding programs. Several cauliflower cultivars resistant to *Xanthomonas campestris* have been selected in India and genetic analysis found that this resistance is dominant and polygenic in several Indian landraces (Bianco and Pimpini 1990). Notable examples of cauliflower resistant lines for this disease are Sn 445, Pua kea and MGS2-3 (Sharma et al. 1972). Pusa Shubhra was developed using Pua kea and MGS2-3 lines and recommended for commercial cultivation (Singh et al. 1993). Walker and Larsen (1980) identified two forms of pathotype-specific major resistance genes for clubroot diseases in cabbage while Weisaeth (1977) reported recessive resistance to this disease in the fodder turnip cultivar "Bohmerwaldkohl" determined by more than one gene. Resistance found in this cultivar has been transferred to many vegetable brassicas. Many of the important cabbage cultivars in the US are now resistant to cabbage yellows disease (Dickson and Wallace 1986). Early purple cauliflower was reported resistant to yellows. Sclerotinia rot caused by *Sclerotinia sclerotiorum* is a major disease of cauliflower in the early stage of development. Several cultivars with moderate resistance have been reported i.e. EC131592, Janavon, EC103576, Kn-81, early Winter Adam`s White head, EC162587, EC177283 (Kapoor 1986; Baswana et al. 1991; Sharma et al. 1995, 1997; Singh and Kalda 1995). Pusa Snowball K-25 has been developed from EC103576 as a resistant source and Pusa snowball-1 is released for commercial cultivation. Resistance to downy mildew disease has been identified in many cauliflower lines, e.g., Igloo, snowball Y, MGS2-3, 1-6-1-4 (Kontaxis et al. 1979; Chatterjee 1993). Pusa Hybrid 2 and Pusa Snowball K-25, resistant to downy mildew were released for commercial cultivation in India.

Alternaria brassicae or *Alternaria brassicola* also causes brown to black, small elongated spots in leaves, stems, and inflorescences of brassica plants. In cauliflower it causes curd blight and resistance to this disease was found in Indian cauliflower lines MGS2-3, Pua kea and 246-4 (Sharma et al. 1991), IIHR 142, IIHR 217 (Pandey et al. 1995) and snowball KT-9 (Sharma et al. 1991). Commercial cultivar Pusa Shubhra released in India is resistant to curd blight (Singh et al. 1993).

2.2.4.3 Insect Resistance

Several insects belonging to specialist and generalist groups attack vegetable brassicas and damage leaves, roots, stems and other parts of the plants making the crop unsuitable for human consumption. Breeding resistant cultivars of *B. oleracea* to insect pests to reduce major economic loss was also a major objective among vegetable *Brassica* breeders. Several studies identified resistant sources for different insect pests (Radcliffe and Chapman 1966; Dickson and Eckenrode 1980; Lamb 1989). Resistance to *Pieris rapae, Plutella xylostella* and *Mamestra brassicae* has been identified in some cabbage lines of the USA (Dickson and Eckenrode 1975, 1980; Dickson and Wallace 1986). Radcliffe and Chapman (1966) have reported the differences in cultivar response to infestation of various insects in the field condition. Stoner (1990, 1992) reported that glossy leaves and high concentrations of glucosinolates are preferred leaf traits for oviposition of some insects. Red cabbage cultivars are less preferred for oviposition than green cultivars (Stoner 1992). Dickson and Eckenrode (1980) reported a glossy leaved cauliflower PI 234599 to be resistant to lepidopteron insects, diamondback moth, cabbage looper, and imported cabbage worm. Cabbage head borer resistance had been reported in many cultivars and F_1 hybrid cauliflowers. The notable examples are Katiki, Early Patna, and F_1s aa × ES102, aa × Katiki (Lal et al. 1991, 1994). Resistance to cabbage aphids have been identified in cauliflower lines NY 13816, NYIR 9605 and others. Picoaga et al. (2003) found resistance to *M. brassicae* in Galician kale landraces.

2.2.4.4 *Quality Traits*

Brassica vegetables are known widely for the presence of rich nutrients and chemical components such as calcium, carotene, vitamin C, vitamin E and a high amount of antioxidants besides providing lots of vegetable fiber (Cao et al. 1996; Nilsson et al. 2006). During the last few decades extensive efforts have been made for breeding nutritionally high quality *B. oleracea* cultivars. Glucosinolates, sulfur-containing secondary metabolites is commonly found in the members of Brasicaceae family. Three types of glucosinolates have been named depending on the amino acid precursor, i.e. aliphatic derived from methionine, aromatic derived from tyrosine or phenylalanine and indolyl derived from tryptophan (Giamoustris and Mithen 1996). Of the three types, the most common in *Brassica* crops are aliphatic glucosinolates. Upon tissue damage, the breakdown product of these glucosinolates gives isothiocyanates, thiocyanates, epithionitriles, oxazolidone and nitriles (Fenwick et al. 1983; Halkier and Gershenzon 2006). Glucosinolates are known for their role in plant defense against fungi, nematodes, herbivores, and weeds (Rosa et al. 1997). Recently, isothiocyanates derived from glucosinolates breakdown products have been found to be anticancer, which induces Phase II detoxifying enzymes such as Quinine reducate, Glutathione-S-transferase, and others (Zhang et al. 1992; Brown et al. 2002). Sulforaphane, the isothiocyanates derived from glucoraphanine found in broccoli is an important subject of study for the last decade. Breeders are trying to increase the glucoraphanine content in broccoli so that this functional food can prevent the risk of cancer in the human body. Genes involved in glucosinolate biosynthesis pathway have been cloned and characterized in *Arabidopsis thaliana* and in brassica crops (Li and Quiros 2002; Halkier and Gershenzon 2006; Bisht et al. 2009).

2.2.4.5 Transfer of Cytoplasmic Male Sterilty into B. oleracea

Cytoplasmic male sterility is widely used in many crops species for using as a pollination control mechanism for hybrid seed production. Since, there is not any naturally occurring cytoplasmic male sterility in *B. oleracea*, cytoplasmic male sterility from *B. nigra, B. napus* and related Cruciferae species radish were transferred by many researchers (Pearson 1972; Bannerot et al. 1974; Dickson 1975; McCollum 1981; Chiang and Crete 1987; Hoser-Krauze 1987; Crisp and Tapsell 1993). Cytoplasmic male sterility "ogura" found in radish by Ogura (1968) was first transferred to broccoli by backcrossing (Bannerot et al. 1974; McCollum 1981). This was later transferred from broccoli to cauliflower by Dickson (1975) and Hoser-Krauze (1987). The transfer of this cytoplasmic male sterility was done in heat tolerant Indian cauliflower such as MS-91, MS-51 and is being used in F_1 hybrid seed production (Sharma et al. 2005). In cabbage *B. nigra* cytoplasmic male sterility was transferred by Pearson (1972), while Chiang and Crete (1987) transferred *B. napus* cytoplasmic male sterility into it. Crisp and Tapsell (1993) later transferred *B. napus* cytoplasmic to make a sterility system from cabbage to cauliflower.

2.3 *Brassica rapa*

Many types of *B.rapa* vegetables were traditionally used as vegetables in East Asian countries including China, Japan and Korea. Chinese cabbage [*B.rapa* L. ssp. *pekinensis* (Lour) Olsson] is the most common type of vegetable in *B. rapa* eaten as fermented food "Kimchi" in Korea and as a fried vegetable in China. Pak choi [ssp. *chinensis* (L) Makino] is used as a fried vegetable in China and as a fresh vegetable of "Ssam" in Korea. Different from other *B.rapa* vegetables, turnip [ssp. *rapifera* (Metzy)] is used as a root vegetable. Root or the whole part of the turnip plant is consumed as "Kimchi" in Korea, as a fried vegetable in China and pickled in Japan.

2.3.1 *History and Evolution of Different Types*

The first record of brassica crop was mentioned in ancient Chinese literature, edited in the 5th century BC, as turnip (*B.rapa* ssp. *rapifera*) and mustard (*B. junceae*). From then to the 7th century, turnip was grown in north China and pak-choi in South China. In the 10th century, at Yang-chou city located in central China, a vegetable was found having wrinkled, hairy leaves as large as fans and not so fibrous as pak-choi in texture. These characteristics are like the primary form of Chinese cabbage (*B.rapa* L. ssp. *pekinensis*). This may suggest that Chinese cabbage might have originated from natural hybridization between turnip and pak-choi. To prove this, artificial crosses were made between turnip and pak-choi collected from North and South China, respectively, from 1960 to 1962. The characteristics of the hybrids obtained were similar to that of Chinese cabbage, having large thin, hairy leaves with fairly clear wings along petioles. In the literature of the 12th century, the primary form introduced to the North developed thick petioles. After that many farmers tried to select more heading types, a premature form of Chinese cabbage was reported in a horticultural book in the 14th century. The typical form of Chinese cabbage was found in the 17th century at a local record of Shiun-Tian-Fu. Li (1981) classified Chinese cabbage into 11 types following morphological, ecological and economical characteristics using local cultivars in China.

In 2007, the cultivation area in China of Chinese cabbage was 26 billion m^2 followed by cabbage (8.9) and Pak-choi (5.3). The major cultivation provinces of Chinese cabbage were Shandong and Hebei, located in North China, covering 21%. However, in South east China, one of the most popular vegetables is pak-choi. Especially in Zhejiang province, where the cultivation area of Pak-choi was 0.9 billion m^2 while that of the Chinese cabbage was 0.3.

Pak-choi is a unique brassica vegetable in China. Thus traditional local varieties and high quality hybrid cultivars were cultivated together throughout the country. One popular inbred variety is "Aikang qing" possessing wide petioles and a high ratio of petiole weight and resistance to TuMV, which made this variety popular in most of China from the early 1980s (Fig. 2-1). Another popular inbred variety is "Shanghai qing", which is very popular in Shanghai and later was cultivated throughout China (Fig. 2-1). This variety is taller than the "Aikang qing" and also shows a high quality of wide petioles and a high ratio of petiole weight. Although not as popular as the other two varieties, "Siyue man" (Wuyue man) possesses bolt resistance, and can be cultivated in spring (Fig. 2-1). Even though the taste of "Siyue man" is inferior, bolt resistance has made this variety popular in the northern provinces of China for winter and spring cultivation. High quality hybrid cultivars were also imported from other countries. One such popular cultivar in the Zhejiang province "Huaguan" was imported from Japan. This cultivar required a short period for harvesting and had a good quality. However its high price prevented further spreading of this cultivar. The official record of introducing Chinese cabbage in Japan was in 1875 at

Figure 2-1 The shape of "Aikang qing"(left), "Shanghai qing"(middle) and "Siyue man" (The photo was presented by a Chinese researcher Tianhwa Fu).

Color image of this figure appears in the color plate section at the end of the book.

Naito Shinjuku Testing Station. After the Russia-Japan war in 1905, Japanese soldiers brought seeds of Chinese cabbage and simultaneously several trading companies started to import various types of Chinese cabbage seeds. After 1920, Chinese cabbage seed production was established in Japan and this made it possible to develop various cultivars (Watanabe 1981).

In Korea, ancient literature of the 13th century mentioned the introduction of Chinese cabbage as a medicinal herb imported from China. In literature of the 17th century, cultivation of Chinese cabbage was mentioned as a vegetable using imported seeds. In the early 19th century, semi-heading type of the local variety was imported and produced a high quality Chinese cabbage in the north province of "Kae-sung" in Korea. This local variety became popular and got a name of "Kae-sung" Chinese cabbage. Using the seed propagation method established in 1850, the "Kae-sung" variety was cultivated in the central province of Korea. Especially in the province of Seoul , smaller but harder heading type was derived and got the name of "Seoul" Chinese cabbage (Fig. 2-2). In 1906, a government research station of Horticultural Demonstration Station was established that started research on vegetable breeding to overcome starvation. As Chinese cabbage and radish (*Raphanus sativus* L.) are the main vegetables for "Kimchi", which is basic side dish in Korean cuisine, research focused on increasing yield. When Dr. Woo Jang-choon returned from Japan, hybrid breeding was started on Chinese cabbage and radish. Dr. Woo contributed to commercial breeding by developing valuable materials and educating students. In the 21st century, 880 varieties of Chinese cabbage, 1,040 varieties of radish, and 18 varieties of turnip were registered by the Korea Seed and Variety Service.

Figure 2-2 The shape of "Seoul" Chinese cabbage.

Color image of this figure appears in the color plate section at the end of the book.

2.3.2 Breeding for Yield and Morphological Traits

Two major local varieties were planted in Korea before the 20th century. The characteristics of the first one, "Kae-sung" Chinese cabbage, are about 40 cm height, semi-heading type, about 10 leaf numbers with dark green colored curly leaf blade, and less hairy. The "Seoul" Chinese cabbage is shorter than the "Kae-sung" Chinese cabbage, semi-heading type, and has about 20 leaf numbers with a light green color. At first, the seeds of "Kae-sung" Chinese cabbage were imported from China, as the seed production failed in the Kae-sung province, where this variety was popular. After development of seed production using autumn harvested plants, this variety was favored and was planted as far down as the Seoul province. Farmers in Seoul continuously selected plants showing good characteristics for making "Kimchi" from the "Kae-sung" variety and developed the "Seoul" variety which was planted in many provinces located at the south of Seoul.

Breeding for yield increase was the most important purpose of the Horticultural Demonstration Station during the 20th century. The first step was introducing various exotic varieties and evaluation of their yield and quality. The planting period of heading Chinese cabbage was longer than semi-heading or non-heading type of local varieties. Farmers planted low quality semi- or non-heading type to produce twice a year before introducing early harvesting heading type in 1960. After this variety, introduction of various heading types (semi or non heading type), which had been traditionally planted in Korea over 800 years. Nineteen foreign varieties were imported and seed production experiments were done on five selected varieties from 1953 to 1959. Late bolting high yielding heading type Chinese cabbage variety "Yaki" was developed from a Japanese cultivar for spring production of Chinese cabbage. Dr. Woo started making hybrid cultivars using self-incompatibility (SI) in Chinese cabbage. In 1960, he developed the first hybrid cultivar "Wonye 1ho" possessing characteristics of disease resistance, tight heading, and high yield (Fig. 2-3). The amount of seed production was 38 kg/ha using SI in a remote place to prevent pollen contamination. This hybrid cultivar development and seed production methods were given to several private seed companies that had the facilities and labor required for seed production. After this cultivar, numerous other varieties and inbreds were developed and distributed to private seed companies that boosted up seed industries in Korea.

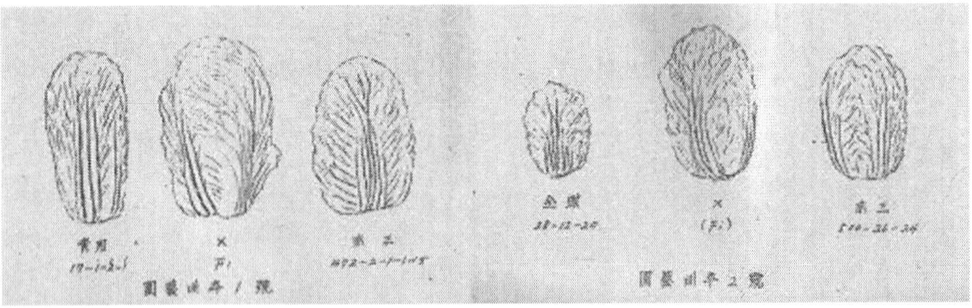

Figure 2-3 Hand drawings of "Wonye 1ho" (left) and "Wonye 2ho" with parental materials (right).

2.3.3 Heterosis Breeding

The main purpose of the hybrid cultivar was high yield and year round production of Chinese cabbage after 1960. The first hybrid cultivar was "Wonye 1ho" for autumn cultivation followed by "Wonye 2ho", distributed from the Horticultural Demonstration Station to seed companies and farmers (Fig. 2-3). The main characteristic of these cultivars was big head size thus requiring a long planting period. The sowing time was before August 5 at the central province at that time, which witnesses viral disease and soft rot due to high temperature. In late autumn, the temperature drops sharply thus sowing time could not be delayed using these cultivars.

The best way of obtaining good hybrid cultivar is making crosses between inbreds and evaluating horticultural traits. However, this specific combining ability test needs a lot of labor and money. If a breeder have 100 inbreds, 5,050 times of bud pollinations are required only for making seeds.

In 1971, a breeder developed one method of making a combination between two parental groups. The horticultural traits of each group were classified by inspecting two parents of "Wonye 2ho" distributed from the

National Horticultural Research Institute (NHRI) in 1960. The A group has characteristics of few leaf numbers, big outer leaves, and incomplete heading, thus can be easily discarded as inferior inbreds. The B group has characteristics of short height, profuse leaf numbers, and similar leaf size, fast growth and complete hard heading. As the head is small and tends to grow ugly, these inbreds were also discarded. However, when hybrid seeds were produced involving these two groups, the horticultural traits were excellent as compared to other crosses using both nice looking inbreds as parents. The first commercial hybrid developed by this crossing method was "Dea-hyung-Garak", an autumn cultivar launched in 1971 and has been planted until recently.

One parental material of "Dea-hyung-Garak" was derived from the early harvesting "Cheong-Bang" that possesses viral disease tolerance and was distributed from a seed company. The other parental material was derived from a Japanese cultivar "Hirathuka 1go" possessing characteristics of early harvesting and strong SI. In the early 1970s, cultivars could be harvested at least 80 days after sowing when the big and half-folding head matured. "Dea-hyung-Garak" could be harvested 70 days after sowing when the medium sized rose shape heading was formed. This beautiful head shape attracted customers, reduced viral disease infection by late sowing and resulted in a short planting period that attracted the farmers. Thus, this cultivar replaced other big semi-folding headed cultivars and became popular.

One seed company continued developing the rose-shape heading variety while other seed companies focused on semi-folded heading type (Fig. 2-4). As a result of continuous breeding in the commercial seed companies and the government research stations, farmers could now select what they wanted from among various high quality hybrids of Chinese cabbage.

Figure 2-4 A rose shaped head (A) and semi-folded head (B) of hybrid cultivars in Chinese cabbage.

Color image of this figure appears in the color plate section at the end of the book.

DNA markers such as randomly amplified polymorphic DNA (RAPD), restriction fragment length polymorphism (RFLP), amplified fragment length polymorphism (AFLP), simple sequence repeats (SSR), and sequence tagged site-polymerase chain reaction (STS-PCR) facilitate the major technique for genetic distance estimation in crops. Among them, RAPD markers are commonly used in a large number of species (Williams et al. 1990). This may be due to the fact that the RAPD facilitates automation, and the low cost of operation; besides, the procedure is simple, speedy, and shows optimum degree of polymorphism in addition to wide genome coverage (Dias et al. 2003).

As the RAPD markers are non-locus specific and low reproducible (Kesseli et al. 1994), several studies failed to use these markers in determining relationships between markers and traits or heterosis (Joyce et al. 1999; Liu et al. 1999; Kwon et al. 2002). However, many studies reported the successful results with RAPD markers. The RAPD distance calculated between species explains the general combining ability and the specific combining ability of the phenotypic character with a global coefficient of determination of 81.6% between two species of *Eucalyptus*. From 26 samples, 415 reliable RAPD bands were obtained using 15 primers. Among them, the effective RAPD markers for predicting hybrid performance were those presenting common frequencies in the two species of *Eucalyptus* (Baril et al. 1997).

Even though many papers have recently reported using DNA markers to predict the combining ability, commercial breeders prefer using the traditional grouping method developed in 1971 to make a combination.

In the 1960s, breeders used the bud pollination method in SI plants of vegetable Brassica crops for development of inbred lines. They counted the number of seed set per flower after self pollination to find out the strength of SI. When an inbred line set seeds less than five per flower, it was considered to possess strong SI and was selected as parental material. The selected parental Inbreds were propagated using artificial bud pollination in a net house to prevent contaminating bees or insects.

The most difficult problem in breeding "Dea-hyung-Garak" was selecting inbreds showing strong SI derived from "Hirathuka 1go". After failure of the first screening using tens of inbreds, the breeders used two big size net houses to screen hundreds of inbreds and finally selected one inbred. Because the "Hirathuka 1go" was a hybrid between Chinese cabbage and cabbage to introduce soft rot disease resistance in Chinese cabbage, the breeders figured that the "Hirathuka 1go" was not yet genetically fixed.

Self-incompatibility (SI) prevents self-fertilization and promotes out-crossing in hermaphrodite plants (Nettancourt 1977). In brassica crops, SI is sporophytically controlled by a single, highly polymorphic *S*-locus consisting of a series of multiple alleles (Bateman 1955). Physical mapping and sequencing analysis of *S*-haplotypes in *Brassica* species have identified three highly polymorphic major SI genes at the *S*-locus. The *SRK* and *SP11/SCR* genes determine the *S*-specificity of the stigma and the pollen, respectively (Schopfer et al. 1999; Takasaki et al. 2000; Takayama et al. 2000; Shiba et al. 2002). The *SLG* gene encodes an abundant, soluble, secreted glycoprotein that enhances the SI recognition reaction perhaps by stabilizing the SRK (Roberts et al. 1979; Dixit et al. 2000; Takasaki et al. 2000). The *SRK* gene consisted of a highly polymorphic *SLG* domain, which is very similar to the *SLG* sequence, a trans-membrain domain and conserved protein kinase domain.

A gene encoding a small cysteine-rich protein, termed *S* locus cysteine-rich protein (*SCR*) was identified between *SLG* and *SRK* genes in the course of genome analysis of the S_8 haplotype of *B. campestris* (Schopfer et al. 1999). The analysis of the S_8 haplotype around the *SLG/SRK* region by two independent groups clearly demonstrated that *SCR* is allelic to *SP11*. Thus, the gene encoding this small cysteine-rich protein was referred to as *SP11/SCR* (Schopfer et al. 1999; Takayama et al. 2000).

Both S_9- and S_8-SP11/SCR contain eight conserved cysteine residues and relatively conserved putative signal peptides, but most regions of the mature protein exhibited low homology. Using a primer designed from the conserved signal peptide sequence, it was possible to amplify allelic *SP11/SCR* genes from most of the pollen-dominant *S*-haplotypes class-I, (Watanabe et al. 2000) and the pollen-recessive *S*-haplotypes class-II (Shiba et al. 2002).

The variation of SI activity due to both genetic and environmental factors was observed. Some cabbage lines showed different levels of SI activity between genotypes/lines, and from plant to plant even within the same lines (Wallace 1979). The method of flower bud pollination was used at first to examine the SI genotype. The result of flower pollination was reliable but it needed a long time and involved the risk of contamination

The SI activity was reported to change by temperature and humidity. Thus, a method of investigating the activity of SI without interference of environment was developed in brassica vegetables (Wallace 1979; Lee et al. 1982). In vitro pollination with 98% relative humidity proved to be the most reliable as compared to greenhouse pollination. Counting the number of penetrated pollen tubes using decolorized aniline blue as a fluorescent stain was a simple and reliable method of assigning the activity of SI allelic phenotype.

There are some reports that the SI activity was not only controlled by the *S*-haplotype but also by other genes. A single nucleotide deletion or alternative transcripts of the *SKR* genes located at the *S*-locus resulted in a low SI activity (Göring et al. 1993; Tantikanjana et al. 1993). The activity of SI of *S*-haplotypes was identified with seed set analysis in 88 lines of radish. The activity of SI appeared to differ depending on the genetic background of inbred lines. It was expected that existence of other factors related to the SI activity is possible (Seo et al. 2003).

The *M* (modifier) gene was located at a locus that is not linked, and recessive mutation (*m*) is epistatic to the *S* locus. It functions in the regulation of water availability at the papilla cell surface, which would be required for the SI reaction as a component of the *SRK* signaling pathway (Ikeda et al. 1997). A recessive mutation of the *M* gene eliminates the SI response in the stigma. Positional cloning of *M* revealed that it encoded a membrane-anchored cytoplasmic serine/threonine protein kinase, designated *M* locus protein kinase (MLPK). Transient expression of MLPK restores the SI response of *mm* papilla cells, suggesting that MLPK is a positive mediator of brassica SI signaling. The plasma membrane localization of MLPK suggests that it might stimulate the function of SRK (Murase et al. 2004).

The seeds of inbred lines having strong SI were propagated with over 3% CO_2 treatment at night following bee pollination during the daytime. The CO_2 treatment required a specific plastic house, which can seize high concentration of CO_2, a CO_2 supply system, bees as pollinators, and ventilation labor for the active pollination of bees (Lee 1979). With this method, the strong SI activity of parental lines was nullified for a while, which made it possible to produce seeds of parental lines in Chinese cabbage. With the help of high concentration CO_2 treatment, the single crossed cultivars, showing high uniformity, have been commonly used for breeding hybrid seeds of brassica vegetables in Korea.

2.3.4 Breeding for Quality

The first cultivar possessing yellow inner leaf was "Norang spring" launched in 1988. This cultivar showed semi-folding heading type and disappeared about 20 years ago because the more favored variety "Dea-hyung-Garak" with a rose shaped heading type was developed. This variety was most popular for around two decades among growers and in the markets. However, when "Norag spring" was launched, the yellow inner leaf color distinguished this from others and customers wanted to buy it. The taste of this cultivar was good and customers believe that the yellow inner leaf of Chinese cabbage tastes like roasted sesame (Fig. 2-5). Furthermore, this cultivar is bolting resistant and not sensitive to calcium deficiency compared to the pre-existing cultivar "High land summer". After "Norang spring" became popular, the characteristic of the yellow inner leaf was introduced not only into other spring cultivars but also into autumn and winter cultivars. Now in the 21st century, even high school students in Korea think that the natural color of Chinese cabbage is yellow.

2.3.5 Breeding for Abiotic Stress Tolerance

Chinese cabbage has tolerance to low temperature but is sensitive to high temperature. Thus for autumn cultivation sowing in late summer and harvesting in late autumn are most natural and can produce high quality Chinese cabbage. However the demand for "Kimchi", especially made of

Figure 2-5 Yellow inner leaf colored spring Chinese cabbage.

Color image of this figure appears in the color plate section at the end of the book.

Chinese cabbage, occurs all year round, forcing breeders to develop heat tolerant cultivars, which can grow enough head in hot weather. In 1973, a heat tolerant summer cultivar "Nae-Seo-beak-ro" was developed by a commercial seed company and this made it possible to grow Chinese cabbage in summer at high lands. One parent of this cultivar was collected from the National Horticultural Research Institute (NHRI), renamed as the Horticultural Demonstration Station, and the other parent was an inbred derived from the Japanese spring cultivar "Mu-so" developed by a seed company. "Nae-Seo-beak-ro" could grow well and make heads at high temperature, could endure high humidity in the monsoon, and it showed resistance to viral disease, soft rot and downy mildew. Using this cultivar in highlands, farmers could harvest considerably high quality Chinese cabbage in summer. After this cultivar, many summer cultivars possessing high heat and disease tolerance were developed by the seed companies.

In the four seasons, the temperature in spring is considerably conducive for growing Chinese cabbage. However, the low temperature of early spring causes bolting that reduces the quality so that Chinese cabbage cannot be used for "Kimchi". Thus breeders tried to develop a bolting tolerant cultivar. In the early 1970s, seed import of a Japanese cultivar "Chun-Chu", popular as a high land summer cultivar in Korea, was impossible and this caused a social problem. At that time, a heat tolerant late bolting cultivar "High Land Summer" was developed by a seed company in 1976 to solve this problem. One parent of this cultivar was an inbred derived from the Japanese cultivar "Kyoto 2go" collected from NHRI and the other parental inbred was derived from the Japanese cultivar "Chun-Chu". As the name represents, "High Land Summer" was developed for summer cultivation in highlands. In highland summer cultivation, unexpected low temperature caused bolting in Chinese cabbage. To select late bolting inbreds, breeders lowered the temperature to 5~10°C during 20 days when the plant reached six true leaf stage and then grew at a natural high temperature of summer. Using this method, they could select enough number of high quality inbred lines to make various crossings and successfully developed a hybrid cultivar possessing a rose heading shape, heat and disease tolerance, and good taste. As this cultivar possessed a late bolting characteristic, farmers noticed that this cultivar can be planted in spring. Later this "High Land Summer" cultivar was widely used as a spring culture nationwide. The "High Land Summer" cultivar is the second long-lived popular cultivar in Korea lasting over 20 years and planted until recently.

Winter culture was possible only in Jeju Island during the 1980s using the Japanese cultivar "Winter". In the late 1980s, the south-coast province imported the new Japanese cultivar "Snow Wind" for winter production of Chinese cabbage. In 1988, a Korean seed company launched a winter cultivar "Ha-ru-bang" for Jeju Island. However, this cultivar could not endure the cold winter conditions and disappeared soon. In 1992, a new winter cultivar "Dong-pung" meaning "east wind" was developed by another seed company. This cultivar grew well in cold winter with a strong root, possessing strong resistance to calcium deficiency and bolting, and tasted good. As this cultivar was highly resistant to cold temperature, cultivation in the south-coast area could produce high quality Chinese cabbage in winter using barren land and the produce could easily be transferred to many big cities in Korea (Fig. 2-6). Thus, winter cultivation in Jeju Island requiring high transportation cost could be minimized.

With the help of developing heat tolerant, late bolting and cold tolerant hybrid cultivars, respectively, Korean farmers could harvest high quality Chinese cabbage while conveniently using barren land all year round. Furthermore, Koreans can enjoy tasty fresh Chinese cabbage at a low price in hot or cold seasons.

Figure 2-6 Winter cultivation area in South-coast province of Korea.

Color image of this figure appears in the color plate section at the end of the book.

2.3.6 Breeding for Biotic Stress: Diseases

In the 1970s, the most severe disease that caused reduction was viral disease in Chinese cabbage. One commercial breeder selected inbreds and F_1 hybrids using special fields where severe viral diseases occurred every year. He launched a cultivar of "Peyongang sin 2ho", which could endure severe viral disease occurring in the southern province of Korea. The Asian Vegetable Research and Development Center (AVRDC), now renamed as The World Vegetable Center, collected worldwide Chinese cabbage germplasm and inoculated them with turnip mosaic virus (TuMV) during 1977–1979. Using five strains of TuMV for artificial inoculation, only "Peyongang sin 2ho" was found resistant to every strain.

When the nursery cultivation method was established, insect-mediated viral disease could be controlled in the early growing stage of Chinese cabbage. These days farmers who want to sell the product, transplant young seedlings in their production field to harvest high quality Chinese cabbage safely. Thus, the problem of viral disease has disappeared after using these seedlings.

In the 20th century, clubroot disease occurred in all areas at every cropping season. The first occurrence of clubroot was reported at Gyeonggi province in 1992 (Lee et al. 2001). From 2002, in every area of the Korean peninsula, including Jeju Island, clubroot disease was found to occur in Chinese cabbage.

The first cultivar showing clubroot resistance was "CR singsing" launched in 1995. Farmers in the Kangwon province, famous for alpine summer culture, faced a severe problem of clubroot mass occurrence because the cropping condition is suitable for the propagation of the clubroot germ. The clubroot resistance of "CR singsing" was introduced from the Japanese cultivar, "CR Shinkki". At first, "CR singsing" was resistant in alpine summer culture; however after two years of cultivation, this variety became susceptible to clubroot.

Later in 2003, another variety, "CR green", was developed by another seed company. This cultivar proved to possess not only clubroot resistance but also the powdery mildew resistance gene. However, it could not be widely cultivated because of the tough leaf texture, which reduced the efficiency of "Kimchi" producing factories. However, clubroot resistance of this cultivar was broken after two years of cultivation.

Recently, breeders are trying to introduce various clubroot resistance genes into many cultivars for spring, summer, autumn and winter cultivation. The resistance sources were introduced from Japan and distributed from the National Institute of Horticultural & Herbal Science (NIHHS), renamed NHRI. In NHRI, researchers are trying to find germplasm showing strong resistance to clubroot race 4 after a single spore isolation. Over 200 clubroot-inoculated Chinese cabbage roots were collected from Korea and each race was identified using the WCD method. Among 13 identified races, race 4 was found most frequently and showed severe symptoms (Cho et al. 2001). Thus single spore isolate of race 4 was used to select resistance of *B. rapa* germplasm. Using nutrient solution for cultivating inoculated plants (Park et al. 2008), an accession of IT033820 was selected as resistance turnip and distributed to seed companies.

Until now, resistant hybrid cultivars are turning sensitive in some parts of Korea and this forced breeders and researchers to develop highly resistant but tasty cultivars in Chinese cabbage. This phenomenon occurs and causes severe problems worldwidely, thus joint research between nations would provide short-cuts to researchers and breeders (Fig. 2-7 and 2-8).

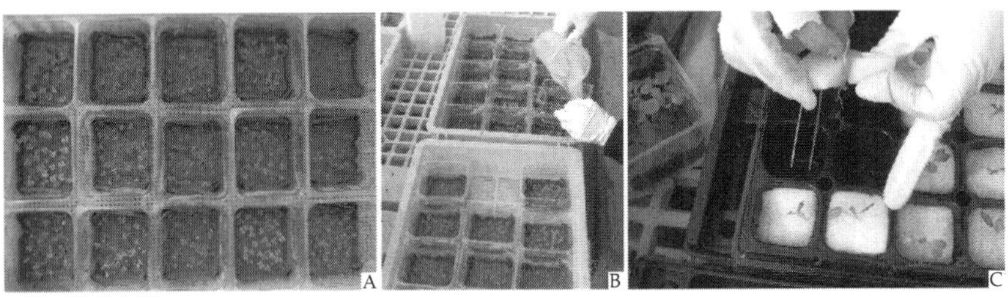

Figure 2-7 Procedure of inoculation method of clubroot in Chinese cabbage (A; Seed germination, B; Input 20 mL of inoculm, C; Trans planting 15 day-old seedlings using 50 pot plug tray and urethane sponge.

Color image of this figure appears in the color plate section at the end of the book.

Figure 2-8 Hydroponic culture of clubroot inoculated Chinese cabbage and the symptoms. (A; 2 weeks after transplanting, B; 6 weeks after transplanting, C; Symptoms of clubroot, left; sensitive, middle; medium sensitive, right; resistance).

Color image of this figure appears in the color plate section at the end of the book.

2.4 Limitations of Classical Breeding and Need for Molecular Breeding

Although classical vegetable *Brassica* breeding has been successful in many aspects such as breeding for plant morphotypes including shape, size, firmness, uniformity, colors and others, improvement in majority of the areas could not be achieved due to many factors. Genetic analysis to determine the number of genes and gene action could largely be achieved for easily scorable qualitative traits, while the majority of economically important traits such as head and curd size of cabbage and cauliflower are quantitative in nature. Breeding for such kind of traits was largely complex due to polygenic inheritance and the influence of the environment. Further phenotypic evaluation is the hardest part of any breeding program, which not only demands extensive labor but also lots of time. F_1 hybrid breeding in cabbage and cauliflower has relied on the self-incompatibility gene having multiple alleles with a complex inheritance pattern. This system in most of the cases was unstable and many of the F_1 hybrids were of low quality due to breakdown of self-incompatibility. However, extensive efforts have been made to transfer available cytoplasmic male sterility system from *B. nigra, B. napus* and *Raphanus* into *B. oleracea* for hybrid breeding.

Many of the resistance sources identified in vegetable brassicas for diseases and insect-pests could not be transferred into suitable cultivars because most of the resistances identified were not completely resistance (Ordas and Cartea 2008). Notable examples are a complex inheritance pattern of resistance to black spot and clubroot disease identified in cabbage, cauliflower and kales. However, quantitative inheritance complicates the transfer of resistance genes in commercial cultivars. The case is similar in clubroot resistance breeding in *Brassica* vegetables. Many race specific resistances have been identified for clubroot in different sources, but exploitation of these genes in breeding could not be done due to incomplete resistance.

Breeding for quality of product by increasing the nutrient contents (vitamin C, carotenoids, antioxidants, storability) and functional foods (glucosinolates) have been major objectives in recent years along with increasing yield, resistance to diseases and pests and other agronomic traits. However, many nutritional compounds such as carotenoids and secondary metabolites such as glucosinolates are synthesized in complex biosynthetic pathways involving many genes. Conventional breeding for desired level of particular compounds is very complex and difficult.

Recent advances in molecular biology, particularly the development of molecular markers and transformation technology, have been very promising for easy identification and manipulation of qualitative and quantitative traits in economically important crop plants including *Brassica* vegetables (Snowdon and Friedt 2004). Molecular markers can locate and determine the number of genes controlling the traits of interest in the genome and using the linked markers, manipulation of traits can be done without much difficulty. This not only makes it easier but also saves the time required for selection of a particular trait without waiting for the actual time of expression of the traits. At the early stage, using linked markers one can select the genome and manipulate in breeding. Several molecular markers have been developed in *Brassica* species and molecular linkage maps have been constructed. Highly saturated maps with co-dominant markers such as simple sequence repeat (SSR) would help in identification of some important recessive genes governing resistances to pests and diseases and quality traits in heterozygous conditions and hence enhance the selection process at the early stage without waiting for the time of character expression.

Agronomically important traits that were difficult to breed conventionally, have been mapped in the genome. Various genes and quantitative trait loci (QTL) controlling resistance against diseases and insect-pests have been mapped in vegetable *Brassica* genome using molecular markers (Snowdon and Friedt 2004). Molecular markers and genetic maps have been widely used to assist breeders in marker-aided selection in *Brassica* species. The genetic maps developed by using molecular markers are useful for the study of genome structure, evolution, mapping and tagging of important genes. In *B. oleracea* and *B. rapa*, linkage maps have been extensively used to identify economically important complex polygenic traits and efforts are being made to transfer these genes using the linked markers to desirable commercial background. Notable examples are mapping of disease resistance and male cytoplasm male sterility restorer genes (Snowdon and Friedt 2004). Linked markers to these economically important traits could be used to transfer this trait into commercial cultivars though marker-assisted selection, molecular manipulation of QTL/gene(s), and cloning and characterization of important genes. It is due to the construction of molecular genetic maps, that we can learn about the complex structure of *Brassica* genome including duplication, and intra- and intergenomic conservation of different chromosomal blocks. Brassicas are the close relatives of model plant *Arabidopsis thaliana,* whose complete genome sequence is available. Most of the genes have been characterized in *Arabidopsis* and molecular mapping and comparative alignment of *Brassica* genome with that of *A. thaliana* chromosomes showed conservation of chromosomal blocks. Further, at the gene level, conservation of 75–80% sequence identified has been reported between coding regions of different brassicas and *A. thaliana*. Comparative alignment of the *Brassica* chromosome containing genes/ QTLs of interest would be helpful for identification of candidate genes from *A. thaliana*. This would help faster cloning and characterization of candidate genes of interest and manipulation in brassica breeding. Taking advantage of this mapping and tagging of glucosinolates genes has been done in *Brassica juncea* (Bisht et al. 2009). Gene-specific markers for high and low allele have been developed and are being used in *B. juncea* breeding for development of canola quality *B. juncea*.

Transgenic technology has been used to develop male sterility in *B. napus* and *B. juncea* using barnase and barstar system (Mariani et al. 1990; Jagannath et al. 2001; Bisht et al. 2007). These systems can be used for developing male sterile line in brassica vegetables for hybrid seed production because most of the cytoplasmic male sterility transferred from *B. nigra, B. napus* and radish were associated with morphological abnormalities. Many *Bacillus thuriengiensis* genes, e.g., *Cry 1A, cry 1Ac, cry 2A,* have been transformed in *Brassica* species for developing insect-pest resistance (Sharma et al. 2005; Ordas and Cartea 2008). Likewise, many genes such as chitinase, glucanase, permantis, etc. can be transformed into *B. oleracea* for producing fungal resistant brassica crops. Many of the insect or disease resistance traits are not found in *Brassica* vegetables. In such cases the trait of economic importance should be transferred from related species through wide hybridization and the embryo-rescue technique. This technique has been used by many to transfer cytoplasmic male sterility from *B. napus* and radish (Inomata 1977; Ayotte et al. 1987). Microspore culture is also a very important technique for rapidly producing homozygous lines with diverse phenotypic variation from wide-crosses because vegetable brassicas are highly cross-pollinated because of self-incompatibility. Doubled haploid lines produced though microspore culture in *Brassica* vegetables have been used for development of linkage maps, mapping and tagging of many important genes and traits QTLs, and other genetic studies. Thus, breeding of brassica vegetables for desired plant type and quality will be greatly enhanced by using the above molecular tools and techniques. This will not only speed the breeding process but also help in precision breeding of vegetable brassicas.

Aknowledgements

This work was supported by grants for the Technology Development Program for Agriculture and Forestry (Grant no. 607002-05), Ministry of Agriculture, Forestry and Fisheries, Republic of Korea.

References

Acciarri N, Schiavi M, Vitelli G, Maestrelli A, Forni E, Giovanessi L (1997) Breeding of green curded cauliflower for fresh market and freezing. International Symposium on Brassicas. Acta Hort 459: 403–410.

Ahluwalia KS, Swarup V, Chatterjee SS (1977) Inheritance of qualitative characters in Indian cauliflower. Veg Science 4: 67–80.

Allgayer H (1928) Genetische Untersuchungen mit Gartenkohl *(Brassica oleracea)* nach Krcuzungsversuchen yon Richard Frcudenberg. Z Ind Abst u Vererb 47: 191–260.

Arus P, Orton TJ (1983) Inheritance and linkage relationships of isozyme loci in *Brassica oleracea.* J Hered 74: 405–412.

Ayotte R, Harney PM, Souza-Machado V (1987). The transfer of triazine resistance from *Brassica napus* L. to *B. oleracea* L. 1. Production of F_1 hybrids through embryo rescue. Euphytica 38: 615.

Bannerot HO, Boulidard L, Cauderon Y, Tempe T (1974) Cytoplasmic male sterility transfer from *Raphanus* to *Brassica.* In: Eucarpia Cruciferae Conference, Scottish Hort Res Inst, Mylnefield, Dundee, Scotland, p 52.

Baril CP, Verhaegen D, Vigneron PH, Bouvet JM, Kremer A (1997) Structure of the specific combining ability between two species of *Eucalyptus.* I. RAPD data. Theor Appl Genet 94: 796–803.

Baswana KS, Rastogi KB, Sharma PP (1991) Inheritance of stalk rot resistance in cauliflower (*Brassica oleracea* var. *botrytis* L.). Euphytica 57: 93–96.

Bateman AJ (1955) Self-incompatibility system in angiosperms. III. Cruciferae. Heredity 9: 52–68.

Bianco VV, Pimpini F (1990) Orticoltura. Pàtron Editore, Bologna.

Bisht NC, Jagannath A, Burma PK, Pradhan AK, Pental D (2007) Retransformation of a male sterile *barnase* line with the *barstar* gene as an efficient alternative method to identify male sterile–restorer combinations for heterosis breeding. Plant Cell Rep 26: 727–733.

Bisht NC, Gupta V, Ramchiary N, Sodhi YS, Mukhopadhyay A, Arumugam N, Pental D, Pradhan AK (2009) Fine mapping of loci involved with glucosinolate biosynthesis in oilseed mustard (*Brassica juncea*) using genomic information from allied species. Theor Appl Genet 118: 413–21.

Borchers EA (1966) Characteristics of male sterile mutant in cauliflower (*Brassica oleracea* L.). Proc Am Soc Hort Sci 88: 406–410.

Branca F (2008) Cauliflower and broccoli. In: J Prohens , F Neuz (eds) Handbook of Plant Breeding. Vegetables I. Springer-Verlag, Berlin, Heidelberg.

Brown AF, Yousef GG, Jeffery EH, Klein BP, Walling MA, Kushadand MM, Juvik JA (2002) Glucosinolate profile in broccoli: variation in levels and implications in breeding for cancer chemoprotection. J Am Hort Sci 127: 807–813.

Cao G, Sofic E, Prior RL (1996) Antioxidant capacity of tea and common vegetables. J Agri Food Chem 44: 3426–3431.

Carlsson M, Von Bothmer R, Merker A (2004) Screening and evaluation of resistance to downy mildew (*Peronospora parasitica*) and clubroot (*Plasmodiophora brassicae*) in genetic resources of *Brassica oleracea*. Hereditas 141: 293–300.

Chand J (1980) Studies on selection for yield and quality in cauliflower. PhD Thesis, PG School, IARI, New Delhi, India.

Chatterjee SS (1993) Cole crops. In: TK Bose ,MG Som , J Abir (eds) Vegetable Crops. Naya Prokash, Calcutta, India, pp 125–223.

Chiang MS (1972) Inheritance of head splitting in cabbage (*B.olearcea* var *capitata* L.). Euphytica 2: 507–509.

Chiang MS (1983) Inheritance of clustered flowers in cabbage. J. Hered 74: 116.

Chiang MS, Crete R (1976) Diallel analysis of the inheritance of resistance to race 6 of *Plasmodiophora brassicae* in cabbage. Can J Plant Sci 56: 865–868.

Chiang MS Crete R (1987) Cytoplasmic male sterility in *Brassica oleracea* induced by *Brassica napus* cytoplasm, female fertility and restoration of male fertility. Can J Plant Sci 67: 891–897.

Cho WD, Kim WG, Takahashi K (2001) Occurrence of clubroot in cruciferous vegetable crops and races of the pathogen in Korea. Plant Pathol 9: 64–68.

Crisp P (1982) The use of an evolutionary scheme for cauliflowers in the screening of genetic resources. Euphytica 31: 725.

Crisp P, Angell SM (1985) Genetic control of green curd colour in cauliflower. Annals of Applied Biology 107: 601–603.

Crisp P, Tapsell CR (1993) Cauliflower (*Brassica oleracea* L.). In: G Kalloo ,BO Bergh (eds) Genetic Improvement of Vegetable Crops. Pergamon Press, Oxford, UK, pp 157–178.

Crisp P, Crute IR, Sutherland RA, Angell SM, Bloor K, Burgess H, Gordon PL (1989) The exploitation of genetic resources of *Brassica oleracea* in breeding for resistance to cluboot (*Plasmodiophora brassicae*). Euphytica 42: 215–226.

Crisp P, Walkey DGA, Bellman E, Roberts E (1975) A mutation affecting curd colour in cauliflower (Brassica oleracea L. var botrytis DC). Euphytica 24: 173–176.

Crute IR, Gray PC, Buczacki ST (1980) Variation in *Plasmodiophora brassicae* and resistant to clubroot disease in *Brassicas* and allied crops. Plant Breed Abstr 50: 91–104.

Dadlani NK (1977) Studies on selection for yield and quality in cauliflower. PhD Thesis, PG School, IARI, New Delhi, India.

de Moel CP, Everaats AP (1990) Growth, development, and yield of white cabbage in relation to time of planting. Acta Hort 267: 279–288.

De Nettancourt D (1977) Incompatibility in angiosperms. Springer-Verlag, Berlin pp 28–56.

Detjen LR, McCue CA (1933) Cabbage characters and their heredity. Del Agri Exp Stn Bull 180: 1–127.

Dias JS, Ferreira ME, Williams PH (1993) Screening of Portuguese cole landraces (*Brassica oleracea* L.) with *Peronospora parasitica* and *Plasmodiophora brassicae*. Euphytica 67: 135–141.

Dias LAS, Marita J, Cruz CD, Barros EG, Salomão TMF (2003) Genetic distance and its association with heterosis in cacao. Braz Arch Biol Technol 46: 339–347.

Dickson MH (1975) G1117A, G1102A and G1106A cytosterile broccoli inbreds. Hort Sci 10: 535.

Dickson MH (1977) Inheritance of resistance to tip burn in cabbage. Euphytica 26: 811–815.

Dickson MH, Eckenrode CJ (1975) Variation in *Brassica oleracea* resistance to cabbage looper and imported cabbage worm in the greenhouse and field. J Econ Entomol 68: 757–760.

Dickson MH, Eckenrode CJ (1980) Breeding for resistance in cabbage and cauliflower to cabbage looper, imported cabbage worm and diamond back moth. J Am Soc Hort Sci 105: 782–785.

Dickson MH, Lee CY (1980) Persistent white curd and other curd characters of cauliflower. J Am Soc Hort Sci 105: 533–535.

Dickson MH, Wallace DH (1986) Cabbage breeding. In: MJ Bassett (ed) Breeding Vegetable Crops. AVI Publ Co, Westport, Connecticut, USA, ppp 395–432.

Dixit R, Nasrallah ME, Nasrallah JB (2000) Post-transcriptional maturation of the S receptor kinase of *Brassica* correlates with co-expression of the S-locus glycoprotein in the stigmas of two *Brassica* strains and in transgenic tobacco plants. Plant Physiol 124: 297–311.

Dixon GR (2006) Vegetable Brassicas and Related Crucifers. CABI Publ, Wallingford, Oxfordshire, UK.

Ellis PR, Hardman JA (1975) Laboratory methods for studying non-preference resistance to cabbage root fly in cruciferous crops. Ann Appl Biol 79: 253–264.

Ellis PR, Hardman JA (1988a) Investigations of the resistance in cruciferous crops to cabbage root fly at Wellesbourne. Report of the 4th Meeting of EUCARPIA/IOBC/WPRS Working Group Breeding for Resistance to Insects and Mites. IOBC/ WPRS Bull XI/3: 63–64.

Fenwick GR, Heaney RK, Mullin WJ (1983) Glucosinolates and their breakdown products in food plants. CRC Crit Rev Food Sci Nutr 18: 123–201.

Foster RE, Walker JC (1946) Improvement of ascorbic acid content in yellows-resistant cabbage. Phytopathology 36: 398.

Gangopadhyay KK, Gill HS, Sharma SR (1997) Heterosis and combining ability studies in early group of Indian cauliflower involving self-incompatible lines. Veg Sci 24: 26–28.

Gray AR, Crisp P (1979) Autumn cauliflower. Report of the National Vegetable Research Station for 1978, pp 57–58.

Giamoustaris A, Mithen R (1996) Genetic of aliphatic glucosinolates. Side-chain modification in *Brassica oleracea*. Theor Appl Genet 93: 1006–1010.

Göring DR, Glavin TL, Schäfer U, Rothstein SJ (1993) An *S* receptor kinase gene in self-compatible *Brassica napus* has a 1-bp deletion. Plant Cell 5: 531–539.

Gray AR (1982) Taxonomy and evolution of broccoli (*B. oleracea* L. var. *italic* Plenck). Econ Bot 36: 397.

Halkier BA, Gershenzon J (2006) Biology and biochemistry of glucosinolates. Annu Rev Plant Biol 57: 303–333.

Hodgkin T (1981) The inheritance of node number and rate of node production in Brussel sprouts. Theor Appl Genet 59: 79–82.

Hodgkin T (1995) Cabbages, kales, etc.. In: J Smartt ,NW Simmonds (eds) Evolution of Crop Plants. 2nd edn. Longman, Burnt Mill, Harlow, UK, pp 76–82.

Hoser-Krauze J (1987) Influence of cytoplasmic male sterility source on some characters of cauliflower (*Brassica oleracea* var. *botrytis* L.). Genet Polonica 28: 101.

Ikeda S, Nasrallah JB, Dixit R, Preiss S, Nasrallah ME (1997) An aquaporin-like gene required for the *Brassica* self-incompatibility response. Science 276: 1564–1566.

Inomata N (1977) Production of interspecific hybrids between *Brassica campestris* and *B. oleracea* by culture *in vitro* of excised ovaries. I. Effects of yeast extracts and casein hydrolysable on the development of excised ovaries. Jpn J Breed 27: 295.

Jagannath A, Bandyopadhyay P, Arumugam N, Gupta V, Burma PK, Pental D (2001) The use of a Spacer DNA fragment insulates the tissue-specific expression of a cytotoxic gene (*barnase*) and allows high-frequency generation of transgenic male sterile lines in *Brassica juncea* L. Mol Breed 8: 11–23.

Joyce TA, Abberton MT, Michelson-Yeates TPT, Forster JW (1999) Relationship between genetic distance measured by RAPD-PCR and heterosis in inbred lines of white clover (*Trifolium repens* L.). Euphytica 107: 159–165.

Jyoti S, Vashistha RN (1986) Gene effects studies of curd weight in mid season cauliflower (*Brassica oleracea* var. *botrytis* L.). Haryana J Hort Sci 15: 263–266.

Kakizaki Y (1930) A dominant wldte-flowered mutant of *Brassica oleracea* L. Jpn J Genet 6: 55–60.

Kale PN, Chatterjee SS, Swarup V (1979) Heterosis and combining ability in Indian cauliflower (*Brassica oleracea* L. var. *botrytis* L.). Veg Sci 6: 11–20.

Kapoor KS (1986) Disease resistance in cauliflower (*Brassica oleracea* var. *botrytis*) against *Sclerotinia sclerotiarum* (lib) De Bari. Veg Sci 13: 285–288.

Kesseli RV, Paran H, Michelmore RW (1994) Analysis of a detailed genetic linkage map of *Lactuca sativa* (Lettuce) constructed from RFLP and RAPD markers. Genetics 136: 1435–1446.

King GJ (1990) Molecular genetics and breeding of vegetable brassicas. Euphytica 50: 97–112.

Kontaxis DG, Mayberry KS, Rubatzky VE (1979) Reaction of cauliflower cultivars to downy mildew in Imperial Valley. Calif Agri 33: 19.

Kotoswki F (1926) Effect of self fertilization in cabbage and onion. Mem Hort Soc (New York) 8: 281–284.

Kristofferson KB (1921) Untersuchungen iber die $F_1$and F_2, generation yon einem spontanen Bastard zwischen Weisskraut and Griinkohl. Sveriges Utsad Tidsk 31–52.

Kristofferson KB (1924) Contributions to the genetics of *Brassica oleracea*. Hereditas 5: 297–364.

Kristofferson KB (1927) Contributions to the genetics of *Brassica oleracea*. II. Hereditas 9: 343–348.

Kwan CC (1934) Inheritance of some plant characters in cabbage, *Brassica oleraceae* var. *cantata*. J Agri Ass'n China 126: 81–127.

Kwon SJ, Ha WG, Wang HG, Yang SJ, Choi HC, Moon HP, Ahn SN (2002) Relationship between heterosis and genetic divergence in 'Tongil'-type rice. Plant Breed 121: 487–492.

Lal OP, Gill HS, Sharma SR, Singh R (1991) Field evaluation in different cultivars of cauliflower against the head borer (*Hellula undalis* Fabricius) (Lepidoptera: pyralidae). J Entomol Res 15: 277–281.

Lal OP, Sharma SR, Singh R (1994) Field resistance in different hybrids of cauliflower against Bihar hairy caterpillar (*Spilosoma obliqua* Walker) (Lepidoptera: arctidae). J Entomol Res 18: 45–48.

Lal G, Chatterjee SS, Swarup V (1979) Genetics of important characters in Indian cauliflower. Genet Agr 33: 181–190.

Lamb RJ (1989) Entomology of oilseed *Brassica* crops. Annu Rev Entomol 34: 211–229.

Laurens F, Thomas G (1993) Inheritance of resistance to clubroot *(Plasmodiophora brassicae* Wor.) in kale *(Brassica oleracea* ssp. *acephala)*. Hereditas 119: 253–262.

Lázaro A, Aguinaglade I (1998) Genetic diversity in *Brassica oleracea* L. (Cruciferae) and wild relatives ($2n = 18$) using isozymes. Ann Bot 82: 821–828.

Lee JT, Ru GL, Lee JN, Kweon HJ, Lee EH, Kim WB, Park HY (2001) Selection of alternative crops on Chinese cabbage in alpine. Kor J Hort Sci Technol 19(suppl II): 63.

Lee SS, Yoon JY, Yoon HM (1982) An in vitro pollination technique for detecting self-incompatibility in Cruciferous crops. Kor J Bred 14(3): 308–313.

Li CW (1981) The origin, evolution, taxonomy and hybridization of Chinese cabbage, 1–10. In: NS Talekar ,TD Griggs (eds) Chinese Cabbage. AVRDC, Taiwan.

Li G, Quiros CF (2002) Genetic analysis, expression and molecular characterization of *BoGSL-ELONG*, a major gene involved in the aliphatic glucosinolate pathway of *Brassica* species. Genetics 162: 1937–1943.

Liu ZQ, Pei Y, Pu ZJ (1999) Relationship between hybrid performance and genetic diversity based on RAPD markers in wheat, *Triticum aestivum* L. Plant Breed 118: 119–296.

Magruder R (1937) Improvement in the leafy cruciferous vegetables. US Dept Agri Yrbk 1937: 289–299.

Mahajan V, Gill HS, Sharma SR, Singh R (1996) Combining ability studies in Indian cauliflower (*Brassica oleracea* var. *botrytis* L.) Group III. Veg Sci 23: 166–170.

Malinowski E (1929) Genetics of *Brassica*. Biblio. Genet 5: 1–26.

Manzanares-Dauleux MJ, Divaret I, Baron F, Thomas G (2000) Evaluation of French Brassica oleracea landraces for resistance to *Plasmodiophora brassicae*. Euphytica 113: 211–218.

Mariani C, DeBeuckeleer M, Truettner J, Leemans J, Goldberg RB (1990) Induction of male sterility in plants by a chimaeric ribonuclease gene. Nature 347: 737–741.

Murase K, Shiba H, Iwano M, Che F-S, Watanabe M, Isogai A, Takayama S (2004) A membrane-anchored protein kinase involved in *Brassica* self-incompatibility signaling. Science 303: 1516–1519.

McCollum GD (1981) Induction of an alloplasmic male sterile *Brassica oleracea* by substituting cytoplasm from "Early Scarlet globe" radish (*Raphanus sativus*). Euphytica 30: 855.

Nasrallah JB, Kao TH, Chen CH, Goldberg ML, Nasrallah ME (1987) Amino-acid sequence of glycoproteins encoded by three alleles of the *S* locus of *Brassica oleracea*. Nature 326: 617–619.

Nasrallah JB, Yu S-M, Nasrallah ME (1988) Self-incompatability genes of *Brassica oleracea:* expression, isolation and structure. Proc Natl Acad Sci USA 85: 5551–5555.

Natti JJ, Dickson MH, Atkin JD (1967) Resistance of *Brassica oleracea* varieties to downy mildew. Phytopathology 57: 144–147.

Nieuwhof M (1961) Male sterility in some cole crops. Euphytica 10: 351–356.

Nieuwhof M (1968) Effect of temperature on the expression of male sterility in Brussels sprout (*B. oleracea* L.var. *gemmifera* DC.). Euphytica 17: 265–273.

Nieuwhof M, Garretsen F (1961) The solidity of the cauliflower curd. Euphytica 10: 301–306.

Nilsson J, Olsson K, Engqvist G, Ekvall J, Olsson M, Nyman M, Akesson B (2006) Variation in the content of glucosinolates, hydroxycinnamic acids, carotenoids, total antioxidant capacity and low-molecular-weight carbohydrates in *Brassica* vegetables. J Sci Food Agri 86: 528–538.

Ogura H (1968) Study on the new male sterility in Japanese radish with special reference to the utilization of this sterility towards the practical raising of hybrid seeds. Mem Fac Agri Kagoshima Univ 6: 39.

Ordas A, Catea ME (2008) Cabbage and kale. In: J Prohens, F Neuz (eds) Handbook of Plant Breeding. Vegetables I. Springer—Berlin, Heidelberg, Germany.

Pandey SC, Naik G, Ramkishan, Sridhar TS (1995) Breeding resistant varieties in cauliflower and cabbage. In: Mandal et al. (eds) Agroecosystem Management. Indian Inst of Hort Res, Bangalore, Karnataka, India.

Park HG, Kwon OH, Kim HT, Na JH, Park Y, Park JY, Park CS, Song KH, Yang DH, Om YH, Yoo IW, Yoon J-Y, Lee BS, Lee SS, Suh HD, Jeong SY, Oh D-G, Cheong JW, Cho YH, Cho Y-S, Cho YC (2008) The Recent History of Vegetable Seed Industry in Korea. Seoul National Univ Press, Korea.

Park S, Yoon MK, Kwak JH (2008) Development of an effective inoculation method for large quantaties of clubroot disease using hydroponics in Chinese cabbage. J Kor Soc Hort Sci 26(4): 449–453.

Pearson OH (1929) A dominant white flower color in *Brassica olearceae* L. Am Nat 68: 561–565.

Pearson OH (1934) Dominance of certain quality characters in cabbage. Proc Am Soc Hort Sci 31: 169–176.

Pearson OH (1972) Cytoplasmically inherited male sterility characters and flavor component from the species cross (*B. nigra* L. Koch) into *Brassica oleracea*. L. J Am Soc Hort Sci 97: 397.

Pease MS (1926) Genetic studies in *Brasica oleracea*. J Genet 16: 363–385.

Piao Z, Ramchiary N, Lim YP (2009) Genetics of clubroot resistance in *Brassica* species. J Plant Growth Regul: doi 10.1007/s00344-009-9093-8.

Picoaga A, Cartea ME, Soengas O, Monetti L, Ordás A (2003) Resistance of kale populations to lepidopterous pests in northwestern Spain. J Econ Entomol 96: 143–147.

Pink DAC and Walkey DGA (1986) Resistance to turnip mosaic and cauliflower mosaic viruses in horticultural brassicas. Asp Appl Biol 12: 69–74.

Provvidenti JJ (1980) Evaluation of Chinese cabbage cultivars from Japan and the Peoples Republic of China for resistance to turnip mosaic virus and cauliflower mosaic virus. J Am Soc Hort Sci 105: 271–573.

Quiros CF (2001) DNA-based marker maps in Brassica. In: Phillips RL, Vasil IK(eds) DNA-based Markers in Plants. Kluwer Academic, London, pp 201–238.

Radcliffe JJ, Chapman RK 1(966) Varietal response to various insect attack on various cruciferous crops. J Econ Entomol 59: 120–125.

Rakow G (2004) Species origin and economic importance of *Brassica*. In: EC Pua ,CJ Douglas (eds) Biotechnology in Agriculture and Forestry, vol 54: *Brassica*. Springer—Berlin, Heidelberg, Germany, pp 1–7.

Rasmusson J (1932) Results from a cross cabbage × savoy cabbage. Hereditas 16: 241–248.

Roberts IN, Stead AD, Ockendon DJ, Dickinson HG (1979) A glycoprotein associated with the acquisition of the self-incompatibility system by maturing stigmas of *Brassica oleracea*. Planta 146: 179–183.

Rosa EAS, Heaney RK, Fenwick GR, Portas CAM (1997) Glucosinolates in crop plants. Hort Rev 19: 99–215.

Ruffio-Chable V, Bellis H, Herve Y (1993) A dominant gene for male sterility in cauliflower (*Brassica oleracea* var. *botrytis*). Phenotype expression, inheritance and use in F1 hybrid production. Euphytica 67: 9–17.

Sampson DR (1967) New light on the complexities of anthocyanin inheritance in *Brassica oleracea*. Can J Genet Cytol 8: 404–413.

Samson S (1930) Le *Brassica oleracea*. Quelques caracttres hereditaires du *Brassica oleracea*. Rev Inst Agri d'Oka 4: 92–95.

Sandhu JS, Singh AK (1977) Inheritance of maturity period and curd weight in cauliflower (*Brassica oleracea* var. *botrytis* L.). Haryana J Hort Sci 6: 161–165.

Schopfer CR, Nasrallah ME, Nasrallah JB (1999) The male determinant of self-incompatibility in Brassica. Science 286: 1697–1700.

Seo JP, Park KW, Kim HJ, Yang SY, Nou IS (2003) Identification of *S*-haplotypes and its relationship with activity of self-incompatibility in radish (*Raphanus sativus* L.). Kor J Breed 35(5): 313–318.

Sharma BR, Swarup V, Chatterjee SS (1972) Inheritance of resistance to black root in cauliflower. Can J Genet Cytol 14: 363.

Sharma BR, Dhiman JS, Thakur JC, Singh A, Bajaj K (1991) Multiple disease resistance in cauliflower. Adv Hort Sci 5: 30–34.

Sharma SR, Gill HS, Kapoor KS (1988) Heterosis and combining ability studies in late cauliflower (*Brassica oleracea* var. *botrytis* L.). Veg Sci 15: 55–63.

Sharma SR, Kapoor KS, Gill HS (1995) Screening against Sclerotinia rot (*Sclerotinia sclerotiarum*), downy mildew (*Peronospora parasitica*) and black rot (*Xanthomonas compestris*) in cauliflower (*Brassica oleracea* var. *botrytis* subvar. *cauliflora* DC). Indian J Agri Sci 65: 916–918.

Sharma SR, Gill HS, Kapoor KS (1997) Inheritance of resistance to white rot in cauliflower. Indian J Hort 54: 86–90.
Sharma SR, Singh PK, Chable V, Tripathi SK (2005) A review of hybrid cauliflower development. J New Seeds 6: (2)151–193.
Shiba H, Kimura N, Takayama S, Hinata K, Suzuki A, Isogai A (2000) Alteration of the self-incompatibility phenotype in Brassica by transformation of the antisense *SLG* gene. Biosci Biotechnol Biochem 64: 1016–1024.
Shiba H, Iwano M, Entani T, Ishimoto K, Shimosato H, Che F-S, Satta Y, Ito A, Takada Y, Watanabe M, Isogai A, Takayama S (2002) The dominance of alleles controlling self-incompatibility in *Brassica* pollen is regulated at the RNA level. Plant Cell 14: 491–504.
Shinohara S (1984) Vegetable Seed Production Technology of Japan. Elucidated with Respective Variety Development Histories, Particulars, vol I. SAACEO, Tokyo, Japan, pp 18–59.
Singh AK, Singh J, Saini SS (1978) Inheritance of curd quality in cauliflower (*Brassica oleracea* var . *botrytis* L.). Hort Res 18: 1–6.
Singh DP, Swarup V, Chatterjee SS (1975) Genetical studies in Indian cauliflowers (*Brassica oleracea* L. var. *botrytis* L.). I. Heterosis and combining ability in maturity group I. Veg Sci 2: 1–7.
Singh DP, Swarup V, Chatterjee SS (1976) Genetic studies in Indian cauliflower (*Brassica oleracea* L. var. *botrytis*). Heterosis and combining ability in maturity group II. Veg Sci 3: 41–46.
Singh H, Kalda TS (1995) Screening of cauliflower germplasm against sclerotinia rot. Indian J Genet 55: 98–102.
Singh R, Gill HS, Sharma SR (1993) Breeding cauliflower for downy mildew resistance (Abstract). In: Golden Jublee Symp–Hort Res–Changing Scenario,, May 24–28, Indian Society of Horticultural Science, Bangalore, Karnataka, India, p 67.
Snogerup S (1980) The wild forms of the *Brassica oleracea* group (2n=18) and their possible relations to the cultivated ones. In: S Tsunoda ,K Hinata , C Gómez-Campo (eds) Brassica Crops and Wild Allies. Jpn Sci Soc Press, Tokyo, Japan, pp 120–132.
Snogerup S, Gustafsson M, von Bothmer R (1990) *Brassica* sect. *Brassica (Brassicaeae)*. I.Taxonomy and variation. Willdenowia 19: 271–365.
Snowdon RJ, Friedt W (2004) Molecular markers in Brassica oilseed breeding: current status and future possibilities. Plant Breed 123: 1–8.
Stoner KA (1990) Glossy leaf wax and plant-resistance to insects in *Brassica oleracea* under natural infestation. Environ Entomol 19: 730–739.
Stoner KA (1992) Density of imported cabbageworms (Lepidoptera: Pieridae), cabbage aphids (Homoptera: Aphididae), and flee beetles (Coleoptera: Chrysomelidae) on glossy and trichome bearing lines of *Brassica oleracea*. J Econ Entomol 85: 1023–1030.
Swarup V, Pal AB (1966) Gene effects and heterosis in cauliflower-I. Indian J Genet 26: 269.
Swarup V, Chatterjee SS (1972) Origin and genetic improvement in Indian cauliflower. Econ Bot 26: 381–394.
Swarup V, Brahmi P (2005) Cole crops. In: BS Dhillon, RK Tyagi, S Saxena, GJ Randhawa (eds) Plant Genetic Resources: Horticultural Crops. Narosa Publ House, New Delhi, India, pp 75–88.
Takasaki T, Hatakeyama K, Suzuki G, Watanabe M, Isogai A, Hinata K (2000) The S receptor kinase determines self-incompatibility in Brassica stigma. Nature 403: 913–916.
Takayama S, Shiba H, Iwano M, Shimosato H, Che F-S, Kai N, Watanabe M, Suzuki G, Hinata K, Isogai A (2000) The pollen determinant of self-incompatibility in *Brassica campestris*. Proc Natl Acad Sci USA 97: 1920–1925.
Thompson KF, Taylor JP (1966) Non-linear dominance relationships between S-alleles. Heredity 22: 519–527.
Trick M, Flavell RB (1989) A homozygous S genotype of *Brassica oleracea* expresses two S-like genes. Mol Gen Genet 218: 112–117.
Tschermak EV (1916) Ueber den gegenwartigen Stand der Gemusezuchtung. Z Pflanzenzucht 4: 65–104.

Van der Meer OP (1985) Male sterility in cole crop—a serial story. EUCARPIA Cruciferae Newsl 10: 58.

Vashistha RN, Neog SJ, Pandita ML (1985) Gene effects studies of curd compactness in cauliflower. Haryana Agri Univ J Res 15: 406–409.

Viani P (1929) Trattato di Orticoltura. Battiato editore, Catania, pp 914–935.

Voorrips RE, Visser DL (1993) Examination of resistance to clubroot in accessions of *Brassica oleracea* using a glasshouse seedling test. Neth J P1ant Pathol 99: 269–276.

Voorrips RE, Kanne HJ (1996) Genetic analysis of resistance to clubroot (*Plasmodiophora brassicae*) in *Brassica oleracea*. 2. Quantitative analysis of root system measurements. Euphytica 93(1): 41–48.

Walker JC (1930) Inheritance of Fusarium resistance in cabbage. J Agri Res (Washington DC) 40: 721–745.

Walker JC (1944) Progress in combination of yellows and mosaic resistance with ascorbic acid content in cabbage. Phytopathology 34: 1012–1013.

Walker JC, Foster RE (1946) The inheritance of ascorbic acid content in cabbage. Am J Bot 83: 758–761.

Walker JC, Pound GS (1947) Improvement of cabbage for disease resistance. Phytopathology 87: 23.

Walker JC, Williams PH (1965) Inheritance of powdery mildew resistance in cabbage. Plant Dise Rep 49: 198–201.

Walker JC, Larsen RH (1980) Bull Washington Agri Exp Stn 547: 12–16.

Walker JC, Williams PH, Pound GS (1965) Internal tipburn in cabbage: Its control through breeding. Wisc Agri Exp Stn Bull, Madison, WI, USA, p 258.

Walkof C (1963) A mutant annual cabbage. Euphytica 12: 77–80.

Wallace DH (1979) Procedures for identifying S-alleles genotypes of *Brassica*. Theor Appl Genet 54: 245–265.

Watanabe E (1981) Development of major Chinese cabbage in Japan. In: NS Talekar ,TD Griggs (eds) Chinese Cabbage. AVRDC, Taiwan, pp 11–27.

Watanabe M, Ito A, Takada Y, Ninomiya C, Kakizaki T, Takahata Y, Hatakeyama K, Hinata K, Suzuki G, Takasaki T, Satta Y, Shiba H, Takayama S, Isogai A (2000) Highly divergent sequences of the pollen self-incompatibility (*S*) gene in class-I *S* haplotypes of *Brassica campestris* (syn. *rapa*) L. FEBS Lett 473: 139–144.

Watts LE (1963) Investigations into the breeding system of cauliflower *Brassica oleracea* var. *botrytis* (L.). I. Studies on self-incompatibility. Euphytica 12: 323–340.

Watts LE (1964) Studies on maturity in F_1 and F_2 generation of cauliflower from crosses between summer, autumn and winter types. J Hort Sci 39: 84–91.

Weisaeth G (1977) Hodelolsortene 'Respla' og 'Resista' foredlainger med Klumratresistens. Gartneryrket 67: 937–939.

Williams JGK, Kubelik AR, Livak KJ, Rafalski JA, Tingey SV (1990) DNA polymorphisms amplified by arbitrary primers are useful genetic markers. Nucleic Acids Res 18: 6531–6535.

Williams PH, Staub J, Sutton JC (1972) Inheritance of resistance in cabbage to black rot. Phytopathology 62: 247–252.

Wills AB (1977) A preliminary gene list in Brassica oleracea.Eucarpia Cruciferae News: 22–24.

Yarnell SH (1956) Cytogenetics of vegetable crops.II. Crucifers Bot Rev 22: 81–166.

Yoshikawa H (1993) Studies on breeding of clubroot resistance in cole crops. Bull Natl Res Inst Veg Ornam Plants Tea Jpn Ser A 7: 1–165.

Zhang Y, Talalay P, Cho CG, Posner GH (1992) A major inducer of anticarcinogenic protective enzymes from broccoli: isolation and elucidation of structure. Proc Nat Acad Sci USA 89: 2399–2403.

3

Diversity Analysis and Molecular Taxonomy of Brassica Vegetable Crops

Guusje Bonnema,[1,3]* *Dunia Pino Del Carpio,*[1,4] and *Jianjun Zhao*[1,2]

ABSTRACT

Brassica crops have great economic value, harbor an enormous diversity and have adapted to cultivation in almost every part of the world. The triangle of U depicts the relationships between the three diploid species *B. rapa, B. oleracea* and *B. nigra* and the three amphidiploid species (*B. napus, B. juncea and B. carinata*) that resulted from interspecific hybridizations. In this chapter, first the different *Brassica* crops and radish are introduced, and their taxonomic position is described. Thereafter the literature on morphological and genetic diversity of vegetable *Brassica's* that belong to the diploid species *B. rapa* (A genome), *B. oleracea* (C genome) and the amphidiploid species *B. juncea* (A and B genomes), and radishes that belong to the related species *Raphanus sativus,* is reviewed. Since these species are not only cultivated for vegetable crops, but also for fodder, oil and condiments, these crop types are also included in this chapter. For the four different species that are reviewed in this chapter, overviews of the origin, the crop types, and the diversity of cultivated and wild forms are given. In the final paragraph the use of genetic and phenotypic diversity assessments for germplasm conservation and plant breeding are discussed.

Keywords: Brassica rapa, Brasica oleracea, Brassica napus, Brassica juncea, Raphanus sativus, genetic diversity, morpho-type, centers of diversity

[1]Laboratory of Plant Breeding, Wageningen University, 6700AA, Wageningen, the Netherlands.
[2]Horticultural College, Hebei Agricultural University, 071001, Baoding, China.
[3]Centre for Biosystems Genomics, 6700 AB Wageningen, the Netherlands.
[4]Graduate School Experimental Plant Sciences, 6708PB Wageningen, the Netherlands.
*Correspondening author: *Guusje.Bonnema@wur.nl*

3.1 Introduction

Brassicas have great economic value and encompass a great diversity of plants. They are grown as vegetables, fodder, sources of oils and condiments. The morphological variation present within *Brassica* species is enormous and from several species practically every part of the plant is used by man. This includes the leaves in crops like heading cabbages and the leafy types that do not form heads (pak choi, komatsuna, etc.), the terminal and axillary buds in cauliflower, broccoli, broccoletto, the seedpods in seed stalk mustard, the swollen stems in taitsai and kohlrabi, the swollen roots in turnips and swede and the seeds in all oil crops. In this chapter, the focus will be on three important Brassica vegetable crops that belong to *Brassica rapa, Brassica oleracea* and *Brassica juncea,* and on *Raphanus sativus*, the radishes. Since *B. rapa, B. juncea* and *R. sativus* also include morphotypes that are cultivated as oil crop and/or condiments, the diversity analyses also extends to the oil crops. For both *B. rapa* and *B. juncea* the diversity within and between morphotypes and the taxonomic relationships between the vegetable and oil types are discussed. Besides the genetic diversity in *R. sativus*, the phylogenetic relationships between radish and *B. rapa/B. oleracea* and *B. nigra* are also discussed. Table 3-1 gives an overview of the different subspecies within the three *Brassica* species and *R. sativus* classified according to the organs that are used for consumption and Fig. 3-1 depicts the different crop types within *B. rapa*. In the different paragraphs for each species the origin of the species is discussed, the crop types are introduced and an overview of studies on the genetic diversity within that species is presented. In the subchapters on *B. oleracea, B. rapa* and *R. sativus* diversity of wild related species that have played a role in the evolution of the different crop types is also discussed. Before we describe the different *Brassica* and *Raphanus* species, an overview is given of the taxonomic position of these genera/species in the Brassicaceae family.

3.2 Taxonomy of the Brasssicaceae

The Brassicaceae is a large plant family (338 genera and 3,700 species) of major scientific and economic importance. The taxonomy of this group is difficult, which is partly due to convergent evolution in nearly every morphological feature used to define tribes and genera. Koch and Kiefer (2006) state that rather than simply considering *Brassica* and *Arabidopsis thaliana* as dicotyledonous model organisms, the whole Brassicaceae family can be used as an example to understand phylogeographic patterns (evolution both in space and time) in a broader context. Even though there is a continuous increase in the number of phylogenetic studies available for various cruciferous taxa of different taxonomic levels

Table 3-1 Overview of different subspecies within three *Brassica* species and *Raphanus sativus* classified according to the organs that are used for consumption.

Organs used	*B. rapa*	*B. oleracea*	*B. juncea*	*R. sativus*
Leafy type—Heading	Chinese cabbage (*pekinensis*)	Chinese kale (*alboglabra*)	var. *capitata* and var. *involuta*	
		Kales (*accephala*)		
		White/red headed cabbage (*capitata*)		
		Savoy headed cabbage (*sabauda*)		
		Tronchuda cabbage (*costata*)		
Leafy types—Loose leaves	Pak choi (*chinensis*)		Large, green or purple leaves (*rugosa* and *latipa*)	
	Wutacai (*narinosa*)		Dissected leaves (var. *crispifolia*, var. *foliosa*, var. *leucanthus*, var. *linearifolia* and var. *multisecta*)	
	Komatsuna (*perviridis*)		strumiferous mustards (var. *strumata*)	
Leafy type—Tilling	Mizuna &Mibuna (*nipposinica* & *japonica*)		var. *gemmifera* and var. *multiceps*	
Axiliary buds		Brussels sprouts (*gemmifera*)		
Swollen stem	Caixin or Caitai (*parachinensis*)	Kohlrabi (*gongylodes*)	Swollen stem mustards (var. *Tumida*, var. *Tsatsai*)	
	Zicaitai (*purpuraria*)		Swollen shoot mustards (var. *crassicaulis*)	
	Broccoletto, Broccoli raab, Cima di rapa (*ruvo*)			

Table 3-1 contd...

Table 3-1 contd...

Organs used	***B. rapa***	***B. oleracea***	***B. juncea***	***R. sativus***
Swollen root	Turnips (*rapa*) (swollen hypocotyl/root)		Root mustard (var. *megarrhizza*, var. *napiformis*)	Small & large rooted (var. *sativus*, var. *hortensis*) Black radish (var. *niger*)
Inflorescence	Caixin or Caitai (*parachinensis*)	Broccoli (*italica*)	Seed stalk mustard (*utilis*)	Serpent, seedpod or rat-tailed radish (var. *caudatus*.)
	Zicaitai (*purpuraria*)	Cauliflower (*botrytis*)		
	Broccoletto, Broccoli raab, Cima di rapa (*ruvo*)			
Oil seed	Turnip rape (*oleifera*)		Oilseed mustard	Oilseed and fodder (*oleifera*)
	Brown and yellow sarson (*dichotoma* & *tricolaris*)		var. *juncea*, var. *gracilis*	

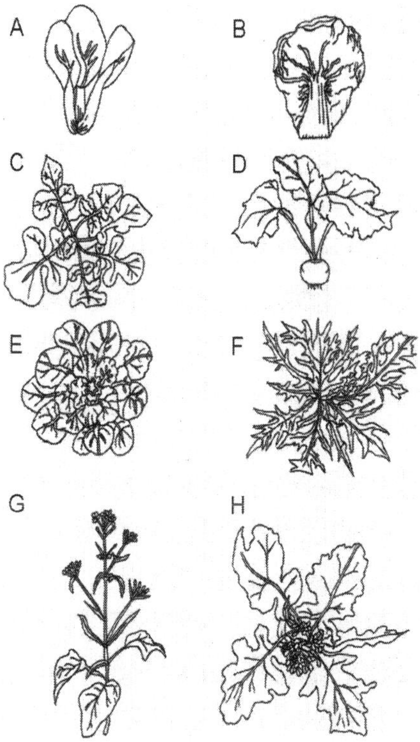

Figure 3-1 Drawings of *Brassica rapa* plants that represent the different morphotypes. A. ?pak choi, B. Chinese cabbage, C. winter oilseed, D. turnip, E. wutacai, F. mizuna, G. yellow sarson, H. broccoletto. Drawings were made by Dunia Pino Del Carpio from pictures of plants of the *B. rapa* core collection grown in Wageningen.

(see review Al-Shehbaz et al. 2006 and references therein), few studies give a comprehensive understanding of the whole family. Three family-wide molecular phylogenetic analyses, which resulted in a proposed new classification scheme to organize genera into tribes, greatly improved the understanding of the evolutionary relationships in the Brassicaceae, compared to early attempts that were mainly based on traits like fruit morphological characteristics (Al-Shebaz et al. 2006; Bailey et al. 2006; Beilstein et al. 2006). Based on these studies, most of the 338 genera were placed into 25 tribes that provide a reference for future phylogenetic studies (discussed in Schranz et al. 2007). The phylogenies that are based on a comprehensive phylogenetic analysis of 746 nrDNA internal transcribed spacer (ITS) sequences, representing 24 of the 25 currently recognized tribes, 146 genera, and 461 species of Brassicaceae, confirmed the monophyly of most of these new tribes (Bailey et al. 2006). All the above studies support the lineages I and II as first defined by Beilstein et al. (2006). Lineage I contains Camelineae (to which *Arabidopsis* belongs) and lineage II contains the agronomically important Brassiceae tribe. The tribe Brassiceae is a monophyletic group and comprises around 240 species and 49–54 genera. This tribe includes the economically important Brassica crops and radish (*Raphanus*). In the paragraph describing genetic diversity of *R. sativus*, the relationship between radish and the *Brassica* species will be discussed.

3.3 Diversity of *Brassica rapa* Vegetables and Oilseeds

3.3.1 *Original Centers of Diversity of* Brassica rapa

Being the first domesticated *Brassica* species, *Brassica rapa* has been cultivated for many centuries (over 4,000 years) from the highlands near the Mediterranean region to Scandinavia, to Germany and into Central Europe, and eventually to Central Asia (Gomez-Campo 1999; Dixon 2006). *B. rapa* was introduced into China through western Asia or Mongolia. The introduction into Japan could have occurred via China or Siberia. In Fig. 3-2, a map of Eurasia is depicted with the primary center and secondary centers of diversity for *B. rapa* are indicated.

A number of earlier studies based on morphology, geographic distribution, isozymes and molecular data indicate that *B. rapa* originated from two independent centers (Denford and Vaughan 1977; Song et al. 1988b; Gomez-Campo 1999). Europe is proposed as one center of origin

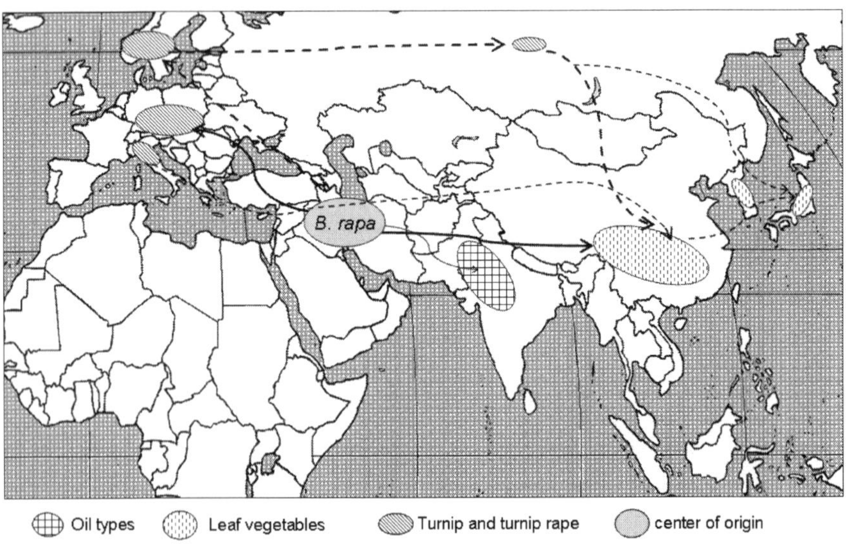

Figure 3-2 A map of Eurasia is depicted with primary center and secondary centers of diversity for *B. rapa* indicated. Two possible routes are depicted for the introduction of *B. rapa* into China.

for turnip rape and turnip types, which were further developed in Russia, Central Asia and the Near East. Eastern Asia is proposed as another center of origin for Asian leafy vegetables. Other cultivar groups of *B. rapa* most likely originated from different morphotypes within the two centers of origin and subsequently evolved separately. Song et al. (1988b) considered that "sarson" and "toria" types in India were derived from the European turnip rape and have evolved separately. In studies of Zhao et al. (2005) and Warwick et al. (2008), the spring oilseed types including the yellow sarson types from India formed a subgroup separate from European and Asian groups, suggesting that the Indian subcontinent may represent the third center of origin and that the separate breeding tradition in this region led to the development of the sarson types.

Turnip is a very old *B. rapa* subspecies that was probably directly domesticated from the wild progenitor that arrived from the Iranian region into Europe (Reiner et al. 1995). In Europe, turnip has been cultivated since 2500–2000 BC and its cultivation spread to Asia after 1000 BC (De Candolle 1886). Many distinct types were known to the Romans at the beginning of the Christian era and some of those varieties bore Greek place names, indicating turnips were earlier cultured in the Roman Empire and Ancient Greece. Turnips were introduced to China and cultivated before Christ based on the Chinese book of poetry "Shih Ching" (Keng 1974). Nowadays Chinese turnips are often replaced by other vegetables and only consumed in the Xinjiang district. The broccoletto's originated from Italy, and form a clearly separate group somewhat related to European turnip and oil types (Zhao et al. 2005).

A large group of *B. rapa* is formed by the leafy vegetables differentiated into several subspecies or cultivar groups mainly from China and Japan. Chinese cabbage is native to China and two main hypotheses regarding the origin of Chinese cabbage exist: one is the hybridization hypothesis suggesting that Chinese cabbage originated from hybridization between turnip (or turnip rape) and pak choi (Li 1981). The loose-leaved type is the ancestral form and gradually developed into the heading form, which was selected for as an adaptation to cool temperatures. The other evolutionary hypothesis was proposed by Tan (1979), who suggested that Chinese cabbage was formed during the introduction of ?pak choi from southern to northern China. Cultivated forms of Chinese cabbage are mentioned later than pak choi in ancient Chinese literary records. During the Tang dynasty (AD 659), the loose-leaved Chinese cabbage was mentioned in the book of "Xin Xiu Ben Cao". The pictures of semi-heading types appeared in the book of "Yin Shan Zheng Yao" in the Yuan dynasty (1330 AD). The heading Chinese cabbage originated between the Yuan and the Ming dynasty (1368–1644 AD), and became especially popular during the Ming dynasty. The origin of other Chinese leafy *B. rapa* types has been discussed but is unclear up to now (Cao et al. 1997; Chen et al. 2000; Guo et al. 2002). Japanese vegetables are likely to be derived directly or indirectly from different types of ?pak choi, but have diverged through geographic isolation and intensive selection (Song et al. 1988b, 1990). In our experiment based on amplified fragment length polymorphism (AFLP) profiles (Zhao et al. 2005), Chinese shuicai accessions that resemble Japanese mizuna group in the ?pak choi cluster and do not group with the Japanese mizuna's that form a distinct cluster.

It is believed that European forms and Asian types of oleiferous *B. rapa* have different origins, involving the Mediterranean area and the region of central Asia, Afghanistan and the adjoining Indian subcontinent. There is a lot of evidence that European oilseed *B. rapa* is genetically very close to the turnip type (Reiner et al. 1995). Domestication is believed to have occurred in the early middle ages in Europe. From the Indian oil crops of *B. rapa*, brown sarson appears to be the ancestral type from which toria and yellow sarson types were selected (Gomez Campo 1999). The history of Chinese oleiferous *B. rapa* domestication in China needs to be further clarified although evolutionary pathways have been proposed by some Chinese researchers, in which the common point is that it possibly derived from Chinese ?pak choi (Liu 1984; Cao et al. 1997). This is also supported by our result based on AFLP profiles, where Chinese oleiferous *B. rapa* (Chinese turnip rape) clusters with ?pak choi (Zhao et al. 2005). The group of winter oil types from Pakistan is not directly related to either East Asian or European types based on AFLP analysis (Zhao et al. 2005). The STRUCTURE analysis based on allele frequencies confirmed that the Pakistan oil types shared genetic background with the Asian ?pak choi group and/or European turnip group (Zhao et al. 2007), which is possibly related to their breeding history. We considered that the winter oils from Pakistan may be an intermediate type that developed in Central Asia in a directional differentiation from the ancestral European type into leafy Asian types. Alternatively winter oils could be the ancestral type of the cultivated *B. rapa* and have further developed into turnips in Europe and into pak choi in Asia. At present we are investigating which theory is correct.

3.3.2 *Description of* Brassica rapa *Crop Types*

Brassica rapa is one of the agronomically important vegetable species and to a minor extent also an oilseed species in the genus *Brassica*. *B. rapa* vegetables are consumed worldwide and provide a large proportion of the daily food intake in many regions of the world. There is wide variation in the plant organs that are consumed, which has resulted in the selection of different morphotypes depending on local preferences. Based primarily on the organs used and secondly on their morphological appearance, a number of major cultivar groups (Table 3-1; Fig. 3-1), which have been given subspecies names in the past, can be distinguished (Specht and Diederichsen 2001; *http://www.plantnames.unimelb.edu.au/Sorting/Brassica_rapa.html*).

Chinese cabbage: *B. rapa* L. subsp. *pekinensis* (Lour.) Hanelt

Chinese cabbage is native to China and is characterized by larger leaves and heads of different shape with winged petioles. It is mainly cultivated north of the Yangze River of China, in Korea and in Japan. The Chinese cabbage is used in autumn and winter for cooking or stewing dishes, but also for fresh salad and salt-pickled. In Korea, Chinese cabbage is used as the major component of "kim-chi", the traditional preserved side dish and salad. At present the Chinese cabbage is commonly found in markets throughout the world.

Pak choi: *B. rapa* L. subsp. *chinensis* (L.) Hanelt

Pak choi does not form a head, is characterized by green-white, enlarged midribs. It is widely cultivated in southern and central China, and there is potential for the crop to be successfully grown in Southeastern Asia, including the Philippines, Malaysia, Indonesia and Thailand.

Wutacai: *B. rapa* L. subsp. *narinosa* (L.H. Bailey) Hanelt; *Brassica chinensis L.* var. *rosularis* Tsen & Lee

Wutacai (flat Chinese cabbage) forms a subgroup of ?pak choi-like cultivars that differ from typical pak choi types by their flat rosettes and many dark-green leaves. This crop is mainly cultivated in Southeastern China, and is more cold tolerant and resistant to bolting. It is used in cooked dishes and salads, but is also used for ornamentals.

Caixin (or Caitai): *B. rapa* L. var. *parachinensis* (L.H. Bailey) Hanelt

Caixin is an early flowering non-heading vegetable with similar leafy features to pak choi. It is mainly cultivated in southern and central China, and distributed in Southeastern Asian countries nowadays. The edible parts of this crop are the young inflorescences and stems that can be harvested 40–80 days after sowing.

Zicaitai: *B. rapa* L. var. *purpuraria* (L.H. Bailey) Kitam

Zicaitai is characterized by the purple red stem and non-heading phenotype, and is mainly cultivated in southern and central China. This flowering purple-stemmed Chinese cabbage has tender early inflorescences, stems and shoots, which are edible. This vegetable is tolerant to low temperature and the purple color intensifies as the temperature decreases.

Taicai (or Tai tsai): *B. rapa* L. ssp. *chinensis* Makino var. *tai-tsai* Hort

Taicais are non-heading cabbage cultivars with irregularly notched leaves of different blade shapes. The tender leaves, stems, and even the conical-shaped succulent taproots are edible. These types are mainly distributed throughout eastern China and are widely cultivated in the Shandong and Jiangsu provinces.

Mizuna and mibuna: *B. rapa* L. subsp. *nipposinica* (L.H. Bailey) Hanelt; *B. rapa* L. var. *japonica*

Mizuna and mibuna types are a small group of Japanese leafy vegetables with many serrated leaves or long narrow leaves. This crop is mainly cultivated and consumed in Japan. The shuicai cultivars from China resemble mizuna or mibuna, and Chinese fennie (tilling) vegetable with strong stooling leaves also belongs to this group (Cao et al. 1997).

Komatsuna: *B. rapa* L. subsp. *perviridis* Bailey

Komatsuna is a type of neep greens, which is consumed for its young leaves, stalks and flower shoots. It is mainly grown in Japan and is also known as Japanese mustard spinach. This vegetable can be stir-fried, pickled, boiled and added to soups or used fresh in salads. It is most often grown in the spring and autumn, as it cannot endure extreme heat or cold for more than a short time.

Turnip: *B. rapa* L. em. Metzg. subsp. *rapa*

The turnip types are a group of cultivars grown for their enlarged hypocotyl and taproot, which can be subdivided in vegetable and fodder turnips. Manifold shapes and colors are typical characteristics of turnips. The turnip preferably grows in misty and cold regions.

Broccoletto, Broccoli raab, Cima di rapa: Broccoletto group; *B. ruvo* L.H. Bailey

Broccoletto is a main Italian group of vegetable *B. rapa* of which the young compact inflorescences are consumed. Broccoletto has a strong stem and short internode length. The edible parts of this type are the small flower heads that appear when the plants are about 20 cm tall. The edible part is quite similar to that of Chinese caixin, which is also utilized in the early flower stage.

Turnip rape: *B. rapa* L. subsp. *oleifera* (DC.) Metzg

Turnip rape is an oil type of *B. rapa*, which is mainly cultivated in Europe, China, India, Pakistan, Bangladesh and Canada, and there is potential for the crop to be successfully grown in the United States, South America and Australia. It falls into different subgroups based on its growth habit. Summer oil types (summer or spring turnip rape) are mainly cultivated in Canada, northern Europe and Bangladesh. Pakistan winter turnip rape is characterized by dissected leaves and rosette seedlings. Cultivars of winter oil type are still cultivated for oil and biomass production in Scandinavian countries, they are more cold tolerant than oilseed rape and have a high growth rate under low temperature. In China, three main oleiferous *B. rapa* ecotypes, viz. spring, winter and semi-winter turnip rape, were developed in adaptation to different climates, soil conditions, cultivation methods and farmer preferences (He et al. 2003), and have been grown as food oil and vegetable crop. Recently, the cultivated oleiferous *B. rapa* in both China and Canada was substantially replaced by the recently introduced *B. napus* cultivars, which have a higher yield and better adaptation.

Sarsons: *B. rapa* L. subsp. *dichotoma* and *trilocularis* (Roxb.) Hanelt

Indian oleiferous *B. rapa* includes three ecotypes, viz. brown sarson (*Dichotoma*), toria (*Dichotoma*) and yellow sarson (*trilocularis*). Brown sarson has long roots, with a limited lateral spread, enabling its successful cultivation under drier conditions. Toria has similar traits to brown sarson in morphology, and is believed to be selected from these types (Gomez Campos 1999). The yellow sarson is characterized by very early flowering, self-compatibility and its yellow seeds.

3.3.3 Phenotypic Diversity of **Brassica rapa** *Vegetables and Oilseeds*

The *B. rapa* species is morphologically very diverse as mentioned above: genetic variation can be observed for many traits when different accessions are grown together under uniform environmental conditions. Within *B. rapa* species, a number of diversity studies have been conducted based on morphology, geographic distribution, isozymes and molecular data.

In the earlier researches on morphological diversity in *B. rapa*, which have been reviewed by Prakash and Hinata (1980) and Gomez-Campo

(1999), a number of cultivar subspecies was recognized and the three main types (leaf, turnip and oil) were classified. A study of morphological diversity within the turnip group (120 landraces) of a Northwestern Spain collection (Padilla et al. 2005), suggested that the classification of landraces provides valuable information about their characteristics and the variability between landraces could be used for breeding programs. In our recent experiment (Pino del Carpio et al. unpublished data), the hierarchical cluster analysis of 26 morphological traits from 163 accessions of *B. rapa* showed three distinct groups (1-the European turnip group, 2-the Chinese cabbage group, 3-the pak choi, brocoletto, Japanese turnip and winter turnip rape group) and two admixtured groups. The further clustering analysis based on high-throughput profiling data of metabolic traits showed four main groups that relate to morphotypes and corresponded very well with the four groups that were defined using STRUCTURE based on AFLP and simple sequence repeat (SSR) markers. Phenotypic characteristics from accessions belonging to the different morphological groups can also be associated with classification based on genomic fingerprints, as was concluded from a study in our laboratory (Zhao et al. 2005). For example, leaf serration is a character that is associated strongly with the grouping in a phenetic tree. Turnips vary widely in shape and color, but as these characteristics are not associated with specific AFLP patterns they could not be differentiated between groups. However the European and Japanese turnips, which are different in leaf edge shape (European Turnips: dissected leaves; Japanese Turnips: no or mildly serrated leaves), also formed two distinct clusters based on AFLP data. The self-incompatibility trait also showed an interesting distribution in our experiment A (Fig. 3-3; Zhao et al. 2005). All of the pak chois of group 1 (PC1) and summer turnip rapes from group-Oil2 which includes yellow sarson, are self-compatible, while most pak choi accessions from group 2 (PC2), turnips from turnip-group 1 (T1), winter turnip rapes (Oil1), brocolletto (Bro), mizuna and komatsuna genotypes are self-incompatible. Incompatibility clearly differentiates subgroups that belong to the same cultivar groups as defined by similar use or phenotype, as it separates PC1 from PC2 and Oil1 from Oil2.

3.3.4 *Genetic Diversity of* **Brassica rapa** *Vegetables and Oilseeds*

In general, a high level of diversity was observed in the *B. rapa* gene pool based on isozyme and molecular data. Das et al. (1999) reported that 69% of 319 AFLP markers and 80% of 125 RAPD markers were polymorphic in nine accessions of *B. rapa*. McGrath and Quiros (1992) assessed 20 accessions including turnip, oil and vegetable types at isozyme and restriction fragment length polymorphism (RFLP) level, in which the highest level

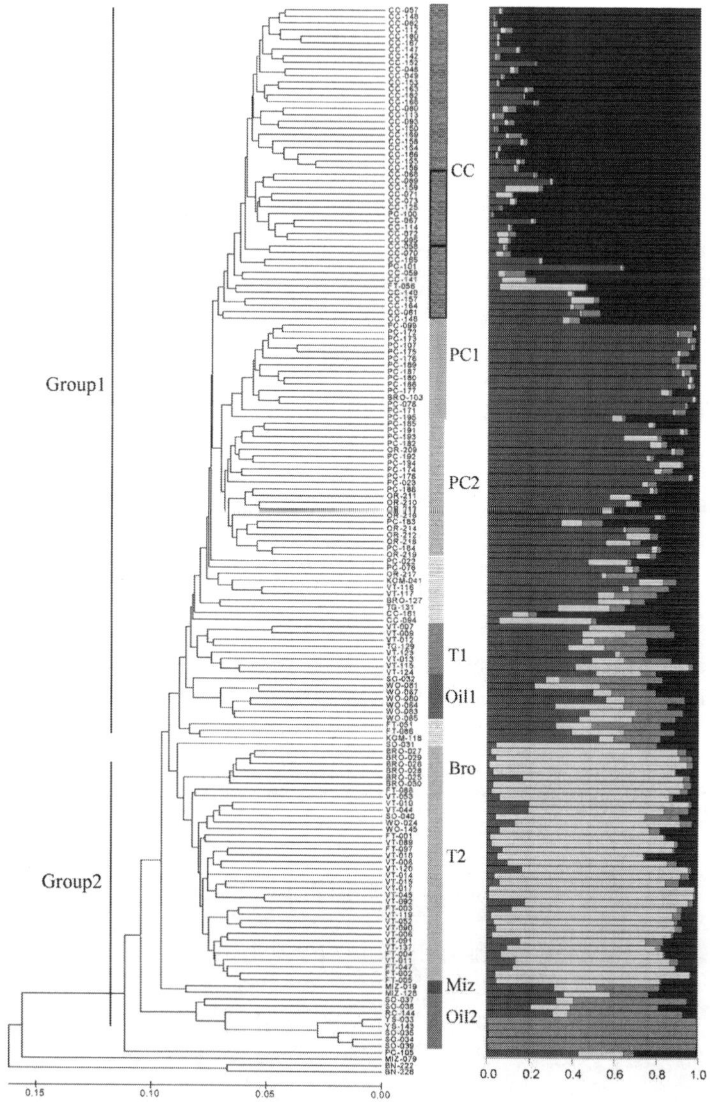

Figure 3-3 UPGMA phenetic tree (left part) and structure image (right part) of 160 *B. rapa* accessions based on AFLP data. The phenetic tree indicates a loose population structure based on UPGMA using MEGA 2.1 software (Zhao et al. 2005), which shows two main subgroups (East Asia and Europe group) and a small subgroup (Sarson types from India). The population structure analysis is based on allele frequencies using the program STRUCTURE (Zhao et al. 2007), which confirmed the UPGMA result and suggested four subgroups. The structure picture was shown in accordance with the order of accessions in the phenetic tree. Accessions in structure are represented by a bar, which is partitioned into several segments with different gray shades according to the individual's estimated membership fractions of the four clusters.

of heterozygosity was shown in central Asian and Indian types compared to European and Asian varieties. Genetic diversity among Nordic turnips comprising 31 germplasm collections, including varieties, elite stocks, landraces and older turnip slash-and-burn types, was studied using nine isozyme loci and 26 alleles (Persson et al. 2001). Genetic variation of 18.7% was observed within the used accessions, and the total heterozygosity value was 0.358. The level of heterozygosity was reduced in the landraces, but it was high within the variety group. In a recent study of Warwick et al. (2008), quite high diversity levels (62–79%) among 93 *B. rapa* accessions were revealed based on 307 AFLPs. Since the number and type of accessions, the evaluation system and the geographic origin are different for different reports, it is difficult to compare the genetic diversity levels. The AFLP fingerprints of a large collection of 160 accessions have shown that there is considerable genotypic variation in the *B. rapa* gene pool (Zhao et al. 2005). The amount of genetic variation present within groups of cultivars, as identified in the phenetic tree, contributed considerably to the high polymorphism rate observed. Low bootstrap values for many of the groups in the phenetic tree suggested that only a few genes are involved in causing the extreme morphologies. Within different cultivar groups, Chinese cabbage, pak choi, and turnip displayed high amounts of polymorphisms (75–82%), but two yellow sarson and two mizuna accessions had remarkably similar AFLP profiles, which may be caused by the limited number of accessions selected from those morphotypes. The high level of genetic diversity within *B. rapa* groups is possibly caused by their wide distribution in geographical regions with different growth habitats and consumer preferences. This could also explain why the different morphotypes could emerge independently in the different regions. Using SSR markers, Ofori et al. (2008) reported on the effect of crop improvement on genetic diversity in three turnip rape cultivars (32 plants per cultivar) representing three diffferent breeding periods, and calculated that genetic diversity within cultivars accounted for 83% of total variation while between cultivar variation was 17%, suggesting that quality breeding did not significantly reduce the genetic diversity within cultivars.

3.3.4.1 Wild B. rapa and Its Relationship to the Different B. rapa Morphotypes

Wild *B. rapa* ssp. *silvestris* (L.) Janchen occurs mainly as a weed in arable fields and as a ruderal escape. It is regarded as the species from which cultivated turnip and turnip rape originated. It is difficult to distinguish wild *sylvestris* from escaped cultivars of the same species (oleiferous *B. rapa*) on the basis of phenotype alone (Andersen et al. 2009). The relationship between cultivated and wild *B. rapa* was studied in an earlier RFLP study (Song et al. 1990), showing that the Chinese leafy types group was distinct from wild European populations and a wild *B. rapa* accession from India was positioned between European and East Asian types. Crouch et al. (1995) reported the genetic diversity within five wild and seven cultivated *B. rapa* accessions using 123 RFLP fragments and principal components analysis, suggesting that two wild Sicily and Algeria accessions were genetically distant from other forms, and two wild Argentina and California accessions were likely to be recent escapes from cultivation. A genetic diversity study with 131 inter-simple sequence repeat (ISSR) markers in 15 cultivated and 17 wild accessions of *B. rapa* from Scandinavian countries was reported by Andersen et al. (2009): the similar genetic variation was observed between the two groups and diversity within wild populations was large (80%). The authors considered that most of the diversity in *B. rapa* can be preserved by maintaining a rather small population with a representative geographical distribution.

3.3.4.2 Relationship of Genetic Diversity with Population Structure and Geographical Distribution

The genetic diversity within *B. rapa* was also related to their wide distribution in geographical regions with different growth habitats and consumer preferences. In a study by Song et al. (1988b) based on RFLP markers, the cultivar types of *B. rapa* were classified into a European turnip group including turnip and turnip rape and an East Asian group including leafy types. Based on genetic diversity analysis at isozyme and RFLP level, McGrath and Quiros (1992) concluded that crop types of *B. rapa* used for similar purposes are derived from geographically differentiated populations. In a recent AFLP analysis of 32 *B. rapa* accessions from the Eurasian continent (Takuno et al. 2007), the grouping of *B. rapa* was also related to geographic origins. The authors suggested that cultivars in East Asia were probably derived from a cultivar group that originated from Europe or central Asia and migrated to East Asia. Cluster analysis in most of the diversity studies indicated that subspecies within *B. rapa* showed a clear separation of two main phenetic groups: European turnip and turnip rape types; Asian turnip, leafy types and Chinese turnip rape. In the collection studied by Zhao et al. (2005), the oil type accessions from Canadian germplasm were missed. Those accessions were included in the recent diversity analysis by Warwick et al. (2008), and clearly were separated from the Indian sarson group and clustered with the European turnip group. Based on allelic frequency analyses (Zhao et al. 2007), it was also likely that part of the genetic background is shared by many accessions belonging to different crop types, suggesting that these accessions that are genetically related possibly share part of their breeding history. These data provide insight into possible application of favorable alleles from accessions from the same morphotype and from different morphotypes of the gene pool for specific crop improvement programs, like introduction of resistance genes from different crop types. With knowledge of the allelic distribution, breeders can estimate the amount of backcrosses needed to retrieve the original crop type.

Leafy *B. rapa* vegetables distributed in East Asian countries show remarkable morphological diversity with local climatic zones. Turnips grown around the world can be classified into several groups by geographical distribution: Teltow turnips; West European turnips with dissected leaves; Asia Minor and Palestine turnips; Russian turnips; Asiatic turnips of Afghan type; Japanese turnips with entire leaves; and European entire-leaved turnips (Sinskaia 1928; Dixon 2006). Limited information is available on the genetic relationship between all these turnip types as measured by molecular marker screens in relation to the different geographical origins. Ren et al. (1995) reported that two Chinese turnip accessions were clustered in the Chinese cabbage group based on random amplified polymorphic DNA (RAPD) analysis among 30 *B. rapa* accessions. We assessed the genetic diversity in a large collection of *B. rapa* by AFLP fingerprinting, and the results showed that genetic similarities are more related to geographic origin (East Asia versus Europe) than to morphological groupings, suggesting either an independent origin and/or a long and separate domestication and breeding history in both regions (Fig. 3-3; Zhao et al. 2005). It was illustrated that European and Japanese turnips formed distinct groups, while Chinese turnips clustered with pak choi types (experiment B). It would be interesting to see whether Chinese turnips are closely related to the Chinese turnip rape. This geographical distinction of the turnips can also be seen in morphological and physiological characters such as leaf shape and flowering time and might either be due to a long separation of breeding of the different turnip types or even indicate an independent origin.

3.4 Diversity of *Brassica oleracea* Vegetables

3.4.1 *Origin of* Brassica oleracea

Several genetic studies in *Brassica oleracea* looked into understanding the origin and the center of domestication of this *Brassica* species and are based on the relationship between the cultivated and the wild relatives. An indication of the close genetic relationship between cultivated and wild relatives is that all $n = 9$ *Brassica* species can be crossed easily and form hybrids and fertile offspring. Wild relatives of *B. oleracea* occur as perennial plants in rocky cliffs on the coast of the Mediterranean region

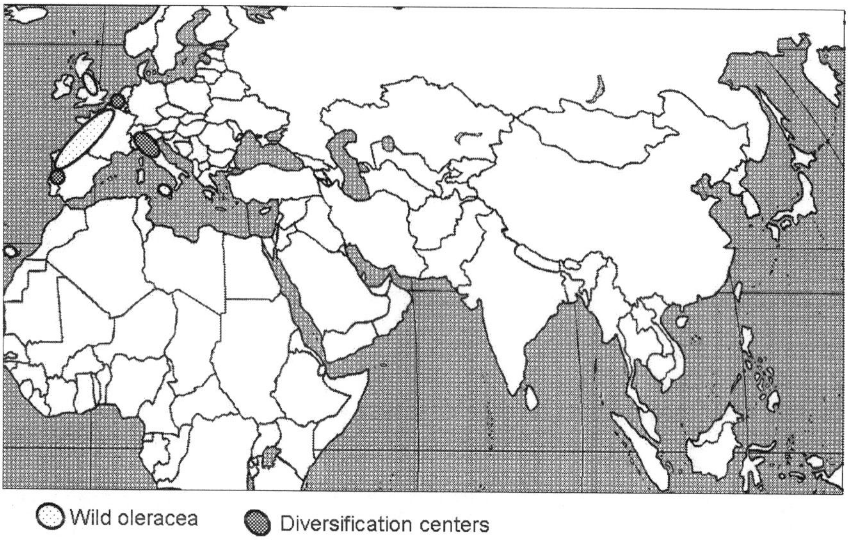

Figure 3-4 A map of Eurasia is depicted with the geographical distribution of wild *oleracea* $n = 9$ species, and the centers of origin for the different *B. oleracea* crop types.

and in northern Spain, western France and southern and Southwest Britain (Tsunoda 1980; Fig. 3-4).

The adaptations of these wild forms to different habitats have led to different cultivated forms as we know them today. The multiple origins of the different *oleracea* morphotypes can be derived from cross-breeding between closely related *Brassica* species growing in geographical proximity to each other. The introgressions from wild species have also been an important factor, which has increased the variability and adaptability of cultivated *B. oleracea*. Several theories have been drawn about the origin of the different forms, for example the origin of stem kales could be found in the *B. rupestris-incana* complex, characterized by the formation of a tall central stem and a large inflorescence before branching (Snogerup 1980). The origin of the bushy kales can probably be found in *B. cretica* given the common branching, shrubby habit and fleshy leaves. The white colored flowers may also indicate an origin derived from *B. alboglabra* and *B. cretica* subsp. *nivea* (Gómez-Campo 1999).

A significant aspect in *B. oleracea* is the development of cultivated forms in different countries for example broccoli, originated in Italy, diversified into different forms and Brussels sprouts developed near Belgium (Gómez-Campo 1999).

Cauliflowers are generally regarded as being derived from broccoli (Crisp 1982; Gray 1982). Crisp (1982) worked on hybridization between cauliflower and broccoli and identified one mutation in a gene in broccoli as being responsible for the cauliflower phenotype. Purugganan et al. (2000) indicated that specific alleles of *B. oleracea CAULIFLOWER* gene (*BoCAL*) were selected by early farmers during the domestication of modified inflorescence structures in *B. oleracea*.

3.4.2 Brassica oleracea *Crop Types*

The vast array of types developed within cultivated *Brassica oleracea* has led to the acceptance of subspecies and variety (cultivar) descriptions based upon specialized morphology of edible parts. We based our description of groups on the classification made by Snogerup (1980) (Table 3.1).

Brussels Sprouts: *B. oleracea* L. var. *gemmifera* DC

Biennial plants with simple erect stems and auxiliary buds, which later develop into compact miniature cabbage heads or sprouts. At the top of the stem there is a rosette of small leaves, which are generally petiolate and subcircular.

Cauliflower: *B. oleracea* L. var. *botrytis* L.

Cauliflower is a biennial or annual herb, with unbranched stems; the leaves are arranged in a rosette, which surrounds the compact terminal flower head (curd). The curd consists of a dome of white to cream or yellow floral meristems that grow on short peduncules.

Broccoli: *B. oleracea* L. var. *italica* Plenck

Broccolis are heading types, which develop a large single terminal inflorescence. The main head consists of groups of fully differentiated green or purple flower buds, which are less densely arranged than cauliflower and with longer peduncles.

Chinese kale: *B. oleracea* L. var. *alboglabra* (L.H.Bailey)

Chinese kale is cultivated since ancient times without apparently obvious wild progenitors. It is an annual herb during the vegetative stage which reaches up to 1–2 m at the end of the reproductive stage. The leaves are crisp and thick.

Other kales: *B. oleracea* L. var. *acephala* DC

Many groups can be distinguished: curly kale, collard, marrow stem kale, palm tree kale, Portuguese kale and thousand headed kale. Kales are old cultivated cole crops closely related and with morphological resemblance to the wild forms of *B. oleracea.*

Kohlrabi: *B. oleracea* L var. *gongylodes* L.

The Kohlrabi is a biennial crop in which the edible consumed part is the thickened stem.

Chinesse cabbage: White headed cabbage (*B. oleracea* L.var. *capitata* L. f. *alba* DC), red headed cabbage *B. oleracea* L var. *capitata* L.f.rubra (L) Thell, savoy headed cabbage *B. oleracea* L. var. *sabauda* L.

All cabbages are characterized by the formation of tight compact heads, the head shapes vary from spherical, to flattened to conical. The leaves are smooth, curled or savoyed and the leaf color can be red-purple in red headed cabbage and green to yellow green in other cabbages.

3.4.3 *Genetic Diversity of* **Brassica oleracea**

Screening for genetic diversity has been widely used in *Brassica oleracea* to reveal the phylogeny of the different types (Astarini et al. 2004), to discuss relationships between populations (Geraci et al. 2001) and to analyze the diversification and origin of *Brassica* crops.

The genetic diversity studies in *B. oleracea* fall into different categories, some are focused in groups (wild and cultivated or both) and/or accessions from different regions and make use of different techniques like isozymes, RFLP, RAPD and DNA sequence analyses. Morphological analyses of different traits have been mostly directed to the preservation and characterization of landraces and cultivated types.

3.4.3.1 Wild B. oleracea and n = 9 Brassica Species

As indicated above wild *oleracea* has been largely studied to compare population attributes. A large study with isozyme and RAPDs compared the population attributes among wild *B. oleracea* (Lanner-Herrera et al. 1996). From a germplasm of 44 wild populations of *B. oleracea* collected in Spain, France and Great Britain, 18 populations were selected for analysis.

The comparison of clustering with isozyme and RAPD of the same populations resulted in an isozyme based tree that displayed greater distances among populations than the RAPD based tree. In this study genetic distances did not appear to be related to geographic distances. The lack of clustering of populations from the same or proximal regions suggested an independent genetic differentiation of populations that occurred throughout the entire range of wild *B. oleracea,* a manifestation of the geographical isolation of wild *Brassica* populations.

Lanner-Herrera et al. (1997) analyzed the relationship between 10 wild *Brassica* species represented by 34 populations with RFLP markers. The UPGMA analysis showed that the species tended to cluster according to the geographic region: *B. cretica* and *B. hilarionis* comprise a cluster that could be called eastern Mediterranean; *B. oleracea, B. bourgeaui*, and *B. montana* define an Atlantic—western Mediterranean cluster; *B. incana, B. rupestris*, and *B. villosa* form an Italian group; and a *B. insularis—B. macrocarpa* association may be called central Mediterranean.

Other studies based on the classification of wild *Brassica* $n = 9$ taxa established the separation of *Brassica* species in groups from the West (*B. oleracea, B. bourgeaui, B. incana, B. montana* and *B. alboglabra*), central Mediterranean groups (*B. rupestris, B. villosa, B. villosa* subsp *drepanensis* and *B. macrocarpa*) and taxa that grow in the eastern Mediterranean (*B. cretica* and *B. hilarionis*)(Lazaro and Aguinagalde 1996, 1998).

Some of the studied species show a variable position in the clustering with others. For example, *B. incana*, which belongs to the rupestris group appeared in a study of Lazaro and Aguinagalde (1996) more closely related to *B. oleracea*, probably because of a former introgression by human influence. In *B. cretica*, the considerable morphological differentiation and events like genetic drift and founder effects can explain the extreme variation between isolated populations and differential relationship with other wild species.

Lanner (1998) also focused on the study of a specific small, variable region in the chloroplast sequence to compare 34 populations representing *Brassica bourgenaui, B. cretica, B. hilarionis, B. incana, B. insularis, B. macrocarpa, B. montana, B. oleracea, B. rupestris* and *B. villosa* (Lanner 1998). This study investigated the relationships that may be detected in a phylogenetic analysis compared with previous phenetic analyses of Lanner-Herrera et al. (1997) and Song et al. (1990). Three groups are well distinguished: one including *B. bourgenaui, B. oleracea, B. cretica, B. incana, B. hilarionis, B. insularis* and *B. montana*, which subdivides into *Brassica bourgenaui-B. oleracea* and *B. cretica-B. incana* clades. A second group includes *B. insularis* and *B. macrocarpa* and a third group comprises *B. rupestris* and *B. villosa*.

Several conclusions were drawn with respect to the relationship between the wild *Brassicas*. The majority of populations of the same species had the same chloroplast haplotype, with divergence observed in two species *B. montana* and *B. cretica*. *B. insularis* also showed variability, which demonstrated the distinction between populations from Tunisia and those from Sardinia and Corsica. The presence of *B. macrocarpa* genetic material in a population of *B. insularis* was considered as evidence of introgression from *B. macrocarpa* into *B. insularis*.

The comparison with previous RFLP studies (Song et al. 1990; Lanner-Herrera et al 1997) agrees and provides additional evidence of the close relationship between *B. villosa* and *B. rupestris* and a more distant relationship to *B. macrocarpa* separated in a different clade.

In the case of the haplotype similarity between *B. cretica* and *B. incana*, the results differ from RFLP studies, which showed that *B. incana* was associated with *B. rupestris* and *B. villosa*. The consensus indicates that the wild species have close relationships and have not diverged significantly since their separation from a common ancestor.

The most divergent species *B. rupestris, B. villosa* and *B. macrocarpa* were also found as a different group with isozymes (Lazaro and Aguinagalde 1998) and this study identifies the Sicilian region as a center of genetic diversity.

The study of diversity with SSRs in the chloroplast genome, including domesticated and wild *B. oleracea* and nine inter-fertile related wild species, was performed by Allender et al. (2007). The results corroborate to the high level of diversity in *B. cretica* with the presence of multiple chloroplast haplotypes as shown already in the study of Lanner et al. (1996). However, the lack of diversity found within *B. oleracea* was remarkable in the study of chloroplast sequences. Panda et al. (2003) also found no variation in chloroplast PCR-RFLP profiles between a cultivated accession and nine wild populations from Spain, France and the UK.

The lack of diversity in this study pointed towards a single center of domestication in the eastern Mediterranean region (Gomez-Campo 1999) for *B. oleracea*.

3.4.3.2 Cultivated Brassica oleracea

Several examples can be found for the study of genetic diversity within types of cultivated *B. oleracea*. Allen et al. (1986) searched for markers at the enzyme level to distinguish the origin of cauliflower stocks. The investigation was restricted to cauliflowers available in the United Kingdom, and included annuals and biennials together with Australian types. The isozymes examined showed evolutionary divergence of the cauliflower types during the selection for different times of development. The results also revealed that European annuals and biennials are parents of the Australia cauliflowers.

The usefulness of different marker systems to examine genetic variation within types has also been compared in several studies. The genetic variability of kale plants (*Brassica oleracea* L. var. *acephala* D.C.) was studied by means of enzymatic polymorphism using polyacrilamide gel electrophoresis and a DNA polymorphism assay based on RAPD (Sawaza et al. 1997). Fifteen clones of kale *B. oleracea* var. *acephala* were studied and differences among clones were observed by isoenzymes and RAPD. However, the dendrogram obtained from both markers were dissimilar, suggesting that the isozymes provided less information than the RAPD about the genome diversity.

A comparison between two marker systems was done in the study of wild cabbage in Dorset (UK), where Raybould et al. (1999) investigated if the restricted pollen and seed flow was the cause of population substructure and isolation by distance. Ten populations were screened with microsatellite primers designed for oilseed rape and compared with data of isozyme loci. Isozymes were highly polymorphic within the population, very similar to the results from British populations investigated by Lanner-Herrera et al. (1996) with isozymes studying wild accessions collected in Spain, France and the United Kingdom. The oilseed rape SSR primers were effective at detecting genetic variation but the number of alleles detected was low with these specific primers. The results showed Dorset populations were isolated by distance with restricted but with significant gene flow among them in Dorset.

Several *B. oleracea* cultivars are restricted to localized areas and/or countries, and these have been subjected to the evaluation of genetic diversity. One case study is on the endangered Italian cultivar Broccoli Fiolaro, which is a local broccoli variety restricted to an area in Northeastern Italy. The study by Vischi et al. (2008) highlights the genetic distinction of the Fiolaro cultivar from other Italian cultivars and the distribution of polymorphism in a geographical pattern. Other studies within types combined morphological data with molecular marker results. In the study of a Turkish kale population the fact that populations clustered according to geographical distribution but not to their morphological characteristics was the most significant result (Okumus and Balkaya 2007).

In a broader study, Lamboy et al. (1994) evaluated the genetic diversity among 56 different *B. oleracea* accessions comprising 14 botanical varieties. The results present that isozymes are partially useful because specific alleles were not limited to a group of accessions but were dispersed all throughout the morphological groups. In this study, the use of RAPD and RFLP markers was suggested for the identification of accessions or botanical varieties. However, Van Hintum et al. (1996) evaluated a selection of 11 white cabbages and nine Brussels sprouts using isozymes, and could classify the groups as expected; all the Brussels sprouts clustered together as did the white cabbages.

Broccoli and cauliflower have been found to separate into different groups with RAPD markers (Hu and Quiros 1991). In the same study, the level of polymorphism detected in cauliflower cultivars was significantly higher than in broccoli. However, when diversity had been evaluated based on database-derived SSR markers, it was found to be low in cauliflower, intermediate in broccoli (Calabrese) and greatest in cabbage (Tonguç and Griffths 2004).

In an RFLP study, Song et al. (1988a, b) revealed that *B. oleracea* accessions could be divided into three groups: a cabbage group including collard, kohlrabi, Portuguese cabbage, borecole and Savoy cabbage; a broccoli group including marrow stem, Jersey kale and Brussels sprouts; and a group including thousand head kale.

The differential grouping of broccoli, cabbage and in particular the distinctness of cauliflower is in accordance with the results from Nienhuis et al. (1993) who also grouped these botanical varieties into different clusters based on RFLP polymorphisms.

Portuguese tronchuda cabbages and kales, two peculiar vegetable types, are of genetic interest since they are only located in particular regions, and can only be found in Portugal or in regions with Portuguese influence. Extensive studies on landraces of tronchuda cabbages and kales indicated similarity based on RFLP patterns, resulting in tronchuda and kales clustering within the same group (Dias et al. 1991). In a study of tronchuda cabbages and kales, which included several other *B. oleracea* types, the clustering tree diverges in groups including broccoli and cauliflower, a group with a mixture of kales and cabbages originally from central-North Europe, and one group formed by Portugese coles. The results of these studies suggest the existence of three major regions of domestication of *B. oleracea* in Europe: Italy, central-North Europe and Portugal (Dias 1995).

3.4.4 Phenotypic Diversity of Brassica oleracea

Based on the morphological appearance of different varieties, *Brassica oleracea* has been statistically classified into groups.

Dias et al. (1993) conducted a study using morphological characters to determine relationships among Portuguese cole landraces using 58 accessions belonging to Portuguese tronchuda cabbage (*B. oleracea* var. *tronchuda* Bailey syn. Var. *costata* De Candolle, Couve Tronchuda), Portuguese galega kale (*B. oleracea* var. *acephala* De Candolle, Couve Galega) and other coles such as Algave cabbages (*B. oleracea* var. *capitata* L., Couve do Algave or *B. oleracea* var. *sabauda* L., Couve Repolho-lombarda do Algave). The accessions were evaluated on 46 morphological characters from seedling stage to ripe silique under field conditions and in two consecutive years. After analysis using clustering by UPGMA, eight main groups were found. Five groups corresponded to tronchuda cabbage landraces and three groups corresponded to cabbages, miscellaneous coles and galega kales. Groups of landraces were primarily associated with morphological differences among accessions and secondly with geographical origin.

Cartea et al. (2003) evaluated the morphological and agronomic characteristics of 15 kale populations. The aim was to determine relationships and also identify morphological characteristics that could be incorporated into breeding programs. The cluster analysis resulted in the identification of five groups, two groups including most of the populations and three small independent groups. Groups of landraces were associated primarily with the geographical origin and secondarily with morphological differences among populations. Clusters separated according to the length of their vegetative phase, North and early populations (Cluster A) and South and late populations (Cluster B). The results of this work are in disagreement with the morphological classification of landraces of *B. oleracea* from Portugal performed by Dias et al. (1993). Although it should be noted that Dias et al. (1991) found that with molecular data (RFLPs) in the same collection the clusters were defined on the basis of their geographical origin and not on the basis of their morphological differences.

As reported above, morphological cluster analysis within crop types has been related often with geographical origin. Balkaya et al. (2005) studied similarities and differences regarding morphological variation of 95 white head cabbage populations as genetic resources collected from different eco-geographical regions of Turkey. The study reported the existence of considerable morphological variability. However, cluster groups were not associated with the geographical origin of the genotypes.

The other important aspect in phenotypic classification is the identification of variability in the adaptation of *B. oleracea* cultivars and landraces to different seasons. For example 21 landraces of cabbage (*B. oleracea* var. *capitata* L.) and two of tronchuda cabbage (*B. oleracea* var. *tronchuda* Bailey) from Galicia (Northwestern Spain) along with five commercial hybrids of cabbage and one commercial variety of tronchuda were evaluated in two planting dates for an early and late harvest (Padilla et al. 2007). The results showed that head forming landraces seemed better adapted to the autumn/winter season than commercial hybrids, with higher yields, better early vigor, and less plants lost after transplanting. However, for characters related to the uniformity of the crops (appearance, days to harvest and synchrony of production), commercial hybrids available in local markets did better than landraces.

3.5 Diversity of *Brassica juncea* Vegetable and Oil Mustards

3.5.1 Origin of Brassica juncea

Brassica juncea is a predominantly self-pollinated allopolyploid species with the A genome of *B. rapa* and the B genome of *B. nigra* and is widely grown as an oilseed in India and Pakistan, as a vegetable in China and Japan, as a condiment in North America and Europe, and more recently as Canola quality oil in Canada and Australia. The cultivation history in China could date back to the 6th century (Liu 1996), while Gomez-Campo (1999) even mentioned records of Su's work (10–61AD) "Tu-Jin-BinCao" a mustard that was a popular crop at the time. Its economic importance and the enormous morphological variability have attracted many researchers, including botanists, taxonomists and breeders since the 17th century. Several studies of comparative morphology, biochemical and molecular diversity provide evidence that the origin of *B. juncea* is not monophyletic but polyphyletic (Vaughan 1977; Prakash and Hinata 1980; Song et al. 1988a; Jain et al. 1994). The region of the Middle East is a likely center of origin because of the occurrence of the wild forms of *B. rapa* and *B. nigra* (Gomez-Campo 1999). Two secondary centers of variation are proposed. North West India harbors the oilseed mustard, and Southwest China harbors the vegetable types (Dixon 2006)(Fig. 3-5). Parallel variation can be observed between *B. rapa* and *B. juncea* in East Asian vegetable forms like the loose head savoy, heading types, stalked types and swollen root types (see Table 3.1), which suggests independent interspecific hybridizations between *B. rapa* types and *B. nigra* that resulted in the different *B. juncea* types (Song et al. 1988a). In *B. juncea* the B-chromosomes show homeologous regions derived from *B. nigra* while two of the three *B. rapa* homologous segments for an 8.7 Mb *Arabidopsis* chromosome 4 segment show an altered structure as observed using GISH (Lysak et al. 2005). Earlier Axelsson et al. (2000) concluded that parental genomes were stably maintained based on an RFLP study. Similarly Panjabi et al. (2008) used intron polymorphism markers and concluded that A and B genomes of *B. juncea* were highly collinear with

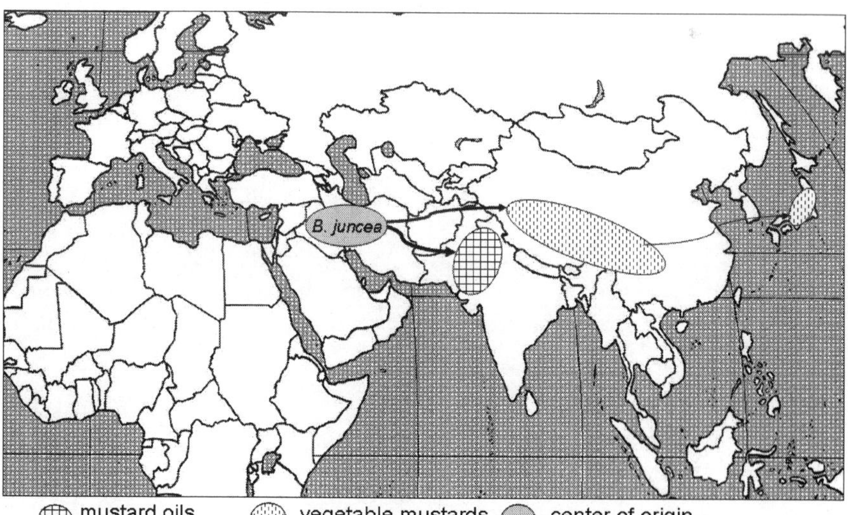

Figure 3-5 A map of Eurasia is depicted with primary center and secondary centers of diversity for *B. juncea* indicated. In India the mustard oils evolved, while in China the vegetable mustards evolved. At present *B. juncea* is widely grown as seed oil in India and Pakistan, as a vegetable in China and Japan, as a condiment in North America and Europe and recently as Canola quality oil in Canada and Australia.

A and B genomes of *B. napus* and *B. nigra*; comparison to the A genome of *B. rapa* was however not made. In a study based on restriction patterns of the chloroplast genomes in *Brassica* amphidiploids, evidence was presented that *B. juncea* has the chloroplast genome identical to *B. rapa* (Erickson et al. 1983). In another study, 40 chloroplast mutations that were phylogenetically informative were used to construct a parsimonious tree that showed that the *B. juncea* chloroplast genome was similar to the *B. rapa* genome (Palmer et al. 1983). Since only a few accessions were studied in both reports, this conclusion needs to be viewed with caution since *B. juncea* could have originated from several independent hybridization events between the parental species (Song et al. 1988a).

3.5.2 *Description of* Brassica juncea *Crop Types*

Brassica juncea is an erect annual to biannual herb, normally unbranched, sometimes with an enlarged taproot. Rosette leaves are very variable in shape and size, some curved and overlapping, some open and spreading, leading to heading and non-heading types. Leaves are entire, divided, with thin or thick petioles and with flat or tumor like surfaces. The fast array of types developed within the cultivated *B. juncea* has led to the acceptance of subspecies and variety (cultivar) descriptions based upon specialized morphology of edible parts. Nomenclature is rather complex, since different authors use different subspecies names. Here nomenclature as described by Gladis and Hammer (1992) is followed, with sometimes references to other nomenclatures.

Leaf Mustards (*B. juncea* var. *rugosa* Bailey; var. *latipa*, var. *crispifolia*, var. *foliosa*, var. *leucanthus*, var. *linearifolia* and var. *multisecta*)

Large plants have broad, large, green or purple leaves (*rugosa* and *latipa*) and plants have dissected leaves (var. *crispifolia*, var. *foliosa*, var. *leucanthus*, var. *linearifolia* and var. *multisecta*).

Head and Curled Leaf Mustards (*B. juncea* var. *capitata* and var. *involuta*)

Plants have leaves with broad petioles and midribs; leaves overlap or form a globose head.

Strumiferous Mustards (*B. juncea var. strumata*)

Plants have large leaves and well developed petioles with cushion- like perturbances at the top.

Tillering Mustards (*B. juncea* var. *gemmifera* and var. *multiceps*)

Plants have many lateral buds on a shortened stem which grow as tillers before flowering.

Swollen Stem Mustards (*B. juncea* var. *tumida;* also named *B. juncea* var. *tsatsai* Mao)

Plants have swollen stems with tumor like perturbances just below the petioles, with large leaves that are entire or dissected and red or green.

Swollen shoot mustards (*B. juncea* var. *crassicaulis*)

Plants have oblong or ovate entire or dissected leaves and shoots are swollen. This crop is distributed in Mongolia, Manchuria and northern China and acclimated to the cold unlike other *B. juncea* types.

Root Mustard (*B. juncea* var. *megarrhizza* also called *B. juncea* var. *napiformis* Paill. et Bois)

Plants have fleshy cylindrical or conical taproots with purple or green, entire or dissected leaves.

Oilseed Mustard (*B. juncea* var. *juncea* also named *B. juncea* var. *gracilis* Tsen et Lee)

This crop with pinnate leaves is distributed in India, central Asia and Europe, with the majority cultured for oil, even though greens are eaten from young plants.

Multi Seed Stalk Mustard (*B. juncea* var. *utilis* Li)

This crop with pinnate leaves is distributed in India, central Asia and Europe, with the majority cultured for oil, even though greens are eaten from young plants.

3.5.3 *Genetic Diversity of* Brassica juncea

Many studies have been conducted that assess the genetic diversity and phenotypic variation among *Brassica juncea* germplasm, however the choice of germplasm with regard to type (oil, mustard or vegetable), geographical origin (India, Pakistan, China, Canada and Australia) and numbers, hamper direct comparisons between these studies. Another complication is the nomenclature of the different types, since different authors use different subspecies names.

Fifty-two mustard (*B. juncea*) accessions, mainly Pakistani landrace mustard types plus a few vegetable cultivars from Japan and a few oilseed cultivars from India and Pakistan were grown in the field in Japan and characterized for 35 agro-morphological characteristics from seedling emergence to maturity (Rabbani et al. 1998b). Since morphological variation within the Pakistani landraces and oil types was low, while variation between these groups and the Japanese vegetable types was high, the authors concluded that mustard in Pakistan has a narrow genetic base, which was confirmed with genetic studies based on DNA polymorphisms and based on SDS-PAGE analysis of total seed proteins (Khan et al. 2008; Rabbani et al. 1998c, 2001). The genetic diversity in a similar collection of mustard (*B. juncea*) that included 41 landraces from Pakistan, six oilseed cultivars or lines from Pakistan, India, China and Australia and five Japanese vegetable cultivars was very low for the oilseed types as measured using 198 polymorphic RAPD fragments, indicating high genetic similarity (Rabbani et al. 1998c). Variability within the cluster of Japanese vegetable mustard was much higher than that of Pakistani accessions. A dendrogram constructed using UPGMA illustrated that Pakistani accessions and the six oilseed cultivars did not form distinct clusters similar to the morphological study. A similar conclusion was drawn in another study of genetic diversity among Indian mustard genotypes from Pakistan comprising 37 germplasm collections, five advanced breeding lines and three improved cultivars based on RAPD markers (Khan et al. 2008). The authors concluded that this low level of genetic diversity was due to intensive selection pressure for uniform phenotypes or due to a narrow genetic base of Pakistani germplasm (Khan et al. 2008). Jain et al. (1994) also used 595 RAPD markers to study the genetic diversity and genetic relationships in Indian mustard (*B. juncea*) among 12 Indian and 11 exotic *B. juncea* genotypes from Russia, Australia, and each one from Poland, the UK, Pakistan, and the USA. A dendrogram constructed using UPGMA cluster analysis of Jaccard genetic similarity coefficients revealed two groups: group A contained seven exotic accessions from Russia, Australia and Poland, while group B comprised the 12 Indian accessions in one subcluster and four exotic accessions that were in a distinct subcluster. The pedigree of Russian and Australian germplasm of group

A was not described; however some Australian accessions had Russian names, suggesting a common origin. The authors state that grouping of the genotypes according to geographic diversity is understandable given the polyphyletic rather than monophyletic origin of *B. juncea*. Another interesting result from the study of Jain et al. (1994) is that they recorded high percentages of heterosis in crosses between Indian and exotic genotypes, while crosses within Indian or exotic genotypes often exhibited negative heterosis. In yet another study, AFLPs were used to assess the genetic diversity among 21 established natural and nine synthetic varieties and lines of oilseed *B. juncea* from Asia, Australia, Canada, Eastern Europe and Russia to identify new sources of variation for hybrid breeding (Srivastava et al. 2001). Clustering using the UPGMA method based on 778 polymorphic out of 1,251 total fragments resulted in three distinct clusters. The Indian and Chinese genotypes plus previously developed synthetics formed one cluster, the newly developed synthetics formed yet another cluster and the lines from Australia, Canada, Eastern Europe and Russia formed the third cluster. Genetic diversity and the relationship of 47 local mustard accessions from Sri Lanka were estimated using 589 fluorescent amplified fragment length polymorphism (FAFLP) markers and several agro-morphological characters (Weerakoon et al. 2008). FAFLP and agro-morphological based clusters were in poor agreement, and resulted in grouping of the 47 accessions into different numbers of clusters. The FAFLP analysis seemed more accurate since it clearly separated the six canola varieties into a single cluster, while with agro-morphological analysis they were scattered among mustard accessions (Weerakoon et al. 2008).

The genetic diversity of selected breeding lines and cultivars of canola quality *B. juncea* was also assessed using AFLPs (Burton et al. 2004). Seventy-seven lines from three of the world's major breeding programs, two from Canada and one from Australia, plus 15 mustard quality *B. juncea* accessions from India, China, Russia and Australia were included in this study, which resulted in the partitioning into five main groups; the mustards formed two groups, Indian mustards and the mustards from the other origins, while the canola quality accessions formed three groups, two of which corresponded to two breeding programs, while the third included lines from all three breeding programs. Variation within breeding programs was larger than variation between them and implications for further breeding strategies were discussed.

A number of studies applied different marker types (RAPD, ITS, AFLP) to reveal genetic relationships among small sets of vegetable mustards (Ren et al. 1995; Qiao et al. 1998; Fu et al. 2006; Qi et al. 2007, 2008). In all these studies, the accessions formed separate clusters but often the different vegetable types like root, expanded stem, leaf and seed stalk mustards grouped together and did not form separate groups. Thus phenotypically similar accessions often possess different genetic backgrounds, suggesting that the diverse morphotypes were selected multiple times and that high selection pressure during domestication resulted in abundant morphological variation in genotypes with similar genetic backgrounds (Qi et al. 2008). No obvious grouping was detected according to geographical origin. The above described studies focus on either oil, mustard or vegetable *B. juncea*; it would be very interesting to compare diversity between Indian and exotic mustard oilseeds on the one hand with Asian vegetable mustards and these mustard oilseeds on the other hand. This comparison was made in a study of Wu et al. (2009), where 95 oilseed and vegetable mustard accessions from China, France, India, Pakistan and Japan were assessed for genetic diversity using 326 scorable fragments (with 161 polymorphic) generated using SRAP markers. They clearly demonstrated that genetic diversity within vegetable mustards was much higher than within oilseed mustards, and also that winter oilseed mustards are genetically more diverse than spring oilseed mustards. The vegetable mustards, winter oilseed mustards and spring oilseed mustards formed three distinct groups, with spring and winter oilseed mustards genetically closer to each other compared to the vegetable mustards, even though the vegetable and winter oilseed mustards share the same geographical origin (China, and even provinces within China). Based on these data they also conclude, like Song et al. (1988a), that oilseed and vegetable types evolved independently and further deduce that based on the higher level of genetic diversity, vegetable types were domesticated earlier than oilseed-types. Unlike the obvious grouping of oilseeds and vegetable mustards, similar vegetable mustards, based on the organs used, did not cluster together, similar to results of other studies (Ren et al. 1995; Qiao et al. 1998; Fu et al. 2006; Qi et al. 2007, 2008). This suggests that selection for different forms affected only a few genes that directed the morphological changes in plant organs. Similar results were found in a study of genetic diversity of the vegetable *B. rapa* types (Zhao et al. 2005).

3.6 Diversity of *Raphanus sativus*

3.6.1 *Taxonomic Relationship of Vegetable* Brassicas *with* Raphanus sativus

Phylogenetic analyses in the subtribe Brassiniae based on RFLPs of chloroplast DNA, indicated a clear division of this subtribe into two main lineages that were designated as *rapa/oleracea* and *nigra* (Warwick and Black 1993, 1997). The genus *Brassica* was polyphyletic as species within this genus were placed in two groups and *Raphanus* was placed in the *rapa/oleracea* lineage. In another study, sequence analysis of two chloroplast noncoding regions from seven species also showed that *R. sativus* was more related to the *rapa/oleracea* lineage than to the *nigra* lineage (Yang et al. 2002). Lu et al. (2008) analyzed nucleotide sequence variations in the mitochondrial *trn K/MatK* region of cultivated and wild radish and related *Brassica* species and also concluded that *Raphanus* accessions formed one cluster that was located in the *rapa/oleracea* lineage, which had diverted from the *nigra* lineage. They further mentioned that *Brassica barreleiri* was the phylogenetically closest *Brassica* species to *Raphanus*. Flannery et al. (2006) developed plastid SSR markers based on 11 gene regions and characterized the plastid genome in several *Brassica* species and Brassicaceae using eight of these plastid SRRs. The analyses failed to separate *B. rapa* and *B. napus* individuals in clearly identifiable groups, but identified a monophyletic group that contained *B. rapa, B. oleracea, B. napus* and *B. nigra*, with a sister group *Raphanus*. These results did not support the hypothesis that *Raphanus* is more related to a *rapa/oleracea* lineage than to a *nigra* lineage (as suggested by Warwick and Black 1993, 1997 and Yang et al. 1998). Many comparative mapping studies suggest that the *Brassica* genomes descent from a common hexaploid ancestor, and are variants of three rearranged ancestral genomes (Parkin et al. 2005; Schranz et al. 2007). A study of comparative chromosome painting using *Arabidopsis* BAC contigs that cover 8.7 Mb of chromosome 4 (At4-b contig) to meiotic chromosomes of 21 species in the Brassicaceae to look for genome triplication and compare that with phylogenetic data obtained from chloroplast sequences (5′-*trnL-trnF* region), concluded that taxa with triplicated At4-b were in the single major Brassicaceae clade with three groups (Lysak et al. 2005). The largest group contained the *rapa/oleracea* lineage with *Raphanus*, and the *nigra* lineage. In none of the three-copy species, the same structural rearrangements were found. Song et al. (1988a, 1990) studied phylogeny of *Brassica* and related genera based on nuclear RFLPs. They presented evidence for the existence of two basic evolutionary pathways for the diploid species: one pathway gave rise to *B. nigra, B. fructulosa, Sinapis arvensis* and more distantly to *R. sativus*, the other gave rise to *B. oleracea* and *B. rapa. R. sativus* represented an intermediate and Song et al. (1990) postulated that it might have been derived from introgression or hybridization between species belonging to the different lineages. Molecular phylogenetic studies based on the internal transcribed spacer (ITS) region of 18S–25S rDNA of *Brassica, Rorippa, Arabidopsis* and allied genera showed that within the tribe Brassiceae, *Raphanus* is more closely related to *B. nigra* than to the *B. oleracea/B. rapa* clade (Yang et al. 1999). Also Thormanm et al. (1994) used both RFLP markers, detected by genomic DNA and cDNA, and RAPD markers to evaluate the genetic relationships among 18 accessions from six cultivated *Brassica* species and one accession from *R. sativus*. Based on the RFLP results, *B. niger* was placed closer to *R. sativus* than to any other *Brassica* species, while based on the RAPD analyses, *B. niger* was closer to *B. juncea* and *B. carinata*, but the position of *R. sativus* was still closer to *B. niger* than to the group with *B. rapa, B. napus* and *B. oleracea*. These results are in agreement with the nuclear data obtained in several studies (Song et al. 1988a, 1990), but are in conflict with the analyses based on mitochondrial and chloroplast DNA. To explain this discrepancy, the authors postulated the possibility that *Raphanus* is a hybrid between the *B. nigra* and *B. oleracea*/ *B. rapa* lineages with the latter as the maternal parent.

3.6.2 *Origin of* Raphanus sativus

Raphanus sativus ($2n = 18$) is an old and globally cultivated oilseed, vegetable and forage crop, which is morphologically extremely variable. As center of domestication of the species, the East Mediterranean region and adjacent Near East are often mentioned (reference book Mansfeld Encyclopedia) (Fig. 3-6). The radish genus includes besides the cultivated species several wild species, like *R. raphanistrum*, *R. landra*, and *R. martimus*, all distributed in the coastal area of the Mediterranean Sea. *R. sativus* var. *raphanistroides* is also a wild species, which is distributed in the Far East, mainly in coastal areas of Japan, Korea and China. It is not clear what the ancestor of the cultivated radish is, even though *R. raphanistrum* is often mentioned.

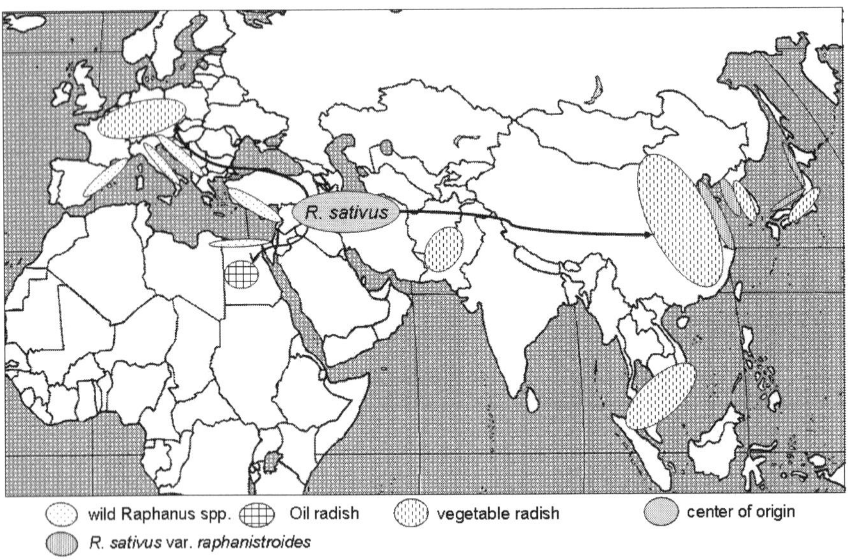

Figure 3-6 A map of Eurasia is depicted with primary center and secondary centers of diversity for *R. sativus* indicated. *R. raphanistrum*, *R. landra* and *R. martimus* are distributed in coastal areas of the Mediterranean Sea, while *R. sativus* var. *raphanistroides* is distributed in coastal areas of Japan, Korea and China.

3.6.3 Description of **Raphanus sativus** *Crop Types*

The many different crop types within cultivated radish have led to the acceptance of subspecies and variety (cultivar) descriptions based upon specialized morphology of edible parts. Nomenclature is rather complex, since different authors use different subspecies names. Here the three main radish types are described; for nomenclature we refer to the following website: *http://www.plantnames.unimelb.edu.au/Sorting/Raphanus.html.*

Oil radish, R. sativus convar. oleifer (Stokes) Alef.

The radish oil crop has been cultivated since ancient times in East Asia, Egypt and perhaps also in India and since the 19th century in some European countries as well. The oil was used for food but also as lamp oil. In West and central Europe, the crop gained importance as green forage, silage and green manure plant. In Japan and Egypt cultivars are grown that are cultivated for their edible leaves, and are considered belonging to this group.

Serpent, Seedpod or Rat-tailed Radish; *R. sativus* convar. *caudatus* (L.f.)

This crop is cultivated in Malaysia, Indonesia, Sri Lanka and India, mainly along the western coast of Punjab, for their extremely long young siliques, which are eaten fresh or in pickles, or sometimes as a cooked vegetable.

Radish, *R. sativus* convar. *sativus* L.

This is an assemblage of different cultivar groups of both radish and small radishes, cultivated worldwide. The two main groups are the radish group (also named *R. sativus* var. *hortensis*), and the small radish group (also named *R. sativus* var. *sativus*). There is enormous variation in shape, size, and color of the tubers, leaves and siliques but also in seasonal development with hundreds of local landraces and cultivars. Radishes are used fresh in salads, but also in cooked or stewed dishes and salted dried or fermented. Small radishes are mainly used as a fresh vegetable, and are also variable, but less compared to the true radishes. Some reports also list *R. sativus* var. *niger* (black radish).

3.6.4 Genetic Diversity of *Raphanus sativus*

3.6.4.1 Cultivated Radish

Genetic diversity of 30 radish accessions from Pakistan was investigated both at the phenotypic and genotypic level with morphological characteristics and also at the DNA level using RAPDs (Rabbani et al. 1998a). Dendrograms were generated based on the morphological data and 202 RAPD bands. Based on morphological traits, the accessions were classified into four major groups that correspond to the four different forms of cultivated radish, large rooted, pod-type and oilseed-seed type. However, based on RAPDs, the accessions were grouped according to their collection sites and not based on their phenotypic differences. A number of other studies in both cultivated and wild radish also showed close relationships between genetic distances and geographical distances (Huh and Ohnishi 2001, 2002; Liu et al. 2008; Wang et al. 2008). Thirty-five late bolting radish cultivars were fingerprinted with RAPD, ISSR and SRAP markers, which resulted in similar estimates of polymorphism rate and genetic similarity coefficients between pair of genotypes (Liu et al. 2008). Clustering using UPGMA showed clusters that were in accordance with geographical origins (Korea, Japan and China) and main characteristics. In another study, 65 accessions of cultivated radish collected from 21 Eurasian and African countries were studied using AFLPs (Wang et al. 2008). In a neighbor-joining tree, the accessions formed groupings according to their geographical origin (Europe, Middle-East, South Asia and East Asia). The branching pattern in the neighbor-joining tree provides some information on the distribution route followed by radish, from the Mediterranean to East Asia via the Middle East and Central Asia or South Asia, similar to the introduction from the Mediterranean area along the silk route to China as suggested by Kitamura (1958) and Kumazawa (1963). Sixty eight European cultivated radish varieties, including open-pollinated varieties, inbred lines, 2*n* and 4*n* hybrid varieties of garden radish *(R. sativus* var. *sativus* DC. Convar. *radicula*) and black radish (*R. sativus* var. *niger*) were fingerprinted using AFLP and ISSR markers. Cluster analyses separated the two weedy species (*R. raphanistrum*) from the cultivated varieties, while within the cultivated germplasm black radish and French breakfast radish types with elongated roots formed separate clusters (Muminovic et al. 2005).

3.6.4.2 Wild and Weedy Radish and Their Relationship to Cultivated Radish

To clarify the relationship between wild and cultivated radish species (both European *R. landra* and *R. raphanistrum* and the Far Eastern *R. sativus* var. *raphanistroides*) and to identify the ancestral species of radish, nine accessions from three wild species and nine accessions of different cultivated varieties together with closely related *Brassica* species were investigated for differences in the *trnK/matK* sequence (Lu et al. 2008). Both wild and cultivated *Raphanus* accessions formed a cluster and belonged to the *rapa/oleracea* group, while the *nigra* group was more distantly related, similar to the findings of Warwick and Black (1993, 1997) and Warwick and Sauder (2005). Multiple origins were suggested for the cultivated radish, since each of the three variety types—European small radish *R. sativus* var. *sativus*, Spanish Black radish *R. sativus* var. *niger* and East Asian big long radish *R. sativus* var. *hortensis*—belonged to different clusters of the phylogenetic tree. This could be ascribed to the frequent hybridization between cultivated and wild radish species, which was also suggested based on chloroplast SSR data and mitochondrial DNA types (Yamagishi and Terachi 2003; Yamagishi and Sasaki 2004). The European wild species are distributed in all three clusters, so that the authors could not conclude which wild species is the ancestor of the cultivated radish (Lu et al. 2008). Allozyme variation was studied within and among open-pollinated cultivars of radish and wild radish populations and it was concluded that domesticated radish retained a population structure similar to that of wild populations, which again suggests frequent hybridizations between wild and cultivated radish (Ellstrand and Marshall 1985). Yamagishi and Sasaki (2004) analyzed 36 radish accessions for variation in the mitochondrial 3'region of *ATP6*, and found that 75% of the varieties were assigned to type I or type II, while seven varieties contained both type I and type II and further varieties with type III or IV were rare. East Asian wild radish was related to European wild radishes, which made the authors suggest that the East Asian wild radish originated in Europe as a weed, as suggested by many others. Kim et al. (2007) identified a novel mitochondrial genome type and developed molecular markers to classify radish cytoplasm types. In addition to the *orf-138-atp8* Ogura mitochondrial DNA organization of cytoplamic male sterility type, they detected several sub-stoichiometric organizations linked to the *atp8* gene. Using the classification system based on PCR amplification, the authors could assign a total of 90 radish cultivars or accessions into three different mitotypes. Lee et al. (2009) recently investigated the organization *atp6-nad3-rps12* of the mitochondrial DNA of a novel cytoplasmic male sterility (CMS) type (DCGMS). They identified an SNP that lead to a frame-shift mutation, which was diagnostic for the DCGMS type.

Several reports address the genetic variation of wild, so-called weedy radish (*R. raphanus*) and *R. raphanistrum*. Panetsos and Baker (1968) analyzed several populations of *Raphanus* growing in California for both morphological and cytological traits and concluded that the populations of the highly successful "wild" (weedy) *R. sativus* originated by hybridization between the cultivated forms of radish with the introduced weed species *R. raphanistrum*. Campbell and Snow (2009) concluded that feral radish likely evolved through gene-flow from *R. raphanistrum* to cultivated radish based on experiments with radish populations over three generations. Based on chloroplast DNA analyses (*trnL-rpl32*), bidirectional hybridization between the progenitor species was suggested for the creation of wild/feral radish in California (Ridley et al. 2008). By studying plant growth traits, it was suggested that loss of the rosette and early flowering were key adaptations in the success of wild radish as a major agricultural weed (Sahli et al. 2008). Huh and Ohnishi (2002) assessed the genetic diversity and genetic structure of East Asian wild radish populations (*Raphanus sativus* var. *hortensis* f. *raphanistroides*), a natural and a cultivated population of *Raphanus sativus* var. *hortensis* from Kazakhstan and a *R. raphanistrum* population from Russia using AFLPs. They showed that the variation within populations was very large (97%) and that there was a close genetic relationship between *R. sativus* from Kazakhstan and *R. raphanistrum*, which confirmed the assumption that *R. raphanistrum* played a role in the origin of *R. sativus*, which probably took place somewhere in the eastern Mediterranean as described previously. In another study, the same authors (Huh and Ohnishi 2001) studied the diversity and population structure of Japanese and Korean populations of wild radish (*R. sativus* var. *hortensis* f. *raphanistroides*) that mainly grow on beaches using allozyme diversity. The genetic diversity within populations was high, which they explained by the insect pollinated outcrossing breeding system, gene-flow from cultivated radishes, large population sizes and high fecundity. In a dendrogram, the Japanese and Korean clusters were clearly distinguished with a close relationship between geographical origin and genetic distance, also within the Korean and Japanese clusters.

3.7 Use of Genetic Diversity Assessments for Germplasm Conservation and for Practical Breeding

The conservation of plant genetic resources in genebanks involves the selection of accessions to be conserved and the maintenance of these accessions for current and future users. For both these issues, knowledge is needed about the distribution of genetic diversity within and between accessions to be sampled from the gene pool, but also about the changes in variation of these samples as a result of regenerations (van Hintum et al. 2007). The goal of genebanks is to preserve the levels of variability in traits such as shape, color, taste, pest and disease resistances and seasonal growing type among other properties that can be valuable sources for breeding. Evaluation based solely on morphological characteristics can be insufficient for an accurate characterization of accessions, since it may under- or over-estimate genetic diversity between accessions. This is clearly illustrated by the fact that especially in *Brassicas*, convergent evolution in almost every trait studied hampers classifications. This is obvious in the studies that profile molecular markers for different vegetable mustards (*Brassica juncea*), since the different vegetable types clustered together and similar types were distributed over different clusters (Ren et al. 1995; Qiao et al. 1998; Fu et al. 2006; Qi et al. 2007, 2008). In *Brassica rapa*, turnips from the East and West also formed distinct clusters when grouped according to AFLP profiles (Zhao et al. 2005). Molecular marker profiles can assist in variety identification (Hu and Quiros 1991), genebank management like elimination of duplicated entrees and checking for uniformity and stability within accession (Kresovich et al 1992; Farnham 1996; Phippen et al. 1997; van Hintum et al. 2007), taxonomic studies (Demeke et al. 1992; Margale et al 1995; Lanner-Herrera et al 1996) and development of minimal sets of markers like RAPD primers, ISSR or SRAP primers to be able to distinguish all cultivars (Srivastava et al. 2001; Tonguç and Griffths 2004; Louarn et al. 2007; Wang et al. 2008). The latter can be used in law suits, where claims on origin of accessions/cultivars can be tested. All these studies corroborate to the usefulness of molecular marker methods to evaluate changes over regeneration processes and the genetic differentiation among accessions.

Genetic diversity assessments can also be used to improve breeding programs. Because *Brassica* vegetables and also oilseeds have been cultivated for many centuries in different parts of the world, the variation within the species has increased as a result of ongoing breeding. Different *Brassica* crop breeding programs have different objectives and priorities since each vegetable type is characterized by its own characteristics and the growing areas diverge. The relationship between accessions revealed by diversity studies, as mentioned above, indicated that different cultivar groups are often region specific, and have emerged independently in the different geographical regions. The market demands are considered by breeders in designing the most desirable ideotype, like Chinese cabbage with ovate or cylindrical heads is favored in different geographical regions and within an ecological region, breeding for specific types (like the ?pak choi related rosette shaped wutacai and early flowering caixin) is necessary to meet specific uses.

Molecular analysis has shown that there is considerable genotypic variation in the *B. rapa* gene pool, while several studies on *B. juncea* mustard oilseeds and *R. sativus* in Pakistan, but also on canola quality *B. junceas* in Canada and Australia, conclude that genetic variation is limited (Rabbani et al. 1998a, b, c; Burton et al. 2004; Khan 2008). The genetic diversity studies, using both molecular markers and phenotypic descriptors, can assist breeders in the choice of starting material for breeding programs, in broadening the base of genetic variation in their gene pools, and in selecting parents for breeding programs or in generating heterotic pools for hybrid breeding programs.

Acknowledgements

We thank Dr. Agnieszka Ludwikow for reviewing the chapter and Dr. Ningwen Zhang for critically reading the manuscript. This work was co-financed by the Centre for BioSystems Genomics (CBSG) which is part of the Netherlands Genomics Initiative/Netherlands Organisation for Scientific Research. Dunia Pino Del Carpio was funded by an IOP genomics program "Brassica nutrigenomics"and Jianjun Zhao by the programme Strategic Scientific Alliances (PSA) of the Royal Netherlands Academy of Arts and Sciences (KNAW) and the Chinese Ministry of Science and Technology.

References

Allen BW, Goodenough PW, Lee JSC, Rutherford PP (1986) Evolution of cauliflower types grown in Great Britain as indicated by the isoenzyme composition of the cauliflower curds. Euphytica 35: 25–31.

Allender CJ, Allainguillaume J, Lynn J, King GJ (2007) Simple sequence repeats reveal uneven distribution of genetic diversity in chloroplast genomes of *Brassica oleracea* L. and (n = 9) wild relatives. Theor Appl Genet 114(4): 609–618.

Al-Shehbaz IA, Beilstein MA, Kellogg EA (2006) Systematics and phylogeny of the Brassicaceae (Cruciferae): an overview. Plant Syst Evol 259: 89–120.

Andersen N, Poulsen G, Andersen B, Kiaer LP, D'Hertefeldt T, Wilkinson M, Jorgensen R (2009) Processes affecting genetic structure and conservation: A case study of wild and cultivated *Brassica rapa*. Genet Resour Crop Evol 56: 189–200.

Astarini IA, Plummer JA, Lancaster RA, Yan G (2004) Fingerprinting of cauliflower cultivars using RAPD markers. Aust J Agri Res 55: 117–124.

Axelsson T, Bowman CM, Sharpe AG, Lydiate DJ, Lagercrantz U (2000) Amphidiploid *Brassica juncea* contains conserved progenitor genomes. Genome 43: 679–688.

Bailey CD, Koch MA, Mayer M, Mummenhoff K, O'Kane SL, Warwick SL, Windham MD, Al-Shehbaz IA (2006) Toward a global phylogeny of the Brassicaceae. Mol Biol Evol 23: 2142–2160.

Balkaya A, Yanmaz R, Apaydin A, Kar H (2005) Morphological characterization of white head cabbage (*Brassica oleracea* var. *capitata* subvar. *alba*) genotypes in Turkey. NZ J Crop Hort 33: 333–341.

Beilstein MA, Al-Shehbaz IA, Kellogg EA (2006) Brassicaceae phylogeny and trichome evolution. Am J Bot 93: 607–619.

Burton WA, Ripley VL, Potts DA, Salisbury PA (2004) Assessment of genetic diversity in selected breeding lines and cultivars of canola quality *Brassica juncea* and their implications for canola breeding. Euphytica 136: 181–192.

Campbell LG, Snow AA (2009) Can feral weeds evolve from cultivated radish (*Raphanus Sativus*, Brassicaceae)? Am J Bot 96: 498–506.

Cao JS, Cao SC, Miao Y, Lu U (1997) Cladistic operational analysis and study on the evolution of Chinese cabbage groups (*Brassica campestris* L.). Acta Horti Sin 24: 35–42.

Cartea ME, Picoaga A, Soengas P, Ordás A (2003) Morphological characterization of kale populations from Northwestern Spain. Euphytica 129: 25–32.

Chen YP, Cao JS, Miao Y, Ye WZ (2000) Analysis of genetic polymorphism in vegetable crops of *Brassica campestris* by RAPD markers. J Zhejiang Univ (Agrii Life Sci) 26: 131–136.

Crisp P (1982) The use of an evolutionary scheme for cauliflowers in the screening of genetic resources. Euphytica 31: 725–734.

Crouch JH, Lewis BG, Lydiate DJ, Mithen R (1995) Genetic diversity of wild, weedy and cultivated forms of *Brassica rapa*. Heredity 74: 491–496.

Das S, Rajagopal J, Bhatia S, Srivastava PS, Lakshmikumaran M (1999) Assessment of genetic variation within *Brassica campestris* cultivars using amplified fragment length polymorphism and random amplification of polymorphic DNA markers. J Biosci 24: 433–440.

De Candolle A (1886) Origin of Cultivated Plants. 2nd edn. Hanfner, New York, USA.

Demeke T, Adams RP, Chibbar R (1992) Potential taxonomic use of random amplified polymorphic DNA (RAPD)—a case-study in *Brassica*. Theor Appl Genet 84: 990–994.

Denford KE, Vaughan JG (1977) Comparative-study of certain seed isoenzymes in 10 chromosome complex of *Brassica-campestris* and its allies. Ann Bot 41: 411–418.

Dias JS (1995) Genetic relationships of Portuguese coles and other close related *Brassica* genotypes using nuclear RFLPs. Genet Resour Crop Evol 42: 363–369.

Dias JS, Lima MB, Song KM, Monteiro AA, Williams PH, Osborn TC (1991) Molecular taxonomy of Portuguese tronchuda cabbage and kale landraces using nuclear RFLPs. Euphytica 58(3): 221–229.

Dias JS, Monteiro AA, Lima MB (1993) Numerical taxonomy of Portuguese Tronchuda cabbage and Galega kale landraces using morphological characters. Euphytica 69: 51–68.

Dixon GR (2006) Vegetable *Brassicas* and related *Crucifers*. Crop Production Science in Horticulture Series, 14, CABI Publ, Wallingford, UK.

Ellstrand NC, Marshall DL (1985) The impact of domestication on distribution of allozyme variation within and among cultivars of radish, *Raphanus sativus* L. Theor Appl Genet 69: 393–398.

Erickson LR, Straus NA, Beversdorf WD (1983) Restriction patterns reveal origins of chloroplast genomes in *Brassica* amphiploids. Theor Appl Genet 65: 201–206.

Farnham MW (1996) Genetic variation among and within the United States collard cultivars and landraces as determined by randomly amplified polymorphic DNA markers. J Am Soc Horti Sci 121: 374–379.

Flannery ML, Mitchell FJG, Coyne S, Kavanagh TA, Burke JI, Salamin N, Dowding P, Hodkinson TR (2006) Plastid genome characterisation in *Brassica* and Brassicaceae using a new set of nine SSRs. Theor Appl Genet 113: 1221–1231.

Fu J, Zhang MF, Qi XH (2006) Genetic diversity of traditional Chinese mustard crops *Brassica juncea* as revealed by phenotypic differences and RAPD markers. Genet Resour Crop Evol 53: 1513–1519.

Geraci A, Divaret I, Raimondo FM, Chèvre AM (2001) Genetic relationships between Sicilian wild populations of *Brassica* analysed with RAPD markers. Plant Breed 120: 193–196.

Gladis T, Hammer K (1992) The Brassica collection in Gatersleben: *Brassica juncea, Brassica napus, Brassica nigra,* and *Brassica rapa*. Feddes Rep 103: 469–507.

Gomez-Campo C (1999) Developments in Plant Genetics and Breeding, vol 4: Biology of Brassica coenospecies. Elsevier, Amsterdam, Laussanne, New York, Oxford.

Gray AR (1982) Taxonomy and evolution of broccoli (*Brassica oleracea* var. *italica*). Econ Bot 36: 397–410.

Guo JX, Zhou NY, Ma RC, Cao MQ (2002) Genetic diversity in *Brassica rapa* revealed by AFLP molecular markers. J Agri Biotechnol 10: 138–143.

He YT, Chen BY, Fu TD, Li DR, Tu JX (2003) Origins and evolution of *Brassica campestris* L. in China. Acta Genet Sin 30: 1003–1012.

Hu J, Quiros CF (1991) Identification of broccoli and cauliflower cultivars with RAPD markers. Plant Cell Rep 10: 505–511.

Huh MK, Ohnishi O (2001) Allozyme diversity and population structure of Japan and Korean populations of wild radish, *Raphanus sativus* var. *hortensis* f. raphanistroides (Brassicaceae). Genes Genet Syst 76: 15–23.

Huh MK, Ohnishi O (2002) Genetic diversity and genetic relationships of East Asian natural populations of wild radish revealed by AFLP. Breed Sci 52: 79–88.

Jain A, Bhatia S, Banga SS, Prakash S, Lakshmikumaran M (1994) Potential use of random amplified polymorphic DNA (RAPD) technique to study the genetic diversity in Indian mustard (*Brassica-Juncea*) and its relationship to heterosis. Theor Appl Genet 88: 116–122.

Keng H (1974) Economic plants of ancient north-china as mentioned in Shih-Ching (book of poetry). Econ Bot 28: 391–410.

Khan MA, Rabbani MA, Munir M, Ajmal SK, Malik MA (2008) Assessment of genetic variation within Indian mustard (*Brassica juncea*) germplasm using random amplified polymorphic DNA markers. J Intgr Plant Biol 50: 385–392.

Kim S, Lim H, Park S, Cho KH, Sung SK, Oh DG, Kim KT (2007) Identification of a novel mitochondrial genome type and development of molecular markers for cytoplasm classification in radish (*Raphanus sativus* L.). Theor Appl Genet 115: 1137–1145.

Kitamura S (1958) Varieties of radish and their transition. In: I Nishiyama (ed) Japanese Radish. Jpn Sci Soc Press, Tokyo, Japan, pp 1–19 (in Japanese).

Koch MA, Kiefer C (2006) Molecules and migration: biogeographical studies in cruciferous plants. Plant Syst Evol 259: 121–142.

Kresovich S, Mcferson JR (1992) Assessment and management of plant genetic diversity—considerations of intraspecific and interspecific variation. Field Crop Res 29: 185–204.

Kumazawa S (1963) Horticulture. Yokendo, Tokyo, Japan, pp 23–50 (in Japanese).

Lamboy WF, McFerson JR, Westman AL, Kresovich S (1994) Application of isozyme data to the management of the United States national *Brassica oleracea* L. genetic resources collection. Genet Resour Crop Evol 41 (2): 99–108.

Lanner C (1998) Relationships of wild *Brassica* species with chromosome number 2n=18, based on comparison of the DNA sequence of the chloroplast intergenic region between trnL (UAA) and (GAA). Can J Bot 76: 228–237.

Lanner C, Bryngelsson T, Gustafsson M (1996) Genetic validity of RAPD markers at the intra- and inter-specific level in wild *Brassica* species with n=9. Theor Appl Genet 93: 9–14.

Lanner-Herrera C, Gustafsson M, Falt AS, Bryngelsson T (1996) Diversity in natural populations of wild *Brassica oleracea* as estimated by isozyme and RAPD analysis. Genet Resour Crop Evol 43: 13–23.

Lanner-Herrera C, Bryngelsson T, Gustafsson M (1997) Relationships of wild *Brassica* species with chromosome number 2n=18, based on comparison on RFLPs. Genome 40: 302–308.

Lazaro A, Aguinagalde I (1996) Phylogenetic relationships between the wild taxa of the *Brassica oleracea* L. group (2n=18) using random amplified polymorphic DNA assay. Sci Hort 65: 219–227.

Lazaro A, Aguinagalde I (1998) Genetic diversity in *Brassica oleracea* L. (Cruciferae) and wild relatives (2n=18) using isozymes. Ann Bot 82: 821–828.

Lee YP, Kim S, Lim H, Ahn Y, Sung SK (2009) Identification of mitochondrial genome rearrangements unique to novel cytoplasmic male sterility in radish (*Raphanus sativus* L.). Theor Appl Genet 118: 719–728.

Li JW (1981) The origins and evolution of vegetable crops in China. Sci Agri Sin 14: 90–95.

Liu HL (1984) Origin and evolution of genus *Brassica* oilseed rapes. Acta Agron Sin 10: 9–18.

Liu LW, Zhao LP, Gong YQ, Wang MX, Chen LM, Yang JL, Wang Y, Yu FM, Wang LZ (2008) DNA fingerprinting and genetic diversity analysis of late-bolting radish cultivars with RAPD, ISSR and SRAP markers. Sci Hort 116: 240–247.

Liu PY. 1996. Chinese Mustard. China Agri Press, Beijing, pp 46–56 (in Chinese).

Louarn S, Torp A, Holme I, Andersen S, Jensen B (2007) Database derived microsatellite markers (SSRs) for cultivar differentiation in *Brassica oleracea*. Genet Resour Crop Evol 54: 1717–1725.

Lu N, Yamane K, Ohnishi O (2008) Genetic diversity of cultivated and wild radish and phylogenetic relationships among *Raphanus* and *Brassica* species revealed by the analysis of trnK/matK sequence. Breed Sci 58: 15–22.

Lysak MA, Koch MA, Pecinka A, Schubert I (2005) Chromosome triplication found across the tribe Brassiceae. Genome Res 15: 516–525.

Margale E, Herve Y, Hu J, Quiros CF (1995) Determination of genetic-variability by RAPD markers in cauliflower, cabbage and kale local cultivars from France. Genet Resour Crop Evol 42: 281–289.

McGrath JM, Quiros CF (1992) Genetic diversity at isozyme and RFLP loci in *Brassica-campestris* as related to crop type and geographical origin. Theor Appl Genet 83: 783–790.

Muminovic J, Merz A, Melchinger AE, Lubberstedt T (2005) Genetic structure and diversity among radish varieties as inferred from AFLP and ISSR analyses. J Am Soc Hort Sci 130: 79–87.

Nienhuis J, Slocum MK, De Vos DA, Muren R (1993) Genetic similarity among *Brassica oleracea* L. genotypes as measured by restriction fragment length polymorphisms. J Am Soc Hort Sci 118: 298–303.

Ofori A, Becker HC, Kopisch-Obuch FJ (2008) Effect of crop improvement on genetic diversity in oilseed *Brassica rapa* (turnip-rape) cultivars, detected by SSR markers. J Appl Genet 49: 207–212.

Okumus A, Balkaya A (2007) Estimation of genetic diversity among Turkish kale populations (*Brassica oleracea* var. acephala L.) using RAPD markers. Rus J Genet 43: 411–415.

Padilla G, Cartea ME, Rodriguez VM, Ordas A (2005) Genetic diversity in a germplasm collection of *Brassica rapa* subsp *rapa* L. from Northwestern Spain. Euphytica 145: 171–180.

Padilla G, Cartea M, Soengas P. Ordás A (2007) Characterization of fall and spring plantings of Galician cabbage germplasm for agronomic, nutritional, and sensory traits. Euphytica 154: 63–74.

Palmer JD, Shields CR, Cohen DB, Orton TJ (1983) Chloroplast DNA evolution and the origin of amphidiploid *Brassica* species. Theor Appl Genet 65: 181–189.

Panda S, Martín JP, Aguinagalde I (2003) Chloroplast and nuclear DNA studies in a few members of the *Brassica oleracea* L. group using PCR-RFLP and ISSR-PCR markers; a population genetic analysis. Theor Appl Genet 106: 1122–1128.

Panetsos CA, Baker HG (1968) Origin of variation in wild *Raphanus Sativus* (Cruciferae) in California. Genetica 38: 243.

Panjabi P, Jagannath A, Bisht NC, Padmaja KL, Sharma S, Gupta V, Pradhan AK, Pental D (2008) Comparative mapping of *Brassica juncea* and *Arabidopsis thaliana* using intron polymorphism (IP) markers: homeologous relationships, diversification and evolution of the A, B and C Brassica genomes. BMC Genom 9: 113.

Parkin IAP, Gulden SM, Sharpe AG, Lukens L, Trick M, Osborn TC, Lydiate DJ (2005) Segmental structure of the *Brassica napus* genome based on comparative analysis with *Arabidopsis thaliana*. Genetics 171: 765–781.

Persson K, Falt AS, Von Bothmer R (2001) Genetic diversity of allozymes in turnip (*Brassica rapa* l. Var. *Rapa*) from the Nordic area. Hereditas 134: 43–52.

Phippen WB, Kresovich S, Candelas FG, McFerson JR (1997) Molecular characterization can quantify and partition variation among genebank holdings: A case study with phenotypically similar accessions of *Brassica oleracea* var capitata L (cabbage) 'Golden Acre'. Theor Appl Genet 94: 227–234.

Prakash S, Hinata K (1980) Taxonomy, cytogenetics and origin of crop *Brassica*, a review. Opera Bot 55: 1–57.

Purugganan MD, Boyles AL, Suddith JI (2000) Variation and selection at the CAULIFLOWER floral homeotic gene accompanying the evolution of domesticated *Brassica oleracea*. Genetics 155: 855–862.

Qi XH, Zhang MF, Yang JH (2007) Molecular phylogeny of Chinese vegetable mustard (*Brassica juncea*) based on the internal transcribed spacers (ITS) of nuclear ribosomal DNA. Genet Resour Crop Evol 54: 1709–1716.

Qi XH, Yang JH, Zhang MF (2008) AFLP-based genetic diversity assessment among Chinese vegetable mustards (*Brassica juncea* (L.) Czern.). Genet Resour Crop Evol 55: 705–711.

Qiao AM, Liu PY, Lei JJ (1998) RAPD analysis of sixteen varieties of mustard. Acta Bot Sin 40: 915–921.

Rabbani MA, Murakami Y, Kuginuki Y, Takayanagi K. (1998a) Genetic variation in radish (*Raphanus sativus* L.) germplasm from Pakistan using morphological traits and RAPDs. Genet Resour Crop Evol 45: 307–316.

Rabbani MA, Iwabuchi A, Murakami Y, Suzuki T, Takayanagi K (1998b) Phenotypic variation and the relationships among mustard (*Brassica juncea* L.) germplasm from Pakistan. Euphytica 101: 357–366.

Rabbani MA, Iwabuchi A, Murakami Y, Suzuki T, Takayanagi K (1998c) Genetic diversity in mustard (*Brassica juncea* L.) germplasm from Pakistan as determined by RAPDs. Euphytica 103: 235–242.

Rabbani MA, Qureshi AA, Afzal M, Anwar R, Komatsu S (2001) Characterization of mustard [*Brassica juncea* (L.) Czern. & Coss.] germplasm by SDS-page of total seed proteins. Pak J Bot 33: 173–179.

Raybould AF, Mogg RJ, Clarke RT, Gliddon CJ, Gray AJ (1999) Variation and population structure at microsatellite and isozyme loci in wild cabbage (*Brassica oleracea* L.) in Dorset (UK). Genet Resour Crop Evol 46: 351–360.

Reiner H, Holzner W, Ebermann R (1995) The development of turnip-type and oilseed-type *Brassica rapa* crops from the wild type in Europe—An overview of botanical, historical and linguistic facts. In: Int Rapeseed Congr, Rapeseed Today and Tomorrow 4, Cambridge, UK, pp 1066–1069.

Ren JP, McFerson JR, Li RG, Kresovich S, Lamboy WF (1995) Identities and relationships among Chinese vegetable brassicas as determined by random amplified polymorphic DNA markers. J Am Soc Hort Sci 120: 548–555.

Ridley CE, Kim SC, Ellstrand NC (2008) Bidirectional history of hybridization in California wild radish, *Raphanus Sativus* (Brassicaceae), as revealed by chloroplast DNA. Am J Bot 95: 1437–1442.

Sahli HF, Conner JK, Shaw FH, Howe S, Lale A (2008) Adaptive differentiation of quantitative traits in the globally distributed weed, wild radish (*Raphanus raphanistrum*). Genetics 180: 945–955.

Sawaza HE, Nagai H, Sodek L. (1997) Characterization of genetic variability of kale plants by enzymatic polymorphism and RAPD. Bragantia 56: 9–19.

Schranz ME, Song BH, Windsor AJ, Mitchell-Olds T (2007) Comparative genomics in the Brassicaceae: a family-wide perspective. Curr Opin Plant Biol 10: 168–175.

Sinskaia EN (1928) The oleiferous plants and root crops of the family Cruciferae, Bull Appl Bot Genet. Plant Breed 9: 1–648.

Snogerup S (1980) The wild forms of the *Brassica oleracea* group (2n=18) and their possible relations to the cultivated ones. In: S Tsunoda , K Hinata , C Gómez-Campo (eds) Brassica Crops and Wild Allies. Jpn Sci Soc Press, Tokyo, Japan, pp 121–132.

Song KM, Osborn TC, Williams PH (1988a) Brassica Taxonomy Based on Nuclear Restriction Fragment Length Polymorphisms (RFLPs). 1. Genome Evolution of Diploid and Amphidiploid Species. Theor Appl Genet 75: 784–794.

Song KM, Osborn TC, Williams PH (1988b) Brassica Taxonomy based on nuclear restriction fragment length polymorphisms (RFLPs). 2. Preliminary analysis of subspecies within *Brassica Rapa* (Syn. *Campestris*) and *Brassica Oleracea*. Theor Appl Genet 76: 593–600.

Song KM, Osborn TC, Williams PH (1990) *Brassica* taxonomy based on nuclear restriction fragment length polymorphisms (RFLPs). 3. Genome relationships in *Brassica* and related genera and the origin of *Brassica oleracea* and *B rapa* (Syn. *campestris*). Theor Appl Genet 79: 497–506.

Specht CE, Diederichsen A (2001) *Brassica*. In: P Hanelt (eds) Mansfeld's Encyclopedia of Agricultural and Horticultural Crops, vol 3. Springer, Berlin, Germany, pp 1435–1465.

Srivastava A, Gupta V, Pental D, Pradhan AK (2001) AFLP-based genetic diversity assessment amongst agronomically important natural and some newly synthesized lines of *Brassica juncea*. Theor Appl Genet 102: 193–199.

Takuno S, Kawahara T, Ohnishi O (2007) Phylogenetic relationships among cultivated types of *Brassica rapa* L. Em. Metzg. As revealed by AFLP analysis. Genet Resour Crop Evol 54: 279–285.

Tan QM (1979) The origin, distribution and domestication of Chinese cabbage. Agri Sci China 4: 68–75.

Thormann CE, Ferreira ME, Camargo LEA, Tivang JG, Osborn TC (1994) Comparison of RFLP and RAPD markers to estimating genetic-relationships within and among cruciferous species. Theor Appl Genet 88: 973–980.

Tonguç M, Griffiths PD (2004) Genetic relationships of *Brassica* vegetables determined using database derived sequence repeats. Euphytica 137: 193–201.

Tsunoda S (1980) Eco-physiology of wild and cultivated forms in *Brassica* and allied genera. In: S Tsunoda , K Hinata , C Gómez-Campo (eds) *Brassica* Crops and Wild Allies. Jpn Sci Soc Press, Tokyo, Japan, pp 110–120.

Van Hintum TJL, Boukema IW, Visser DL (1996) Reduction of duplication in a *Brassica oleracea* germplasm collection. Genet Resour Crop Evol 43: 343–349.

Van Hintum TJL, de Wiel CCMV, Visser DL, van Treuren R, Vosman B (2007) The distribution of genetic diversity in a *Brassica oleracea* gene bank collection related to the effects on diversity of regeneration, as measured with AFLPs. Theor Appl Genet 114: 777–786.

Vaughan JG (1977) A multidisciplinary study of the taxonomy and origin of Brassica crops. Bio Sci 27: 35–40.

Vischi M, Fiori M, De Paoli E, Padovan S, Guarda M, Olivieri A (2008) Broccoli 'Fiolaro' (*Brassica oleracea*) an endangered typical Italian cultivar. A genetic analysis by SSR markers. Plant Genet Resour 6: 215–221.

Wang N, Kitamoto N, Ohsawa R, Fujimura T (2008) Genetic diversity of radish (*Raphanus sativus*) germplasms and relationships among worldwide accessions analyzed with AFLP markers. Breed Sci 58: 107–112.

Warwick SI, Black LD (1993) Molecular Relationships in Subtribe Brassicinae (Cruciferae, Tribe Brassiceae). Can J Bot 71: 906–918.

Warwick SI, Black LD (1997) Phylogenetic implications of chloroplast DNA restriction site variation in subtribes Raphaninae and Cakilinae (Brassicaceae, tribe Brassiceae). Can J Bot 75: 960–973.

Warwick SI, Sauder CA (2005) Phylogeny of tribe Brassiceae (Brassicaceae) based on chloroplast restriction site polymorphisms and nuclear ribosomal internal transcribed spacer and chloroplast trnL intron sequences. Can J Bot 83: 467–483.

Warwick SI, James T, Falk KC (2008) AFLP-based molecular characterization of *brassica rapa* and diversity in canadian spring turnip rape cultivars. Plant Genet Resour 6: 11–21.

Weerakoon SR, Weerasena OVDSJ, Peiris PKD, Somaratne S (2008) Differentiation of Sri Lankan mustard (*Brassica Juncea*) accessions using FAFLP markers and agro-morphological data. Res J Biotechnol: 239–243.

Wu X, Chen B, Lu G, Wang H, Xu K, Gao G, Song Y (2009) Genetic diversity in oil and vegetable mustard landraces revealed by SRAP markers: its implications in evolution and domestication. Genet Resour Crop Evol: DOI 10.1007/s10722-009-9420-8.

Yamagishi H (2004) Assessment of cytoplasmic polymorphisms by PCR-RFLP of the mitochondrial orfB region in wild and cultivated radishes (*Raphanus*). Plant Breed 123: 141–144.

Yamagishi H, Terachi T (2003) Multiple origins of cultivated radishes as evidenced by a comparison of the structural variations in mitochondrial DNA of *Raphanus*. Genome 46: 89–94.
Yamagishi H, Sasaki J (2004) Differentiation of the cytoplasm of Japanese cultivated radishes (*Raphanus sativus*) into four types in the 3' region of mitochondrial atp6. Breed Sci 54: 189–195.
Yamagishi H, Tateishi M, Terachi T, Murayama S (1998) Genetic relationships among Japanese wild radishes (*Raphanus sativus* f. raphanistroides Makino), cultivated radishes and R-raphanistrum revealed by RAPD analysis. J Jpn Soc Hort Sci 67: 526–531.
Yang YW, Tseng PF, Tai PY, Chang CJ (1998) Phylogenetic position of *Raphanus* in relation to *Brassica* species based on 5S rRNA spacer sequence data. Bot Bull Acad Sin 39: 153–160.
Yang YW, Lai KN, Tai PY, Ma DP, Li WH (1999) Molecular phylogenetic studies of *Brassica, Rorippa, Arabidopsis* and allied genera based on the internal transcribed spacer region of 18S–25S rDNA. Mol Phylogenet and Evol 13: 455–462.
Yang YW, Tai PY, Chen Y, Li WH (2002) A study of the phylogeny of *Brassica rapa, B-nigra, Raphanus sativus,* and their related genera using noncoding regions of chloroplast DNA. Mol Phylogenet Evol 23: 268–275.
Zhao JJ, Wang XW, Deng B, Lou P, Wu J, Sun RF, Xu ZY, Vromans J, Koornneef M, Bonnema G (2005) Genetic relationships within *Brassica rapa* as inferred from AFLP fingerprints. Theor Appl Genet 110: 1301–1314.
Zhao JJ, Paulo MJ, Jamar D, Lou P, van Eeuwijk F, Bonnema G, Vreugdenhil D, Koornneef M (2007) Association mapping of leaf traits, flowering time, and phytate content in *Brassica rapa*. Genome 50: 963–973.

4

Molecular Linkage Maps: Strategies, Resources and Achievements

Danuta Babula-Skowrońska,[1]* *Agata Cieśla,*[1] and *Jan Sadowski*[1,2]

ABSTRACT

This chapter provides a detailed overview of marker systems, genetic maps and the status of genome mapping for diploid species of the *Brassica* genus. Progress in molecular biology techniques development and a high level of DNA polymorphisms, disclosed both among various *Brassica* accessions and subspecies, allowed construction of comprehensive genetic maps. We describe different marker systems used in the *Brassica* chromosomal maps construction with their specific advantages and disadvantages, thus the choice of a particular marker system must be adapted to the particular problem to be solved. On the other hand, the presence of the same markers on different *Brassica* maps allows comparison of their localization, giving insight into the structure and evolution of genomes of the diploid and amphidiploid species. Construction of *Brassica* genetic maps appeared to be of great significance for increasing breeding efficiency as it permitted detection and analysis of QTLs, marker- or gene-assisted selection, gene flow and fingerprinting for clone identification of economically important genes. Furthermore, the maps facilitated detection of a target region by searching public mapping database. Importantly, the *Brassica* species are closely related to *Arabidopsis thaliana,* therefore the knowledge obtained from the *A. thaliana* genome sequencing has opened the way for comparative genomics, evolutionary studies and gene structure analysis.

Finally, it needs to be emphasized that all genetic mapping experiments are based on polymorphism of genetic markers, thus the final pattern

[1]Institute of Plant Genetics, Polish Academy of Sciences, Strzeszyńska 34, 60-479, Poznań, Poland.

[2]Department of Biotechnology, Institute of Molecular Biology and Biotechnology, Faculty of Biology, Adam Mickiewicz University, Umultowska 89, 61-614 Poznań, Poland.

*Corresponding author: *dbab@igr.poznan.pl*

of duplication/triplication of the *Brassica* genomes is incomplete and requires further verification. Nevertheless, the present knowledge of genomic relationship among diploid *Brassica* species, collected due to genetic maps development, made the understanding of the *Brassica* genomes evolution possible. Finally, the studies of chromosomal location and expression of duplicated genes in the *Brassica* species will provide insight into the role of polyploidization/diploidization cycles in divergence of gene functions leading to the phenotypic plasticity of vegetable *Brassica* species.

Keywords: DNA markers, linkage maps, comparative maps, segmental duplications, paleopolyploids

4.1 Introduction

The first genetic linkage associations of genes controlling morphological traits in *Brassica* were detected over 80 years ago (Pease 1926). Further research describing linkage groups for the *Brassica* species concerned only a limited number of morphological and isozyme markers (Yarnell 1956; Stringam 1977; Sampson 1978; Arus and Orton 1983). Despite identification and rough mapping of genes for morphological characters and isozyme markers, little was known about their organization in the *Brassica* genomes (James and Williams 1980). Development of molecular biology techniques and a high level of DNA polymorphisms disclosed both between various *Brassica* accessions and subspecies allowed construction of comprehensive *Brassica* genetic maps (Figdore et al. 1988; Song et al. 1988; McGrath et al. 1990). The first detailed restriction fragment length polymorphism (RFLP) maps of *Brassica* have been published in the beginning of the 1990s (McGrath et al. 1990; Song et al. 1991; for a review see Babula et al. 2007). Many maps of *Brassica* have been constructed on the basis of the F_2 or doubled haploid (DH) lines derived from different species, subspecies or varieties. This facilitated mapping of the genes or quantitative trait loci (QTLs) of economically important traits, identification of introgressions between different genomes, marker-assisted breeding and gene pyramiding aimed at crop improvement. The presence of the same markers on different *Brassica* maps allows comparison of their localization, giving insight into the structure and evolution of genomes of the diploid and amphidiploid species.

Up to date more than 1,000 different markers have been located on 15 genetic maps of the *Brassica* genus species using different populations and marker systems (Lim et al. 2007). This chapter provides a detailed overview of genetic maps and the status of genome mapping for diploid species of *Brassica* genus.

4.2 Evolution of Marker Types: From RFLPs to SNPs

Along with accumulating knowledge about primary genome structure, new types of markers, namely molecular markers, have appeared. Since the polymerase chain reaction (PCR) technique was first introduced in 1983 by Kary Mullis, a range of new PCR-based markers have been used in genetic map construction. The DNA markers have shed new light on plant genome research and breeding. The molecular markers have been widely used for identification and analysis of agronomically important genes (Plieske and Struss 2001) and for genetic analysis of the *Brassica* genomes structure (Williams et al. 1990; Table 4-1). Importantly, the *Brassica* species are closely related to *Arabidopsis thaliana,* therefore the knowledge obtained from the *A. thaliana* genome sequencing has opened the way for comparative genomics, evolutionary studies and gene structure analysis (Snowdon and Friedt 2004). As a result of the genome sequencing of the model plant *A. thaliana* (The *Arabidopis* Genome Initiative 2000), many techniques of marker production used in the *Brassica* research have become more informative. Each marker system has specific advantages and disadvantages, thus the choice of a particular marker system must be adapted to the particular problem to be solved. Molecular markers widely used in *Brassica* research can be classified into three groups as listed below.

1. Hybridization-based DNA markers, such as restriction fragment length polymorphism (RFLP)
2. PCR-based DNA markers, such as random amplified polymorphic DNA (RAPD), amplified fragment length polymorphism (AFLP), sequence-characterized amplified region (SCAR), cleaved amplified polymorphic sequences (CAPS), simple sequence repeat (SSR) and sequence tagged site (STS), and sequence-related amplified polymorphism (SRAP)
3. DNA markers based on single nucleotide polymorphism (SNP; for details about molecular markers see Gupta et al. 1999).

4.2.1 Restriction Fragment Length Polymorphism

Restriction fragment length polymorphism (RFLP) is one of the earliest DNA-based techniques used in genetic studies. The polymorphism is analyzed in Southern blot hybridization assay. It was first used for human genome mapping and later it was adopted for plant genome mapping, including *Brassica* (Table 4-1; for a reviev see Gupta et al. 1999). There were two main types of probes used in the RFLP technique. In the first, unknown genomic sequences were used as hybridization probes; they were fragments of genomic DNA obtained after digestion with restriction endonucleases. The enzyme used most often was *Pst*I and the restriction fragments obtained were referred to as *Pst*I-fragments. With progress in

Table 4-1 Characteristics of genetic markers applied in the *Brassica* studies.

Properties	RFLP	RAPD	AFLP	SCAR	CAPS	SSR	STS	SRAP	SNP
Type of polymorphism	point mutations, insertions, deletions	point mutations, insertions, deletions	point mutations, insertions, deletions	point mutations, insertions, deletions	point mutations, insertions, deletions	changes in microsatellites sequence length	point mutations, insertions, deletions	point mutations, insertions, deletions	point mutations
Frequency of occurrence in genome	high	high	high	high	high	mean	rare	high	very high
Genome location	coding regions	whole genome	whole genome	whole genome	whole genome	repeated DNA	whole genome	coding regions	whole genome
Type of inheritance	codominant	dominant	dominant	dominant/ codominant	codominant	codominant	dominant/ codominant	dominant/ codominant	codominant
Possibility of alleles detection	yes	no	no/yes (rare)	yes	yes	yes	no	yes	yes
DNA sequence knowledge required	no	no	yes/no	yes	yes	no	yes	yes	yes
References	Song et al. 1990; Kianian and Quiros 1992a; Lan et al. 2000; Babula et al. 2003	Williams et al. 1990; Nozaki et al. 1997	Vos et al. 1995; Sebastian et al. 2000; Pradhan et al. 2003	Piao et al. 2004; Varshney et al. 2004	Piao et al. 2004; Varshney et al. 2004	Dietrich et al. 1994; Plieske and Struss 2001	Plieske and Struss 2001; Choi et al. 2007	Li and Quiros 2001; Rahman et al. 2007	Snowdon and Friedt 2004; Rahman et al. 2007

genome sequencing, the expressed sequence tags (ESTs) could be used as hybridization probes in RFLP analysis.

RFLP analysis was widely used in the *Brassica* genetics. However, taxonomists were the first beneficieries of RFLP. Random nuclear sequences selected from *Pst*I genomic DNA library were used to estimate the phylogenetic relationships between vegetable subspecies of *Brassica oleracea* and *Brassica rapa* (Song et al. 1990). The RFLP markers have been found to be very useful in genetic map construction of the *Brassica* genomes due to the high level of DNA polymorphism in these species, and the codominant inheritance of RFLP markers. Since the 1980s many RFLP-generated maps have been constructed for different *Brassica* crops. The first genome restriction fragment polymorphisms for the *Brassica* species were estimated by screening *Eco*RI-, *Hind*III-, or *Eco*RV-digested total genomic DNA (Figdore et al. 1988). The first linkage map based on RFLP for *B. oleracea* was constructed by Slocum et al. (1990). These studies proved that with the help of RFLP markers, it was possible to detect genome rearrangements such as tandem duplications and a strong conservation of some parts of the linkage loci organization. RFLP markers were used in linkage map construction for different vegetable *Brassica* as shown by Landry et al. (1992), Camargo et al. (1997) or Hu et al. (1998). RFLP markers were used also in integration of linkage maps based on intraspecific and interspecific mapping populations of *B. oleracea* (Kianian and Quiros 1992a). It significantly increased the level of polymorphism and introduced RFLP analysis as an efficient method for evolutionary considerations. The RFLP together with AFLP markers were used in map construction of the *B. oleracea* and *B. juncea* chromosomes (Sebastian et al. 2000; Pradhan et al. 2003). This combined approach offered the advantages of both marker types.

The sequencing project of the *A. thaliana* genome has opened a new era in the *Brassica* genetics. In the absence of complete knowledge about the *Brassica* genome, ESTs and genome sequence of related organisms have offered an opportunity to examine a large number of coding sequences (Table 4-2). EST library from the *B. rapa* ssp. *pekinensis* flower buds was used for characterization of the coding-sequence divergence of closely related *B. rapa* and *A. thaliana* genomes (Tiffin and Hahn 2002). These studies showed that the sequences, which were not significantly similar to those of *A. thaliana*, did not demonstrate essential similarity to other sequences deposited in public databases. This observation suggests that some of the similar sequences that were not detected for being significantly changed was a result of distinct selection in comparison to those of *A. thaliana*. It shows that *A. thaliana* ESTs can be used in evolutionary studies of *Brassica*, but the most popular application of this type of hybridization probes is map construction. ESTs refer to coding sequences so they can be used for identification of coding regions in a genome and for discovering

Table 4-2 Number of the *Brassica* ESTs clones available in EST database in NCBI (updated May 2009).

Species	Number of ESTs
Brassica rapa	194,741
Brassica nigra	1,814
Brassica oleracea	63,201
Brassica carinata	2,500
Brassica juncea	5,393
Brassica napus	656,151

new genes. Thus, their localization on the genetic maps can bring valuable information. RFLP markers based on unknown sequences as well as ESTs were used in the construction of the comparative map between *B. oleracea* and *A. thaliana* (Lan et al. 2000). Andy Paterson's group has used ESTs from *A. thaliana* in *Brassica* mapping and this approach allowed determination of the structural changes in the genome of these two species (Lan et al. 2000). ESTs can be also very useful for localization of the disease resistance loci on the genetic maps in economically important *Brassica* species, and regarding the conservation of ESTs, the same probes can be used in genome studies even in distantly related species like *Arabidopsis*, *Gossypium* and *Sorghum* (Paterson et al. 1996; Lan et al. 2000).

ESTs from the *A. thaliana* genome sequencing project served as probes for RFLP analysis in map construction for *B. oleracea*. EST-based probes corresponded to 12 main functional categories of known genes from *A. thaliana* (Babula et al. 2003). A large number of RFLP markers were used for construction of the sequence-tagged linkage map for *B. rapa*. ESTs clones from four different tissue-specific libraries of *B. rapa* and sequences representing genes from *A. thaliana* generated by PCR, were used as hybridization probes (Kim et al. 2006). All markers used in this experiment made the resultant map very detailed.

4.2.2 Random Amplified Polymorphic DNA

Interest in the development of genetic maps has prompted the search for markers that would allow fast and cheap genome analysis. One of the markers proposed was the random amplified polymorphic DNA (RAPD; Williams et al. 1990). This is a simple marker system based on the amplification of genomic DNA with single oligomer primers of random sequence. The primers permit detection of polymorphisms when DNA sequence is not known. It is possible to analyze a wide range of species using a universal set of primers (for more details about RAPD markers see Williams et al. 1990).

RAPD markers have become very useful in the *Brassica* species research taking into account the high level of polymorphism within each species, and the various genomes in basic diploid and natural amphidiploid species. *Brassica* genomes' analysis revealed that with the use of a blend of arbitrary primers it is possible to identify genome-specific and species-specific RAPD markers (Quiros et al. 1991; Table 4-1). RAPD markers proved to be important in investigation of genetic identities and relationships among botanical varieties of *B. oleracea* (Kresovich et al. 1992). RAPD together with RFLP markers were used to analyze the relationships between selected Brassicaceae species. A range of 18 different botanical varietes of *Brassica* species and one variety of *Raphanus sativus* was chosen as material for taxonomic analysis. Results revealed the worth of RAPD markers for this type of analysis (Thormann et al. 1994). Lazaro and Aguinagalde (1998) used RAPD markers to evaluate genetic diversity in *B. oleracea* and wild relatives. As a result of these studies, the *Brassica* populations were grouped into three main taxonomic branches. RAPDs were used for *B. rapa* linkage map construction and for selected QTLs analysis. This investigation revealed that some of the RAPD markers used were linked with agronomic traits such as bolting time and self-incompatibility, which makes RAPD markers useful in marker-assisted selection (Nozaki et al. 1997).

4.2.3 *Amplified Fragment Length Polymorphism*

Amplified fragment length polymorphism (AFLP) is a technique resembling that of RFLP. The main difference between these two approaches is that AFLP is based on PCR amplification instead of southern hybridization for detection of restriction fragments, like in RFLP technique (for details about AFLP markers see Vos et al. 1995; Mueller and Wolfenbarger 1999).

AFLP markers have been used in many different studies, e.g. phylogenetic relationships, systematic studies or epidemiological hospital monitoring in a wide range of taxa (Muller and Wolfenbarger 1999). The AFLP analysis has become a tool successfully adapted in the *Brassica* genomics. Because of its many advantages, application of AFLP markers permits fast construction of genetic linkage maps in different members of the *Brassica* genus like in the case of *B. juncea* (Pradhan et al. 2003), *B. oleracea* (Sebastian et al. 2000) or *B. napus* (Lombard and Delourme 2001) (Table 4-1).

Literature data show some interesting features of AFLP markers. In construction of the *B. juncea* linkage map, differences between *Eco*RI- and *Pst*I-based AFLP distributions were observed (Pradhan et al. 2003). *Eco*RI-based AFLP analysis was more disposed to form some clusters than *Pst*I-based AFLP markers (this was also noted in tomato and maize mapping). In connection with these observations, it is presumed that *Eco*RI-based AFLPs are more suitable for map saturation, while *Pst*I-based AFLP markers have a better genome distribution. AFLP markers were also used for integration of the genetic maps across different populations in *B. oleracea* (Sebastian et al. 2000). They were useful in filling gaps in the integrated map, but it was noticed that a significant number of AFLP markers used was located near the centromer regions. This finding can be explained by the repetitive DNA presence in these regions (Sebastian et al. 2000).

The main disadvantage of AFLP markers is that they mostly represent the dominant type of inheritance. Also methylation of genomic DNA can result in false positive polymorphism when the restriction enzymes are sensitive to methylation (Li and Quiros 2001). AFLP technique has been widely used to target specific loci but the assay is too complicated to be adapted for high-throughput selection, marker-assisted selection (MAS) or map-based cloning. This particular marker, AFLP, can be converted into a PCR-based one, such as CAPS or SCAR (Piao et al. 2004).

4.2.4 Sequence-Characterized Amplified Region and Cleaved Amplified Polymorphic Sequence

These two types of PCR-based markers were especially useful in the *Brassica* genome analysis when special agriculturally important genes were searched (Table 4-1). Clubroot disease caused by *Plasmodiophora brassicae* is a very dangerous infection of the *Brassica* crops including Chinese cabbage (*B. rapa* ssp. *pekinensis*) (Piao et al. 2004) (for more information about clubroot disease in the *Brassica* crops see Chapters 2 and 5). The source of resistance in *Brassica* has been identified as *CRb* gene; it has been mapped using AFLP markers. A few AFLP markers, which mapped near the resistance gene, were selected for conversion into sequence-characterized amplified region (SCAR) and cleaved amplified polymorphic sequence (CAPS) markers. AFLP markers were sequenced from both ends and specific primers served for development of SCAR markers. SCAR markers did not give polymorphic products, so the products obtained were digested by restriction enzymes. With the help of this kind of markers a local map around *CRb* gene was constructed.

B. juncea known as Indian mustard is a very important crop in China, however, it is often affected by white rust, a dangerous crucifers disease caused by *Albugo candida*. Breeding for resistance against this pathogen has not been successful although sources of resistance are available. The gene responsible for white rust resistance was mapped for *B. rapa* (Kole et al. 1996, 2002) and *B. juncea* (Prabhu et al. 1998, Mukherjee et al. 2001; Varshney et al. 2004), but RFLP and RAPD markers used for the target gene are not appropriate for marker-assisted selection (MAS). Therefore, it was necessary to develop markers more tightly linked with pathogen resistance gene and produce SCAR/CAPS markers (Varshney et al. 2004). CAPS markers were generated from RAPD markers by the way described above.

SCAR and CAPS markers are very useful in map-based cloning (MBC) and marker-assisted selection (MAS). The markers of these types can be also used for screening plants for specific genes in breeding programs. SCAR and CAPS markers are quite fast in the search for resistance genes and importantly, many markers of this type are codominant.

4.2.5 Simple Sequence Repeats

The simple sequence repeats (SSRs), known as microsatellites, are short DNA sequences in which a motif of one to six bases is tandemly repeated. Their origin is closely related to the errors appearing during DNA replication and recombination process (Goldstein and Schlötterer 1999). Subsequently, additional sets of these repeated sequences are added to the strand. Many studies indicate that SSRs are abundant and their distributions differ not only among species but also between transcribed and non-transcribed regions both in plants and mammals (Katti et al. 2001; Morgante et al. 2002; Toth et al. 2000). The analyses of plant DNA sequences revealed a frequency of one SSR every 6–7 kb, which is equivalent to the corresponding data for mammals, while the studies based on oligonucleotide hybridization suggested the presence of one SSR every 65 to 80 kb (Panaud et al. 1996; Cardle et al. 2000). On the other hand, analysis of the microsatellite motifs distribution in the *B. oleracea* genomic shotgun sequences brought the frequency of one SSR every 4 kb (Iniguez-Luy et al. 2008).

SSRs have many characteristic features such as: a high level of polymorphism resulting from length variability, codominant Mendelian inheritance, almost random distribution in eukaryotic genomes, locus-specificity, conservation among closely related species and relative easiness of isolation and characterization (Weber and May 1989; Beckmann and Soller 1990; Queller et al. 1993). Hence, they appeared as a basic marker type for construction of the genetic maps (for example Dietrich et al. 1994; Dib et al. 1996; McCouch et al. 2002; Piquemal et al. 2005; Suwabe et al. 2006). Furthermore, SSR markers were used in linkage and association analyses (Abdurakhmonov et al. 2005), phylogenetics (Plieske and Struss 2001; Chung and Staub 2003; Flannery et al. 2006) and population studies (Rosenberg et al. 2002). Because of the duplicated nature of many genomes, the microsatellite sequences could occur in multiple copies with similar flanking regions forming families (Zhang 2004). Thus, many primer pairs could amplify two or more loci. On the other hand, identification of the microsatellite motifs within gene sequences made it possible to distinguish their duplicated copies (Babula-Skowronska et al., unpublished).

This wide range of SSR use in genetic studies initiated development of microsatellite discovery programs for the majority of agronomically important crops (Condit and Hubbell 1991; Akkaya et al. 1992). Currently, in the public domain 1,545 SSRs for *Brassica* are available, which have been isolated from different *Brassica* genomes or transferred from *A. thaliana* (Szewc-McFadden et al. 1996; Westman and Kresovich 1998; Plieske and Struss 2001; Saal et al. 2001; Lowe et al. 2002; Suwabe et al. 2002; Table 4-3). Identification of SSRs in the *Brassica* genomes is based on probe hybridization to genomic or cDNA libraries followed by DNA sequencing (Kresovich et al. 1995; Plieske and Struss 2001; Suwabe et al. 2002; Lowe et al. 2004). Recently, SSRs were identified for *B. rapa* based on the bacterial artificial chromosome (BAC)-end sequences selected from the KBrH library (Choi et al. 2007). A dynamic increase in the number of SSR sequences in the *Brassica* database allowed development of many genetic markers, which were added to the framework RFLP maps or became the basis of new SSR-based linkage maps (Sebastian et al. 2000; Piquemal et al. 2005; Choi et al. 2007; Gao et al. 2007). As SSR markers are highly transferable with high reliability among related species, they are used for integration of genetic maps. The first experiments concerned integration data from two different segregating populations of *B. oleracea* (Sebastian et al. 2000; Babula-Skowronska et al., unpublished) and integration of A-genome maps in *B. napus* and *B. rapa* (Choi et al. 2007). Moreover, SSRs developed from the *B. oleracea* sequences were used for inspection of genetic relationships within the clade (Tonguc and Griffiths 2004) and in chloroplast genomes of this species and wild relatives (Allender et al. 2007), while the SSRs from *B. napus* were used for finding their distribution in Brassicaceae such as *B. oleracea*, *B. rapa*, *B. nigra*, *Diplotaxis* ssp., *B. tournefortii*, *Sinapis alba*, *Raphanus sativus* and *Eruca sativa* and for estimation of the extent of genetic diversity among 32 varieties of *B. napus* (Plieske and Struss 2001). These data indicated high variation potential of the species of U triangle, but low transferability of microsatellite markers into distantly related Brassicaceae species. Also, the distribution, putative function, and mutational mechanisms of SSRs' evolution have been studied (Ayele et al. 2005; Li et al. 2002; Iniguez-Luy et al. 2008).

4.2.6 Sequence Tagged Sites

Sequence tagged sites (STS) belong to PCR-based markers, characterized by a high specificity. There are supposed to be only a single locus in a genome for particular STS, if properly selected. Markers of this type are often used to increase density of genetic maps. The reference linkage map for the *Brassica* A genome developed in the multinational *B. rapa* sequencing project, also includes some STS markers (Choi et al. 2007; Table 4-1). The primer pairs for

Table 4-3 Summary of the *Brassica* SSR markers based on the published information and data available in the public domain (*www.brassica.info*).

Source of of SSR data	References	Number of SSR					Total number of SSR
		B. rapa	***B. nigra***	***B. oleracea***	***B. napus***	***A. thaliana***	
BBSRC Microsatellite Program	Lowe et al. 2002; Lowe et al. 2004	89	113	95	101		397
HRI Microsatellites	Smith and King 2000			6			6
	Kresovich et al. 1995; Szewc-McFadden et al. 1996				24		24
	Lagercrantz et al. 1993				5		5
AAFC BAG	Plieske and Struss 2001; Saal et al. 2001				81		80
	Suwabe et al. 2002	38					38
	Uzanova and Ecke 1999				8		8
	Bell and Ecker 1994					30	30
INRA Versailles						100	100
Celera AgGen *Brassica* Consortium	Piquemal et al. 2005						372
FITO (Wisconsin)	Iniguez-Luy et al. 2008						587
PBC	Choi et al. 2007				470		470

STS markers were designed from the mapped genomic sequences (Teutonico and Osborn 1994) and BAC-end sequences.

To overcome limitations connected with RFLP and AFLP markers, such as high costs and time consuming procedures or not enough markers associated for genes crucial in breeding programs, a way should be found to transform RFLPs and AFLPs into PCR-based markers, such as STS. This was found in studies on finding markers associated with *Phoma* resistance genes in the *Brassica* B genome (Plieske and Strus 2001).

4.2.7 Sequence-Related Amplified Polymorphism

Sequence-related amplified polymorphism (SRAP) is a simple PCR-based marker system developed by Li and Quiros (2001) (Table 4-1). It is based on amplification of genomic sequences with two specific primers (for details see Li and Quiros 2001). SRAP marker technique allows for coding sequence investigation and easy sequencing of selected bands. Markers of this type have a good distribution in the genome because they are based on open-reading frames (ORFs), so they have been used in the *B. oleracea* map construction together with AFLP markers (Li and Quiros 2001). SRAP markers were placed in the nine main linkage groups of *B. oleracea* resulting in no significant difference in genome coverage obtained for AFLP. SRAP markers revealed relatively high frequency of codominant type of inheritance, which is a very important attribute. Using the resultant map, SRAP markers were used for tagging of glucosinase desaturation gene (*BoGLS-ALK*). SRAP133 marker was located inside this gene that is important for MAS. A MAS allows fast screening of numerous segregating populations. SRAP markers have also many other applications. They can generate high level of polymorphisms in many other crops and they were used to find markers for male sterility in Chinese cabbage, *cms* fertility restorer gene in rapeseed and virus resistance gene in celery (Li and Quiros 2001). SRAP markers have also found application in cDNA fingerprinting. In *B. rapa* some tissue-specific bands were identified with markers of this type, which corresponded to genes expressed only in the tissue investigated (Li and Quiros 2001).

SRAP markers were also used in investigation of markers for seed coat color gene in *B. rapa*. The color of the *Brassica* seeds varies from dark brown to bright yellow. The yellow color of the seeds is most desirable as the seed coat in this color is thinner and yellow seeds contain more oils, more proteins with lower fiber content, which is important for its use as fodder. Thus, it would be important to identify genes responsible for seed coat color. Three such genes were identified in *B. rapa*. SRAP marker linked with the seed coat color gene (*Br1/br1*) was found. On the basis of this SRAP marker, the SNP and SCAR markers were obtained. The latter can be useful in MAS for monitoring genotypes in the early stage of development and in selection of yellow seeds coat color alleles configuration expected (Rahman et al. 2007).

4.2.8 *Single Nucleotide Polymorphism*

Single nucleotide polymorphisms (SNPs) resulting from point mutations in DNA sequence occur abundantly in the genome. Analysis of SNPs allows identification of allelic variation of particular genes. SNPs are widely used in human disease diagnostics, they have a great potential in genetic mapping and genome analysis in many different organisms also in vegetables *Brassica* (Snowdon and Friedt 2004; Table 4-1).

SNP markers were used in the comparative genomics study of *B. rapa* and *A. thaliana* to screen a specific region responsible for clubroot resistance in *Brassica* (Suwabe et al. 2006). SNPs were also used to analyze the seed color genes in F_2 and BC_1 of *B. rapa* breeding materials (Rahman et al. 2007).

In future, SNPs are expected to be an important marker system in the *Brassica* genetic studies. Knowledge of genome sequences of *Brassica* could be used to develop locus-specific SNP markers. Before it happens, it is necessary to develop tools and methods for SNP detection and interpretation in polyploid genomes of *Brassica*. New high-throughput methods like matrix-assisted laser desorption/ionization time of flight mass spectrometry (MALDI-TOF) may be adapted to SNPs detection. It allows reduction in the time and costs of SNP analysis (Snowdon and Friedt 2004).

4.3 Development of Public Marker Resources for *Brassica*

Together with the start of the Multinational *Brassica* Genome Project (MBGP *Brassica*) the information platform at the web pages has been made accessible (*www.brassica.info, http://brassica.bbsrc.ac.uk*). The platform facilitates exchange of information from the *Brassica* genome analyses at the genetics, genomics, proteomics and metabolomics levels between different laboratories. Besides the above data, the platform also includes the sets of genetic markers mapped in the hitherto published *Brassica* linkage maps. Currently, a total of 2,665 *Brassica* sequence tagged markers including 584 RFLPs, 1,545 SSRs, 536 SNP/InDel, intron polymorphism and ACGM plus over 1,600 additional AFLP and RAPD markers have been accumulated. Additionally, the detailed description for microsatellite sets has been placed at the microsatellite information exchange *http://www.brassica.info/resource/markers/ssr-exchange.php*. Here, the basic information concerning primer sequences, repeat classes and resources has been given. The majority of SSRs have been isolated and described by BBSRC Microsatellite Program,

Celera Consortium (commercial), INRA Versailles and AAFC Consortium (commercial). This characterizes predominantly SSRs from *B. rapa*, *B. nigra*, *B. oleracea* and *B. napus*. The remaining SSRs have been developed and characterized by several small research groups (Lagercrantz et al. 1993; Bell and Ecker 1994; Kresovich et al. 1995; Szewc-McFadden et al. 1996; Uzanova and Ecker 1999; Saal et al. 2001; Plieske and Struss 2001). Many of the isolated SSRs have been localized on the *Brassica* genetic maps (for example Sebastian et al. 2000; Piquemal et al. 2005; Choi et al. 2007; Gao et al. 2007; D. Babula-Skowronska et al., unpubl). Increased mapping data covering markers for the existing *Brassica* linkage maps, dramatically enriched this database.

4.4 Nomenclature of Genetic Markers and Genes

The marker names, accessions and orientation of the linkage groups in the *Brassica* species were arbitrary, because different nomenclatures for markers, genes and their paralogs were used by different researchers (Østergaard and King 2008). Hence, many markers with different names located on the different linkage maps might represent the same locus. The lack of standardized nomenclature for linkage groups and common probes in the *Brassica* species can hinder the exchange of genetic information and comparison of the previously published maps. Therefore, the Steering Committee for the Multinational *Brassica* Genome Project recommended a new nomenclature system for the *Brassica* linkage groups, markers and genes (*http://brassica.info/resource/maps/lg-assignments.php*; Østergaard and King 2008). The "single letter marker type designator codes" system for marker names has been adopted after De Vicente et al. (2004). According to this proposition, the single letter marker assay type designate: a-AFLP, e-EST, f-RFLP, m-SSR, p-CAPS, r-RAPD and t-STS, c-SCAR, s-SNP, x-other and u-unspecified. Østergaard and King (2008) proposed a system of naming the linkage groups assigned to the three diploid *Brassica* genomes. Here, the consensus nomenclature of the chromosome/linkage group (A1–A10, C1–C9 and B1–B8 in *B. rapa*, *B. oleracea* and *B. nigra*, respectively) has been assigned to the canonical diploid *Brassica* genomes in the "triangle of U" as: A, B and C. It should facilitate integration of genetic linkage maps constructed by different research groups using different mapping populations and different sets of molecular markers. The standard systematic gene nomenclature system for the *Brassica* genus takes the form: <GENUS 1 LETTER> [<species 2 letters>]<GENOME 1 LETTER> | <X>.<NAME 3–6 LETTER CODE>.<locus assignment 1 letter>. An example is the orthologue of the *Arabidopsis* ABA INSENSITIVE 1 (*ABI1*) gene isolated from the C genome of *B. napus*: *BnaC.ABI1.a*.

4.5 Mapping Populations

4.5.1 Mapping Populations in Vegetable Brassicas Genetics

Genetic mapping requires a suitable molecular marker system, software for computational data analysis and specific types of plant mapping populations. There are several types of mapping populations that are widely used in genetic linkage mapping: F_2, backcrosses, doubled haploids (DHs), recombinant inbred lines (RILs) and near-isogenic lines (NILs). Of crucial importance for mapping populations development is selection of parents. The parents of the mapping population must show sufficient variation in the traits of interest at both the DNA sequence and phenotype level. The choice of appropriate mapping population must be made in view of the specific aim of genetic map construction and the types of genetic markers which are to be used. If one wants to map specific loci, short-term mapping populations like F_2 or backcrosses may be used, but if one wants to pursue a global genome mapping, long-term mapping, populations like DHs, NILs or RILs should be selected (for more details about the type of mapping populations see Hittalmani et al. 2002 and *http://www.scribd.com/doc/6229849/Mapping-Population* , Singh and Prasanna). The number of lines of mapping population is also very important and depends on the type of population and traits which are to be mapped. In genetic map construction for vegetable Brassicas many different mapping populations have been used (Table 4-4). According to enormous morphological diversity in the genus *Brassica*, there are many possibilities for parental selection. Intraspecific populations reveal high numbers of segregating loci like in *B. oleracea* but sometimes only in interspecific populations high polymorphism level is observed (Kianian and Quiros 1992a). Many details about the main existing mapping populations for the genus *Brassica* are available at *http://www.brassica.info/resource/plants/mapping-populations.php*.

4.5.2 Vegetable Brassicas Mapping Populations

4.5.2.1 F_2 Population

F_2 population usually is created by self-pollination or sib mating of F_1. F_2 lines are the products of single meiotic cycle and consequently, the ratio for the dominant marker is 3:1 and for the codominant marker it is 1:2:1. F_2 population can be developed relatively easily in a short time so it is a good population for preliminary mapping. This population cannot be used for quantitative traits loci (QTL) mapping because each individual is genetically different, which prevents their use in replicated trials over locations and years. Also exact replicas of lines cannot be obtained as it is a short-term population.

Table 4-4 Selected populations used in the *Brassica* genetic and breeding (based on *http://www.brassica.info/CropStore/populations.php, with modifications*).

Species	Population name	Population type	Number of lines	Parental lines	Marker type	Distortion of marker segregation	Authors
B. rapa	Turnip x mock pak-choi	F_2	61	turnip (*B. rapa* ssp. *rapifera* cv "Yorii Spring") and mock pak-choi (*B. rapa* ssp. *parachinensis*, cv "Kwan Hoo Choi"	Isozymes RFLP	6 loci with distorted segregation ratios, towards of maternal (turnip) allele	McGrath and Quiros 1991
	'Michihili' x Spring broccoli	F_2	95	Chinese cabbage 'Michihili' (*B. rapa* ssp. *pekinensis*) and Spring broccoli (*B. rapa* ssp. *utilis*)	RFLP	8 loci with distorted segregation ratios	Song et al. 1991
	'Horizon' x 'R500'	F_2	104	Canola-type cultivar "Horizon" and yellow sarson type "R500" (*B. rapa* ssp. *tricolaris*)	RFLP	85 loci with distorted segregation ratios, towards the "Horizon" allele	Chyi et al. 1992
	Per x R500	F_2 (F_3)	100	*B. rapa* biennial cultivar Per and annual cultivar R500	RFLP	33 (23 %) loci with distorted segregation ratios; 30% of loci towards the "R500" allele, 40% of loci towards the Per allele	Teutonico and Osborn 1994
	Per x R500	(F_6) RIL	87	Winter turnip rape (Per)(*B. rapa* ssp. *oleifera*) and a spring yellow sarson (R500)(*B. rapa* ssp. *tricolaris*)	RFLP	28 loci with distorted segregation ratios; 57 % of loci towards the „R500" allele	Kole et al. 1997
	BraAGF2	F_2	94	Two DH lines of *Brassica rapa* ssp. *pekinensis*, Nou 7 (a clubroot-resistant line) and G004 (susceptible Chinese cabbage line)	RFLP SSR RAPD	15% of the total loci with distorted segregation rations	Suwabe et al. 2003; 2006; 2008
	JWF3p	F_2	134	Developed from Chinese cabbage F_1 cultivar Jangwon (*B. rapa* ssp. *pekinensis*)	RFLP SSR		Kim et al. 2006

	R-o-18 x B162	F_2	114	*B. rapa* accessions B162 and Yellow-Sarson line (R-o-18)	AFLP SSR	16% of the total loci with distorted segregation rations	Soengas et al. 2007
	BraCKDH	DH	78	Two different inbred lines of Chinese cabbage (*B. rapa* ssp. *pekinensis*), "Chiifu-401-42" and "Kenshin-402-43"	AFLP SSR RAPD ESTP STS CAPS	38.3% of the total loci with distorted segregation rations, generally towards male parent "Kensin" allele	Choi et al. 2007
B. nigra	BniTIF2	F_2	83	B1164 selected from theTurkey population and B1157 selected from the India population	Isozymes RFLP RAPD	Towards one of parental alleles in the cluster	Truco and Quiros 1994
	Catania x RC	BC	88	A rapid cycling accession (CrGC) and wild Italian accession (Catania)	RFLP	9% of the total loci with distorted segregation rations, half of the deviations were skewed towards a recurrent parent allele	Lagercrantz and Lydiate 1995; Lagercrantz 1998
B. oleracea		F_2	96	"Packman" broccoli (*B. oleracea* ssp. *italica*) and "Wisconsin Golden Acres" cabbage (*B. oleracea* ssp. *capitata*)	RFLP		Slocum et al. 1990
	C x CA	F_2	>60	Collard (*B. oleracea* ssp. *acephala*) and cauliflower (*B. oleracea* ssp. *botrytis*)	RFLP	9.7% of the total loci with distorted segregation ratios	Kianian and Quiros 1992a
	C x B	F_2	>60	Collard (*B. oleracea* ssp. *acephala*) and broccoli (*B. oleracea* ssp. *italica*)	RFLP	5.2% of the total loci with distorted segregation ratios	Kianian and Quiros 1992a
	U x CA	F_2	>60	Wild kale from USSR(*B. oleracea* ssp. *acephala*) and cauliflower (*B. oleracea* ssp. *botrytis*)	RFLP	12.2% of the total loci with distorted segregation ratios	Kianian and Quiros 1992a

Table 4-4 contd...

Table 4-4 contd...

Species	Population name	Population type	Number of lines	Parental lines	Marker type	Distortion of marker segregation	Authors
	K x I	F_2	52	Kohlrabi (*B. oleracea* ssp. *gongylodes*) and *B. insularis*	RFLP	59% of the total loci with distorted segregation ratios, towards maternal parent (kohlrabi) allele	Kianian and Quiros 1992a
		F_2	120	cabbage line (No. 86-16-5) resistant to race 2 of *Plasmodiophora brassicae* (Woronin) and a rapid cycling line (CrGC No. 85)	RFLP QTL Morphological RAPD SCAR	Towards one or the other parental RFLP allele	Landry et al. 1992; Cheung et al. 1997
	BolAGDH	BC_1	296	A12DHd (*B. oleracea* ssp. *alboglabra*) and GDDH33 (*B. oleracea* ssp. *italica*)	RFLP RAPD	7 loci with distorted segregation ratios	Ramsay et al. 1996
	BolAGDH	DH	169	A12 (rapid cycling Chinese kale line; *B. oleracea* ssp. *alboglabra*) and GD (calabrese; *B. oleracea* ssp. *italica*)	RFLP		Bohuon et al. 1996
		F_2	124	Cabbage (*B. oleracea* ssp. *capitata*) and broccoli (*B. oleracea*. ssp. *italica*)	RFLP RAPD	22% of the total loci with distorted segregation ratios	Camargo et al. 1997
		F_2	56	RCB (self compatible) and *B. oleracea* var. Green Comet	RFLP		Lan et al. 2000
		F_2	247	RCB and *B. oleracea* var. Cantanese	RFLP		Lan et al. 2000
		F_2	250	RCB and *B. oleracea* var. Pusa Katki	RFLP		Lan et al. 2000

		F_2	246	RCB and *B. oleracea* var. Bugh Kana	RFLP		Lan et al. 2000
	BolNGDH	DH	97	DJ7032 (cauliflower line, Nedcha; *B. oleracea* ssp. *botrytis*) and DJ3753 (Brussels sprout hybrid Gower; *B. oleracea* ssp. *gemmifera*)	AFLP RFLP		Sebastian et al. 2000
		F_2	143	Broccoli "Early-Big" (*B. oleracea* ssp. *italica*) and cauliflower "An-Nan early" (*B. oleracea* ssp. *botrytis*)	AFLP		Li et al. 2003; Gao et al. 2007

Populations of this type were widely used in the initial mapping experiments for vegetables Brassicas. From amongst *B. rapa* Chinese cabbage cultivars, Michihili and Spring broccoli were selected as female and male parents, respectively to generate the mapping population (Song et al. 1991; Table 4-4). A set of 95 F_2 lines was analyzed to form a map- based on RFLP markers.

F_2 populations were often exploited to construct maps and to map genes or markers linked to agronomically important traits in *B. oleracea*. For example F_2 population composed of 96 individuals derived from a cross between "Packman" broccoli (*B. oleracea* var. *italica*) and "Wisconsin Golden Acres" cabbage (*B. oleracea* var. *capitata*) was used as a base for the first RFLP linkage map for *B. oleracea* (Slocum et al. 1990; Table 4-4). Four different F_2 populations were used for analysis of linkage arrangements and evolutionary implications in *B. oleracea* (Kianian and Quiros 1992a; Table 4-4). Three intraspecific hybrids were used to get F_2 populations: collard (B115) x cauliflower (B265), collard (B115) x broccoli (B008) and wild kale (B1661) x cauliflower (B265), each of the populations covered a set of 60 individuals. The fourth F_2 population, consisting of 52 lines was developed by interspecific cross between kohlrabi (B225) and *B. insularis* (B364). Combination of these populations was used for studying segregation of RFLP loci and map association. On this basis it was possible to observe distortions in the segregation ratios, gene distance as well as duplications of genes and whole genome allowing evolutionary implications. Sometimes F_2 populations are used for mapping after selfing, in the form of F_3 generations ($F_{2:3}$). $F_{2:3}$ population is used in particular applications such as mapping of recessive genes or sometimes QTLs, but like F_2 population, it is also not "immortal" (Singh and Prasanna *http://wwwscribdcom/doc/6229849/Mapping-Population*). F_3 population can also be used when plants from F_2 generation are not available (e.g., when F_2 plants are too old). Such a conversion of F_2 population into F_3 generation was applied to chromosomal mapping in *B. oleracea* with expressed sequence tags (ESTs) from *A. thaliana*) (Babula et al. 2003). F_3 generation was derived from F_2 population produced by the above-mentioned cross between collard (B115) and cauliflower (B265). The F_3 generation (originated from F_2 population developed from Chinese cabbage "Jangwon" *B. rapa* ssp. *pekinensis*) was also used to create sequence-tagged linkage map of *B. rapa* (Kim et al. 2006; Table 4-4).

4.5.2.2 Backcross Mapping Population

Backcross mapping population is obtained in the process, which begins with the cross between the chosen female and male parents. After obtaining F_1 generation, one single plant is selected to pollinate the plant used as a female parent. As a consequence, the first backcross generation comes into existence (BC_1). Backcross populations can be developed in a relatively short time although they are also not "immortal" populations. They are applied for marker-assisted breeding and in QTL mapping (Hittalmani et al. 2002; *http://www.brassica.info/resource/plants/mapping-populations.php*).

Backcross population was supposed to be promising for fine-mapping QTL in *B. oleracea* (Bohuon et al. 1996). Considering the presence of short fragment of chromosome introgressed into the recipient genome, QTLs may be mapped with very fine exactness. The starting point for the special population design, was a well known cross between two doubled haploid lines of *B. oleracea*, A12DHd (A12) and GDDH33 (GD) (more details about the genetic cross between A12DHd and GDDH33 in Section 4.5.2.3; Bohuon et al. 1996; Table 4-4). A12 was used as a female parent (recipient of the fragment of chromosome) and GD was used as a donor of introgressed genetic material). One F_1 individual was chosen to pollinate three A12 plants to obtain the first backcross generation (BC_1). In this backcross, BC_1 population has a cytoplasm from A12. This process was repeated for the second backcross (BC_2) generation. The 176 individuals from BC_1 were used for segregation analysis of RFLP and genetic map construction in order to reveal appropriate linkage between markers and identify the region of the genome with introgressed chromosomal fragment. The resultant map was compared with the previous maps such as those created by Bohuon and colleagues and particular individuals from BC_1 were selected to be the parents of BC_2. The approach with marker-assisted selection proved to be a successful way for generation of backcross mapping population (Ramsay et al. 1996). The same A12xGD backcross population was used for mapping QTLs for flowering time control in *B. oleracea* (Rae et al. 1999).

4.5.2.3 Doubled Haploids

Double haploids (DHs) are developed by culturing microspores or anthers. DHs result from one meiotic cycle similarly as F_2 population, but the expected ratio for both types of markers is 1:1 (dominant and codominant). This is an "immortal" population because it can be replicated in different places and years without any genetic change. This stability guarantees the possibility of comparisons of the results obtained in different experiments. DH populations are ideal for QTL and global genome mapping (Hittalmani et al. 2002; Singh and Prasanna *http://www.brassica.info/resource/plants/mapping-populations.php*).

Embriogenetic nature of the *Brassica* microspores allowed a relatively rapid production of DH lines (Bohuon et al. 1996). With reference to many maps for *B. oleracea* based on F_2 populations, DH mapping populations were indispensable for data comparison and integration. In *Brassica* and Oilseeds Research Department, John Innes Centre in Norwich, UK, a DH population was produced for *B. oleracea*, which became one of the most widely used populations in the *B. oleracea* research, known as A12xGD or BolAGDH. A plant from DH lines of *B. oleracea* var. *alboglabra* (A12DHd) was pollinated by pollen from a single plant from the microspore-derived DH line of *B. oleracea* var. *italica* (GDDH33) to obtain a set of genetically identical F_1 offspring. From a cross between two highly polymorphic lines a set of 196 DH individuals was obtained. This population was used for comparative analysis of conserved fragments C genome between *B. oleracea* and *B. napus*, based on RFLP probes (Bohuon et al. 1996).

DH populations were used to make the first integrated linkage map for *B. oleracea* based on two different F_1-derived DH populations: the above-mentioned A12xGD and the second one named NxG. NxG (also known as BolNGDH) obtained by the cross between F_1-derived DH lines DJ7032 (*B. oleracea* var. *botrytis*) and DJ 3753 (*B. oleracea* var. *gemmifera*) (Sebastian et al. 2000; Table 4-4; See Section 4.7.3). The integrated map is much more saturated than the initial maps. The resultant map can be used for identification of homeologous loci and collinear chromosomal segments. Two types of genetic markers were used for integration of data from two DH populations A12xGD and NxG. RFLP, AFLP and microsatelite markers were analyzed for both populations and after separate maps construction, integration was achieved by using JOINMAP software.

DH populations were frequently produced to study a particular feature, not the whole genome. A DH population was used for identification of two loci for resistance to clubroot in *B. rapa* (Suwabe et al. 2003; Table 4-4). Two DH lines were used as parents, "G004" and "A9709". Line G004 was a clubroot resistance line developed by crossing "Homei P09" (a line with a high ability to regenerate from microspores) and "Siloga S2" (a line derived by selfing of the clubroot resistant cultivar of European fodder turnip). The second parental line A9709 was clubroot-susceptible derived from a cross between two cultivar lines of Chinese cabbage: "Homei P09" and "Nozaki 2". A cross between two parental lines "G004" and "A9709" gave an F_2 population, which was used for identification of the clubroot resistance loci.

DH populations were also used in map construction for oilseed crops: Indian mustard (*B. juncea*) and canola (*B. napus*). For *B. juncea* genome analysis, DH populations were obtained through a microspore culture from F_1 generation derived from the cross between Varuna and Heera (Pradhan et al. 2003). Parental lines differed significantly because they represented two distinct gene pools. Varuna was the crop that was largely widespread in India and Heera was the East European mustard line with canola quality. The DH population obtained was a set of 123 lines. It was used to produce high density genetic linkage map for Indian mustard based on RFLP and AFLP markers. The same DH population was used in the studies of QTL directly related to the yield (Ramchiary et al. 2007). This population was also used in comparative mapping and evolutionary analysis of *B. juncea* and *A. thaliana* (Panjabi et al. 2008). Also, for *B. napus* many DH populations are available. One of the first DH populations for *B. napus* was constructed to map loci connected with controlling vernalization requirement and flowering time (Ferreira et al. 1994, 1995). TNDNDH population was obtained by crossing *B. napus* line Ningyou7 with Tapidor and then some F_1 individuals were used to obtain DH lines (Qiu et al. 2006). Another DH population for *B. napus* was produced by Parkin and colleagues (Parkin et al. 1995; 2005). It was the DH population from the cross between the winter breeding line CPB87/5 and the newly resynthesized *B. napus* line SYN1 (interspecific cross between *B. rapa* and *B. oleracea*). This population was used for identification of the A and C genomes of amphiploid genome *B. napus*.

4.5.2.4 Recombinant Inbred Lines

Recombinant inbred lines (RILs) can be produced in two ways: by prolonged self-pollination and by sib-mating of the offspring of individual members of F_2 population to obtain fully homozygous lines (Broman 2005). After many rounds of crosses, RILs revealed the 1:1 genetic segregation ratio like DHs. RILs are obtained after many meiosis events so they are mosaics of parental genome, which makes them excellent for detection of closely linked markers. RILs have great potential in QTL analysis because of their genome structure and "immortal" character. They can be used over many seasons in different environmental conditions, which is significant for QTL mapping (Kole et al. 1997). For more details about RILs see Broman (2005) and Hittalmani et al. (2002).

The mapping populations of these types were used in construction of genetic linkage map for *B. rapa*. The F_2 populations were derived from the cross between *B. rapa* cultivars, Per (a winter turnip rape) and R500 (a spring yellow sarson), which differed in many features (e.g., seed color, leaf pubescence, erucic acid content, vernalization requirement, flowering time, resistance to white rust, winter survival and freezing tolerance). RILs used for map construction were the F_6 generation with high ratio of homozygosity (Kole et al. 1997; Table 4-4).

4.5.2.5 Near-Isogenic Lines

Near-Isogenic Lines (NILs) are generated by repeated self-pollination or by backcrossing of the F_1 plants to the recurrent parents (Singh and Prasanna). NILs can be used to reveal particular gene effects and to display potential interactions between genes and environment. They are often used for QTL mapping (Yano and Sasaki 1997).

NILs are not frequently used for genetic studies in vegetable Brassicas like *B. oleracea* (Ping et al. 2007) or *B. rapa* (Muangprom and Osborn 2004). They were exploited for genome analysis in *B. napus* (Foisset et al. 1995; Li et al. 2006).

4.6 Mapping Software Used

The construction of genetic maps is based on the segregating data collected for experimental or natural populations. Together with the development of molecular techniques, the rapid increase in the number of loci has prompted development of specialized computer programs. Currently, different computer software packages are available for estimation of recombinant frequencies and map construction, among them the most popular are MapMaker (Lander et al. 1987; Lincoln et al. 1992) and JoinMap (Stam 1993). For the *Brassica* species most maps have been constructed using both the programs. The first maps of *Brassica* were developed by using the MapMaker program (for example Song et al. 1991; Chyi et al. 1992; Kole et al. 1997). However, construction of high-density and consensus maps, by consolidation of all segregating data derived from different mapping populations, required more advanced approaches, such as JoinMap, adapted by the majority of research groups (for example, Sebastian et al. 2000). Both programs use a similar approach for constructing maps and the same types of mapping populations: BC_1, F_2, RILs, doubled haploids, and the outbreeders full-sib family. They are based on efficient algorithms that allow simultaneous multipoint linkage analysis of numerous loci. Hence, the loci identification and order within the linkage groups obtained by these two programs are similar. The only major difference noted was the length of the maps. For example, Qi et al. (1996) reported that the length and observed genome coverage with loci mapped was greater when obtained with MAPMAKER than with JOINMAP, which resulted from different algorithms used in the map distance calculation. Furthermore, JoinMap allows integration of the segregating data from different populations into a single consensus map. Construction of the genetic map is performed in a few steps. Initially, linkage analysis is performed to detect which loci are linked. Linkage groups are assembled using two-point data. This is done by calculating all the pairwise linkage distances among markers in the data set prepared. The basic number of the linkage groups should be equivalent to the number of chromosomes in a given species. Next, the order of the markers within a linkage group is determined using three-point analysis. Three-point analysis is used to find the most likely locus orders within a linkage group. Determination of the locus ordering on the linkage map requires the use of criteria and algorithms to find the optimal sequence of loci. The criteria include the maximum likelihood (Lander et al. 1987), the minimum sum of adjacent recombination fractions (SARF), the maximum sum of adjacent LOD scores (SALOD)(Liu and Knapp 1990), the minimum number of crossovers and the "least square locus order" (Stam 1993). Map distances in centiMorgans are calculated from recombination frequencies using the Haldane (Haldane 1919) or Kosambi (Kosambi 1944) mapping functions.

4.7 Map Construction-Map Types

Molecular marker-based genetic linkage maps in poorly characterized species such as Brassicas are fundamental tools for different research purposes. They are useful in genome structure analysis, studies of genomic relationship, diversity and genome evolution of the cultivated *Brassica* species, in identification of quantitative trait loci (QTL) and genetic markers closely linked to important traits such as disease resistance, abiotic stress, and morphological traits and in breeding programs based on marker-assisted selection (MAS). Another goal of map construction is to support the genome sequencing project. Hence, individual maps are often constructed with a specific goal in mind. The first molecular maps of the *Brassica* species were constructed in the 1980s. Currently, over 1,000 different molecular markers have been mapped onto at least 15 genetic maps of the *Brassica* genus derived from a wide range of crosses (Lim et al. 2007). The detailed description of the genetic maps available for diploid species of *Brassica* genus is given below and in Table 4-5.

4.7.1 Preliminary Maps

The preliminary genetic maps of *Brassica* were mainly based on anonymous sequences that allowed their rapid saturation. The first linkage groups of *B. rapa* were published by McGrath and Quiros (1991) (Table 4-5). The map was constructed on the basis of a set of RFLP probes located previously on the *B. oleracea* chromosomes (McGrath et al. 1990) to compare conservation of gene synteny between these species. The authors identified 49 loci using 31 RFLP probes and four isozymes, from which 31 loci were genetically linked to eight groups, covering 262 cM of the *B. rapa* map. This analysis showed extensive gene duplication; 35% of the probes used detected duplicated loci and 19% disclosed a single unique locus. Similar duplication degree of genome was obtained for *B. oleracea* (McGrath et al. 1990). The *B. rapa* map was constructed on the basis of the distinct subspecies cross, hence distorted loci, intraspecific gene duplications or deletions were observed (Table 4-4). These changes indicated the genome divergence within this species and could be assigned to morphological diversity within species. The first detailed maps of *B. rapa* were constructed by Song et al. (1991)

Table 4-5 Summary of the *Brassica* genetic maps.

Species	Type of marker	Source of probes	Total length (cM)	Number of probes	Number of markers	Average number of loci per probe	Average interval between markers (cM)	Authors
B. rapa	RFLP Isozyme	Genomic clones isolated from *B. oleracea* and *B. napus*, cDNA clones from *B. napus* seedlings, napin seed storage protein gene (Crouch et al. 1983)	262	35	49	1.5	5.3	McGrath and Quiros 1991
	RFLP	Three genomic DNA libraries from: "Early White" cauliflower (*B. oleracea* ssp. *botrytis*), "Wisconsin Golden Acre" cabbage (*B. oleracea* ssp. *capitata*) and "WR 70 days" Chinese cabbage (*B. rapa* ssp. *pekinensis*)	1,850	188	280	1.3	6.6	Song et al. 1991
	RFLP	A *Pst*I library of "Westar" (*B. napus* L.)	1,876	269	360	1.3	5.2	Chyi et al. 1992
	RFLP	A *Pst*I genomic DNA library and a cDNA library from the *B. napus* cultivar Westar; *Eco*RI genomic DNA library from *B. rapa* cultivar Tobin, cloned genes from *A. thaliana*	1,785	116	139	1.2	13.5	Teutonico and Osborn 1994
	RFLP	A *Pst*I genomic DNA library and a cDNA library from *B. napus* cultivar Westar; *Eco*RI genomic DNA library from *B. rapa* cultivar Tobin (Teutonico and Osborn 1994); cloned genes from *B. napus* and *A. thaliana*	890	102	144	1.4	6.0	Kole et al. 1997

SSR RFLP RAPD	SSR markers describer early by Suwabe et al.(2002, 2004); RFLP (Harada et al. 1988);	1,006		262		3.7	Suwabe et al. 2006
SNP and/or indel markers	RAPD primers, single-copy genes from *A. thaliana* for SNPs and indels	743		273		3.6	Suwabe et al. 2008
RFLP SSR RAPD	*B. rapa* EST clones from four different tissue-specific libraries; GST representing genes from *A. thaliana* chromosomes 4 and 5 flowering-time related genes; SSR primers (Suwabe et al. 2002; Lowe et al. 2004)	1,287	396	545	1.3	2.4	Kim et al. 2006
AFLP SSR RAPD ESTP STS CAPS	AFLP with use *Eco*RI/*Mse*I and, *Pst*I/*Mse*I enzymes; SSR, STS and ESTP were identify from BAC-end sequences of a KBrH library of the inbred line "Chiifu-401-42" (Park et al. 2005); CAPS assay based on monomorphic products of STS and ESTP; RAPD primers from the Operon Technologies (USA) and WA (Japan)	1,182		556		2.8	Choi et al. 2007
AFLP SSR	SSR primers from Kim et al. 1999, Suwabe et al. 2002, Lowe et al. 2004 and AAFC *Brassica*/*Arabidopsis* Genomics Initiative	664	316	246	0.8	2.9	Soengas et al. 2007

Table 4-5 contd...

Table 4-5 contd...

Species	Type of marker	Source of probes	Total length (cM)	Number of probes	Number of markers	Average number of loci per probe	Average interval between markers (cM)	Authors
B. nigra	Isozymes RFLP RAPD	Genomic and cDNA clones isolated from the *B. napus* libraries, clones of known function, RAPD primers from the Operon Technologies (USA)	677	76	124	1.6	5.5	Truco and Quiros 1994
	RFLP	Genomic *Pst*I clones from *B. napus*, and *B. oleracea*, *Brassica* cDNA clones, *A. thaliana* cDNA clones	885	158	288	1.9	3.1	Lagercrantz and Lydiate 1995
	RFLP	Genomic *Pst*I clones from *B. napus*, *B. rapa* and *B. oleracea*	778	158	288	1.9 (2.6)*	2.7	Lagercrantz and Lydiate 1996
	RFLP	*A. thaliana* DNA fragments: genomic *Pst*I clones, anonymous cDNAs, genomic clones, YAC-end and gene sequences	751	160	284	1.8	2.6	Lagercrantz 1998
B. oleracea	RFLP	Genomic DNA clones derived from three different subspecies of *Brassica*	820	197	258	1.3	3.5	Slocum et al. 1990
	Phenotypic RFLP Isozyme	Clones isolated from a *B. napus* cDNA library; genome-specific probes isolated from the *B. oleracea* and *B. napus* genomic libraries	747	62	108	1.7	6.9	Kianian and Quiros 1992a
	RFLP	Probes from a library *B. napus*: embryo cDNA clones, seedling-specific cDNA, cruciferin and napin cDNA	1,112	149	201	1.3	5.5	Landry et al. 1992

RFLP RAPD STS SCAR Isozymes	cDNA clones from *B. napus*, genomic DNA from *B. oleracea*	1,606	218	310	1.4	5.7	Cheung et al. 1997
RFLP	Libraries of cloned *Brassica Pst*I fragments (Sharpe et al. 1995)	875	167	303	1.8	2.9	Bohuon et al. 1996
RFLP RAPD Isozymes	*EcoR*I genomic cDNA library from *B. rapa* cv. Tobin (pTG clones); *Pst*I genomic DNA library of *B. napus* cv. Westar (pWG clones); *Pst*I genomic DNA library from *Brassica;* S-glycoprotein and S-locus related protein; phosphoglucose isomerase	747	138	138	1.0	5.4	Ramsay et al. 1996
RFLP RAPD Self-incompatibility locus	*Pst*I genomic DNA library and cDNA library from *B. napus* cv. Westar (pWG clones) and *EcoR*I genomic cDNA library from *B. rapa* cv. Tobin (pTG clones)	921	125	159	RFLP 1.1 RAPD 2.7 per primer pair	5.6	Camargo et al. 1997
RFLP RAPD STS SCAR Phenotypic Isozymes	cDNA clones from *B. oleracea* and *B. napus* library (Harada et al. 1988; from C. Quiros)	1,606		310		5.1	Cheung et al. 1997
RFLP Isozymes	Genomic *B. oleracea* and *B. napus* libraries, cDNA library, gene specific clones	1,738		167		10.4	Hu et al. 1998

Table 4-5 contd...

Table 4-5 contd...

Species	Type of marker	Source of probes	Total length (cM)	Number of probes	Number of markers	Average number of loci per probe	Average interval between markers (cM)	Authors
	RFLP	*Brassica* RFLP probes; *A. thaliana* genome clones	875		82		13.3	Bohuon et al. 1998
	RFLP	*Arabidopsis* ESTs, *Brassica Pst*I genomic clones, *Arabidopsis* cDNA clones cloned RAPD-PCR products	862	369	367	1.50	2.35	Lan et al. 2000
	RFLP AFLP SSR	*Brassica* genomic DNA clones	889		433		3.3	Sebastian et al. 2000
	RFLP AFLP SSR	*Brassica* genomic DNA clones	893**		547		2.6	Sebastian et al. 2000
	RFLP	*A. thaliana* EST and full cDNA clones	1806;	95;	212;	2.2;	8.5	Babula et al. 2003;
			855	160	306	1.9	2.8	Kaczmarek et al. 2009
	SRAP SSR Gene-markers	Genomic and cDNA clones, broccoli BAC clones, *B. oleracea* genes	703	192	1257		0.6	Gao et al. 2007

* including monomorphic bands.
** integrated map of *B. oleracea*.

and Chyi et al. (1992) (Table 4-5). Song et al. (1991) constructed the map of *B. rapa* containing 280 RFLP markers and covering a genetic distance of 1,850 cM. A more saturated map based on diversity of parental lines was constructed by Chyi et al. (1992) (Tables 4-4 and 4-5). This map included 360 RFLP markers distributed along 10 linkage groups and covering a total of 1,876 cM. Both Song et al. (1991) and Chyi et al. (1992) observed numerous sequence duplications, which were either dispersed in the whole *B. rapa* genome or arranged in the duplicated regions. Song et al. (1991) found three linkage groups involved in the most significant duplicate linkages and similar structural rearrangements among them. These structural rearrangements within duplicated blocks were responsible for evolution of the *B. rapa* genome. Although, the next map of *B. rapa* developed by Teutonico and Osborn (1994) included 139 RFLP loci only, it has been used for comparative analysis of *B. rapa, B. napus, B. oleracea* and *A. thaliana* (Table 4-5). The loci localized on this map included three simply inherited traits such as leaf hairs (pubescence), yellow seeds and low seed erucic acid. All these loci identified were mapped into the 10 main linkage groups and one additional small group, covering a genetic distance of 1,785 cM. Like the other authors in earlier reports, Teutonico and Osborn (1994) noticed a high level of loci duplication in *B. rapa*, forming small repeated chromosomal regions. The lack of larger duplicated regions could result from accumulation of local rearrangements after duplication or limited polymorphism for the used probes. The F_2 population as the basis of this map was later developed to recombinant inbred (RI) population from the same F_1 plant (Kole et al. 1997; Table 4-4). It permitted construction of the first framework map for *B. rapa*. In the initial stage, this map included 144 RFLP loci corresponding to genomic clones, cloned genes from *B. napus* and *A. thaliana* and three QTLs controlling seed color, leaf pubescence and resistance to white rust. They covered a genetic distance of 890 cM with an average distance between loci of 6 cM (Table 4-5). Comparison of the F_2 and RI maps indicated a similar loci order along the linkage groups in these maps. Only some minor deviations in the loci order and interval distances in both maps were observed. Similarly as in the previous studies, the level of duplicated loci was high. It confirmed the complexity of the *B. rapa* genome structure.

Despite the importance of *B. nigra* as one of the progenitors of the tetraploid *B. juncea* and *B. carinata* species, only a few maps of this species have been constructed. The first of them was published by Truco and Quiros (1994). This map included 124 markers, and 122 of them were linked into 11 linkage groups with the total map distance of 677 cM and the average map density of 5.5 cM per marker (Table 4-5). This map provided information on the genome structure and complexity revealing a high degree of sequence duplication. Truco and Quiros found that 40% of the markers mapped were duplicated. Many of them were organized in intragenomic chromosomal segments and others were dispersed over the whole genome or formed tandem arrangements. It confirmed the hypothesis about the polyploid origin of the *Brassica* species (Prakash and Hinata 1980).

Slocum et al. (1990) constructed the first detailed map of *B. oleracea* including 258 RFLP markers arranged into nine linkage groups and covering a genetic distance of 820 cM (Table 4-5). The majority of the RFLP markers detected were duplicated, and these duplicated copies were located on different linkage groups or organized in tandem duplications in the same linkage group. A comparison of the loci order along linkage groups indicated a high degree of their collinearity, although some rearrangements within these duplicated regions were observed. These data supported previous cytological data indicating existence of duplicated regions in the *B. oleracea* genome. Also preliminary molecular data from analysis of morphological, isozyme and RFLP markers in series of monosomic alien chromosome addition lines of *B. oleracea* indicated a high level of gene duplication (McGrath et al. 1990). Additionally, they detected some rearrangements such as insertion or deletion in the two parental accessions ("Packman" broccoli and "Wisconsin Golden Acres" cabbage) (Table 4-4). These data were confirmed in a wide range of the *Brassica* accessions. The rearrangements identified and chromosome restructuring have occurred during the *B. oleracea* evolution. They were considered as basic mechanism for gene evolution and differential gene expression (Ohno 1970). Next, using the same F_2 mapping population, Kennard et al. (1994) analyzed location and variation of 22 morphological traits such as leaf dimensions, heading habit, internode distance and flower development in the *B. oleracea* genome. Using the earlier mapped RFLP markers in this population Slocum et al. (1990) detected most of the major genes and many minor genes controlling the morphological traits analyzed. They also observed possible duplications of trait loci by comparing them with duplicated markers identified in the earlier analysis (Slocum et al. 1990). However, no preferential associations were observed between the loci controlling morphological traits and the pairs duplicated loci. Kianian and Quiros (1992a, 1992b) constructed a composite map for *B. oleracea* based on four different populations (see Section 4.7.3). These four component maps were analyzed for the location, organization, and variation of multigene families including sequences homologous to rRNA, napin, cruciferin, self-incompatibility, isocitrate lyase and malate synthase in the *B. oleracea* genome (Kianian and Quiros 1992b). The presence of the same loci allowed comparison of segregating data and indicated the chromosomal aberrations in these populations. In parallel, another map of *B. oleracea* based on a cross between the cabbage line resistant to race 2 of *Plasmodiophora brassicae* and a rapid cycling line was developed (Landry et al. 1992). This map consisted of 201 loci including genes controlling leaf morphology and biennial flowering and covered a total genetic length of 1,112 cM. Furthermore, two dominant QTLs for resistance to race 2 of the clubroot disease were detected. This map was updated by Cheung et al. (1997) (Table 4-5). They added 136 marker loci so together with the earlier mapped loci the map included 310 loci arranged into nine main linkage groups and four smaller ones covering a genetic distance of 1,606 cM. It covered the greatest distance with more markers when compared to those used in the earlier *B. oleracea* maps, providing additional information on the *B. oleracea* genome organization. This map was compared with the *B. napus* map revealing homoeology between C genomes of *B. oleracea* and *B. napus*.

4.7.2 High-Density Maps

Fast development of different marker systems has advanced construction of the high-density genetic maps. The maps have been constructed de novo or through the extension of the already published maps with the help of additional markers, as discussed below. The new markers provided a more accurate picture of the genome organization. These extended maps have become an excellent tool for marker-assisted breeding and accurate studies of genome evolution.

Presently, some high-density maps are available for *B. rapa*. The high-density map (JWF3) of *B. rapa* published by Kim et al. (2006) includes 545 markers assigned to 10 linkage groups covering a total map length of 1,287 cM with an average interval of 2.4 cM (Table 4-5). Most of the markers were detected by RFLP analysis using the sequenced expressed sequence tag (EST) probes derived from four different *B. rapa* tissue-specific libraries. Furthermore, five homologues of *FLC* (MADS-box flowering-time regulator) genes have been located on this map. In addition, 21 SSR markers have been mapped to assign each linkage group to the existing published A-genome linkage groups of *B. rapa* (Suwabe et al. 2002) and *B. napus* (Parkin et al. 1995; Lowe et al. 2004). Assuming the average marker density of 2.4 cM and 529 Mb Johnston et al. (2005) estimated that one marker is per Mb. The authors presented the pattern of intragenomic duplications in *B. rapa* (Kim et al. 2006). They found that many chromosomal regions had two or three copies, which supported the earlier discovery described above. Also, the detailed analysis of *FLC* gene indicated its duplication. In parallel, the SSR-based map of *B. rapa* including SSR, RFLP and RAPD markers was published (Suwabe et al. 2006; Table 4-5). This map included 262 markers with a genetic distance of 1,005.5 cM. The mapped SSR markers allowed assignment of the corresponding linkage groups in the *B, rapa, B. oleracea, B. napus* and *A. thaliana* maps. Earlier, genetic analysis of two loci controlling clubroot resistance (*Crr1* and *Crr2*) in the same F_2 population had been carried out (Suwabe et al. 2003). They found close linkage between these loci and SSR markers which can be used in marker-assisted selection. Moreover, they discovered that the clubroot resistance was under oligogenic control. Hence, three QTLs were identified and mapped on this map. Another *B. rapa* map published by Soengas et al. (2007) was used for genetic analysis of QTLs controlling resistance to black rot caused by *Xanthomonas campestris* pv. *campestris* (Pammel) Dowson. This map was constructed on the basis of 223 AFLP and 23 SSR markers covering a total map distance of 664 cM (Table 4-5). The order and orientation of each linkage group were determined with respect to the previously published maps (see Kim et al. 1999 and Suwabe et al. 2004). Finally, the reference genetic map (CKDH) for *B. rapa* was constructed as support of the *Brassica rapa* Genome Sequencing Project (*Br*GSP, Choi et al. 2007; Fig. 4-1; Table 4-5). This map included the total of 556 markers distributed into nine linkage groups with a total length of 1,182 cM and an average distance of 2.1 cM between two loci. These linkage groups were identified by using common SSR markers with the reference map of *B. napus* (Parkin et al. 1995). Recently an attempt was made at construction of the *B. rapa* high-resolution genetic map via mapping more than 1,000 SSR markers derived from BAC-end sequences, ESTs and BACs (Hong et al. 2008).

Also, for *B. oleracea* some high-density maps based on different mapping populations and set of markers have been constructed (for review see Babula et al. 2007). Bohuon et al. (1996) constructed the *B. oleracea* map with 303 RFLP loci corresponding to 167 informative probes using a highly polymorphic DH population (Table 4-4). These loci were distributed in nine linkage groups corresponding to nine pairs of chromosomes and covered a genetic distance of 875 cM. From among these probes, as many as 54% detected more than one polymorphic locus, confirming the duplicated nature of *B. oleracea* genome. Each individual linkage group was oriented in relation to chromosome arms and numbered according to the corresponding linkage groups of *B. napus* (Parkin et al. 1995). Because of the use of the same set of RFLP probes for construction of linkage maps of *B. oleracea* and *B. napus*, the structure of C genomes could be compared in these species (See Section 4.10). The parental individuals of the mapping population derived by Bohuon et al. (1996) have been used to construct the library of recombinant backcross lines covering the whole genome (Ramsay et al. 1996). Linkage analyses in BC_1 generation allowed the detection of 138 markers arranged into nine linkage groups and covering a genetic distance of 747 cM (Table 4-5). This work demonstrated the use of recombinant backcross lines to QTLs location and marker-based selection. Camargo et al. (1997) constructed the genetic map of *B. oleracea* based on RFLP and RAPD markers to localize a single locus controlling self-incompatibility (SI). The map made by Camargo et al. contained 159 loci and covered a genetic distance of 921 cM (Table 4-5).

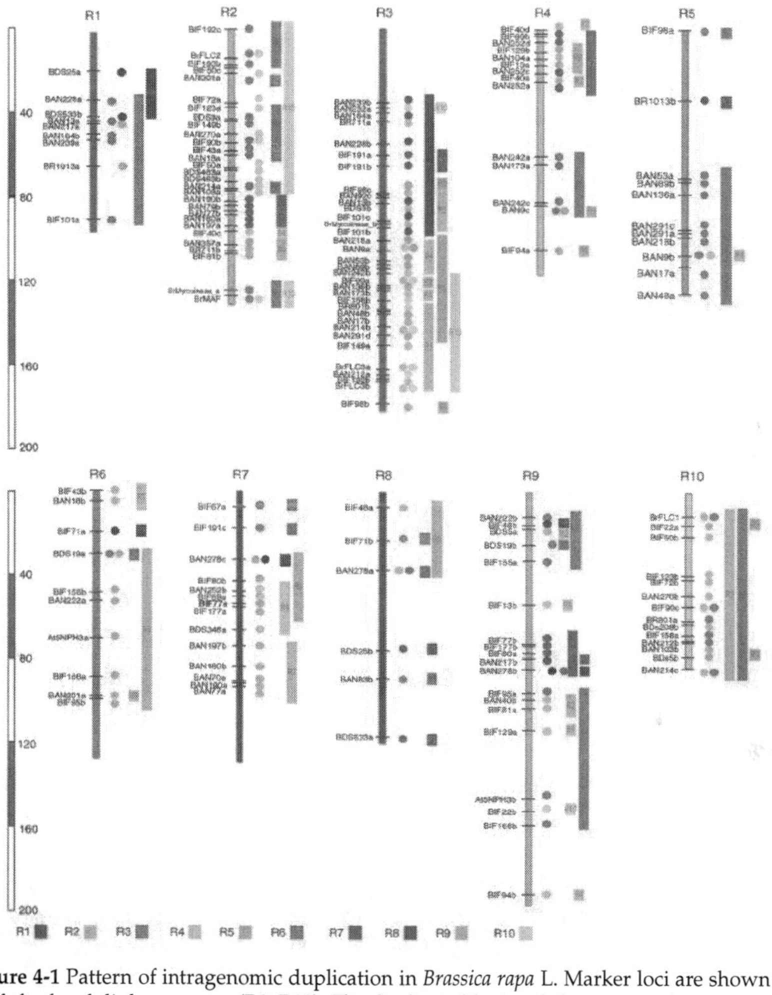

Figure 4-1 Pattern of intragenomic duplication in *Brassica rapa* L. Marker loci are shown on the left of each linkage group (R1–R10). The duplicated loci and chromosomal segments are shown on the right of each linkage group. Each of 10 linkage groups is assigned to another color in gray scale (reprinted with permission from Kim et al. 2006).

The single SI locus was mapped into linkage group 2. Earlier, three linked loci for the same *SLG6* probe had been detected, although no locus was assigned to SI (Kianian and Quiros 1992a). The presence of a single locus was confirmed by analysis of SI reaction within individuals of the population they studied by pollinating with the one parent.

Another method for construction of high-density maps is based on the enrichment of the published maps with additional markers. For example it has been used by Saal et al. (2001), who added 31 SSR markers to the existing genetic map of *B. oleracea* (Kianian and Quiros 1992a, Hu et al. 1998). Earlier, these SSR markers had been identified as C-genome specific in the *B. napus* genome. The presence of these SSR markers in the two species allowed testing the usefulness of rapeseed SSR markers in their mapping in *B. oleracea.*

The availability of the whole *A. thaliana* genome sequence permitted easier transfer of sequence information to the *Brassica* species. This knowledge was used for construction of the detailed linkage maps of the *Brassica* species. These maps were developed on the basis of RFLP markers derived from the *A. thaliana* ESTs or cDNAs corresponding to the known genes by Lan et al. (2000) and Babula et al. (2003). Comparative analysis of the maps made for the two species permitted defining the pattern of chromosomal collinearity between *B. oleracea* and *A. thaliana*. Babula et al. (2003) added 212 loci corresponding to the 95 *A. thaliana* EST probes to the existing *B. oleracea* map (Kianian and Quiros 1992a, Hu et al. 1998; Table 4-5). It covered a genetic distance of 1,806 cM with the average locus density of 9 cM. Lately, this map has been enriched by an additional group of loci corresponding to the *A. thaliana* EST and cDNA clones (Babula et al. 2006; Ziolkowski et al. 2006; Kaczmarek et al. 2009). Finally, this map included 306 loci corresponding to 160 *A. thaliana* probes and covering a genetic distance of 855 cM (Table 4-5). Now, this map is enriched by about over 40 SSR markers to integrate it with the reference map of *B. oleracea* (Sebastian et al. 2000, D. Babula-Skowronska et al., unpubl.). The highest density maps for *B. oleracea* were constructed by Sebastian et al. (2000) and Gao et al. (2007) (Table 4-5). The first of them represented a *B. oleracea* consensus map including 547 markers and covering a total map length of 893 cM (Table 4-5 and See Section 4.7.3). This map was constructed by integration of all segregating data from two different DH mapping populations: the Chinese kale x calabrese (BolAGDH map) and cauliflower x Brussels (BolNGDH map) (Table 4-4). The BolAGDH map was adapted from the study by Bohuon et al. (1996) and enriched with AFLP and SSR markers. It included 433 markers covering a genetic distance of 889 cM with an average locus interval of 3.3 cM. However, the BolNGDH map was constructed by G. King's group. It comprised 223 markers covering a genetic distance of 831 cM with an average locus interval of 4.7 cM. Additionally, each linkage group was assigned by physical mapping to their respective chromosomes using FISH (Howell et al. 2002; Fig. 4-1). This map was adopted by other laboratories for more detailed studies. For example Ryder et al. (2001) conducted analysis of the *B. oleracea* genome organization within the linkage groups O3 and O6 and the corresponding sequences in the *A. thaliana* genome. The selected regions included markers of important genes such as sporophytic self-incompatibility locus *S*, MADS box transcription factors *BoAP1* and *BoCAL*. Comparative analysis of loci from the *B. oleracea* linkage group O6 and the *A. thaliana* BAC map provided evidence of collinearity between these species and locus duplication through identification of two regions A and B. Almost all marker loci within region A of the linkage group O6 corresponded to single orthologous loci on the *A. thaliana* chromosome 1, however region B represented paralog of region A with inversion. It was consistent with those of Lan et al. (2000). On the other hand, no evidence of internal duplication within the region of linkage group O3 was found. Recently, the highest-density genetic map for *B. oleracea* was constructed by Gao et al. (2007) (Table 4-5). This map includes 1,257 markers distributed across nine linkage groups and it length is 703 cM. Each linkage group was assigned to their respective chromosomes using FISH and to chromosomes of *A. thaliana*. Also, on the basis of common SSR markers this map was assigned to the reference map of *B. oleracea* (Sebastian et al. 2000). Because of the presence of genes involved in glucosinolate pathway, resistance to cotyledon stage downy mildew and inflorescence development, it can be useful for marker-assistant selection. Particularly interesting can be the linkage groups: BoLG2 including *BoGSL-ELONG* associated with cotyledon stage downy mildew resistance and BoLG9 comprising the region with the *AOP* gene family members associated with side chain modification of aliphatic glucosinolates.

The first high-density RFLP map of *B. nigra* was constructed by Lagercrantz and Lydiate (1995) (Table 4-5). It was developed on the basis of the backcross generation F_1 with highly a heterozygous parent (Table 4-4). Initially, the segregating data for female (recurrent parent) and male (F_1) were scored separately. Finally, a combined male and female map was constructed. It comprised 288 loci in eight linkage groups with a total map length at 855 cM. This map provided important information about the male and female recombination frequencies and B genome structure indicating the intragenomic duplication and evolution and collinearity level between *B. nigra* and *B. oleracea*. Significant differences detected in the recombination frequencies between the parents were attributed to genetic effect. The data suggested that enhanced male and female recombination frequencies could be in proterminal regions and at centromeres of chromosomes, respectively. Comparison of marker loci along all linkage groups identified revealed the presence of triplicated chromosomal regions covering the whole *B. nigra* genome. Their presence confirmed the duplicated nature of the *B. nigra* genome. On the other hand, some translocations covering chromosomal segments were observed. The map was used to establish B genome structure and compare A, B and C genomes in *Brassica* (Lagercrantz and Lydiate 1996; see Section 4.10). The results pointed out that the whole *B. nigra* genome was represented by duplicated chromosomal segments. Eight segments were identified, each of which had three copies. This finding confirmed the hypothesis that modern *Brassica* species including *B. nigra* derived from a hexaploid ancestor. Finally, this map was updated with an additional set of markers derived from the *A. thaliana* probes (Lagercrantz 1998; Table 4-5). Two hundred eighty four loci were added to the existing *B. nigra* map (Lagercrantz and Lydiate 1995) covering a total map length of 751 cM. Each linkage group with one exception contained the loci detected with probes from all five *A. thaliana* chromosomes. These studies allowed a more accurate definition of the genome architecture of *B. nigra*. As mentioned above, the complexity of the *B. nigra* genome of duplication was detected by Lagercrantz (1998), which supported the hypothesis about a hexaploid ancestor of the *Brassica* species. He suggested that the present-day *A. thaliana* genome structure represents the ancestor.

4.7.3 Composite and Integrated Maps

The composite maps were obtained by alignment of common markers localized on a few different maps. They could be treated as the first integrated maps. As shown above, individual maps were constructed for special purposes; moreover, they were developed on the basis of different marker systems and populations. This fact together with the highly duplicated nature of the *Brassica* genomes and the presence of more numbers of gene copies made it difficult to transfer information between the maps and opened the possibility of erroneous attribution of the corresponding linkage groups in different maps. Despite these difficulties, mapping with the help of multiple populations offers some definite advantages over the mapping based on a single population. In particular, a greater number of loci can be placed onto the map, which gives greater genomic coverage of the genome and permits more accurate estimation of the map distances. Finally, the common polymorphic loci in all studied crosses increase the chances of segregations in other populations. Hence, they can be used as anchor markers to align the corresponding linkage groups from several maps. Recently, Cmap program was adopted to compare and integrate the existing genetic maps of the *Brassica* species and also with the *A. thaliana* genetic map (Lim et al. 2007). The information contained within these maps can be further enriched when these maps are synthesized into a single consensus map to represent a given species.

The first of the *B. oleracea* composite maps was developed by alignment of four linkage maps (Kianian and Quiros 1992a). These component maps were constructed on the basis of three intraspecific (collard x cauliflower; collard x broccoli; wild kale x cauliflower) and one interspecific (kohlrabi x *B. insularis*) F_2 segregating populations (Tables 4-4 and 4-5). Alignment of common linkage groups of these maps by using shared loci allowed generation of a combined *B. oleracea* RFLP map. This map included 108 RFLP, isozyme and morphological loci distributed into 11 linkage groups and covering a total length of 747 cM. Eight of the eleven identified linkage groups were assigned to their respective chromosomes by alignment with the gene synteny groups of the *B. campestris-oleracea* addition lines (McGrath et al. 1990). Comparison of the loci order along individual linkage groups of these four maps indicated frequent rearrangements such as translocations and inversions in the *B. oleracea* genome, even at the subspecies level. Moreover, 56% of the probes mapped on the composite map detected duplicated sequences. Like in the earlier studies, these results indicated that 44% of the *B. oleracea* genome is duplicated, supporting the hypothesis about its paleopoliploidal origin (Quiros et al. 1987; McGrath et al. 1990). The duplicated loci were randomly dispersed or formed tandem arrangement such as gene families and duplicated chromosomal segments. One of these populations derived from a cross between collard and cauliflower was further developed by Hu et al. (1998), Saal et al. (2001) and Babula et al. (2003). Hu et al. (1998) added to the existing map 167 RFLP loci covering a genetic distance of 1,738 cM (Table 4-5). These authors also confirmed the duplicated nature of C genome of *B. oleracea* by revealing duplicated loci within and between chromosomes. This map was aligned with three other, previously published maps (Landry et al. 1992; Ramsay et al. 1996; Camargo et al. 1997). Each of these maps was constructed independently with different sets of probes and segregating populations. Using 1–5 common RFLP markers for each of the nine linkage groups, four maps were aligned. Accurate consolidation of all linkage groups was not possible, but some chromosomal regions were enriched with many markers. Only, six of the nine linkage groups of the Camargo's et al. (1997) map were aligned with the base map. Further, a homology for four linkage groups in the base map and the map of Landry's et al. (1992) was found. Five groups were aligned in the map of Ramsay et al. (1996) and that of Camargo et al. (1997). Each of these groups was assigned to their respective chromosomes using the map constructed with monosomic alien chromosome addition lines by McGrath et al. (1990). In general, the linear order of markers was maintained in all maps, but the distances between markers varied. However, some chromosomal rearrangements were observed. For example, a linkage group in the map of Landry et al. (1992) carrying a clubroot resistance and consisting of markers from two other linkage groups was rearranged. This composite map enriched by a large number of markers could be useful in further targeted research such as gene cloning. The consolidation of some genetic maps was continued by Lan et al. (2000). They constructed a comparative map of *B. oleracea* and *A. thaliana* using selected *A. thaliana* ESTs as probes (Table 4-5; see Section 4.11). In this study they used a composite *B. oleracea* map that was developed on the basis of mapping of the common *A. thaliana* ESTs in four populations. The unique loci mapped in each of the four populations were integrated on the basis of the closest common flanking loci. With the use of 217 *A .thaliana* ESTs, 113 *Brassica PstI* genomic clones, 23 *A. thaliana* cDNA clones and four cloned RAPD-PCR products, 367 polymorphic loci were detected. The "one to one correspondence" and "duplication" models developed in this study confirmed the earlier data concerning the duplicated nature of the *B. oleracea* genome. According to these models, 41% of RFLP loci detected were duplicated. The majority of them were organized in large intra-chromosomal segments. The duplicated segments were observed on the *B. oleracea* chromosomes 1, 2 and 9. Analysis of the arrangement along all linkage groups allowed displaying at least 22 rearrangement breakpoints differentiating *B. oleracea* homologs. Although some regions could be triplicated, these models did not support triplication of the *Brassica* genome.

Sebastian et al. (2000) worked out the first integrated map for *B. oleracea* based on segregating data from two distinct DH mapping populations and using 105 common markers (RFLP, AFLP and microsatellites) for the component BolAGDH and BolNGDH maps (Table 4-5). The consolidation of the segregating data of both maps allowed generation of a high-density map including 547 markers and covering a genetic distance of 893 cM, which is a 99% of the *B. oleracea* genome. The order and orientation of each linkage group was established with respect to original BolAGDH map (Bohuon et al. 1996). Because of the duplicated nature of the *B. oleracea* genome, banding patterns of the parental lines for the multiple-copy clones were checked, so that only homologous loci were compared. The correlation test concerning the position of the homologous markers in both component populations revealed close conservation of marker order with small rearrangements. In two linkage groups (LGs 1 and 5) the marker order was identical and 5 linkage groups (LGs 4, 6, 7, 8, 9) showed single inversion, whereas in the LG 3 and LG 2 two and several inversions were detected, respectively. Similarity of the marker order in the BolAGDH, BolNGDH and integrated maps means that the integrated map can be used as a *B. oleracea* consensus map. At present this map is further integrated with the map constructed by Babula et al. (2003) and updated by Kaczmarek et al. (2009) (D. Babula-Skowronska et al., unpubl). In this process, the SSR markers and low-copy *A. thaliana* ESTs were selected as anchor loci to assign the corresponding linkage groups from both maps.

4.8 Molecular Cytogenetic Maps

The genetic maps, even with high-density genes/markers are never complete. What is more, they present relative distances between markers only, based on the frequency of meiotic recombination. More accurate information about the physical distribution of genes on the chromosomes originates from the cytogenetic maps showing their approximate positions, however they characterize low-density genes. Therefore, to get maximum information from the genetic, cytogenetic and physical maps for the genome sequencing projects, these maps must be integrated. The first step of this process is alignment of the linkage groups with karyotype (Howell et al. 2002; Fig. 4-1). It allows determination of a framework of sequence-anchored BAC clones covering the whole genome, determining of the orientation of linkage groups with respect to chromosome arms and comparison of genetic and physical distance of selected markers. Integration of cytogenetic and genetic maps was carried out for many plant species such as wheat, barley or *A. thaliana* (for example Koornneef and Van der Veen 1983; Künzel et al. 2000; Sandhu et al. 2001).

For the *Brassica* species, karyotypes and various genetic linkage maps have been established by different research groups. Identification and cytological characterization of individual chromosomes in *Brassica* is problematic because of the small size of chromosomes, specific-chromosome landmarks, as well as similar chromosome lengths and arms ratios. Nevertheless, some karyotypes of *Brassica* using in situ hybridization have been developed (Armstrong et al. 1998; Howell et al. 2002; Koo et al. 2004).

Connection of linkage groups of the existing *B. rapa* genetic maps with individual chromosomes is an important phase of the Multinational *B. rapa* Genome Sequencing Project (*Br*GSP). Because of duplicated nature of the *B. rapa* genome, the anchoring of sequence contigs on the physical map and reconstruction chromosomes sequence can be problematic. However, the availability of two distinct maps for this species should facilitate this process (Kim et al. 2006; Choi et al. 2007). The first of them, the high-density map of *B. rapa* constructed by Kim et al. (2006) has been initially attributed to the other existing maps. Using 21 SSR markers designated to A genome of *B. rapa* and *B. napus* (Parkin et al. 1995; Suwabe et al. 2002; Lowe et al. 2004), each linkage group of this map has been assigned to an accepted classification of the *B. rapa* linkage groups (R1–R10). Comparison of the relative length of each linkage group with the length of corresponding cytogenetic chromosome identified by Lim et al. (2005) has revealed compatibility between the linkage groups and the chromosomes, while the nine linkage groups of the reference genetic map for *B. rapa* (CKDH) have recently been assigned to chromosomes by using locus-specific ESTs to identify the corresponding BAC clones in KBrH library (Choi et al. 2007; Hong et al. 2008). These BACs identified used as FISH probes allowed integration of the genetic and physical maps of *B. rapa* ssp. *pekinensis*.

Truco and Quiros (1994) assigned the *B. nigra* linkage groups to chromosomes using two sets of alien addition lines, *B. napus-B. nigra* and *B. oleracea-B. nigra* (Chevre et al. 1991; Quiros et al. 1991; This 1992). Seven of the probes used for construction of the *B. nigra* genetic map were located on the chromosomes of the addition lines *B. napus-B. nigra*. Their location allowed assignment of three linkage groups B1, B3 and B4 to chromosomes 1, 5 and 2, respectively.

In *B. oleracea*, the first efforts to assign linkage groups to chromosomes were taken up by Kianian and Quiros (1992a). They assigned eight of nine linkage groups from a *B. oleracea* composite map to chromosomes on the basis of the data reported by McGrath et al. (1990). More accurate integration of the cytogenetic and genetic maps was carried out by Howell et al. (2002). They selected the *B. oleracea* var. *alboglabra* line A12DHd, which was previously used as a parent of the mapping population for the construction of the reference genetic map (Bohuon et al. 1996; Sebastian et al. 2000), to produce partial karyotype (Armstrong et al. 1998) and a BAC library. Integration of both maps of the *Brassica* C genome was achieved with 22 probes representing 19 loci. They used genomic plasmids, cosmids and BAC clones with genetically mapped inserts as FISH probes for their localization on the cytogenetic map. Because of duplicated nature of the *B. oleracea* genome, it should be verified if individual FISH-based loci corresponded to the same locus on the genetic map. In the study by Howell et al.(2002) individual linkage groups of the *B. oleracea* genetic map were assigned to corresponding chromosomes of the karyotype. The assignment of linkage groups to chromosomes was as follows: Linkage group (LG) O1 to chromosome 8, LG O2 to chromosome 5, LG O3 to chromosome 1, LG O4 to chromosome 2, LG O5 to chromosome 6, LG O6 to chromosome 9, LG O7 to chromosome 4, LG O8 to chromosome 7 and LG O9 to chromosome 3 (Fig. 4-1). Additionally, eight of the nine linkage groups were oriented with respect to chromosome arms. Comparison of the orientation of linkage groups and chromosome arms in the genetic and cytogenetic maps, respectively, proved that four chromosomes (2, 4, 7 and 9) were in the same orientation as corresponding linkage groups, whereas the other four chromosomes (1, 3, 5 and 8) were in the opposite orientation towards these groups. The remaining chromosome 6 was probably also ordered in the opposite orientation. Using these data, Gao et al. (2007) aligned the constructed transcriptome map for *B. oleracea* based on cDNA-SRAP markers to C-genome linkage groups and chromosomes of *B. oleracea* and *B. napus*. A similar strategy of integration of genetic and physical maps of *B. oleracea* has been followed by Ziolkowski et al. (2006) (Fig. 4-2). These authors analyzed four rearrangement breakpoints by BAC-FISH and RFLP mapping of three *A. thaliana* chromosomal regions. For analysis, 95 *Arabidopsis* BACs representing chromosomal segments on the chromosome 1, 2 and 3 were selected. They used BAC contings as FISH probes to identify homologous regions on the *B. oleracea* chromosomes. Additionally, 36 *Arabidopsis* ESTs from the physical mapped segments were chosen to identify the corresponding loci on the *B. oleracea* genetic map. It allowed the assignment of particular linkage groups to individual

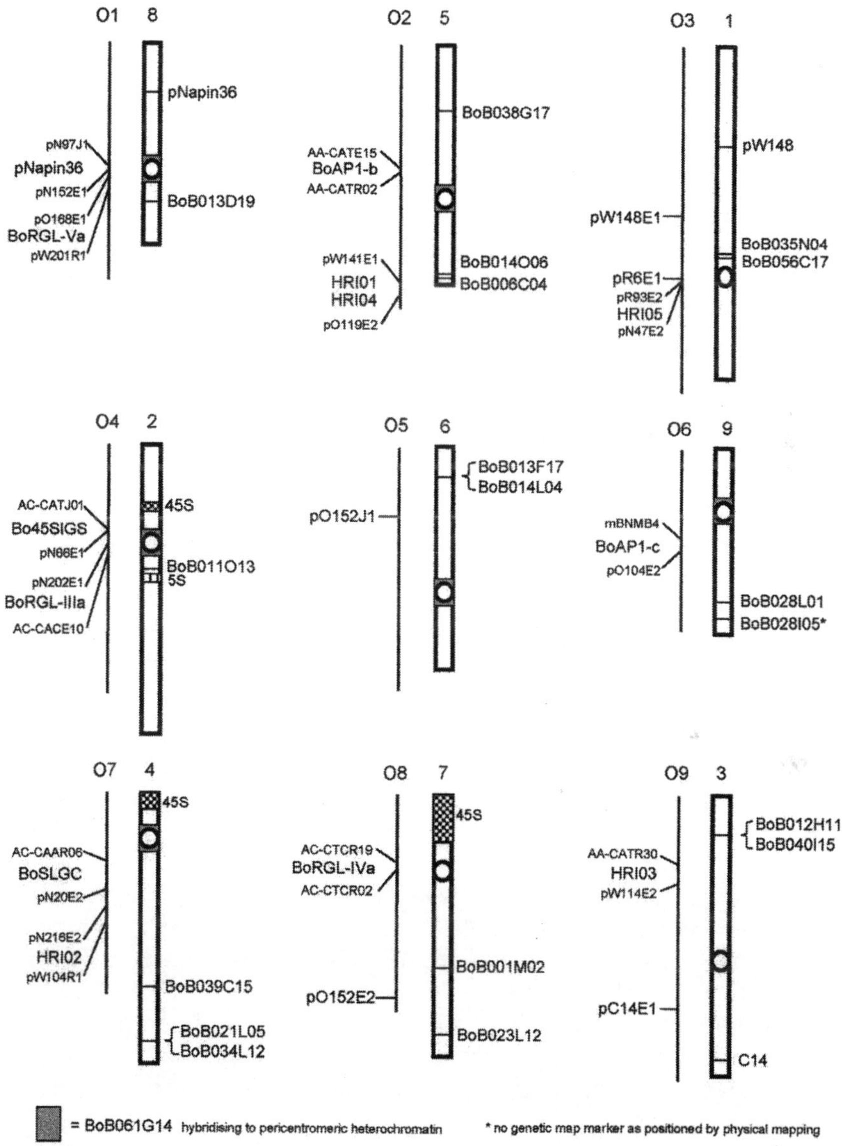

Figure 4-2 Assignment of the linkage groups (O1–O9) of the integrated genetic map to chromosomes (1–9) in *B. oleracea* by using the corresponding RFLP loci and FISH probes, respectively. The linkage groups are represented by vertical bars with marked location of the specific RFLP loci corresponding to FISH probes. The chromosomes are shown on the right of the linkage groups in relative lengths with marked location of the BAC, 5S rDNA and 45S rDNA probes. Open circles represent centromeres. Chromoosomes 1, 3, 5, 6 and 8 are inverted to follow the orientation of linkage groups (reprinted with permission from Howell et al. 2002).

Brassica chromosomes. Additionally, localization of analyzed group of *A. thaliana* ESTs on the *B. oleracea* reference map BolAGDH facilitated the identification of corresponding chromosomes of the basic karyotype (Howell et al. 2002).

4.9 Intragenomic Duplications

Results of many studies of the plant genome structures based on molecular markers and cytogenetic data indicated their ancient polyploid origin (for example Helentjaris et al. 1988; Kianian and Quiros 1992a; Shoemaker et al. 1992). One of the consequences is the presence of a great number of duplicated genes (Bennetzen 2000). The first mentions the duplicated nature of the *Brassica* genomes that appeared in the 1930s when it was postulated that the diploid species represent secondary polyploids derived from ancestral genomes with a fewer number of chromosomes (Catcheside 1934; Röbbelen 1960; Prakash and Hinata 1980). Further, more detailed studies based on RFLP mapping and comparative studies between the *A. thaliana* and *Brassica* species revealed that 40 to 50% of the loci were duplicated within Brassica genomes (McGrath et al. 1990; Slocum et al. 1990; Kianian and Quiros 1992a; Kowalski et al. 1994; Truco and Quiros 1994; Lagercrantz and Lydiate 1996; Lagercrantz 1998; Lan et al. 2000; Babula et al. 2003; Parkin et al. 2005). They could be classified as randomly dispersed single loci, linked-gene families or blocks duplicated loci in non-homologous chromosomes (Kianian and Quiros 1992a). Hu et al. (1998) classified duplications into two major groups: intra- and interchromosomal ones. The origin mechanism of loci duplication is yet unknown. It is suggested that randomly dispersed duplicated loci have arisen by reverse transcription of mRNA or heterogeneous nuclear RNA with subsequent reinsertion into the genome (Helentjaris et al. 1988). The linked-gene families can be a result of unequal crossing-over events. However, the segmental duplication is known to originate from the overlapping of reciprocal translocations followed by hybridization (Gottlieb 1983). Another way of loci duplication probably follows the formation of the interspecific aneuploids by combining homoeologous chromosomes of related genomes after hybridization and amphiploidization.

Analysis of available genetic maps of *Brassica* indicates a partial duplication of genomes forming complex arrangements of chromosomal segments present in at least three copies (Kowalski et al. 1994; Lagercrantz and Lydiate 1996). These triplicated chromosomal regions are associated with ancient polyploidization of the *Brassica* genomes. The reiterated segments are characterized by significant collinearity, although some rearrangements such as inversions, insertions and substitutions have observed (for example Lagercrantz 1998; Babula et al. 2003; Parkin et al.

2005). The sequence-level studies of duplicated segments indicated that some gene copies were deleted within individual regions (O'Neill and Bancroft 2000; Rana et al. 2004). These modifications within genomes can lead to formation of novel phenotypic variations of traits among these species (Paterson et al. 2001). This observation has opened a new approach to phenotypic evolution of polyploidy in plants.

The duplicated structure of the *B. rapa* A genome has been confirmed by some other authors. McGrath and Quiros (1991) found extensive gene duplication including 11 of 31 analyzed probes in the *B. rapa* genome. Of the 11 probes, 3 duplicated loci were located on the linkage groups and the remaining duplicated copies were unlinked. Similar results had been obtained earlier for *B. oleracea* by using the same set of probes (McGrath et al. 1990). Also, Song et al. (1991) observed extensive sequence duplication because almost 37% of probes detected duplicated sequences in *B. rapa*. Many of them formed linked arrangements located on different linkage groups. On the other hand, some small rearrangements such as insertions, deletions and inversions within these reiterated segments were found. Analysis of a high-density map of *B. rapa* with 545 loci allowed revealing a duplication pattern of this genome (Kim et al. 2006). In accordance with it, the majority of chromosomal regions were reiterated two to three times. This duplication pattern of *B. rapa* genome has been confirmed by Choi et al (2007) (Fig. 4-3). They found the duplicated region on R2 and R10 corresponding to the single region on the *A. thaliana* chromosome 5. It is in agreement with the discoveries of Parkin's et al. (2005) and Kim's et al. (2006), who found that many chromosomal segments reiterated 2 or 3 times exhibiting extensive triplication of A genome (Rana et al. 2004). On the other hand, Yang et al. (2006) observed, that almost 88% of genes located around the duplicated *FLC* regions were deleted.

Also, evidence has been found of the duplicated nature of the *B. nigra* B genome. Three different types of arrangements have been identified: scattered through the genome, organized in tandem (gene families) and in blocks of duplicated loci conserved in more than one linkage group. Truco and Quiros (1994) detected segmental duplication located on the linkage groups B1 and B5. On the basis of a limited number of duplicated sequences, they established that the chromosome corresponding to the linkage group B1 could have resulted from an exchange of chromosomes B2 and B4, while, the chromosome corresponding to the linkage group B6 could have been derived from the chromosomes B3 and B5. Lagercrantz and Lydiate (1996) using the *Brassica* RFLP probes suggested that the entire *B. nigra* B genome consists of large, duplicated segments which reiterated three times . Graphically, the whole genome of *B. nigra* is represented by eight sets of triplicated chromosomal segments with almost completely conserved loci order. The triplication has been confirmed by Lagercrantz

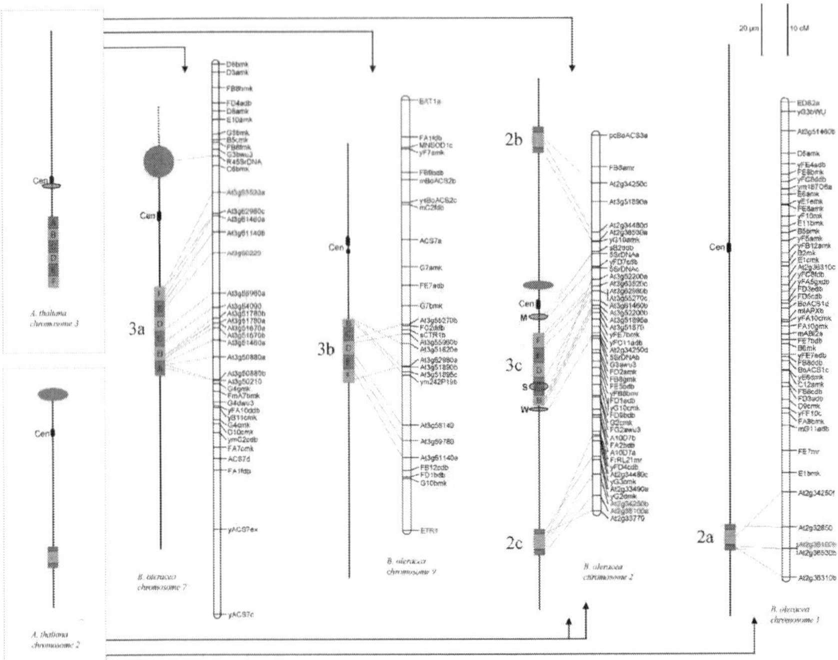

Figure 4-3 Genetic and physical mapping of the selected regions of *A. thaliana* chromosomes 2 and 3 in the *B. oleracea* genome. The idiograms of the *A. thaliana* chromosomes 2 and 3 with BAC blocks representing FISH probes of the analyzed regions are shown on the left. The centromeres are marked by "Cen". The circles represent localization of 45 rDNA and 5S rDNA loci. The idiograms of the *B. oleracea* pachytene chromosomes and corresponding linkage groups on the right show a triplicated nature of the *B. oleracea* C genome. Long-tailed arrows indicate the connection of the *A. thaliana* chromosomes 2 and 3 with their homologous in *B. oleracea* 2a, 2b, 2c and 3a, 3b, 3c, respectively. The horizontal lines between ideograms of chromosomes and linkage groups represent genes selected from the *A. thaliana* regions analyzed used as RFLP probes for genetic mapping in *B. oleracea* (reprinted with permission from Ziolkowski et al. 2006).

(1998) with additional 284 loci detected by the *A. thaliana* probes. These data support the hypothesis that *B. nigra* has descended from a hexaploid ancestor. Many of the collinear segments detected in the *B. nigra* genome were short because of high frequency of rearrangements. Nevertheless, some examples of long collinear segments have been found in this genome. One of them covered at least 40 cM of the top *A. thaliana* chromosome 5 and was present in three homeologous copies distributed over three different linkage groups: G2, G5 and G8. Two of these segments were characterized by high collinearity, while the third one contained large inversion when compared to that of *A. thaliana*. Another example was the 40 cM segment of the *A. thaliana* chromosome 2, which corresponded to three homeologous segments in *B. nigra*. One of these segment included 50 cM region on the *B. nigra* linkage group G6 and the other two on the *B. nigra* linkage groups G1 and G8 were interrupted by segments from the other *A. thaliana* chromosomes. These data indicated that none of the *A. thaliana* chromosomes were significantly over- or underrepresented in duplicate homeologous copies in *B. nigra*. Sadowski and Quiros (1998) corroborated the duplicated nature of the *B. nigra* genome by analysis of gene members of the single-copy *A. thaliana* gene complex. This complex included six genes (including *RPS2* gene) located on the chromosome 4 and they were expanded to 17 loci in the *B. nigra* genome and distributed over five linkage groups. The gene clusters with four to five members were found on the two *Brassica* linkage groups, B1 and B4 and on other partial smaller segments on B2, B3 and B8. Only gene complex on the *B. nigra* linkage group B1 was completely conserved relative to the corresponding gene cluster in *A. thaliana*. Physical analysis of this segment based on PFGE proved its extended size in the linkage group B1 of *B. nigra* compared with the size of this cluster in *A. thaliana* which is in agreement with the genome-size difference between these two species (Arumuganathan and Earle 1991).

Similarly, many studies indicated the duplicated nature of the *B. oleracea* C genome supporting the hypothesis about its secondary poliploidization (Prakash and Hinata 1980). As follows from RFLP analysis, at least 35% of the genes identified in this genome have been duplicated (Slocum et al. 1990; McGrath et al. 1990; Hu et al. 1998; Lan et al. 2000; Babula et al. 2003). These loci have been organized in tandem and segmental arrangements on all chromosomes. Detailed analysis of duplicated segments distribution in the *B. oleracea* genome indicated that at least some of them could be triplicated (Kowalski et al. 1994; Ziolkowski et al. 2006) or represented by two to five copies (Truco et al. 1996; Lan et al. 2000; Babula et al. 2003). Kianian and Quiros (1992a) analyzing the linkage map generated from three intra- and one interspecific populations, observed that the distances and order of markers within duplicated chromosomal segments survived. On the other hand, some rearrangements at the subspecies level were identified. The majority of these chromosomal modifications were represented by small inversions followed by translocations (Kianian and Quiros 1992a; Babula et al. 2003). In addition to these variations, the tandem duplications of individual genes were observed (Babula et al. 2003). The authors suggested that these rearrangements appeared as a consequence of genome evolution and variation in the chromosome number in the *Brassica* species.

The duplication level of the *Brassica* A, B and C genomes detected indicate their complexity as a result of polyploidization during their ancestry (Kianian and Quiros 1992a; Lagercrantz 1998). Within duplicated segments found in the genomes, numerous rearrangements were observed (Gustafsson et al. 1976; Quiros et al. 1988; Kianian 1990). Accumulation of different chromosome rearrangements led to renewed genome diploidization. Some specific chromosome rearrangements have been found in the *Brassica* genomes. Among them, small inversions were the most frequent forms of aberration followed by chromosomal translocations (Ryder et al. 2001; Babula et al. 2003). In addition to inversion and translocations, deletions seem to be another important molding force of the *Brassica* genomes (Song et al. 1991; O'Neill and Bancroft 2000; Babula et al. 2003; Rana et al. 2004; Parkin et al. 2005). On the basis of the above data, the hypothetical models of the *Brassica* species evolution were proposed. One of the hypotheses known as the triplication theory indicated that the *Brassica* genomes were composed of three rearranged variants of an ancestral genome and descended from a common hexaploid ancestor (Lagercrantz and Lydiate 1996). On the basis of comparative genetic mapping between the *Arabidopsis* and *Brassica* species the allopolyploid nature of the putative hexaploid ancestor was proposed (Lagecrantz 1998). Lysak et al. (2005) using the physical comparative chromosome painting between *A. thaliana* and 21 Brassiceae and related species proposed that the ancestral Brassiceae genome became triplicated via allohexaploidy 7.9–14.6 Mya. These analyses indicated significance of the chromosome fusion and fissions in karyotype evolution of the Brassiceae. Other evidence supporting the triplication theory was provided by genetic/physical analysis of the selected chromosomal regions of *A. thaliana* in the *B. oleracea* genome (Ziolkowski et al. 2006). However, some authors did not confirm this theory because of identification of the *Brassica* gene copies in the number smaller or greater than three. Another hypothesis was proposed by Quiros' group, who suggested amphiploidization as the origin of genomes of cultivated *Brassica*. This supposition would imply that the ancestral genome was composed of 5–7 chromosomes, which differentiated into several related genomes with similar chromosome numbers (Truco et al. 1996). Then, they diverged from one another by geographical isolation, leading to changes in chromosomal structure due to the accumulation of translations. Hybridization, amphiploidization and subsequent backcrossing of these genomes led to the appearance of the diploid *Brassica* species. Further, aneuploidy and secondary chromosomal rearrangements modified the genomes resulting in the present-day *Brassica* genomes of 8, 9 and 10 chromosomes in *B. nigra*, *B. oleracea* and *B. rapa*, respectively.

As all mapping experiments are based on polymorphism of genetic markers, the final pattern of duplication/triplication genome is incomplete and requires further verification. Nevertheless, the present knowledge of genomic relationship among diploid *Brassica* species brings us nearer to understanding the evolution of *Brassica* and gene transfer between them.

4.10 Homeology of A, B and C Genomes in the Brassica Genus

Three diploids and three amphidiploid genomes formed from them belong to the cultivated *Brassica* species; each with a characteristic chromosome number: A genome (n = 10) in *B. rapa*, B genome (n = 8) in *B. nigra* and C genome (n = 9) in *B. oleracea*, AB genomes in *B. juncea* (n = 18), AC genomes in *B. napus* (n = 19) and BC genomes in *B. carinata* (n = 17). Results of the phylogenetic studies of diploid and amphidiploid species revealed a relationship between the above species known as the U-triangle (U 1935; see Chapters 6 and 8 in this volume). This relationship was confirmed by more detailed investigation including cytogenetic study (Prakash and Hinata 1980); analyses of nuclear DNA content (Verma and Rees 1974), isozymes (Coulthart and Denford 1982; Quiros et al. 1987), proteins (Vaughan 1977), rRNA genes (Quiros et al. 1987) and chloroplast DNA (Erickson et al. 1983, Palmer et al. 1983); artificial resynthesis of hybrids (Ramanujam and Srinivasachar 1943; Mizushima 1950; Olsson 1960) and RFLP marker analysis (Song et al. 1988; Hosaka et al. 1990). Analysis of the cytological, biochemical and molecular data showed that the three diploid *Brassica* species evolved in two pathways: *B. rapa* and *B. oleracea* from one pathway having a common origin and *B. nigra* from another (Vaughan 1977; Attia and Robbelen 1986; Song et al. 1988; Warwick and Black 1991). The chromosomal collinearity level within the *Brassica* genus was determined by comparative genome mapping using common DNA probes (Slocum 1989; McGrath and Quiros 1991; Truco and Quiros 1994; Bohuon et al. 1996). Comparisons of gene order along chromosomes of the related species allow investigation and corroboration or rejection of the evolution hypotheses of their origin. The majority of comparative genome structure analyses among species of the *Brassica* genus concerned *B. rapa* and *B. oleracea* (A and C genomes, respectively), *B. oleracea* and *B. napus* (C and AC genomes, respectively), *B. rapa* and *B. napus* (A and AC genomes, respectively) and *B. rapa*, *B. nigra* and *B. oleracea* (A, B and C genomes, respectively) (Table 4-6). Nevertheless essential similarity of the diploid and amphidiploid *Brassica* genomes was proved, extensive rearrangements, forming unique combinations of chromosomal regions were observed as expected because of different chromosome numbers and genome evolution in the *Brassica* species. The studies have provided only a crude estimation of the genome structure and organization in the compared species because of insufficient marker saturation in the existing maps. Another problem was the duplicated nature of the *Brassica* genomes making the identification of orthologous regions in these species difficult.

Table 4-6 Characteristics of homoeology between A, B and C genomes of the *Brassica* species.

Species	Genome homeology	Species	Common loci/ probes	Authors
B. rapa (A genome)	C genome	*B. oleracea* (McGrath et al. 1990)	95 cDNA probes	McGrath and Quiros 1991
	A genome	*B. napus*		Hoenecke and Chyi 1991
	A and C genomes	*B. napus* (Ferreira et al. 1994) *B. oleracea* (Camargo 1994)	63 RFLP loci 46 RFLP loci	Teutonico and Osborn 1994
	A genome	*B. napus* (Parkin et al. 1995)	116 SSR loci	Choi et al. 2007
	A genome	*B. juncea* (Axelsson et al. 2000)	29 SSR loci	Choi et al. 2007
	A genome	*B. napus*	41 SSRs 3 SNP/ InDel	Suwabe et al. 2008
B. nigra (B genome)	A and C genomes	*B. rapa* (Sharpe et al. 1995) *B. oleracea* (Bohuon et al. 1996)		Lagercrantz and Lydiate 1996
	A and C genomes	*B. rapa* (McGrath and Quiros 1991) *B. oleracea* (Kianian and Quiros 1992a)		Truco et al. 1996
B. oleracea (C genome)	A genome	*B. rapa*		Slocum 1989
	C genome	*B. napus* (Parkin et al. 1995)	129 RFLP loci	Bohuon et al. 1996
	A, B and C genomes	*B. oleracea* *B. nigra* *B. rapa*	Five genes complex as probe	Sadowski et al. 1996
	A genome	*B. rapa* *B. napus*	117 RFLP probes 7 RAPD primers	Cheung et al. 1997
	C genome	*B. oleracea* (Sebastian et al. 2000; Howell et al. 2002); *B. napus* (Bohuon et al. 1996; Lowe et al. 2004; Piquemal et al. 2005)	22 SSR loci	Gao et al. 2007
	C genome	*B. napus* (Sun et al. 2007)	77 SRAP markers	Gao et al. 2007

Analysis of A and C genomes and their relationships to other *Brassica* genomes has been a subject of many studies. The first genetic comparison of A and C genome structures in *B. rapa* and *B. oleracea*, respectively has been carried out by Slocum (1989) and McGrath and Quiros (1991) (Table 4-6). They compared the genetic maps based on random genomic and cDNA clones. McGrath and Quiros (1991) tested over 100 cDNA clones to recognize the corresponding DNA sequences in both species. About 95% of the probes detected at least one polymorphic locus in these genomes. Using 27 common marker probes, McGrath and Quiros (1991) indicated similar gene copy number forming the conserved block loci of both genomes. Their results as well as that of an earlier study (Slocum 1989) indicated that the same set of single copy loci was observed in both species. Three single loci were located on one linkage group in *B. rapa* and *B. oleracea*. On the other hand, effects of genetic differentiation in each of the genome were observed. The cytogenetic analysis of meiosis in pollen mother cells of *B. rapa* and *B. oleracea* (*B. alboglabra*) performed by Heneen et al. (1995) indicated differences in chromatin condensation. This differentiation was further confirmed by genetic mapping using isozyme and DNA-based markers to discriminate between these genomes. The results of all studies suggested that *B. rapa* and *B. oleracea* originated from a common ancestor, although chromosome reshuffling after speciation of both species has occurred. The similarity of A and C genome structures can allow their sexual recombination forming amphidiploid *B. napus*.

Other studies concerned a comparison of A and C genome structure in *B. rapa* and *B. oleracea*, respectively, with that of the corresponding genome of the amphidiploid *B. napus* species. On the basis of the interspecific hybridization and cytogenetic data, *B. oleracea* and *B. rapa* were proposed as progenitors of the amphidiploid *B. napus* (U 1935). The above thesis was confirmed by RFLP mapping distinguishing A and C genome specific markers. Starting from the 1990s, numerous comparisons of C genome in *B. oleracea* and *B. napus* and A genome in *B. rapa* and *B. napus* revealed their conservation (Lydiate et al. 1993; Teutonico and Osborn 1994) (Table 4-6). Teutonico and Osborn (1994) were the first to compare genetic maps of three species, i.e. *B. rapa*, *B. napus* and *B. oleracea*. The *B. rapa* map constructed by these authors was first compared with the previously published map for *B. oleracea* (Camargo 1994). The comparison revealed many regions with conserved loci order in both maps, confirming a relationship between these species, however, some rearrangements such as deletions or insertions were observed. These rearrangements could affect the distances between these loci in the way reflecting differences in the genomes size of *B. rapa* and *B. oleracea*. Further, this *B. rapa* map was compared with the previously published map for *B. napus* (Ferreira et al. 1994) (Table 4-6). All common RFLP loci with the exception of two were preserved, a common loci order along the corresponding linkage groups in both maps. Similarly, the distances between the common loci were different in these species. As the *B. rapa* map has been constructed on the basis of the cloned genes mapped in *A. thaliana*, a comparison of the two maps was possible. Bohuon et al. (1996) compared the *B. oleracea* high density map constructed by them with C genome linkage groups in the previously published *B. napus* map (Parkin et al. 1995; Table 4-6). Although the two maps contained 132 common probes detecting 270 polymorphic loci in *B. oleracea* and/or *B. napus*, a comparative analysis was carried out by using 129 common RFLP loci. Comparison of the loci order along the corresponding C genome linkage groups of *B. oleracea* and *B. napus* revealed a collinear pattern for 123 of 129 common loci. The rest of the six loci did not show compatibility due to the scoring errors (*flowers* on O3/N13 and *pW197dNP* on N15), their presence in the rearranged region in one of the parents of the *B. oleracea* cross (*pW197*f and *pR36*e on O7) or the detection of different loci by *pR116* on O3/N13. The analysis of the loci arrangement along the corresponding linkage groups on the *B. oleracea* and *B. napus* maps revealed a high level of collinearity, indicating similarity of C genome structure in both species. The total length of the collinearity region was 702 cM and 779 cM in *B. oleracea* and *B. napus*, respectively. This observation has been confirmed by Cheung et al. (1997), who using the same set of RFLP probes and RAPD and STS primers, also indicated partial collinearity of C genome linkage groups in *B. oleracea* and *B. napus* (Table 4-6). The highest level of conservation was found between the *B. oleracea* linkage group 1 and the *B. napus* linkage group 3. These authors estimated that 85% of the *B. oleracea* genome had homoeology with the *B. napus* genome and 48% of the *B. napus* genome was derived from *B. oleracea*. Comparative analysis of the *B. oleracea* and *B. napus* maps has also performed by Ryder et al. (2001). They aligned the homoeologous C genome linkage groups of *B. oleracea* and *B.napus* including O3 and N13 and O6 and N16, respectively (Parkin et al. 1995) (Table 6). The analyzed region of the linkage group N16 was also homoeologous with A-genome chromosomes N6 and N7 (Cavell et al. 1998). In spite of the collinearity of C genomes of *B. oleracea* and *B. napus*, some major genome rearrangements between these species were observed. Within the conserved regions simple inversions or translocation including some loci were detected. Interestingly, the rearrangements were more often found in C genome of *B. napus* than *B. oleracea*. Similarly, comparative analysis of A genome in *B. rapa* and *B. napus* has been made. The first studies based on comparison of A-genome linkage groups of both species have been performed by Hoenecke and Chyi (1991). They observed in general the conserved markers order for most linkage groups in *B. rapa* and *B. napus*. Additionally, some minor and two large rearrangements were identified. The rearrangements detected were a consequence of interspecific hybridization of A and C genome progenitors.

On the other hand, a homoeology between A and C genomes has been found, confirming the results of earlier studies (McGrath and Quiros 1991; Teutonico and Osborn 1994).

Development and general availability of SSR markers facilitated comparison of A, B and C genome structures in the *Brassica* species. Gao et al. (2007) by using 22 common SSR markers aligned nine linkage groups BoLG1-BoLG9 of their *B. oleracea* map to equivalent linkage groups O1–O9 and the corresponding chromosomes in this species (Sebastian et al. 2000; Howell et al. 2002) and the *B. napus* linkage groups N11–N19 (Bohuon et al. 1996; Lowe et al. 2004; Piquemal et al. 2005) (Table 4-6). Additionally, 77 genomic SRAP markers detected and mapped on their map were included in the ultra-dense *B. napus* map constructed by Sun et al. (2007). However, on the basis of 116 common SSR markers, a reference genetic map of *B. rapa* ssp. *pekinensis* (Choi et al. 2007) was aligned to the reference A genome map of *B. napus* (Parkin et al. 1995). The SSR analysis indicated higher transferability between A and C genomes species than between this group and B genome species (Lowe et al. 2004). It confirmed the known evolutionary relationship between different *Brassica* species. This map based on 29 common SSR loci was aligned to A genome of *B. juncea* (Axelsson et al. 2000; Table 4-6). These studies indicated general conservation of the marker order along linkage groups representing A genome, that is R1 to R10, N1 to N10 and J01 to J10 in *B. rapa*, *B. napus* and *B. juncea*, respectively. However, this order was disturbed by several minor rearrangements reflecting differences in the genome structure in the species studied. Recently, Suwabe et al. (2008) constructed an integrated linkage map of A genome for *B. rapa* and *B. napus* using 44 common markers, 41 SSRs and 3 SNP/InDel (Table 4-6). They established that the order of the common markers along the linkage groups of both maps were conserved indicating a high level of A genome collinearity in these species. On the linkage groups A4, A5 and A10 small internal inversions were observed. The above results led to the conclusion that after integration of A and C ancestral genomes to the amphidiploid genome of *B. napus*; the integrity of their chromosomes was disturbed. The first comparative studies of the three diploid genomes were performed by the Lydiate's and Quiros's groups (Lagercrantz and Lydiate 1996; Truco et al. 1996). In the Lydiate's group three genetic maps of *B. rapa* (Sharpe et al. 1995), *B. nigra* (Lagercrantz and Lydiate 1996) and *B. oleracea* (Bohuon et al. 1996) generated by the same set of probes were compared (Fig. 4-4; Table 4-6). This study indicated that the *Brassica* A, B and C genomes were very similar in spite of different chromosome numbers. The same authors found a high level of genomes conservation indicating the genome triplication as a general pattern of the *Brassica* genomes. It suggested that all diploid *Brassica* genomes inherited complete but rearranged duplicated segments of ancestral genomes. Moreover, the fission and fussion processes were

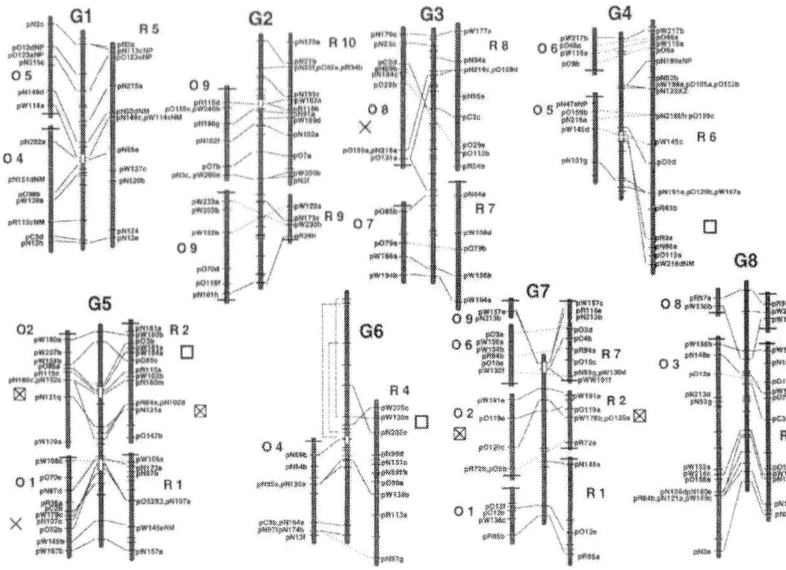

Figure 4-4 Alignment of the *B. nigra* linkage groups with the corresponding segments of the *B. rapa*, and *B. oleracea* linkage groups. The vertical lines with the marker loci on the left and right represent the linkage groups. G1–G8, R1–R10 and O1–O9 correspond to the *B. nigra*, *B. rapa* and *B. oleracea* linkage groups, respectively. Lines drawn between segments indicate the common loci (reprinted with permission from Lagercrantz and Lydiate 1996).

indicated to take place during the genomes divergence and they were identified as responsible for changes in the chromosome number.

Truco et al. (1996) compared the order of loci along the linkage groups in the maps of *B. rapa* (McGrath and Quiros 1991), *B. nigra* (Truco and Quiros 1994), and *B. oleracea* (Kianian and Quiros 1992a) developed by a set of common probes (Table 4-6). The intergenomic conserved regions with extensive reordering among A, B and C genomes, were found. The highest homology was observed for A and C genomes, which confirmed their common origin. The results of this analysis suggested the presence of at least five ancestral chromosomes that could be considered as a foundation to build the present-day genomes by duplicating and reshuffling the existing information.

More detailed analyses of the diploid *Brassica* genomes structure were performed by Sadowski et al. (1996) (Table 4-6). These studies were limited to genetic and physical analysis of a cluster of five genes selected from a 15 kb segment on the *A. thaliana* chromosome 3 in three diploid *Brassica* species. This gene complex was used as a probe to identify common chromosome segments in *B. rapa*, *B. nigra* and *B. oleracea*. Three of the five probes displayed RFLPs in all three populations, whereas two of them segregated only in

B. nigra and *B. rapa*. Analysis of the hybridization patterns indicated that the number of the cluster members varies among species, only one of them was present in the single copy in all three genomes. The genes At1, At2 and At4 were represented by two or three loci in the *Brassica* genomes, mapping to three linkage groups of *B. oleracea* (C1, C6 and Cx) and to two in both *B. nigra* (B1 and B2) and *B. rapa* (A3 and A10), although not all genes were revealed due to the lack of polymorphism. Therefore, physical mapping based on pulse field gel electrophoresis (PFGE) was used in order to detect all five genes on the linkage groups A10, B2 and C1 of *B. rapa*, *B. nigra* and *B. oleracea*, respectively. The genetic and physical analyses indicated conservation of these chromosomal regions in *Brassica* in terms of general gene organization, relative to *A. thaliana*. Moreover, the additional, partial complexes in all three *Brassica* genomes were detected. The most complete secondary cluster was found in *B. nigra* containing four genes. These genes were linked to common sequences as genes coding the storage protein napin in all three genomes.

4.11 Genome Conservation between the *Brassica* and *A. thaliana* Species

Genome structure analyses indicate that many genomes at least partially have remained conserved during evolution (O'Brien et al. 1988; Whitkus et al. 1992; Bennetzen and Freeling 1993). The first studies of the genome structure revealed significant conservation of gene content and order among different plant species that diverged from a common ancestor more than 50 million years ago (Mya)(Crepet and Feldman 1991). Comparative mapping with the use of a common set of markers allows identification of homoeologous or homologous loci within collinear chromosomal regions in related species as well as rearrangements differentiating the genomes compared. Results of numerous studies indicated that mostly small rearrangements often took place in the plant genomes. For example, in rice and *Arabidopsis* 20% and 60% of genes identified were rearranged, respectively (Vision et al. 2000; Bennetzen and Ma 2003). Presently, these changes are mainly attributed to polyploidization and rediploidization processes, whose effects in the form of segmental duplication can be observed even in all diploid flowering plants (Kellogg and Bennetzen 2004). Hence, many genes are represented by some homologs, making it difficult to identify the accurate orthologos in the species compared. Knowledge of the genome conservation and ancient chromosomal rearrangements allows drawing conclusions about the evolution of different species. Therefore, a major challenge to plant biologists is to explain the mechanisms involved in the structural and functional evolution of duplicated genomes. Comparative genome analyses based on cross-hybridization of different plant genomes allow identification of sets of orthologous genes retaining identical gene order in some species. These studies have been initiated between closely related both diploid and polyploid species including the *Brassica* ones (Bonierbale et al. 1988; Helentjaris et al. 1988; Hulbert et al. 1990; Lagercrantz 1998; Babula et al. 2003).

The *Brassica* species represent a group of crops closely related to the model plant *A. thaliana*, whose genome sequence had been already determined. The phylogenetic analyses indicated that these species diverged from their common ancestor 17–18 Mya (Yang et al. 2006). Many detailed studies proved that the sequence homology between the *A. thaliana* genes and their homologues in *Brassica* is in the range from 75% to 90% in coding and < 70% in intronic regions (Cavell et al. 1998; Grant et al. 1998; Wroblewski et al. 2000; Quiros et al. 2001). Thanks to this high sequence homology, the *A. thaliana* genomic and cDNA clones could be used as probes on Southern blots to detect related sequences in the *Brassica* genomes. Their use gave new insight into the structure and evolution of their genomes. Analysis of chromosomal regions in *A. thaliana* and *Brassica* crops permitted detection of the collinear segments with a conserved gene order and a high degree of sequence conservation between these species (for example Sadowski et al. 1996; Lagercrantz et al. 1998; O'Neill and Bancroft 2000; Quiros et al. 2001; Babula et al. 2003; Gao et al. 2004; Ziółkowski et al. 2006). These studies showed that the *Brassica* genomes were a mosaic of *Arabidopsis*-like segments (Lagercrantz 1998; Babula et al. 2003). Moreover, many of the *Brassica* chromosomal regions were duplicated/triplicated with respect to the homologous segments of the *A. thaliana* genome (Lagercrantz 1998; Lan et al. 2000; Babula et al. 2003; Parkin et al. 2005). The existence of multiple chromosomal segments with partially conserved gene order in both genomes indicated that they could originate from the whole-genome duplication events in the *Arabidopsis* and *Brassica* lineages. The first evidence in support of this theory was found by mapping a small number of genes in both the *A. thaliana* and *Brassica* species (Kowalski et al. 1994; Teutonico and Osborn 1994). Kowalski et al. (1994) constructed the comparative map of *A. thaliana* and *B. oleracea* based on the previously mapped and newly selected sets of probes (Chang et al. 1988; Slocum et al. 1990). They detected 11 collinear chromosomal regions with a conserved loci order in the *A. thaliana* and *B. oleracea* maps covering 24.6% and 29.9% of the genome, respectively (Table 4-7). Analysis of distribution of these segments along the linkage groups pointed to triplication of the *B. oleracea* genome. Also, the duplication of the *A. thaliana* genome was suggested. Further and more detailed studies based on the genetic and physical mapping, permitted a more accurate definition of the degree of chromosome collinearity between *A. thaliana* and *Brassica* crops. Recently, in *B. rapa*, a reference high-density map was aligned to the *A. thaliana* genome

Table 4-7 Characteristics of conserved chromosomal regions in the *A. thaliana* and *Brassica* species.

Species	Average length of conserved region (cM)	Number of syntenic groups	Number of common loci	Number of breakpoints	Type of rearrangements	Map coverage (%)	Authors
B. rapa	No data	29	153	No data	Deletions, duplications, inversions	29	Choi et al. 2007
B. oleracea	21.3	11	110	26	Translocations, inversions	29.9	Kowalski et al. 1994
	7.1	34	273	No data	Translocations, Deletions, duplications	> 28	Bohuon et al. 1998; Lukens et al. 2003
	No data	24	186	19	Duplications, Inversions	No data	Lan et al. 2000
	20	42	212	24	Inversions, duplications, translocations	46.5	Babula et al. 2003
	19.3	24	306	14-34	Translocations, duplications	54	Kaczmarek et al. 2009
B. nigra	8	83	284	79-112	Inversion, translocations/ fusions	100	Lagercrantz 1998

(Choi et al. 2007). These studies indicated that 78% of the mapped markers had sequence homology to the *Arabidopsis* genome. A detailed analysis of distribution of these markers along all linkage groups allowed detection of 139 homologous regions identifying 29 syntenic groups of markers between these species. These groups covered 344 cM, representing 29% of the *B. rapa* reference map (Table 4-7). Some of them revealed extensive genomic rearrangements as inversions, duplications, deletions and overlappings of many syntenic regions. Lan et al. (2000) and Babula et al. 2003 (updated by Kaczmarek et al. 2009) constructed a comparative map of *B. oleracea* and *A. thaliana* based on the mapping of *Arabidopsis* ESTs to characterize genome organization of both species (see Section 4.7.2). The EST-based maps are more informative and recognize homologous regions between species. The simple *Arabidopsis/Brassica* "one to one" and "duplication" models accepted by Lan et al. (2000) indicated that 57% of comparative loci have been found in syntenic segments between *Arabidopsis* and *B. oleracea* (Table 4-7). It corroborated extensive synteny between both species. Also, further studies of Babula et al. (2003) showed extensive conservation of many chromosomal segments in *Arabidopsis* and *B. oleracea.* They found 42 regions with conserved gene organization covering 61.6% and 46.5% of the *A. thaliana* and *B. oleracea* maps, respectively (Table 4-7). In some cases these conserved regions included entire chromosomal arms in *Arabidopsis*. The updated map based on 160 *A. thaliana* ESTs and full-length cDNA allowed identification of 24 collinearity regions covering over 54% of the *B. oleracea* genome. These results were confirmed by LineUp algorithm specifying statistical significance of chromosomal homology detected on the basis of the density and order of genes (Hampson et al. 2003). Furthermore, the detected blocks of collinearity between *A. thaliana* and *B. oleracea* were compared with the conserved blocks found under the Ancestral Karyotype developed by Schranz et al. (2006). Lan et al. (2000) and Babula et al. (2003), later Kaczmarek et al. (2009) detected that each of the conserved regions in *A. thaliana* was homologous to at least two regions in *B. oleracea* confirming a secondary polyploid origin of *Brassica*. Simultaneously, Lukens et al. (2003) on the basis of the literature data (Bohuon et al. 1996) defined the most closely related segments in *B. oleracea* and *A. thaliana*. They proposed the criteria for distinction of the orthologous from the paralogous loci based on a comparison of the position of the sequenced *Brassica* loci on a *B. oleracea* genetic map with positions of their putative orthologs within the *A. thaliana* genome. Using these criteria they identified 34 syntenic regions in *A. thaliana* and *B. oleracea* (Table 4-7). The majority of *B. oleracea* linkage groups were strongly associated with a single *A. thaliana* chromosome: O2 and At5, O4 and At2, O5 and At1, O8 and At1, O9 and At5, respectively, although seven of the nine *B. oleracea* linkage groups had collinear regions with more than one *A. thaliana* chromosome.

Rana et al. (2004) detected identical order genes in *B. rapa* and *B. oleracea*, which confirmed that the two genomes were very similar and had a triplicated structure with respect to that of *Arabidopsis*. This similarity was expected, because the lineages of *B. oleracea* and *B. rapa* diverged only by about 4 Mya. Similarly, the comparative map of *A. thaliana* and *B. nigra* constructed by Lagercrantz (1998) allowed identification of 41 short collinear regions, distributed along all black mustard linkage groups (Table 4-7). He also identified some large chromosomal segments with conserved loci order. Analysis of this comparative map confirmed triplication of the *B.nigra* genome, i.e. all collinear regions were represented by three copies in *B. nigra*, while it occurred in its single homologous in *A. thaliana*. Further, the collinearity of chromosomal regions in the *A. thaliana* and *Brassica* genomes was tested at the micro-level including selected gene groups or chromosomal regions in both species. An example is the genetic and physical mapping of gene complex in the diploid *Brassica* genomes originally defined in *A. thaliana* (Sadowski et al. 1996). Combination of the linkage analysis and physical mapping by PFGE proved useful in defining the gene cluster in *Brassica* and permitted identification of the genes present in duplicated regions and determination of the physical sizes of all clusters in three diploid *Brassica*. Five linked genes being in a single copy on the *Arabidopsis* chromosome 3 were used as a probe. This analysis revealed the presence of multiple copies of this gene complex in the three *Brassica* genomes, although the number of cluster members varied among species. Interestingly, each of the three *Brassica* genomes had only a single complete cluster the same as in *Arabidopsis* and the other showed a greater or smaller rearrangement. Furthermore, the size of the whole cluster was larger in the *Brassica* genomes due to the gene structure and the sizes of intergenic regions. Similarly, Sadowski and Quiros (1998) using the map constructed by Truco and Quiros (1994), analyzed genetic and physical organization of the six gene-cluster (including *RPS2* gene coding resistance to *Pseudomonas syringae*) from the *A. thaliana* chromosome 4 in *B. nigra*. The RFLP analysis indicated a simple hybridization pattern obtained with the *RPS2* probe revealing a single locus and complex patterns obtained with the remaining five probes. In total, 17 RFLP loci were identified in the *B. nigra* genome by the six *A. thaliana* probes distributed over five linkage groups. The gene cluster with four to five members were found on the two *B. nigra* linkage groups and other smaller segments on three linkage groups. Only a single almost complete cluster with a total conservation of the *A. thaliana* complex was found. The physical analysis of this conserved segment showed differences in the size of this cluster in *B. nigra* and *A. thaliana*, which is in agreement with the genome-size difference of these two species. A different strategy of combined linkage analysis and physical mapping has been used by Ryder et al. (2001). They studied the organization of two regions in the *B. oleracea* genome selected from the integrated linkage map and the corresponding sequences in the *A. thaliana* physical map. The first of them included 74.8 cM region of the shortest *B. oleracea* linkage group O6 with the sporophytic self-incompatibility locus *S* and two copies of the MADS box transcription factor gene *BoAP1*. The second of them included 30.4-cM region from the longest *B. oleracea* linkage group O3 with a copy of the MADS box transcription factor gene *BoCALa* (Sebastian et al. 2000). They identified two regions within the linkage group O6 with the locus duplication and collinearity with *A. thaliana*. The linear markers organization indicated good conservation with respect to the corresponding region of *A. thaliana*. One of these two regions showed a total collinearity with a single segment on the *A. thaliana* chromosome I. However, the latter region represented an inverted version of the first region. They also found evidence for local rearrangements within an overall pattern of collinearity. Organization of markers from the *B. oleracea* linkage group O3 indicated the lack of internal duplication. However, this linkage group contained sequences homologous to several regions of the *A. thaliana* genome i.e. the *A. thaliana* chromosome I, II, II, IV and V. However, there are markers common to this region and the linkage group O6. Because of imperfections of genetic mapping such as a limited level of polymorphism, the chromosomal collinearity in the *A. thaliana* and *Brassica* genomes was not estimated. Recently, Gao et al. (2007) compared sequences of the 11 *B. oleracea* genes and three BAC-end and 155 cDNA markers to the *A. thaliana* physical map. Their results were consistent with data reported by Li et al. (2003) and partially in agreement with the alignment of C genome of the *B. napus* and *A. thaliana* chromosomes. The differences in C genome structure of *B. oleracea* and *B. napus* might reflect chromosomal changes taking place during the allopoliploization process of *B. napus*. The above mentioned data have been supported by cytogenetic studies and sequence analysis as described in Chapter 6. Besides the chromosomal collinearity between the *A. thaliana* and the *Brassica* species, numerous rearrangements have been found (Table 4-7). Kowalski et al. (1994) identified at least 26 extensive chromosomal rearrangements as translocations and inversions that occurred in *A. thaliana* and *B. oleracea* since the divergence of both genomes. Lan et al. (2000) found at least 19 chromosomal structural rearrangements that differentiate both these species. Some of these chromosomal breakpoints seemed to be common in the *B. oleracea* and *Arabidopsis* genomes pointing to the chromosome organization of their common ancestor. As in the earlier studies, many rearrangements such as inversion and translocation/fusions within the collinear segments are observed in *B. nigra* and *A. thaliana*, which have occurred since the divergence of both species from a common ancestor. Lan et al detected 79 to 112 different rearrangements depending on the method for estimating the number of these rearrangements used.

Finally, they presented the reconstruction of an ancestral genome structure having determined a minimal number of rearrangements.

The genomic duplications observed in the *Brassica* species and accumulation of mutations such as deletions, insertions and inversions including more chromosomal regions or individual genes could lead to rapid changes in genome structure and the divergence of gene function increasing phenotypic plasticity in the *Brassica* species. It was found that in *B. rapa* 62% genes were conserved within homoeologous segments and had their counterpart in *Arabidopsis*. Similarly, in *B. oleracea* 68% genes were conserved. These genes may also show function conservation. The relationship between gene localization within collinearity regions in *B. oleracea* and *A. thaliana* and their function conservation has been confirmed by Babula et al. (2006). Initially, they analyzed chromosomal organization of selected genes and their homologues involved in the ethylene biosynthesis and signaling pathways in the *B. oleracea* genome. Twenty two loci with sequence homology to these genes were placed on the existing *B. oleracea* map. Most of them were located within the collinearity regions in *B. oleracea* and *A. thaliana*. Next, changes in gene expression in response to ozone stress for three *ACS* genes in *B. oleracea* and *A. thaliana* were studied. These analyses confirmed functional conservation of the genes in both species. In recent years many studies suggested that the lack of some genes within homoeologous segments could result from gene loss or insertion that can be the main mechanism responsible for these changes. The variety of functions of each copy of the gene can lead to nonfunctionalization (one of duplicated gene copies is silenced), subfunctionalization (both copies may become partially compromised), or neofunctionalization (one of duplicated gene copies may acquire a novel function).

4.12 Conclusions

The current progress in the *Brassica* genome analysis is mainly associated with the availability of informative genetic maps, knowledge of the whole *Arabidopsis* genome sequence and development of sensitive cytogenetic methods such as BAC-FISH. These findings allowed pre-determination of the genomic structure of *Brassica* indicating the genomes triplication and presence of many rearrangements as a consequence of poliploid origin. Moreover, analysis of markers order along linkage groups of three diploid *Brassica* species disclosed a high degree of chromosomal conservation between A, B and C genomes. It is important in light of the advanced project of *B. rapa* genome sequencing (*Br*GSP), because *B. oleracea* and *B. nigra* will become the main beneficiaries of this knowledge. The first step is the alignment of the reference maps of the *Brassica* species. It is known that each of these maps has been constructed on the basis of different mapping populations, which means they are difficult to compare. Presently, this process is facilitated via mapping SSR sequences. The sequences are easily transferable between different species allowing identification of the corresponding sequences in various genomes. The first studies confirmed the usefulness of SSRs in alignment of some *Brassica* maps, which was discussed in this chapter. Furthermore, SSR programs currently developed in *Brassica* provide a vast number of microsatellite sequences. For example, for high-resolution mapping in *B. rapa*, BACs, ESTs and over 1,000 SSRs have been developed.

Construction of molecular maps is of great significance for increasing breeding efficiency as it permits analysis of QTL detection, marker- or gene-assisted selection, gene flow and fingerprinting for clone identification of economically important genes. Furthermore the maps facilitate detecting of a target region by searching public mapping database. Development of integrated maps, DNA and cDNA libraries and comprehensive databases is of importance for genome sequencing and structure and functional genomics. The studies of expression of duplicated genes in the *Brassica* species provide insight into the role of polyploidization in divergence of gene function leading to the phenotypic plasticity of *Brassica*.

Acknowledgements

This work was supported by a research grant from the Polish State Committee for Scientific Research (grant no. PBZ-MNiSW-2/3/2006/19 awarded to JS).

References

Abdurakhmonov IY, Abdullaev AA, Saha S, Buriev ZT, Arslanov D, Kuryazov Z, Mavlonov GT, Rizaeva SM, Reddy UK, Jenkins JN, Abdullaev A, Abdukarimov A (2005) Simple sequence repeat marker associated with a natural leaf defoliation trait in tetraploid cotton. J Hered 96: 644–653.

Akkaya MS, Bhagwat AA, Cregan PB (1992) Length polymorphisms of simple sequence repeat DNA in soybean. Genetics 132: 1131–1139.

Allender CJ, Allainguillaume J, Lynn J, King GJ (2007) Simple sequence repeats reveal uneven distribution of genetic diversity in chloroplast genomes of *Brassica oleracea* L. and (n = 9) wild relatives. Theor Appl Genet 114: 609–618.

Armstrong, SJ, Fransz P, Marshall DF, Jones GH (1998) Physical mapping of DNA repetitive sequences to mitotic and meiotic chromosomes of *Brassica oleracea* var. *alboglabra* by fluorescence *in situ* hybridization. Heredity 81: 666–673.

Arumuganathan K, Earle ED (1991) Nuclear DNA content of some important plant species. Plant Mol Rep 9: 208–219.

Arus P, Orton TJ (1983) Inheritance and linkage relationships of isozyme loci in *Brassica oleracea*. J Hered 74: 405–412.

Attia T, Röbbelen G (1986) Cytogenetic relationship within cultivated *Brassica* analysed in amphihaploids from the three diploid ancestors. Can J Genet Cytol 28: 323–329.

Axelsson T, Bowman CM, Sharpem AG, Lydiate DJ, Lagercrantz U (2000) Amphidiploid *Brassica juncea* contains conserved progenitor genomes. Genome 43: 679–688.

Ayele M, Haas BJ, Kumar N, Wu H, Xiao Y, Van Aken S, Utterback TR, Wortman JR, White OR, Town CD (2005) Whole genome shotgun sequencing of *Brassica oleracea* and its application to gene discovery and annotation in *Arabidopsis*. Genome Res 15: 487–495.

Babula D, Kaczmarek M, Barakat A, Delseny M, Quiros CF, Sadowski J (2003) Chromosomal mapping of *Brassica oleracea* based on ESTs from *Arabidopsis thaliana*: complexity of the comparative map. Mol Genet Genom 268: 656–665.

Babula D, Misztal LH, Jakubowicz M, Kaczmarek M, Nowak W, Sadowski J (2006) Genes involved in biosynthesis and signalisation of ethylene in *Brassica oleracea* and *Arabidopsis thaliana*: identification and genome comparative mapping of specific gene homologues. Theor Appl Genet 112: 410–420.

Babula D, Kaczmarek M, Ziolkowski PA, Sadowski J (2007) *Brassica oleracea*. In: C Kole (ed) Genome Mapping and Molecular Breeding in Plants, vol 5: Vegetables. Springer, Berlin, Heidelberg, New York, pp 227–285.

Beckmann JS, Soller M (1990) Toward a unified approach to genetic mapping of eukaryotes based on sequence tagged microsatellite sites. Biotechnology (NY) 8: 930–932.

Bell CJ, Ecker JR (1994) Assignment of 30 microsatellite loci to the linkage map of *Arabidopsis*. Genomics 19: 137–144.

Bennetzen JL (2000) Comparative sequence analysis of plant nuclear genomes: microcolinearity and its many exceptions. Plant Cell 12: 1021–1029.

Bennetzen JL, Freeling M (1993) Grasses as a single genetic system: genome composition, collinearity and complementarity. Trends Genet 9: 259–261.

Bennetzen JL, Ma J (2003) The genetic collinearity of rice and other cereals based on genomic sequence analysis. Curr Opin Plant Biol 6: 128–133.

Bohuon EJR, Keith DJ, Parkin IAP, Sharpe AG, Lagercrantz U, Lydiate DJ (1996) Alignment of the conserved C genome of *Brassica oleracea* and *Brassica napus*. Theor Appl Genet 93: 833–839.

Bohuon EJR, Ramsay LD, Craft JA, Arthur AE, Marshall DF, Lydiate DJ, Kearsey MJ (1998) The association of flowering time quantitative trait loci with duplicated regions and candidate loci in *Brassica oleracea*. Genetics 150: 393–401.

Bonierbale MW, Plaisted RL, Tanksley SD (1988) RFLP maps based on a common set of clones reveal modes of chromosomal evolution in potato and tomato. Genetics 120: 1095–1103.

Broman KW (2005) The genomes of recombinant inbred lines. Genetics 169: 1133–1146.

Camargo LEA (1994) Mapping RFLP and quantitative trait loci in *Brassica oleracea*. PhD Thesis, Univ of Wisconsin, Madison, WI, USA.

Camargo LEA, Savides L, Jung G, Nienhuis J, Osborn TC (1997) Location of the self-incompatibility locus in an RFLP and RAPD map of *Brassica oleracea*. J Hered 88: 57–60.

Cardle L, Ramsay L, Milbourne D, Macaulay M, Marshall D, Waugh R (2000) Computational and experimental characterization of physically clustered simple sequence repeats in plants. Genetics 156: 847–854.

Catcheside DG (1934) The chromosomal relationships in the Swede and turnip groups of *Brassica*. Ann Bot 48: 601–633.

Cavell AC, Lydiate DJ, Parkin IA, Dean C, Trick M (1998) Collinearity between a 30-centimorgan segment of *Arabidopsis thaliana* chromosome 4 and duplicated regions within the *Brassica napus* genome. Genome 41: 62–69.

Chang C, Bowman JL, Dejohn AW, Lander ES, Meyerowitz EM (1988) Restriction fragment length polymorphism linkage map for *Arabidopsis thaliana*. Proc Natl Acad Sci USA 85: 6856–6860.

Cheung WY, Champagne G, Hubert N, Landry BS (1997) Comparison of the genetic maps of *Brassica napus* and *Brassica oleracea*. Theor Appl Genet 94: 596–582.

Chevre AM, This P, Eber F, Deschamps M, Renard M, Delseny M, Quiros CF (1991) Characterization of disomic addition lines *Brassica napus*—*Brassica nigra* by isozyme, fatty acid, and RFLP markers. Theor Appl Genet 81: 43–49.

Choi SR, Teakle GR, Plaha P, Kim JH, Allender CJ, Beynon E, Piao ZY, Soengas P, Han TH, King GJ, Barker GC, Hand P, Lydiate DJ, Batley J, Edwards D, Koo DH, Bang JW, Park BS, Lim YP (2007) The reference genetic linkage map for the multinational *Brassica rapa* genome sequencing project. Theor Appl Genet 115: 777–792.

Chung SM, Staub JE (2003) The development and evaluation of consensus chloroplast primer pairs that possess highly variable sequence regions in a diverse array of plant taxa. Theor Appl Genet 107: 757–767.

Chyi Y-S, Hoenecke ME, Sernyk JL (1992) A genetic linkage map of restriction fragment length polymorphism loci for *Brassica rapa* (syn. *campestris*). Genome 35: 746–757.

Condit R, Hubbell SP (1991) Abundance and DNA sequence of two-base repeat regions in tropical tree genomes. Genome 34: 66–71.

Coulthart M, Denford KE (1982) Isozyme studies in *Brassica* I Electrophoretic techniques for leaf enzymes and comparison of *B. napus*, *B. campestris*, and *B. oleracea* using phosphoglucomutase. Can J Plant Sci 62: 621–630.

Crepet WL, Feldman GD (1991) The earliest remains of grasses in the fossil record. Am J Bot 78: 1010–1014.

Crouch ML, Tenbarga KM, Simon AE, Ferl R (1983) cDNA clones for Brassica napus seed storage proteins: evidence from nucleotide sequence analysis that both subunits of napin are cleaved from a precursor polypeptide. J Mol Appl Genet 2: 273-283.

De Vicente C, Thomas M, Adriana A (2004) Descriptors for genetic markers technologies. International Plant Genetic Resources Institute (IPGRI) Annual Report (www.futureharvest.org)

Dib C, Faure S, Fizames C, Samson D, Drouot N, Vignal A, Millasseau P, Marc S, Hazan J, Seboun E, Lathrop M, Gyapay G, Morissette J, Weissenbach J (1996) A comprehensive genetic map of a human genome based on 5,264 microsatellites. Nature 380: 152–154.

Dietrich WF, Miller JC, Steen RG, Merchant M, Damron D, Nahf R, Gross A, Joyce DC, Wessel M, Dredge RD, Marquis A, Stein LD, Goodman N, Page DC, Lander ES (1994) A genetic map of mouse with 4,006 simple sequence length polymorphisms. Nat Genet 7: 220–245.

Erickson LR, Straus, NA, Beversdorf WD (1983) Restriction patterns reveal origins of chroloplast genomes in *Brassica* amphidiploids. Theor Appl Genet 65: 201–206.

Ferreira ME, Williams PH, Osborn TC (1994) RFLP mapping of *Brassica napus* using F_1-derived doubled haploid lines. Theor Appl Genet 89: 615–621.

Ferreira ME, Satagopan J, Yandell BS, Williams RH, Osborn TC (1995) Mapping loci controlling vernalization requirement and flowering time in *Brassica napus*. Theor Appl Genet 90: 727–732.

Figdore SS, Kennard WC, Song KM, Slocum MK, Osborn TC (1988) Assessment of the degree of restriction fragment length polymorphism in *Brassica*. Theor Appl Genet 75: 833–840.

Flannery ML, Mitchell FJ, Coyne S, Kavanagh TA, Burke JI, Salamin N, Dowding P, Hodkinson TR (2006) Plastid genome characterisation in *Brassica* and Brassicaceae using a new set of nine SSRs. Theor Appl Genet 113: 1221–1231.

Foisset N, Delourme R, Barret P, Renard M (1995) Molecular tagging of the dwarf *BREIZH* (*Bzh*) gene in *Brassica napus*. Theor Appl Genet 91: 756–761.

Gao M, Li G, Yang B, McCombie WR, Quiros CF (2004) Comparative analysis of a *Brassica* BAC clone containing several major aliphatic glucosinolate genes with its corresponding *Arabidopsis* sequence. Genome 47: 666–679.

Gao M, Li G, Yang B, Qiu D, Farnham M, Quiros CF (2007) High-density *Brassica oleracea* linkage map: identification of useful new linkages. Theor Appl Genet 115: 277–287.

Goldstein DB, Schlötterer C (1999) Microsatellites: Evolution and Applications. Oxford Univ Press, Oxford, UK.

Gottlieb LD (1983) Isozyme number and phylogeny. In: U Jensen ,DE Fairbrothers (eds) Proteins and Nucleic Acids in Plant Systematics. Springer, Berlin, Heidelberg, New York, pp 209–221.

Grant MR, McDowell JM, Sharpe AG, de Torres Zabala M, Lydiate DJ, Dangl JL (1998) Independent deletions of a pathogen-resistance gene in *Brassica* and *Arabidopsis*. Proc Natl Acad Sci USA 22: 15843–15848.

Gupta PK, Varshney RK, Sharma PC, Ramesh B (1999) Molecular markers and their application in wheat breeding. Plant Breed 118: 369–390.

Gustafsson M, Bentzer B, Von Bothmer R, Snogerup S (1976) Meiosis in Greek *Brassica oleracea* group. Bot Not 129: 73–84.

Haldane JBS (1919) The combination of linkage values and the calculation of distance between loci of linked factors. J Genet 8: 299–309.

Hampson S, McLysaght A, Gaut B, Baldi P (2003) Lineup: statistical detection of chromosomal homology with applications to plant comparative genomics. Genome Res 13: 999–1010.

Harada JJ, Baden CS, Comai L (1988) Spatially regulated genes expressed during seed germination and postgerminative development are activated during embryogeny. Mol Gen Genet 212: 466–473.

Helentjaris T, Weber D, Wright S (1988) Identification of the genomic locations of duplicate nucleotide sequences in maize by analysis of restriction fragment length polymorphism. Genetics 118: 353–361.

Heneen WK, Chen BY, Cheng BF, Jonsson A, Simonsen V, Jorgensen RB, Davik J (1995) Characterization of the A and C genomes of *Brassica campestris* and *B. alboglabra*. Hereditas 123: 251–267.

Hittalmani S, Shashidhar HE, Bagali PG, Huang N, Sidhu JS, Singh VP, Khush GS. (2002) Molecular mapping of quantitative trait loci for plant growth, yield and yield related traits across three diverse locations in a doubled haploid rice population. Euphytica 125: 207–214.

Hoenecke M, Chyi Y-S (1991) Comparison of *Brassica napus* and *B. rapa* genomes based on restriction fragment length polymorphism mapping. In: DI McGregor (ed) Proc 8th Int Rapeseed Congress GCIRC, Saskatoon, Canada, pp 1102–1107.

Hong, CP, Kwon SJ, Kim JS, Yang T-J, Park B-S, Lim JP (2008) Progress in understanding and sequencing the genome of *Brassica rapa*. Int J Plant Genomics 2008: 1–10.

Hosaka K, Kianian SF, McGrath JM, Quiros CF (1990) Development and chromosomal localization of genome specific DNA markers of *Brassica* and the evolution of amphidiploid and n=9 diploid species. Genome 33: 131–142.

Howell EC, Barker GC, Jones GH, Kearsey MJ, King GJ, Kop EP, Ryder CD, Teakle GR, Vicente JG, Armstrong SJ (2002) Integration of the cytogenetic and genetic linkage maps of *Brassica oleracea*. Genetics 161: 1225–1234

Hu J, Sadowski J, Osborn TC, Landry BS, Quiros CF (1998) Linkage group alignment from four independent *Brassica oleracea* RFLP maps. Genome 41: 226–235.

Hulbert SH, Richter TE, Axtell JD, Bennetzen JL (1990) Genetic mapping and characterization of sorghum and related crops by means of maize DNA probes. Proc Natl Acad Sci USA 87: 4251–4255.

Iniguez-Luy FL, Voort AV, Osborn TC (2008) Development of a set of public SSR markers derived from genomic sequence of a rapid cycling *Brassica oleracea* L. genotype. Theor Appl Genet 117: 977–985.

Johnston JS, Pepper AE, Hall AE, Chen ZJ, Hodnett G, Drabek J, Lopez R, Price HJ (2005) Evolution of genome size in *Brassicaceae*. Ann Bot 95: 229–235

James RV, Williams PH (1980) Clubroot resistance and linkage in *Brassica campestris*. Phytopathology 70: 776–779.

Kaczmarek M, Koczyk G, Ziolkowski PA, Sadowski J (2009) Comparative analysis of the *Brassica oleracea* genetic map and the *A. thaliana* genome. Genome 52: 620–633.

Katti MV, Ranjekar PK, Gupta VS (2001) Differential distribution of simple sequence repeats in eukaryotic genome sequences. Mol Biol Evol 18: 1161–1167.

Kellogg EA, Bennetzen JL (2004) The evolution of nuclear genome structure in seed plants. Am J Bot 91: 1709–1725.

Kennard WC, Slocum MK, Figdore SS, Osborn TC (1994) Genetic analysis of morphological variation in *Brassica oleracea* using molecular markers. Theor Appl Genet 87: 721–732.

Kianian SF (1990) Evolutionary trends in the *Brassica oleracea*-cytodeme: chromosomal and linkage relationships. PhD Thesis, Univ of California, Davis, CA, USA.

Kianian SF, Quiros CF (1992a) Generation of a *Brassica oleracea* composite RFLP map: linkage arrangements among various populations and evolutionary implications. Theor Appl Genet 84: 544–554.

Kianian SF, Quiros CF (1992b) Genetic analysis of major multigene families in *Brassica oleracea* and related species. Genome 35: 516–527.

Kim HJ, Chang-Soon J, Kyoo-Won C, Yong Pyo L (1999) AFLP and RAPD mapping of Chinese cabbage (*Brassica rapa* L var *pekinensis*). In: Plant Anim Genome VII Conf, San Diego, CA, USA, p 141.

Kim JS, Chung TY, King GJ, Jin M, Yang T-J, Jin Y-M, Kim H-I, Park B-S (2006) A sequence-tagged linkage map of *Brassica rapa*. Genetics 174: 29–39.

Kole C, Teutonico R, Mengistu A, Williams PH, Osborn TC (1996) Molecular Mapping of a Locus Controlling Resistance to *Albugo candida* in *Brassica rapa*. Genetics 86: 367–369

Kole C, Kole P, Vogelzang R, Osborn TC (1997) Genetic linkage map of a *Brassica rapa* recombinant inbred population. J Hered 88: 553–557.

Kole C, Williams PH, Rimmer SR, Osborn TC (2002) Linkage mapping of genes controlling resistance to white rust (*Albugo candida*) in *Brassica rapa* (syn. *campestris*) and comparative mapping to *Brassica napus* and *Arabidopsis thaliana*. Genome 45: 22–27.

Koo DH, Plaha P, Lim YP, Hur Y, Bang JW (2004) A high resolution karyotype of *Brassica rapa* ssp. *pekinensis* revealed by pachytene analysis and multicolor fluorescence in situ hybridization. Theor Appl Genet 109: 1346–1352.

Koornneef M, JH Van der Veen (1983) Trisomics in *Arabidopsis thaliana* and the location of linkage groups Genetica 61: 41–46.

Kosambi DD (1944) The estimation of map distance from recombination values. Ann Eugen 12: 172–175.

Kowalski SP, Lan TH, Feldmann KA, Paterson AH (1994) Comparative mapping of *Arabidopsis thaliana* and *Brassica oleracea* chromosomes reveals islands of conserved organization. Genetics 138: 499–510.

Kresovich S, Williams JGK, McFerson JR, Routman EJ, Schaal BA (1992) Characterization of genetic identities and relationships of *Brassica oleracea* L. via a random amplified polymorphic DNA assay. Theor Appl Genet 85: 190–196.

Kresovich S, Szewc-McFadden AK, Bliek SM, McFerson JR (1995) Abundance and characterization of simple –sequence repeats (SSRs) isolated from a size-fractionated genomic library of *Brassica napus* L. (rapeseed). Theor Appl Genet 91: 206–211.

Künzel G, Korzun L, Meister A (2000) Cytologically integrated physical restriction fragment length polymorphism maps for the barley genome based on translocation breakpoints. Genetics 154: 397–412.

Lagercrantz U (1998) Comparative mapping between *Arabidopsis thaliana* and *Brassica nigra* indicates that *Brassica* genomes have evolved through extensive genome replication accompanied by chromosome fusions and frequent rearrangements. Genetics 150: 1217–1228.

Lagercrantz U, Lydiate DJ (1995) RFLP mapping in *Brassica nigra* indicates differing recombination rates in male and female meioses. Genome 38: 255–264.

Lagercrantz U, Lydiate D (1996) Comparative genome mapping in *Brassica*. Genetics 144: 1903–1910.

Lagercrantz U, Ellegren H, Andersson L (1993) The abundance of various polymorphic microsatellite motifs differs between plants and vertebrates. Nucl Acids Res 21: 1111–1115.

Lan T-H, DelMonte TA, Rieschmann KP, Hyman J, Kowalski SP, McFerson J, Kresovich S, Paterson AH (2000) An EST-enriched comparative map of *Brassica oleracea* and *Arabidopsis thaliana*. Genome Res 10: 776–788.

Lander ES, Green P, Abrahamson J, Barlow A, Daly MJ, Lincoln SE, Newberg LA (1987) MAPMAKER: an interactive computer package for constructing primary genetic linkage maps of experimental and natural populations. Genomics 1: 174–181.

Landry BS, Hubert N, Crete R, Chang MS, Lincoln SE, Etoh T (1992) A genetic map for *Brassica oleracea* based on RFLP markers detected with expressed DNA sequences and mapping of resistance genes to races 2 of *Plasmodiophora brassicae* (Woronin). Genome 35: 409–420.

Lazaro A, Aguinagalde I (1998) Genetic diversity in *Brassica oleracea* L. (*Crucifera*) and wild relatives (2n=18) using RAPD markers. Ann Bot 82: 829–833.

Li G, Quiros CF (2001) Sequence-related amplified polymorphism (SRAP), a new marker system based on a simple PCR reaction: its application to mapping and gene tagging in *Brassica*. Theor Appl Genet 103: 455–461.

Li G, Gao M, Yang B, Quiros CF (2003) Gene for gene alignment between the *Brassica* and *Arabidopsis* genomes by direct transcriptome mapping. Theor Appl Genet 107:168–180.

Li RJ, Wang HZ, Mao H, Lu YT, Hua W (2006) Identification of differentially expressed genes in seeds of two near-isogenic *Brassica napus* lines with different oil content. Planta 224: 952–962.

Li Y-C, Korol AB, Fahima T, Beiles A, Nevo E (2002) Microsatellites: genomic distribution, putative functions and mutational mechanisms: a review. Mol Ecol 11: 2453–2465.

Lim GAC, Jewell EG, Li X, Erwin TA, Love C, Batley J, Spangenberg G, Edwards D (2007) A comparative map viewer integrating genetic maps for *Brassica* and *Arabidopsis*. BMC Plant Biol 7: 40.

Lim KB, de Jong H, Yang TJ, Park JY, Kwon SJ, Kim JS, Lim MH, Kim JA, Jin M, Jin YM, Kim SH, Lim YP, Bang JW, Kim HI, Park BS (2005) Characterization of rDNAs and tandem repeats in the heterochromatin of *Brassica rapa*. Mol Cells 19: 436–444.

Lincoln S, Daly M, Lander E (1992) Mapping genes controlling quantitative traits with MAPMAKER/QTL 11. Whitehead Institute Technical Report, 2nd edn.

Liu BH, Knapp SJ (1990) GMENDEL: a program for Mendelian segregation and linkage analysis of individual or multiple progeny populations using log-likelihood ratio's. J Hered 81: 407.

Lombard V, Delourme R (2001) A consensus linkage map for rapeseed (*Brassica napus* L.): construction and integration of three individual maps from DH populations. Theor Appl Genet 103: 491–507.

Lowe AJ, Jones AE, Raybould AF, Trick M, Moule CL, Edwards KJ (2002) Transferability and genome specificity of a new set of microsatellite primers among *Brassica* species of the U triangle. Mol Ecol Notes 2: 7–11.

Lowe AJ, Moule C, Trick M, Edwards KJ (2004) Efficient large-scale development of microsatellites for marker and mapping applications in *Brassica* crop species. Theor Appl Genet 108: 1103–1112.

Lukens L, Zou F, Lydiate D, Parkin I, Osborn T (2003) Comparison of a *Brassica oleracea* genetic map with the genome of *Arabidopsis thaliana*. Genetics 164: 359–372.

Lydiate D, Sharpe A, Lagercrantz U, Parkin I (1993) Mapping the *Brassica* genome. Outlook Agri 22: 85–92.

Lysak MA, Koch MA, Pecinka A, Schubert I (2005) Chromosome triplication fund across the tribe Brassiceae. Genome Res 15: 516–525.

McCouch SR, Teytelman L, Xu Y, Lobos KB, Clare K, Walton M, Fu B, Maghirang R, Li Z, Xing Y, Zhang Q, Kono I, Yano M, Fjellstrom R, DeClerck G, Schneider D, Cartinhour S, Ware D, Stein L (2002) Development and mapping of 2240 new SSR markers for rice (*Oryza sativa* L.). DNA Res 31: 199–207.

McGrath JM, Quiros CF (1991) Inheritance of isozyme and RFLP markers in *Brassica campestris* and comparison with *B. oleracea*. Theor Appl Genet 82: 668–673.

McGrath JM, Quiros CF, Harada JJ, Landry BS (1990) Generation of alien chromosome addition lines from synthetic *Brassica oleracea* monosomic alien chromosome addition lines with molecular markers reveals extensive gene duplication. Mol Gen Genet 223: 198–204.

Mizushima U (1950) Karyogenetic studies of species and genus hybrids in the tribe *Brassiceae* of *Cruciferae*. Tohoku J Agri Res 1: 15–27.

Morgante M, Hanafey M, Powell W (2002) Microsatellites are preferentially associated with nonrepetitive DNA in plant genomes. Nat Genet 30: 194–200.

Muangprom A, Osborn T (2004) Characterization of draft gene in *Brassica rapa*, including the identification of a candidate gene. Theor Appl Genet 108: 1378–1384.

Mueller UG, Wolfenbarger LL (1999) AFLP genotyping and fingerprinting. Trends Ecol Evol 14: 389–394.

Mukherjee AK, Mohapatra T, Varshney A, Sharma R, Sharma RP (2001) Molecular mapping of a locus controlling resistance to *Albugo candida* in Indian mustard. Plant Breed 120: 483–487.

Nozaki T, Kumazaki A, Koba T, Ishikawa K, Ikehashi H (1997) Linkage analysis among loci for RAPDs, isozymes and some agronomic traits in *Brassica campestris* L. Euphytica 95: 115–123.

O'Brien SJ, Seuánez HN, Womack JE (1988) Mammalian genome organization: an evolutionary view. Annu Rev Genet 22: 323–351.

Ohno S (1970) Evolution by Gene Duplication. Springer, Berlin, Heidelberg, New York.

Olsson G (1960) Species crosses within the genus *Brassica*. II Artificial *Brassica napus* L. Hereditas 46: 351–396.

O'Neil CM, Bancroft I (2000) Comparative physical mapping of segments of the genome of *Brassica oleracea* var. *alboglabra* that are homoeologous to sequenced regions of chromosome 4 and 5 *Arabidopsis thaliana*. Plant J 23: 233–243.

Østergaard L, King GJ (2008) Standardized gene nomenclature for the *Brassica* genus. Plant Meth 4:10.

Palmer JD, Shield CR, Cohen DB, Orton TJ (1983) Chloroplast DNA evolution and the origin of amphidiploid *Brassica* species. Theor Appl Genet 65: 181–189.

Panaud O, Chen X, McCouch SR (1996) Development of microsatellite markers and characterization of simple sequence length polymorphism (SSLP) in rice (*Oryza sativa* L.). Mol Gen Genet 16: 597–607.

Panjabi P, Jagannath A, Bisht NC, Padmaja KL, Sharma S, Gupta V, Pradhan AK, Pental D (2008) Comparative mapping of *Brassica juncea* and *Arabidopsis thaliana* using Intron Polymorphism (IP) markers: homeologous relationships, diversification and evolution of A, B and C *Brassica* genomes. BMC Genom 9: 113.

Parkin IA, Sharpe AG, Keith DJ, Lydiate DJ (1995) Identification of the A and C genomes of amphiploid *Brassia napus* (oilseed rape). Genome 38: 1122–1133.

Parkin IA, Gulden SM, Sharpe AG, Lukens L, Trick M, Osborn TC, Lydiate DJ (2005) Segmental structure of the *Brassica napus* genome based on comparative analysis with *Arabidopsis thaliana*. Genetics 171: 765–781.

Paterson AH, Lan TH, Reischmann KP, Chang C, Lin YR, Liu SC, Burow MD, Kowalski SP, Katsar CS, DelMonte TA, Feldmann KA, Schertz KF, Wendel JF (1996) Toward a unifed map of higher plant chromosomes, transcending the monocot-dicot divergence. Nat Genet 14: 380–382.

Paterson AH, Lan TH, Amasino R, Osborn TC, Quiros C (2001) *Brassica* genomics: a complement to, and early beneficiary of the *Arabidopsis* sequence. Genome Biol 2: 1011.1–1011.4.

Pease MA (1926) Genetic situation in *Brassica oleracea*. J Genet 16: 363–385.

Piao ZY, Deng YQ, Choi SR,Park YJ, Lim YP (2004) SCAR and CAPS mapping of CRb, a gene conferring resistance to *Plasmodiophora brassicae* in Chinese cabbage (*Brassica rapa* ssp. *pekinensis*). Theor Appl Genet 108: 1458–1465.

Ping L, Jungeng K, Guoyu Z, Bonnema G, Zhyuang F, Xiaowu W (2007) Transcript profiling of domonant male sterile mutant (Ms-cd1) in cabbage during flower bud development. Plant Sci 172: 111–119.

Piquemal J, Cinquin E, Couton F, Rondeau C, Seignoret E, Doucet I, Perret D, Villeger MJ, Vincourt P, Blanchard P (2005) Construction of an oilseed rape (*Brassica napus* L.) genetic map with SSR markers. Theor Appl Genet 111: 1514–1523.

Plieske J, Struss D (2001) Microsatellite markers for genome analysis in *Brassica*. I development in *Brassica napus* and abundance in Brassicaceae species. Theor Appl Genet 102: 689–694.
Prabhu KV, Somers DJ, Rakow G, Gugel RK (1998) Molecular markers linked to white rust resistance in mustard *Brassica juncea*. Theor Appl Genet 97: 865–870.
Pradhan AK, Gupta V, Mukhopadhyay A, Arumugam N, Sodhi YS, Pental D (2003) A high-density linkage map in *Brassica juncea* (Indian mustard) using AFLP and RFLP markers. Theor Appl Genet 106: 607–614.
Prakash S, Hinata K (1980) Taxonomy, cytogenetics and origin of crop Brassicas, a review. Opera Bot 55: 1–57.
Rae AM, Howell EC, Kearsey MJ (1999) More QTL for flowering time revealed by substitution lines in *Brassica oleracea*. Heredity 83: 586–596.
Rahman M, McVetty PBE, Li G (2007) Development of SRAP, SNP and Multiplexed SCAR molecular markers for the major seed coat color gene in *Brassica rapa* L. Theor Appl Genet 115: 1101–1107.
Ramanujam S, Srinivasachar D (1943) Cytogenetic investigations in the genus *Brassica* and the artificial synthesis of *B. juncea*. Ind J Genet 3: 73–88.
Ramchiary N, Padmaja KL, Sharma S, Gupta V, Sodhi YS, Mukhopadhyay A, Arumugam N, Pental D, Pradhan AK (2007) Mapping of yield influencing QTL in *Brassica juncea*: implications for breeding of a major oilseed crop of dryland areas. Theor Appl Genet 115: 807–817.
Ramsay LD, Jennings DE, Bohuon EJR, Arthur AE, Lydiate DJ, Kearsey MJ, Marshall DF (1996) The construction of a substitution library of recombinant backcross lines in *Brassica oleracea* for the precision mapping of quantitative trait loci. Genome 39: 558–567.
Rana D, Boogaart T, O'Neill CM, Hynes L, Bent E, Macpherson L, Park JY, Lim YP, Bancroft I (2004) Conservation of the microstructure of genome segments in *Brassica napus* and its diploid relatives. Plant J 40: 725–733.
Rosenberg NA, Pritchard JK, Weber JL, Cann HM, Kidd KK, Zhivotovsky LA, Feldman MW (2002) Genetic structure of human populations. Science 298: 2381–2385.
Röbbelen G (1960) Beiträge zur analyse des *Brassica* genomes. Chromosoma 11: 205–228.
Ryder CD, Smith LB, Teakle GR, King GJ (2001) Contrasting genome organisation: two regions of the *Brassica oleracea* genome compared with collinear regions of the *Arabidopsis thaliana* genome. Genome 44: 808–817.
Qi X, Stam P, Lindhout P (1996) Comparison and integration of four barley genetic maps. Genome 39: 379–394.
Qiu D, Morgan C, Shi J, Long Y, Liu J, Li R, Zhuang X, Wang Y, Tan X, Dietrich E, Weihmann T, Everett C, Vanstraelen S, Beckett P, Fraser F, Trick M, Barnes S, Wilmer J, Schmidt R, Li J, Li D, Meng J, Bankroft I (2006) Comparative linkage map of oilseed rape and its use for QTL analysis of seed oil and erucic acid content. Theor Appl Genet 114: 67–80.
Queller DC, Strassmann JE, Hughes CE (1993) Microsatellites and kinship. Trends Ecol Evol 8: 285–288.
Quiros CF, Ochoa O, Kianian SF, Douches D (1987) Analysis of the *Brassica oleracea* genome by the generation of *B. rapa-oleracea* chromosome addition lines: characterization by isozymes and rDNA genes. Theor Appl Genet 74: 758–766.
Quiros CF, Ochoa O, Douches DS (1988) Exploring the role of x = 7 species in *Brassica* evolution: hybridization with *B. nigra* and *B. oleracea*. J Hered 79: 351–358.
Quiros CF, Hu J, This P, Chevre AM, Delseny M (1991) Development and chromosomal localization of genome specific markers by polymerase chain reaction in *Brassica*. Theor Appl Genet 82: 627–632.
Quiros CF, Grellet F, Sadowski J, Suzuki T, Li G, Wroblewski T (2001) *Arabidopsis* and *Brassica* comparative genomics: sequence, structure and gene content in the *ABI1-Rps2-Ck1* chromosomal segment and related regions. Genetics 157: 1321–1330.

Saal B, Plieske J, Hu J, Quiros CF, Struss D (2001) Microsatellite markers for genome analysis in *Brassica* II Assignment of rapeseed microsatellites to the A and C genomes and genetic mapping in *Brassica oleracea* L. Theor Appl Genet 102: 695–699.

Sadowski J, Quiros CF (1998) Organization of an *Arabidopsis thaliana* gene cluster on chromosome 4 including the *RPS2* gene, in the *Brassica nigra* genome. Theor Appl Genet 96: 468–474.

Sadowski J, Gaubier P, Delseny M, Quiros CF (1996) Genetic and physical mapping in *Brassica* diploid species of a gene cluster defined in *Arabidopsis thaliana*. Mol Gen Genet 251: 298–306.

Sampson DR (1978) A second gene for hairs in *B. oleracea* and its tentative location in linkage group four. Can J Genet Cytol 20: 101–109.

Sandhu D, Champoux JA, Bondareva SN, Gill KS (2001) Identification and physical localization of useful genes and markers to a major gene-rich region on wheat group 1S chromosomes. Genetics 157: 1735–1747.

Schranz EM, Lysak MA, Mitchell-Olds T (2006) The ABC's of comparative genomics in the Brassicaceae: building blocks of crucifer genomes. Trends Plant Sci 11: 535–542.

Sebastian RL, Howell EC, King GK, Marshall DF, Kearsey MJ (2000) An integrated AFLP and RFLP *Brassica oleracea* linkage map from two morphologically distinct double haploid mapping populations. Theor Appl Genet 100: 75–81.

Sharpe AG, Parkin AP, Keith DJ, Lydiate DJ (1995) Frequent non reciprocal translocations in the amphidiploid genome of oilseed rape (*Brassica napus*). Genome 38: 1112–1121.

Shoemaker RC, Guffy RD, Lorenzen LL, Specht JE (1992) Molecular genetic mapping of soybean: map utilization. Crop Sci 32: 1091–1098.

Singh AK, Prasanna BM Molecuar mapping in crop plants: development and characterization of mapping populations http://wwwscribdcom/doc/6229849/Mapping-Population.

Slocum MK (1989) Analyzing the genomic structure of *Brassica* species and subspecies using RFLP analysis. In: T Helentjaris, E Burr (eds) Development and Application of Molecular Markers to Problems in Plant Genetics. Cold Spring Harbor Laboratory Press, New York, USA, pp 73–80.

Slocum MK, Figdore SS, Kennard WC, Suzuki JY, Osborn TC (1990) Linkage arrangement of restriction fragment length polymorphism loci in *Brassica oleracea*. Theor Appl Genet 80: 57–64.

Smith LB, King GJ (2000) The distribution of BoCAL-a alleles in *Brassica oleracea* is consistent with a genetic model for curd development and domestication of the cauliflower. Mol Breed 6: 603–613.

Snowdon RJ, Friedt W (2004) Molecular markers in *Brassica* oilseed breeding: current status and future possibilities. Plant Breed 123: 1–8.

Soengas P, Hand P, Vicente JG, Pole JM, Pink DAC (2007) Identification of quantitative trait loci for resistance to *Xanthomonas campestris* pv *campestris* in *Brassica rapa*. Theor Appl Genet 114: 637–645.

Song KM, Osborn TC, Williams PH (1988) *Brassica* taxonomy based on nuclear restriction fragment length polymorhisms (RFLPs). 1 Genome evolution of diploid and amphidiploid species. Theor Appl Genet 75: 784–794.

Song KM, Osborn TC, Williams PH (1990) *Brassica* taxonomy based on nuclear restriction fragment length polymorphism (RFLPs). 3 Genome relationships in *Brassica* and related genera and the origin of *B. oleracea* and *B. rapa* (syn. *campestris*). Theor Appl Genet 79: 479–506.

Song KM, Suzuki JY, Slocum MK, Williams PH, Osborn TC (1991) A linkage map of *Brassica rapa* (syn. *campestris*) based on restriction fragment length polymorphism loci. Theor Appl Genet 82: 296–304.

Stam P (1993) Construction of integrated genetic linkage maps by means of a new computer package: JoinMap. Plant J 3: 739–744.

Stringam GR (1977) Linkage relationships of four seedling mutants in turnip rape. J Hered 68: 391–394.

Sun Z, Wang Z, Tu J, Zhang J, Yu F, McVetty PB, Li G (2007) An ultradense genetic recombination map for *Brassica napus*, consisting of 13551 SRAP markers. Theor Appl Genet 114: 1305–1317.
Suwabe K, Iketani H, Nunome T, Kage T, Hirrai M (2002) Isolation and characterization of microsatellites in *Brassica rapa* L. Theor Appl Genet 104: 1092–1098.
Suwabe K, Tsukazaki H, Iketani H, Hatakeyama K, Fujimura M, Nunome T, Fukuoka H, Matsumoto S, Hirai M (2003) Identification of two loci for resistance to clubroot (*Plasmodiophora brassicae* Woronin) in *Brassica rapa* L. Theor Appl Genet 107: 997–1002.
Suwabe K, Iketani H, Nunome T, Ohyama A, Hirai M, Fukuoka H (2004) Characteristics of microsatellites in *Brassica rapa* genome and their potential utilization for comparative genomics in cruiferae. Breed Res 54: 85–90.
Suwabe K, Tsukazaki H, Iketani H, Hatakeyama K, Kondo M, Fujimura M, Nunome T, Fukuoka H , Hirai M, Matsumoto M (2006) SSR-based comparative genomics between *Brassica rapa* and *Arabidopsis thaliana*: the genetic origin of clubroot resistance. Genetics 173: 309–319.
Suwabe K, Morgan C, Bancroft I (2008) Integration of *Brassica* A genome genetic linkage map between *Brassica napus* and *B. rapa*. Genome 51: 169–176.
Szewc-McFadden AK, Kresovich S, Bliek SM, Mitchell SE, McFerson JR (1996) Identification of polymorphic, conserved simple sequence repeats (SSRs) in cultivated *Brassica* species. Theor Appl Genet 93: 534–538.
Teutonico RA, Osborn TC (1994) Mapping of RFLP and qualitative trait loci in *Brassica rapa* and comparison to the linkage maps of *B. napus*, *B. oleracea* and *Arabidopsis thaliana*. Theor Appl Genet 89: 885–894.
The *Arabidopsis* Genome Initiative (2000) Analysis of the genome of the flowering plant *Arabidopsis thaliana*. Nature 408: 796–815.
This P (1992) Characterisation et development de marqueurs moleculaires en vue de l'analyse de lignees d'addition *Brassica napus-Brassica nigra* et *Diplotaxis erucoides-Brassica nigra*. PhD Thesis, Univ of Perpignan, France.
Thormann CE, Ferreira ME, Camargo LEA, Tivang JG, Osborn TC (1994) Comparison of RFLP and RAPD markers to estimating genetic relationship within and among cruciferous species. Theor Appl Genet 88: 973–980.
Tiffin P, Hahn MW (2002) Coding sequence divergence between two closely related plant species: *Arabidopsis thaliana* and *Brassica rapa* ssp. *pekinensis*. J Mol Evol 54: 746–753.
Tonguc M, Griffiths PD (2004) Genetics relationships of *Brassica* vegetables determined using database derived simple sequence repeats. Euphytica 137: 193–201.
Toth G, Gaspari Z, Jurka J (2000) Microsatellites in different eucariotic genomes: survey and analysis. Genome Res 10: 967–981.
Truco MJ, Quiros CF (1994) Structure and organization of the B genome based on a linkage map in *Brassica nigra*. Theor Appl Genet 89: 590–598.
Truco MJ, Hu J, Sadowski J, Quiros CF (1996) Inter- and intra-genomic homology of the *Brassica* genomes: implications on their origin and evolution. Theor Appl Genet 93: 1225–1233.
U N (1935) Genome analysis in *Brassica* with special reference to the experimental formation of *B. napus* and peculiar mode of fertilization. Jpn J Genet 7: 784–794.
Uzunova MI, Ecke W (1999) Abundance, polymorphism and genetic mapping of microsatellites in oilseed rape (*Brassica napus* L.). Plant Breed 118: 323–326.
Varshney A, Mohapatra T, Sharma RP (2004) Development and validation of CAPS and AFLP markers for white rust resistance gene in *Brassica juncea*. Theor Appl Genet 109: 153–159.
Vaughan JG (1977) A multidisciplinary study of the taxonomy and origin of *Brassica* crops. BioScience 27: 35–40.
Verma SC, Rees H (1974) Nuclear DNA and the evolution of allotetraploid *Brassiceae*. Heredity 33: 61–68.
Vision TJ, Brown DG, Tanksley SD (2000) The origins of genomic duplications in *Arabidopsis*. Science 290: 2114–2117.

Vos P, Hogers R, Bleeker M, Reijans M, van de Lee T, Hornes M, Frijters A, Pot J, Peleman J, Kuiper M, Zabeau M (1995) AFLP: a new technique for DNA fingerprinting. Nucl Acids Res 21: 4407–4414.

Warwick SI, Black LD (1991) Molecular systematics of *Brassica* and allied genera (subtribe Brassicinae, Brassiceae): chloroplast genome and cytodeme congruence. Theor Appl Genet 82: 81–92.

Weber JL, May PE (1989) Abundant class of human DNA polymorphisms which can be typed using the polymerase chain reaction. Am J Hum Genet 44: 388–396.

Westman AL, Kresovich S (1998) The potential for cross-taxa simple-sequence repeat (SSR) amplification between *Arabidopsis thaliana* L. and crop brassicas. Theor Appl Genet 96: 272–281.

Whitkus R, Doebley J, Lee M (1992) Comparative mapping of sorghum and maize. Genetics 132: 1119–1130.

Williams JGK, Kubelik AR, Livak KJ,Rafalski JA, Tingey S (1990) DNA polymorphisms amplified by arbitrary primers are useful as genetic markers. Nucl Acids Res 25: 6531–6535.

Wroblewski T, Coulibaly S, Sadowski J, Quiros CF (2000) Variation and phylogenetic utility of the *Arabidopsis thaliana Rps2* homolog in various species of the tribe Brassiceae. Mol Phylogenet Evol 16: 440–448.

Yang TJ, Kim JS, Kwon SJ, Lim KB, Choi BS, Kim JA, Jin M, Park JY, Lim MH, Kim HI, Lim YP, Kang JJ, Hong JH, Kim CB, Bhak J, Bancroft I, Park BS (2006) Sequence-level analysis of the diploidization process in the triplicated *FLOWERING LOCUS C* region of *Brassica rapa*. Plant Cell 18: 1339–1347.

Yano M, Sasaki T (1997) Genetic and molecular dissection of quantitative traits in rice. Plant Mol Biol 35: 145–153.

Yarnell SH (1956) Cytogenetics of the vegetable crops. II Crucifers Bot Rev 22: 81–166

Zhang DX (2004) Lepidopteran microsatellite DNA: redundant but promising. Trends Ecol Evol 19: 507–509.

Ziolkowski PA, Karczmarek M, Babula D, Sadowski J (2006) Genome evolution in *Arabidopsis*/ *Brassica*: conservation and divergence of ancient rearranged segments and their breakpoints. Plant J 47: 67–74.

5

Molecular Mapping of Complex Traits

Małgorzata Kaczmarek,[1]* *Matthew N. Nelson,*[2] and *Wallace A. Cowling*[2]

ABSTRACT

The term "complex traits" relates to any phenotype which escapes from classical Mendelian single-locus type of inheritance (recessive or dominant). Many important biological characters, from grain yield or pathogen resistance to complex human diseases as asthma or diabetes, result from a segregation of numerous quantitative trait loci (QTL). The fact that total genetic variation underlying those phenotypes is divided into portions between a large number of loci greatly complicates attempts to identify regions related to that trait. An addition of environmental effects and interactions between those genes completes the picture. This chapter is devoted to molecular mapping of complex traits in plants. First, we review commonly used approaches to analyze the nature of genetic variation for quantitative traits, from the 1920s, when their basis was established, until today. The following two sections illustrate successes achieved to date on identifying QTLs related to morphology, flowering time control, pathogen resistance, transformation and plant regeneration and nutrient composition in vegetable Brassicas. The last part focuses on recently introduced technologies such as single nucleotide polymorphism (SNP), diversity arrays technology (DArT) or expression microarrays, which have not been extensively used for analysis of QTLs in Brassica species, yet are methods of choice for future projects concerning heterosis. Finally, analysis of the genetic basis of crop domestication and adaptation along with epigenetic regulation of crop development are discussed.

Keywords: Brassica, QTL, molecular mapping, linkage disequilibrium, pathogen resistance, array-based technologies, crop domestication

[1]Institute of Plant Genetics Polish Academy of Sciences, Strzeszyńska 34, 60-479 Poznań, Poland.
[2]School of Plant Biology and Institute of Agriculture, The University of Western Australia, 35 Stirling Highway, Crawley WA 6009, Australia.
*Corresponding author: *mrum@igr.poznan.pl*

5.1 Introduction

Genetic variation underlying quantitative phenotypes such as plant yield, pathogen resistance, etc. result from a segregation of numerous quantitative trait loci (QTL), each explaining a portion of the total variation, and whose expression is modified by interactions with other genes and the environment (Mackay 2001). The development of DNA markers enabled the construction of saturated genetic maps and localization of QTL for various phenotypes in various organisms (Glazier et al. 2002). Owing to their favorable genetic attributes, plants are used as model organisms for the study of quantitative traits. They are particularly suitable for high-resolution mapping and positional cloning. Much of the early work in this field was carried out by crossing the individuals of distinct phenotypes to maximize both QTL effects and DNA marker polymorphisms and constructing large experimental populations. Precise evaluation of mean phenotypic values called for statistically significant data obtained from replicated tests performed in different environmental conditions on either clonally reproduced genotypes or stable mapping populations such as recombinant inbred lines (RILs) or introgression lines (ILs). The creation of nearly isogenic lines (NILs) that differ only at a single region increased the precision of estimates of the effects of individual QTL or the interactions between QTLs (Paran and Zamir 2003). Although NILs can help to define the position of a QTL to a small region and thereby provide a shortlist of candidate genes for further investigation, the identity of the gene(s) underlying a QTL must be confirmed through the genetic and/or functional complementation and gene expression analysis (Paran and Zamir 2003). Two contrasting attributes of QTL analysis are: the relative simplicity of QTL mapping (to date ~ 1,750 papers have been published on QTL mapping in plants alone; *www.ncbi.nlm.nih.gov*; April 2010) versus the difficulty in assigning genetic locus to the QTL (until 2005 only 12 genes underlying QTL loci have been cloned) (Mackay 2001).

Until recently, the genetic bases of quantitative traits were described mainly using statistical methods of heritability analysis. In order to gain information on molecular aspects of what is now called genetic architecture of quantitative traits, the explanation of participation of individual genes in variation underlying those traits is needed. Such a detailed description requires the integration of classical genetics, evolutionary and ecological genetics, molecular population genetics, molecular genetics, and developmental biology, at a level of resolution that has not yet been attained (Mackay 2001).

5.1.1 Theoretical View on the Genetic Basis of Variation Underlying Quantitative Traits

While the number of genetically described loci is limited, even though 39 eukaryotic genome sequencing projects have been completed (www.ncbi.nlm.nih.gov, 8 April 2010), the issue remains "is it because mutations of minute effect escape from typical mutational screens?" (Mackay 2001). This question cannot be answered until the nature of genetic variation for quantitative traits is resolved. There were a number of theories on this matter.

First was a classic thesis called Fisher's infinitesimal theory. His description of a process of adaptation assumed that all inherited traits are shaped by countless number of mutations with tiny effects (Fisher 1930; Farrall 2004).

The continuation of this model was proposed by Mather, who stated that there were two separate categories of genes: "oligogenes" characterized by Mendelian inheritance and "polygenes" which controlled quantitative traits. Mutations regarding the latter category were limited to alleles with subtle effects (Mather 1941; Mather and Harrison 1949; Mackay 2001; Barton and Keightley 2002).

With the expansion of QTL mapping experiments, new evidence appeared that contradicted both those theories. These studies have reported QTLs that explained only 20% of the observed variation in traits. As a consequence of these observations, Robertson and Reeve (1952) stated that there were no separate classes of genes, but what was called oligogenes were subject to a whole range of mutational effects from lethal effects to almost impossible to distinguish isoalleles with quantitative effects. The extension of this theory was known as the "candidate gene" theory, which proposed "segregation of above mentioned isoalleles was a source of quantitative variation in situations when it co-localized with loci at which major mutations affecting the trait occurred" (Mackay 1985; Robertson 1985). Both theories gave a recombination of a central role as supportive of genetic advances but the latter also put mutation in this position (Hill 1982).

Fisher's theory has been modified several times (Kimura 1983; Orr 1998). The latter author proposed an exponential model of QTL effects. In other words, quantitative variation was controlled by the small number of loci with large effects and large number of loci with progressively smaller effects (Farrall 2004). This theory, in contrast to the infinitesimal model, gave hope to the approaches explaining the heritability of quantitative traits.

An observed phenotype, in respect to the quantitative traits discussed here, result from the expression of genes involved in developmental, physiological and/or biochemical pathways, which interact with each other and are under the influence of environmental factors. One way to investigate the function of these genes is the classical genetic approach—mutagenesis.

The multiple constraints of mutational tagging of QTL candidate genes were extensively discussed in excellent reviews by Nadeau and Frankel (2000) and Mackay (2001). In conclusion, it must be admitted, that although the value of this approach cannot be overestimated, however, it cannot fully replace QTL mapping in identification of a subset of loci affecting variation of the trait.

5.1.2 QTL Mapping

To perform QTL mapping one needs two things: a population in which phenotypic variation for the analyzed trait can be observed and DNA markers that detect polymorphisms in the experimental population, which then serve as loci for linkage map construction. The principle of QTL mapping, arguably traceable back to the 1920s (Sax 1923), is that simple traits or markers can be used to investigate the genetic control of more complex traits through linkage analysis. The basis of this approach has been further improved by findings of abundant DNA polymorphism most commonly small insertion/deletions and single nucleotide substitutions (Mackay 2001). Methods of detection of molecular variation have become simplified from the first generation of DNA markers (restriction fragment length polymorphism, RFLP, analyzed using Southern blots; Bonierbale et al. 1988) to high-throughput methods for both polymorphism discovery and genotyping (Kristensen et al. 2001; Gupta et al. 2008) (for more information see Chapter 4.2).

Improvement in mapping population development have also occurred (see Chapter 4.5). Traditionally, populations have been created by crossing two individuals with differences in trait phenotype. QTL mapping then utilized linkage disequilibrium (LD) between QTL alleles and marker alleles generated this way. The observation that LD was also present in many other types of populations (for details see Mackay 2001), have resulted in exploiting them in approaches known as association mapping (Gupta et al. 2005). This strategy has an advantage over linkage analysis, in respect to the most desirable attribute of QTL mapping - precision, which is directly proportional to the number of meioses sampled, correlated with the number of opportunities for recombination between the marker and the trait locus. Association studies, in contrast to artificially constructed populations with usually limited number of individuals or families, benefit from many generations of recombination, which have occurred in nature. Although, the level and distribution of LD has been investigated to the nucleotide level in many plant species (Lijavetzky et al. 2007; Mather et al. 2007; Xing et al. 2007; Fusari et al. 2008; Yan et al. 2009), *Arabidopsis thaliana* with it's genome-wide studies is still the notable exception (Kim et al. 2007). Thus, QTL mapping usually includes: initial saturation of the genome with molecular markers performed using linkage analysis, followed by higher resolution confirmation studies of detected QTL, that end up with identification of candidate genes with recombination or association mapping.

5.1.2.1 Initial Genome Inspection

Key steps in the linkage mapping of QTL in inbreeding organisms were outlined in the previous paragraph. Along with the development of new methods of detecting polymorphism, there has been the improvement in statistical methods for mapping QTL and development of "golden rules" for experimental design and interpretation. The choice is between least-square (LS) (Soller et al. 1976) and maximum-likelihood (ML) (Lander and Botstein 1989; Knapp et al. 1990) methods taking into account both their advantages and disadvantages.

Many QTL mapping procedures prevent simultaneous evaluation of association of many QTLs with markers or marker intervals in the mapping population. To overcome possible favoritism on single marker and interval mapping analysis caused by the presence of multiple linked QTLs, one can use either composite interval mapping (CIM) (Jansen and Stam 1994; Zeng 1994) or the multiple interval mapping (MIM) (Kao et al. 1999) method. The former combines ML interval mapping with multiple regression, using marker cofactors to reduce the above mentioned bias and to increase the power to detect QTLs by decreasing the within marker-class phenotypic variation. The latter has been developed to provide estimates of positions and main and interaction effects of multiple QTLs. The investigator has to take into account that both methods are model dependent, and are influenced by different numbers of marker cofactors and window size (the region to either side of the test interval within which no marker cofactors are fitted) (Mackay 2001).

The need to improve production traits in agriculturally important crops has been addressed by mapping QTLs for traits of economic importance. The numbers of QTLs mapped in these studies are limited by sample size considered as a number of individuals in mapping population (correlated with the number of recombination events tested) and as a number of different alleles present in the two parents lines, which represent only a portion of the existing genetic variation. The methods to overcome this genetic limitation are four-way (Xu 1996) and eight-way cross design (Talbot et al. 1999; Mott et al. 2000), combining multiple line cross experiments (Xie et al. 1998) and utilizing parental strains derived by divergent artificial selection from a large base population (Nuzhdin et al. 1999).

Despite many obvious limitations, which lower the value of QTL mapping, this approach helps to understand the genetic basis of naturally occurring variation, albeit at the resolution of large chromosomal regions.

5.1.2.2 High-Resolution Mapping

High-resolution mapping aims to limit the size of the region exhibiting LD with the QTL to a single gene, by increasing recombination in the QTL region. Since there is a proportional relationship between the number of recombination and the number of generations, the required precision of mapping is achieved by exploiting further generations of the cross than F_2 (Darvasi and Soller 1995). Also, the effects of single QTL can be magnified by constructing strains that differ only in respect to the region adjoining the QTL, from a whole chromosome (chromosome substitution lines) to a smaller interval (ILs and NILs; see Section 4.5 and 5.1). Numerous observations of trait phenotype performed on this type of lines increase the precision with which QTL genotype is specified.

It is gratifying to discover examples of effective fine-scale recombination mapping of QTL in plants, such as the localization of tomato fruit weight QTL, *fw2.2*, to a 1.6 cM interval containing four unique transcripts (Frary et al. 2000) and correspondence of tomato fruit-specific apoplastic invertase gene *Lin5* to the *Brix9-2-5* QTL, which controls fruit glucose and fructose contents, based on co-segregation of recombinants in a 484-bp region within this gene with QTL phenotypes (Fridman et al. 2000).

5.1.3 Gene Identification

Once a QTL is mapped to a region realistic for candidate gene identification, the question arises: "which one is it?". Again, the above mentioned studies on tomato can serve as an example of this approach. Observation of differences in QTL phenotype and polymorphic genotypes in a candidate gene for the same individuals was reported for tomato *Lin5* gene and a fruit sugar content (Fridman et al. 2000), provided evidence that the QTL was correlated with this gene. Likewise, functional complementation, in which the trait phenotype is recovered in transgenic organisms, can also help in gene identification. Functional tests can be carried out through overexpressing or down-regulating the gene(s) of interest (Salvi and Tuberosa 2005). These tests serve best in situations where QTL effects are large and their gene action is of a dominant type. Otherwise, it is necessary to assemble data from different sources, which together specify a candidate gene. Candidates can also be identified from genetic complementation tests, LD mapping, quantitative differences of gene expression and/or protein function, and exploring other species for orthologous QTL (Mackay 2001; Holland 2007). Expression profiling experiments may help in validation of candidate gene only when QTL for trait co-localize with QTL controlling expression of a candidate gene (Morgante and Salamini 2003).

5.1.3.1 Complementation Tests

One of the methods to demonstrate the identity of a causal gene and a QTL is the complementation test. This approach has specific requirements, which are: the possibility of preparing controlled crosses, the availability of mutants for all loci in the QTL region analyzed and clear discrimination of candidate genes. However, a weakness of this analysis is that failure to complement cannot discriminate between two types of genetic interaction: allelic and epistatic. Thus, final proof will include similarities in gene expression patterns (Doebley et al. 1997) and associations of molecular polymorphisms in the candidate gene with differences in the trait phenotype (Long et al. 1996, 1998).

5.1.3.2 Association Mapping

An introduction of association mapping method has been made in Section 5.1.1. It must be added here that besides an increase of precision of QTL mapping this approach has other advantages over linkage mapping in biparental offspring populations. It allows for detection of QTL in a wide range of germplasm rather than only those few that are polymorphic in classic populations. Since the population is natural, it takes less time and resources to assemble than standard mapping populations (Flint-Garcia et al. 2003). This method depends on measuring the linkage disequilibrium between QTL and other closely linked loci in a natural population (Mackay 2001). The extent of linkage disequilibrium is influenced by a number of factors including: the average amount of recombination per generation experienced by that region of the genome, the number of generations since the original mutation, and the population size (Hill and Robertson 1968; Falconer and Mackay 1996; Hartl and Clark 1997). The disadvantage of such collections or populations is that linkage disequilibrium also occurs for reasons other than genetic linkage, such as drift, selection, and admixture of populations. Association mapping has been applied to plants, at the level of both individual genes and the whole genome (Gupta et al. 2005; Mackay and Powell 2007).

Association mapping can be performed either to screen candidate loci in a previously defined interval or to look for associations without a priori linkage information. In a situation where genotypes at all polymorphic sites in the region of interest are known, it is possible to connect one of them with the sites generating the phenotypic effects (QTN-quantitative trait nucleotide) (Long et al. 1998; Long and Langley 1999). This is rarely true, and as a consequence association studies are conducted with only a subset of marker loci. As a corollary, the power to detect an association depends on the number of individuals sampled and the density of polymorphic markers.

5.1.3.3 Comparative QTL Analysis

There are several ways of exploiting synteny between closely related plant species in the course of QTL analysis. In a situation when one recognizes similar localizations of QTL for the same traits in different species, one can perform candidate gene verification in a model organism, and then look for linkage and association for the orthologous locus in other species. Conversely, high-resolution mapping could be carried out in one species and the correspondence of the genes in the interval with the QTL could be evaluated in a model organism (Mackay 2001).

There are several examples of some conservation of developmental pathways (flowering time control or plant architecture) among distantly or closely related plants species. QTL analysis of genes controlling photoperiodic response in rice and *Arabidopsis* has suggested their functional conservation, besides the obvious differences in the way they operate (reviewed in Paran and Zamir 2003).

5.1.4 Identification of Sequence Variation Underlying QTL

The obvious continuation of QTL analysis is identification of causal association between genetic polymorphism in gene sequences and phenotypic variation in traits analyzed (Mackay 2001; Morgante and Salamini 2003). And again, LD mapping is a method of choice. Studies of genotype-phenotype associations at candidate genes performed in humans and *Drosophila* revealed some interesting features. All significant associations have been found in introns and non-coding regions (for review see Mackay 2001), suggesting that variation in regulatory sequences can cause fine quantitative differences in phenotype (Morgante and Salamini 2003). In plants 12 genes corresponding to QTLs have been identified (Salvi and Tuberosa 2005). For two of them: *fw2.2* in tomato (Cong et al. 2002; Nesbitt and Tanksley 2002) and *tb1* in maize (Lukens and Doebley 2001; Hubbard et al. 2002), mutations have been detected in 2.6 kb and 50 kb upstream, respectively. The problems appear when it comes to describing the effects of mutations in non-coding regions, due to their distance from the regulated gene (Stam et al. 2002), laborious differentiation from neutral base pair substitutions and difficulty in predicting their phenotypic effect (Morgante and Salamini 2003). In conclusion, in quantitative control of the trait both the allelic diversity within gene transcription units and in distant regions contribute.

As a closing remark, it should be emphasized that understanding of molecular basis of phenotypic variation should not be restricted only to point mutations affecting genes but should also bear in mind the possible role of epistasis in QTL variation (Yedid and Bell 2002) and epigenetic phenomena namely gene silencing, DNA methylation, RNA interference and heterochromatic DNA (Volpe et al. 2002; Zilberman et al. 2003). Also, it is worth mentioning that long terminal repeats retrotransposons have the ability to modify the expression of adjacent genes and can play a role in molecular basis of hybrid vigor (Kashkush et al. 2003).

5.2 Target Traits

5.2.1 Morphology

Plants from the genus *Brassica* have a long history of cultivation around the world as vegetable, oil, fodder, and condiment crops (for details see Chapters 1,2 and 3). The wide variation in morphology evident today is likely to be in part due to differing selection pressure based on environmental growth conditions and the need of the consumer (Zhao et al. 2007). While QTL analysis has been used in other crop species primarily for disease resistance and flowering time control, the highly variable morphological forms of vegetable Brassicas provide additional opportunities for scientists who seek to understand the genetic basis of plant growth and development.

Many of the early studies observed morphological traits to be controlled by many genes (Pease 1926; Pelofske and Baggett 1979). One of the earliest attempts to identify genetic markers associated with genes affecting morphological and maturity characters was by Figdore et al. (1993; Table 5-1). Although their work focused mainly on the analysis of genetic basis of resistance to the clubroot pathogen, the trait loci controlling head type, color and extrusion were studied in a cross between susceptible cauliflower (*B. oleracea* var. *botrytis*; Early White) and resistant broccoli (*B. oleracea* var. *italica*; CR-7), using RFLP markers. For each of the traits analyzed, several chromosomal regions were found to contribute to each trait. This supported the conclusion that these traits are under polygenic control.

The complexity of genetic control of a subset of morphological traits was evaluated by Kennard et al. (1994; Table 5.1) in an F_2 population developed from a cross between cabbage (*B. oleracea* var. *capitata*) and broccoli (*B. oleracea* var. *italica*). QTL analysis utilizing a single factor ANOVA enabled marker-locus associations to be identified for 16 of the 22 traits. In most cases, the same chromosomal region was detected to be correlated with different yet related traits. An example of this was the complex associations found among leaf dimension traits (i.e. lamina length, petiole length, basal petiole thickness and leaf length) on linkage group 4C, each of them having the same gene action (partial dominance), the same genotypic class having the largest mean (broccoli homozygous and heterozygous). Also, head forming leaf overlap, petiole length, lamina leaf, annual vs. biennial habit were found to be highly associated on linkage group 5C. Previous studies, based on classical genetic methods, have observed heading habit, petiole length and lamina width to be linked (Pease 1926). Similar results were obtained for analysis of heading habit and biennial habit (Pelofske and Bagget 1979). As a corollary, this could be evidence that a common set of genetic loci control these morphological traits in different *B. oleracea* cultivars.

The degree of genetic divergence, as an effect of evolution of *Brassica* genus, has been elucidated by Lan and Paterson (2000; Table 5-1), in respect to curd-related traits using genotypes of different geographical origin: Italy, India and Thailand, which were Cantanese (CAN), Pusa Katki (PK) and Bugh Kana (BK), respectively. These three *B. oleracea* varieties, exhibiting different morphologies, were crossed to a common parent, the rapid-cycling Brassica (RCB), to create three F_2 populations of *B. oleracea.* These populations were genotyped using RFLP markers and a total of 86 QTLs were identified for eight curd-related traits in *B. oleracea.* Using comparative mapping among the three mapping populations, they deduced that at least 67 different genetic loci were involved representing 54 ancestral genes (based on perceived patterns of genome duplication). These results supported previous research which found that the enlarged curd of cauliflower is under compound genetic control (Watts 1964; Kalloo and Bergh 1993; Kennard et al. 1994). Based on the observation of decreased complexity of genetic control of most traits, it was hypothesized that *B. oleracea* var. BK was a progenitor of cauliflower and broccoli. The comparison to the population used by Kennard et al. (1994; Table 5-1), who mapped similar traits, provided additional evidence on the differences in QTLs identified among these three populations. QTLs at nine locations were shared by both studies. Comparative mapping to *A. thaliana* identified several possible candidate genes for the *Brassica* QTLs.

The same RFLP-genotyped *B. oleracea* F_2 populations were used by Lan and Paterson (2001; Table 5-1) to detect QTLs affecting the size of leaves and stems. They found 47 QTLs in three mapping populations, which represented approximately 35 different genetic loci, as inferred by comparison of the three linkage maps. There was some evidence that a proportion of these QTLs may have corresponded to homeologous locations in the *B. oleracea* genome. Mutants in potentially orthologous *Arabidopsis* genes equivalent to QTLs related to ancestral genes were identified.

Sebastian et al. (2002; Table 5-1), using a population of doubled haploid (DH) lines derived from a cross between two contrasting crop types, an annual cauliflower (*B. oleracea* var. *botrytis*) and biennial Brussels sprout (*B. oleracea* var. *gemmifera*), located QTLs controlling 27 morphological and developmental traits, including leaf, flowering, axillary bud and stem characters. The previously published study by Sebastian et al. (2000) reported linkage map construction of *B. oleracea* genome utilizing the same population and RFLP, AFLP (amplified fragment length polymorphism) and simple sequence repeat (SSR) markers. QTL analysis based on multiple-marker regression identified 32 QTLs, affecting 17 morphological traits, up to four per trait. The range of additive genetic variation explained by them was between 6 and 43% . The work of Kennard et al. (1994; Table 5-1) compares closely to the results obtained in this study, concerning co-localization of markers associated with two leaf characters on linkage groups O6 and O7. Additionally, the similarity of mode of gene action provided evidence of common genetic control of these characters. Despite this, homology between those regions could not be evaluated due to incompatibility of marker sets used in both studies. In contrast, a QTL for petiole length found on O3 by Sebastian et al. (2002; Table 5-1) seemed to be homologous to a QTL detected in another *B. oleracea* study (Camargo et al. 1995).

Bettey et al. (2000; Table 5-1) investigated components of seed vigor and pre-emergence seedling growth traits in *B. oleracea* using a segregating mapping population of 105 DH lines. They found some correlation among traits, particularly among germination traits, and between seed weight and pre-emergence seedling growth traits. They concluded that germination and pre-emergence seedling growth were under distinct genetic control. QTL analysis revealed regions on linkage groups O1, O3, O6, O7 and O9 controlling these traits. Interestingly, two of these QTLs shared a similar position with QTLs for flowering time found previously by Bohuon et al. (1998; Table 5-1) in the same mapping population.

Wide variation for morphological traits exists also in *B. rapa* but less research has been conducted in this species compared to *B. oleracea* (see Chapters 2.3 and 3.3). A total of 22 traits including flowering time, seed and pod traits, growth-related traits, leaf morphology, and turnip formation was studied by Lou et al. (2007)(Fig. 5-1; Table 5-1) using several populations derived from three main morphotypes of *B. rapa* (the oilseed-, leafy-, and turnip morphotypes). QTL analysis was performed using both interval mapping and multiple-QTL model mapping methods. Altogether, 19 QTLs affecting morphological traits (excluding flowering time) were detected. Principal component analysis and co-localization of QTLs indicated that some of the loci controlling leaf and seed-related traits (on R07) and those for flowering time and turnip formation (on R02) might be the same.

Most of the above described QTL analyses concentrated on traits related to the morphology and development of shoot characters. The other group of vegetable Brassicas are plants grown for their succulent storage roots. One of the major vegetables within this group is turnip (*B. rapa* L. ssp. *rapifera* (Metzg.) Sinsk.). The first report on QTL analysis for root traits in *Brassica* crops was by Lu et al. (2008; Table 5-1) who detected 18 QTLs for three root traits: taproot thickness, length and weight. Simple interval

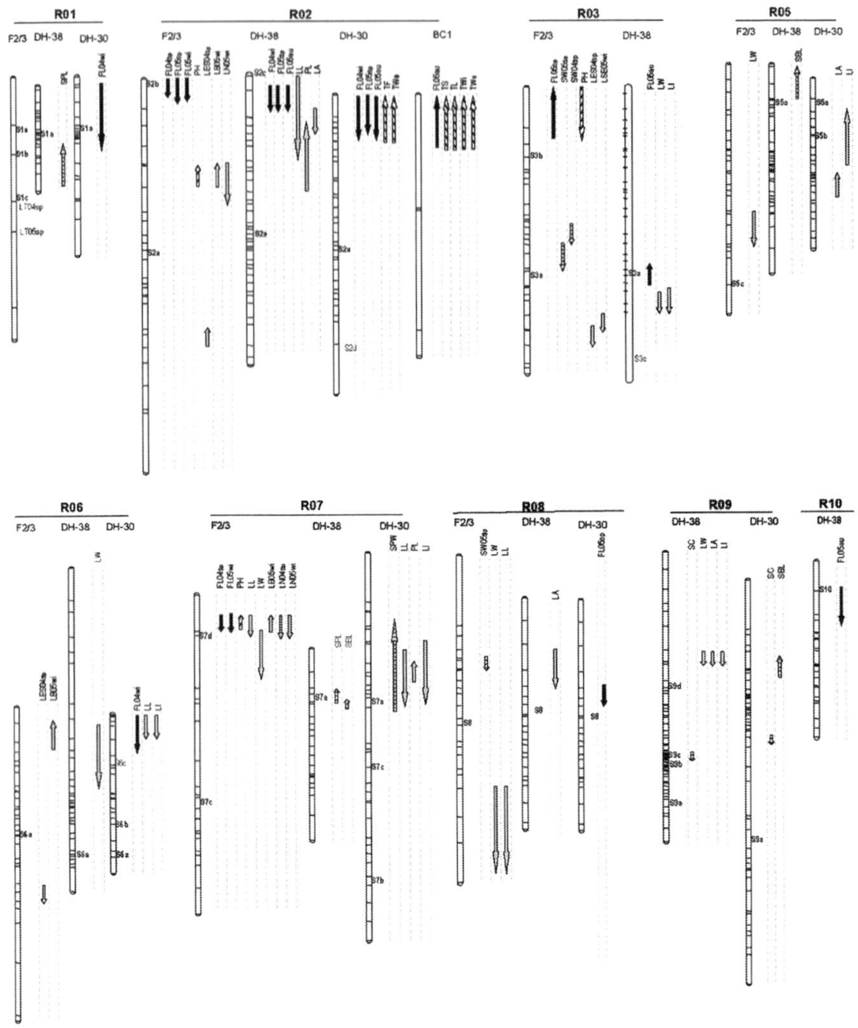

Figure 5-1 Locations of QTL for the traits analyzed in the four mapping populations of *B. rapa*. The linkage groups of different maps are aligned based on common SSR (S1–S10) or common AFLP markers (P Lou et al., unpublished data). The lengths of the arrows indicate the 2-LOD support intervals. The traits (see abbreviations in Table 5-1) are indicated above each column. The direction of the arrow's head indicates the allelic effect: upward, RC-144 increases and CC-156 decreases for F2/3, YS-143 increases and PC-175/VT-115 decreases in DH-38 and DH-30; downward: CC-156 increases and RC-144 decreases, YS-143 decreases and PC-175/VT-115 increases in DH-38 and DH-30. The filling pattern of arrows refers to different groups of phenotypic traits. Flowering time; seed; plant height; leaf; turnip. S1a, BRMS096; S1b, Ra2G09; S1c, BRMS037; S2a, Na12H09; S2b, BrFLC2; S2c KS50030; S2d, BrMAF; S3a, BRMS043; S3b, BRMS042; S3c, BrFLC3; S5a, BRMS034; S5b, BRMS007; S5c, Ra3H10; S6a, BRMS014; S6b, Na12H07; S6c, KS51082; S7a, BRMS018; S7b, Ol12E03; S7c, Ra2A01; S7d, BRMS036; S8, Ra2E12; S9a, BRMS051; S9b, Na10A08; S9c, Ol12F02; S9d, Ol10D08; S10, BrFLC1. (reprinted from Lou et al. 2007 with permission of Oxford University Press).

mapping was conducted on an F_2 population derived from a cross between two parents that has contrasting shoot and root characteristics: turnip (*B. rapa* ssp. *rapifera*) and non-headed Chinese cabbage (*B. rapa* ssp. *chinensis*). Results indicated that there is an extensive genetic variation for taproot thickness, length and weight traits in the analyzed hybrid population and that favorable alleles can be combined to improve storage-root yield.

5.2.2 Flowering Time Control

Morphological types of plant species of the genus *Brassica* could be also differentiated based on their flowering habits: annual versus biennial. The former, represented by broccoli and cauliflower, groups plants which produce flowers during single growing season. Cabbage or kohlrabi exemplifies the latter, which will not flower unless exposed to low temperatures (vernalization). The recognition of genetic factors that control the switch from vegetative to reproductive stage will broaden the understanding of the basis of plant development (see Chapter 2.2.2.5). The modification of these factors will also enable the cultivation of a particular *Brassica* species beyond their present geographical range.

In past decades, two contrasting concepts on genetic basis of control of flowering related traits have been proposed. On the one hand, single gene control has been put forward (Detjen 1926; Horovitz and Perlaska 1954; Walkoff 1963; Landry et al. 1992), on the other hand - potentially many genes have been suggested (Baggett and Wahlert 1975; Pelofske and Baggettt 1979; Baggett and Kean 1989). The genetic control of flowering time was analyzed in an F_2 population derived from a cabbage (*B. oleracea* var. *capitata*) by broccoli (*B. oleracea* var. *italica*) cross with the use of RFLP markers (Kennard et al. 1994). As a result, four regions of the genome controlling annual/biennial habit and flowering time were identified, and epistasis between two genes controlling flowering time was detected.

This approach was improved upon by Camargo and Osborn (1996; Table 5-1), who used in their study replicated F_3 families, derived from a cross between the same *B. oleracea* subspecies as Kennard et al. (1994; Table 5-1), but using different parental genotypes; this provided a better estimate of the genotype effects. Marker interval analysis indicated that flowering time was controlled by three or more loci, supporting other studies that reported oligogenic control. Two QTLs located on separate linkage groups were associated both with annual/biennial habit and flowering time. Comparison with previous analysis revealed the linkage of loci on LG2 and LG6 with a gene encoding S-glycoprotein, involved in self-incompatibility mechanism of *B. oleracea* (Nasrallah et al. 1985) and petiole length affecting QTL (Camargo et al. 1995), respectively. A third region on LG 8 was associated with flowering time only. Consistent with

Table 5-1 An overview of QTL analysis for vegetable Brassica crops.

Trait	Plant Material	Spatial Conditions	Temporal Conditions	Marker	Statistical approach	Software	QTL	Variation explained (%)	QTL position LG/Flanking markers/ distances (cM)	Authors
Resistance to race 2 of *Plasmodiophora brassicae* Leaf morphology (LMOR)	78 F_3 *B. oleracea* var. *capitata* resistant No. 86-16-5 x rapid cycling *B. oleracea* CrGC No. 85	growth cabinet	43 days after inoculation	RFLP	Interval mapping	Mapmaker/ QTL	CR2A CR2B LMOR	58 15	LG6, 2NF11-2ND3 (22); LG1, 3NE4a-3ND3 (12); LG3, 3ND12, 21	Landry et al. (1992)
Resistance to race 7 of *Plasmodiophora brassicae* (R)	180 F_2 clubroot resistant *broccoli B. oleracea* var. *italica* CR7 x susceptible cauliflower *B. oleracea* var. *botrytis* "Early White"	greenhouse	35 days after inoculation	RFLP	Chi-square, R-square, F-tests	-	R	12.5 7.5 10.2	LG1, 14a; LG1, 34-75a (35); LG4, 48; LG9, 177b	Figdore et al. (1993)
Head type (H)							H	16.6 7.3–9.8 9.9–13.9	LG9, 2; LG3, 96a-188(4); LG9, 2-37a(19)	
Head color (C)							C	12.5 16.2 13.9	LG3, 96a-188(4); LG4, 71-59(25); LG9, 79-37a(39)	
Head extrusion (E)							E	12.2 7.6	LG8, 6a-82a(6); LG1, 64	
Heading data (maturity-M)							M	12.7; 15.5 6.6–12.8 4.0–9.4 4.7	LG1, 19-14a(72); LG4, 71-59(25); LG5, 210b-118(40); LG6, 110	

Developmental traits: Flowering traits: Annual/biennial	70–90 F_2 cabbage "Wisconsin Golden Acre" x broccoli "Packman"	greenhouse	37–97 days after sowing	RFLP	Single factor ANOVA, Chi-square tests, Pearson correlation coefficients	SAS GLM procedure, SAS FREQ, SAS UNI-VARIATE, SAS CORR		6.7– 42.7 (single marker loci), 60.1 (multilocus model)	2C, 152 (18); 5C, 67a- 4a (27); 7C, 179-131b (35);	Kennard et al. (1994)
Days to bud									1C, 29b (78); 2C, 152-165 (42); 3C, 133-92b (21); 7C, 148-131b (53); 9C, 43b;	
Days to flower									1C, 84-93a (12); 1C, 29b (78); 2C, 152-165 (42); 3C, 133-113 (3); 3C, 92b (61); 7C, 148-131b (53); 7C, 25b (93);	
Days from bud to flower									3C, 133 (40); 8C, 170 (37);	
Size of cluster								 - 	2C, 165 (60); 4C, 26 (47); 5C, 97a (25); 9C, 43b;	
Leaf traits: Lamina length								12.0 8.0 17.9 19.5	4C, 10b-26 (15); 5C, 67[a]; 6C, 30; 6C, 111;	
Lamina width								10.3 11.3 11.1 12.1	1C, 61a; 4C, 128; 5C, 174; 7C, 25b;	
Petiole length								15.6 12.1	4C, 10b-26 (25); 5C, 92a;	
Basal petiole width								13.5	7C, 25b;	

Table 5-1 contd...

Table 5-1 contd...

Trait	Plant Material	Spatial Conditions	Temporal Conditions	Marker	Statistical approach	Software	QTL	Variation explained (%)	QTL position LG/Flanking markers/ distances (cM)	Authors
Basal petiole thickness							-	11.6	4C, 10b-26 (25);	
Leaf length								16.9	6C, 111;	
								19.4	4C, 10b;	
								18.1	4C, 26;	
								6.7	5C, 67a;	
								7.4	5C, 92a;	
Lamina length/width								13.9	6C, 30-111;	
								9.5	1C, 61a;	
								14.2	4C, 128;	
								15.4	4C, 10b;	
								11.5	4C, 26;	
Lamina/petiole length								7.1	6C, 30;	
								8.1	5C, 92a;	
Stem traits:										
Distance between second and third node									2C, 8(41); 3C, 133-95 (35); 5C, 97a-9a (43);	
Distance between third and fourth node									8C, 170-168 (28); 3C, 133-95 (35); 5C, 9a;	
Distance of twenty internodes									7C, 179; 8C, 115-143 (10); 1C, 29b; 2C, 83b; 3C, 133-95 (35); 4C, 26; 9C, 153a	

Flowering time (FT) Annual/biennial (PF)	92 F_3 families cabbage x broccoli	field	15 surveys at 5–15 days intervals	RFLP	Interval mapping, single-factor ANOVA	MapMaker/ QTL 1.1 Minitab's GLM procedure	FT PF	19.7 29.9 19.2 2.9 3.4	LG2, *ec2h2-slg6*; LG6, *wg6f10-wg7g10*; LG8, *ec3b2-ec3e10*; LG2, *ec2h2-slg6*; LG6, *wg7g10-wg3g9a*	Camargo and Osborn (1996)
Resistance to *Plasmodiophora brassicae* Woron	90 F_2 clubroot resistant kale *B. oleracea* var. *acephala* C10 x susceptible cauliflower *B. oleracea* var. *botrytis* 48.4.7	glasshouse	6 weeks 2 years	RAPD	Student's t-test, Kruskal and Wallis test, One way ANOVA	-	-	-	OPA4-700, OPA16-510; OPB11-740; OPA1-1880; OPL6-780; OPA18-1490; OPE20-1250	Grand-clement and Thomas (1996)
Resistance to *Plasmodiophora brassicae* Woron	91 DH-lines cabbage *B. oleracea* var. *capitata* "Bindsachsener" x *B. oleracea* var. *italica* "Greenia"	phytotron chamber	6 weeks	RFLP AFLP	Multiple-QTL Model (MQM), Kruskal and Wallis test	GENSTAT5 MAPQTL 3.0	*pb-3* *pb-4*	54 6	LG3, 4NE11a; LG1, 2NA8c; LG6, 2NA1b-A20	Voorips et al. (1997)
Flowering time	149 DH lines *B. oleracea* var. *alboglabra* (A12DHd) x *B. oleracea* var. *italica* (GDDH33)	field (two trials)	until the first flower	RFLP	Marker regression, ANOVA, SIM function, Interval mapping	http:/ web. bham.ac.uk/ g.g.seaton/ MQTL Mapmaker/ QTL	FTO2.1 FTO3.1 FTO5.1 FTO5.2 FTO9.1 FTO9.2	9 3 8 32 8 16 - - 47 22 11 37	O2, (78±9); O3, (12); O5, (16±13); O5, (40±10); O9, (36–46); O9, (74–94)	Bohuon et al. (1998)
Resistance to *Plasmodiophora brassicae* Woron pathotype 1 and 3	138 F_2 cabbage *B. oleracea* var. *capitata* x kale *B. oleracea* var. *acephala*	infested field	60 days after trans-planting	RAPD RFLP Isozyme morpho-logical	Interval mapping	QGENE	-	30	LG3, WG6A1- WG1G5 (2.6)	Mori-guchi et al. (1999)

Table 5-1 contd...

Table 5-1 contd...

Trait	Plant Material	Spatial Conditions	Temporal Conditions	Marker	Statistical approach	Software	QTL	Variation explained (%)	QTL position LG/Flanking markers/ distances (cM)	Authors
Flowering time Late flowering (LF) Early flowering (EF)	79 substitution lines *B. oleracea* A12DHd x GDDH33	field (eight trials)	until the first flower	RFLP AFLP	ANOVA, Tukey-Kramer multiple compari-son tests	-	LF EF LF LF1 LF2 LF LF1 LF2 LF3 EF	-	O1, (0–30.3); O1, (30.3–38.1); O2, (65–80); O3, (0–35); O3, (35–75); O5, (0–34.2); O9, (0.0–43.0); O9, (43.0–64.4); O9, (70.8–106.8); O9, (23.1–47.0)	Rae et al. (1999)
Curd-related traits Days to budding (dfb) Days from budding to flowering (dofdbf)	247 F_2 rapid cycling *B. oleracea* (RCB) x *B. oleracea* var. Cantanese (CAN), 250 F_2 RCB x *B. oleracea* var. Pusa Katki (PK), 246 F_2 RCB x *B. oleracea* var. Bugh Kana (BK)	field	180 days after planting on the first flowering day	RFLP	Interval mapping, multiple QTL model	SAS, Microsoft Excel, Mapmaker-QTL	dfb	20.9 39.6 11.0 1.2 9.9 65.3 6.5 87.6 44.4 44.1 13.9 5.9 21.9 27.9 10.0 39.7 7.8 15.9 9.4 43.8 7.3 4.6	BK, C4, EW4D04w+6; BK, C7, EW8C11a+2; BK, C1, EST429a+7; BK, C7, EST151c+0; CAN, C3, EW8F03a+5; CAN, C4, EW9E10+12; CAN, C5, EST195b+3; CAN, C7, WG3D11+14; PK, C3, EST411c+0; PK, C4, EST114a+1; PK, C7, EW6B07a+3; PK, C8, EW8D10b+0; PK, C9, EST131a+17; PK, C4, EW9B02X+1; BK, C4, EW2B12s+4; BK, C7, EW8C11a+1; BK, C8, EW5G04b+10; CAN, C4, EW9E10+12; CAN, C5, EST195b+3; CAN, C7, EW8C11a+4; CAN, C8, EW8E09c+2; CAN, C3, EW8F03a+1	Lan and Paterson (2000)

First-rank branching (rk1) Side branching (sidb) Cluster width (cluw) Curd width (headw)								11.6 63.6 76.6 11.4 7.4 24.6 5.5 15.8 8.2 7.2 14.3 5.9 27.2 7.6 34.8 7.6 9.2 7.9 6.2 6.5 45.9 13.6 30.4 8.1 37.8 11.4 12.5 3.8 6.1 5.9 8.0 6.9 33.4 5.6	CAN, C4, EST453g+3; PK, C3, EST411c+0; PK, C4, EST373a+16; PK, C7, EW7E01a+1; PK, C8, EW8D10b+1; PK, C9, EST131a+13; PK, C5, EW8A11a+2; BK, C7, WG3D11+11; CAN, C1, EST566a+2; CAN, C3, EW8F03a+0; CAN, C4, EST453g+3; CAN, C5, EW8F11c+0; CAN, C7, WG3D11+8; CAN, C8, EW8E09c+0; PK, C4, EST122b+6; PK, C5, EW8A11a+0; PK, C6, WR1D12a+0; PK, C7, EW6B07a+3; PK, C1, EST217a+3; PK, C9, EST131a+14; BK, C7, EW8C11a+1; CAN, C4, EST55b+16; CAN, C7, EW8C11a+5; CAN, C8, EW7C10b+7; PK, C4, EST429E+1; PK, C5, EW8A11a+1; PK, C7, EW8F06+0; PK, C9, EW1G03+0; CAN, C1, EW5H03+0; CAN, C2, EW9F06a+0; CAN, C4, EW9E10+0; PK, C1, EST125+0; PK, C3, EST411c+0; PK, C4, EW9B02x+0	

Table 5-1 contd...

Table 5-1 contd...

Trait	Plant Material	Spatial Conditions	Temporal Conditions	Marker	Statistical approach	Software	QTL	Variation explained (%)	QTL position LG/Flanking markers/ distances (cM)	Authors
								14.9	BK, C4, EW4D04w+7;	
								10.5	BK, C7, EW8C11a+4;	
								11.3	CAN, C1, EW7C10a+0;	
								11.7	CAN, C4, EST453g+4;	
								7.8	CAN, C5, EW7E08a+1;	
								22.1	CAN, C7, WG3D11+6;	
								5.1	CAN, C9, EW1G03a+1;	
								3.7	CAN, C3, EST22b+4;	
								6.2	CAN, C8, EW8E09c+0;	
Apical shoot length (rrl1st)								51.2	PK, C3, EST411c+0;	
								54.3	PK, C4, EST122b+4;	
								7.6	PK, C5, EW8A11a+0;	
								6.4	PK, C7, EW5A12a+0;	
								4.3	PK, C9, EW8E09b+2;	
								22.7	BK, C1, EW2E07a+12;	
								11.3	BK, C4, EW2B12s+10;	
								15.2	BK, C7, EW8C11a+1;	
								7.5	BK, C7, EW6C11a=0;	
								9.1	CAN, C3, EW9E05a+0;	
								11.0	CAN, C4, EST453g+0;	
First branch length (rrl2nd)								7.6	CAN, C5, EST195b+7;	
								6.8	CAN, C7, EW6B07a+0;	
								12.8	CAN, C9, EW1G03a+1;	
								11.0	PK, C1, EW8C11x+3;	
								7.9	PK, C1, WST88c+0;	
								25.5	BK, C1, EW2E07a+15;	
								15.4	BK, C4, EW4D04w+23;	
								15.6	BK, C7, EW8C11a+2;	
								7.5	BK, C7, EW3A04b+23;	
								4.3	BK, C7, EW5C12f+0	

Germination: Unstressed MGT Water stress MGT Heat treatment Conductivity Seedling growth: Seed weight Root weight Shoot weight Shoot: root weight ratio Root growth rate Shoot growth rate Root length Shoot length Shoot: root length ratio	105 *B. oleracea* DH lines A12DHd x GDDH33	-	time taken by shoots to reach 1.5 cm and by roots to reach 4 cm; up to 15 days	RFLP AFLP micro-satellites	ANOVA Interval mapping Kruskal-Wallis test Marker regression	Genstat 5 REML MapQTL 3.0 QTL Java Applet V0.70 (http://web.bham.ac.uk/g.g.seaton/qtl_ns/brass.html)	-	13.7 14.5 16.4 13.7 16.6	O1, AC-CACJ01, (118); O3, AC-CAAE02 , (84.2; O5, pN23E2, (5.0); O6, pO10E2, (72.6); O3, pW102J1, (88.2); O9, pO160E1, (71.2); O6, AC-CATE18, (32.9); O9, pN3E1, (107.8); O6, pR64E2, (11.2); O9, AC-CAGJ05, (11.2); O9, pO160E1, (71.2); O4, pO147e1N, (20.8); O9, AC-CAGJ05, (47.3); O6, pR64E2, (11.2); O9, AC-CAGJ05, (47.3); O1, AC-CACE08, (93.9); O4, pW188E2N, (37); O6, AA-CATJ08, (20.2); O3, LEW6E1, (16.3); O6, pR64E2, (11.2); O7, pr36E2, (5.9); O7, pN64E2, (37); O9, AC-CAGJ05, (47.3); O3, pN213J2, (61.7); O9, pC14 (cos), (88); O7, mBN72AJ1, (67.3)	Bettey et al. (2000)

Table 5-1 contd...

Table 5-1 contd...

Trait	Plant Material	Spatial Conditions	Temporal Conditions	Marker	Statistical approach	Software	QTL	Variation explained (%)	QTL position LG/Flanking markers/ distances (cM)	Authors
Flowering time	60 F_2 *B. juncea;* 60 F_2 *B. oleracea;* 95 F_2 *B. rapa*	glasshouse	until the first fully expanded flower	RFLP	Interval mapping, Two-factor analysis of variance	MapMaker/ QTL 1.1 JMP 3.02		14 36 14 13 13 21 23	J2, (43); J3, (33); J10, (25); J12, (27); J15, (27); J18, (20); O2, (25); O3, (13); O9, (18); R2, (43); R3, (18); R10, (14)	Axelsson et al. (2001)
Flowering time	426 BC_3S_1 plants of *B.rapa* RI line (R500(annual) xPer(biennial)) xR500	growth chamber field	144–160 days after sowing for non-vernalized plants; all vernalized plants were recorded	RFLP	Interval mapping	MapMaker v.2.0 (Haldane function)	*VFR2*	-	LG8, *tg1g9*	Kole et al. (2001)
Plant size	247 F_2 rapid cycling *B. oleracea* (RCB) x *B. oleracea* var. Cantanese (CAN), 250 F_2 RCB x *B. oleracea* var. Pusa Katki (PK), 246 F_2 RCB x *B. oleracea* var. Bugh Kana (BK)	field	until the appearance of the first flower	RFLP	Interval mapping, multiple QTL model	SAS, Microsoft Excel, Mapmaker QTL	Laml Lamw	15.0 33.0 4.5 3.6 13.7 30.3 4.0 20.0 32.4 9.6 36.7 9.3	BK, C4, EW4D04w+43; BK, C7, EW8C11a+0; BK, C6, EW2C08a+14; BK, C9, EST131a+0; CAN, C4, WG3F04b+4; CAN, C7, EW8C11a+4; CAN, C1, EW6F02+1; BK, C1, EW2E07a+16; BK, C7, EW8C11a+0; CAN, C4, EST453g+2; CAN, C7, EW8C11a+4; CAN, C8, EW809c+0	Lan and Paterson (2001)

							Petl	19.5	BK, C7, EW8C11a+2;	
								12.9	CAN, C4, EW7B02+2;	
								25.7	CAN, C7, EW8C11a+2;	
								1.5	CAN, C7, EW3A04b+3;	
								7.3	CAN, C8, WR2F06x+7;	
							Stl	11.1	CAN, C1, EST125+9;	
								6.8	CAN, C1, EW5H03+0;	
								7.0	CAN, C3, EST165d+0;	
								10.8	CAN, C5, EST195b+8;	
								9.7	CAN, C6, EW8A09+6;	
								22.1	CAN, C7, WG3D11+12;	
								14.3	PK, C1, EST373c+4;	
								5.9	PK, C3, EW5F09a+6;	
								11.3	PK, C4, EST122b+2;	
								9.5	PK, C6, WR1D12a+0;	
							Stw	18.7	BK, C2, EW7B04c+16;	
								7.1	BK, C4, EW2B12s+0;	
								32.2	BK, C7, EW8C11a+2;	
								5.6	BK, C8, EW5G04b+13;	
								6.9	BK, C9, EW6B07d+0;	
								5.7	BK, C6, EWc208a+2;	
								11.7	CAN, C4, EST55b+14;	
								11.0	CAN, C5, EST195b+10;	
								33.3	CAN, C7, Ew8C11a+2;	
								32.5	PK, C4, EST133b+7;	
								8.0	PK, C7, EW5A12a+0;	
								12.4	PK, C9, EST131A+15;	
								5.3	PK, C5, EST453d+0;	
							non	13.7	BK, C7, EW8C11a+0;	
								13.7	CAN, C3, EW8F03a+3;	
								16.5	CAN, C7, EW8C11a+4;	
								5.8	CAN, C4, EST55b+6;	
								4.3	CAN, C6, EW5C05c+0;	
								17.7	PK, C4, EW9E10+4;	
								6.1	PK, C9, EW8E11a+0	

Table 5-1 contd...

Table 5-1 contd...

Trait	Plant Material	Spatial Conditions	Temporal Conditions	Marker	Statistical approach	Software	QTL	Variation explained (%)	QTL position LG/Flanking markers/ distances (cM)	Authors
Agrobacterium rhizogenes-mediated transformation efficiency	73 DH lines *B. oleracea* var. *alboglabra* (A12DHd) x *B. oleracea* var. *italica* (GDDH33)	phytotron chamber	35 days after inoculation	RFLP AFLP micro-satelite	Marker regression, interval mapping, ANOVA	QTL CAFE http://web.bham.ac.uk/g.g.seaton/	-	 8.8 5.4 4.3 29.9 8.3 7.8	GFP-fluorescent root production; LG1, (100); LG3, (85); LG7, (62); Adventitious root production; LG3, (34 and 92); LG5, (38); LG7, (66)	Cogan et al. (2002)
Resistance to *Albugo candida*	87 RI lines from *B. rapa* Per (biennial)x R500 (annual)	glasshouse	cotyledon stage	RFLP	Interval mapping	Mapmaker 2.0, Mapmaker/ QTL 1.1	AC2 resistance AC7 resistance	89.8 1.7 96.3	LG4, *wg6c1a-Pub1;* LG2, *wg2d11-ec5a6a;* LG4, *wg6c1a-Pub1*	Kole et al. (2002)
Flowering time	78 BC3S1 *B. rapa* segregating for *FR1;* 100 BC1S1 *B. rapa* segregating for *FR2;* 326 F2 plants *B. rapa* (*fr1/fr1, VFR2/VFR2 _ FR1/FR1,vfr2/ vfr2*) (Kole et al. 2001)	field growth chamber	days after sowing to the first open flower (DTF) number of leaves on the main axis at flowering (LN)	RFLP SSR	Composite interval mapping (CIM), two-factor analysis of variance with means weighted	JoinMap 3.0, QTL Cartographer, Proc MIXED of SAS	*FR1* *VFR1* *FR2*	80.6 14.0 39.0	R2, *BrFLC2*, (11.4); R2, *wg6b10*, (53.9); R3, *BrFLC5*, (24.7)	Schranz et al. (2002)

Developmental characteristics †	86 DH lines cauliflower *B. oleracea* var. *botrytis* x Brussel sprout *B. oleracea* var. *gemnifera*	field	78, 142 and 162 days after sowing	RFLP AFLP SSR	Marker regression, ANOVA	http:/web. bham.ac.uk/ g.g.seaton/	LW	14.5	L.G O6, (16);	Sebastian et al. (2002)
Leaf traits								32.4	L.G O7, (16);	
								44.5	L.G O8, (28);	
									L.G O8, (80);	
							LPL	16.9	L.G O3, (44);	
								10.2	L.G O6, (16);	
								10.6	L.G O7, (26);	
							BPL	45.4	L.G O1, (28);	
									L.G O1, (70);	
								11.3	L.G O2, (6);	
							APL	14.3	L.G O1, (28);	
							WPL	25.6	L.G O1, (0);	
							LAS	37.8	L.G O3, (94);	
							MW	10.5	L.G O2, (30);	
							LN	26.0	L.G O6, (4);	
							WN	17.7	L.G O8, (4);	
							LT	8.3	L.G O1, (32);	
								11.0	L.G O5, (28);	
							LTZ	11.9	L.G O5, (46);	
Flowering traits							VN	14.1	L.G O7, (18);	
								23.0	L.G O8, (24);	
Axillary bud traits							NN[c]	42.9	L.G O8, (22);	
							NL[c]	12.8	L.G O4, (6);	
								14.1	L.G O7, (56);	
							NN[d]	17.9	L.G O8, (14);	
								10.9	L.G O9, (62);	
							AB[d]	10.8	L.G O1, (22);	
Stem traits							PH	9.1	L.G O7, (22);	
							SL	20.6	L.G O1, (34);	
									L.G O1, (68);	
								6.6	L.G O7, (48);	
								11.3	O8, (24)	

Table 5-1 contd...

Table 5-1 contd...

Trait		Plant Material	Spatial Conditions	Temporal Conditions	Marker	Statistical approach	Software	QTL	Variation explained (%)	QTL position LG/Flanking markers/ distances (cM)	Authors
Agrobacterium-mediated transformation		Selected lines from AG DH population of *B. oleracea*, 16 SLs GD33/ A12 for O1, O3 and O7 groups	*in vitro* culture	5-week-old plants	RFLP	Generalized linear model with Poisson error and log-link function, Analysis of variance	-	Transgenic-root production, Adventitious- and transgenic-root production	-	O1, (99.3–108.6); O3, (69.3–81.2)	Cogan et al. (2004)
Regeneration from protoplasts		128 F_2 broccoli *B. oleracea* var. *italica* x rapid-cycling *B. oleracea* (RC)	phytotron chamber	5, 10 and 15 weeks after transfer of microcalli	AFLP micro-satellite	Interval mapping, modified bulked segregant analysis	JoinMap 2.0, PROC GLM SAS, PROC CORR SAS, PlabQTL	5 weeks 5 weeks 10 weeks 10 weeks 15 weeks 15 weeks	19.3% 28.6% 28.8%	O2[a]/C8[b], 16 cM; O9/C7, 11 cM; O2/C8, 17 cM; O9/C7, 11 cM; O2/C8, 17cM; O9/C7, 11 cM	Holme et al. (2004)
Resistance to *Plasmodiophora brassicae* Woron pathotype 1, 2, 4 and 7 (five isolates described near)	Ms6 K92 Pb137–522 K92–16 eH	266 $F_{2/3}$ resistant kale *B. oleracea* var. *acephala* C10 x susceptible broccoli *B. oleracea* var. *italica* HDEM	glasshouse	6–8 weeks	RAPD RFLP ACGM (amplified consensus genetic markers)	Interval mapping (IM), composite interval mapping (CIM), ANOVA, Pearson coefficient	PROC GLM, QTL Cartographer	*Pb-Bo1* *Pb-Bo2* *Pb-Bo5a* *Pb-Bo5b* *Pb-Bo9b* *Pb-Bo1* *Pb-Bo2* *Pb-Bo8* *Pb-Bo1* *Pb-Bo3* *Pb-Bo1* *Pb-Bo4* *Pb-Bo5a* *Pb-Bo1* *Pb-Bo2* *Pb-Bo5a* *Pb-Bo5b* *Pb-Bo9a*	34.0 6.8 4.3 3.0 3.4 77.4 2.5 2.1 80.7 4.7 68 2.8 3.3 20.7 21.0 4.2 6.1 5.5	LG1, T2, (20.41); LG2, S07.1900, (85.41); LG5, PBB7b, (0.01); LG5, ab1.1350, (116.81); LG9, a04.1900, (44.78); LG1, T2, (20.41); LG2, PBB38a, (66.01); LG8, c01.980, (36.49); LG1, T2, (20.41); LG3, aa7.1400, (144.46); LG1, T2, (20.41); LG4, aa9.983, (112.89); LG5, s07.200, (29.27); LG1, ae05.800, (0.01); LG2, r10.1200, (88.21); LG5, PBB7b, (0.01); LG5, a18.1400, (118.19); LG9, aj16.570, (10.01)	Rocherieux et al. (2004)

Agrobacterium tumefaciens susceptibility	59 plants of AGDH population *B.oleracea* 56 F_1 families 5 substitition lines for LG 09	*in vitro* culture	5-days-old seedlings were inoculated and after 50 days evaluated	RFLP	Generalized linear model with logit link and binomial error	JoinMap 3.0, MapQTL 4.0, Genstat 5.0	-	-	O9, pW233, (40.1)	Sparrow et al. (2004a)
Transgenic and adventitious root production after *Agrobacterium rhizogenes* treatment	48 DH plants of NG population *B. oleracea* var. *botrytis* (N) x *B. oleracea* var. *gemmifera* (G)	*in vitro* culture	6-days-old seedlings were inoculated and evaluated after 35 days		Marker regression, Interval mapping, ANOVA	QTL CAFÉ	 1 QTL 2 QTL 2 QTL	21.1–24.54 40.15–24.54 15.7–37.5 - -	LG1, (34–38) [1,2,3]; LG8, (2–8) [1,2,3]; LG9, (64–104)[2,3]; LG9, (0–4) [1,2,3]; LG9L, (58–88) [1,2,3]	Oldacres et al. (2005)
Resistance to *Plasmodiophora brassiceae*	94 $F_{2/3}$ lines of *B. rapa* G004 DH line x nou7 Chinese cabbage	greenhouse	6 weeks from sowing	SSR RFLP RAPD SNP Indel markers	Interval mapping	Mapmaker /exp 3.0, Mapmaker /QTL 1.1	'Waka-yama-01' resistance 'Ano-01' resistance	26.8 18.3 10.5 71.7 15.9	LG7, BRMS-297-BRMS-088, (~40); LG6, BRMS-100-BRMS-096, (~60); LG2, BN288D-WE24-1, (~10); LG7,BRMS-297-BRMS-088, (~40); LG2, BN288D-WE24-1, (~10)	Suwabe et al. (2006)

Table 5-1 contd...

Table 5-1 contd...

Trait	Plant Material	Spatial Conditions	Temporal Conditions	Marker	Statistical approach	Software	QTL	Variation explained (%)	QTL position LG/Flanking markers/ distances (cM)	Authors
Flowering time (FL) Seed-related trait: Seed pod length (SPL) Seed pod width (SPW) Beak length (SBL) Seed color (SC) Seed weight (SW) Growth-related trait: Plant height (PH) Branches (PB) Leaf trait: Leaf edge shape (LES) Leaf trichomes (LT) Leaf number (LN) Leaf lobes (LB) Lamina length (LL) Lamina width (LW) Petiole length (PL) Leaf area (LA) Leaf index (LI) Turnip trait: Turnip formation (TF) Turnip shoots (TS) Turnip length (TL) Turnip width (TWi) Turnip weight (TWe)	178 $F_{2/3}$ lines *B.rapa* rapid cycling line RC-144 x Chinese cabbage line CC-156; 71 DH lines Pak choi PC-175 x Yellow sarson YS-143; 64 DH lines vegetable turnip VT-115 x YS-143; 136 backcross population lines F_1 plant (VT-115xYS-143) xVT-115	F_2: non-heated greenhouse F_3: heated and non-heated greenhouse DH: heated greenhouse and open field BC: open field	harvested siliques before flowering state		Interval mapping (IM), Multiple-QTL model mapping (MQM)	MAPQTL 5.0, Mapchart	FLQTL-1 FLQTL-2 FLQTL-3 FLQTL-4 FLQTL-5 FLQTL-6 FLQTL-7 FLQTL-8 SPQTL-1 SPQTL-2 SPQTL-3 SPQTL-4 SCQTL-1 SWQTL-1 SWQTL-2 PHQTL-1 PHQTL-2 PHQTL-3 LQTL-1 LQTL-2 LQTL-3 LQTL-4 LQTL-5 LQTL-6 LQTL-7 LQTL-8 LQTL-9 LQTL-10 TuQTL-1	13.4 8.9–59.3 19.7 9.3 18.3 14.3–15.1 17.9 11 11.1 20.3 27.2–38.1 25.5 61.7–65.5 10.2–11.6 17.6 15.7 8.9 23.9 10.3–25.8 6.5 20.5–26.4 7.0–24.2 12.5–22.5 9.1 13.7–21.9 10.6 9.4–11.2 10.8–12.6 24.0–40.0	R01, middle; R02, top; R03, bottom; R03, top; R06, middle; R07, H-middle; R10, top; R08, middle; R01, middle; R05, top; R07, H-middle; R09, H-middle; R09, middle; R03, middle; R08, H-middle; R02, H-middle; R03, top; R07, H-middle; R02, H-middle; R02, L-middle; R03, bottom; R05, middle; R06, middle; R06, L-middle; R07, H-middle; R08, H-middle; R08, bottom; R09, H-middle; R02, top	Lou et al. (2007)

Flowering time	134 F_2 *B. oleracea* var. *italica* x *B. oleracea* var. *capitata*	greenhouse, field	days to flowering (DTF) from transplanta-tion to the field up to 130 DTF	RFLP SRAP CAPS AFLP	Composite interval mapping (CIM)	JoinMap 3.0, MapQTL 2.0, QTL Carto-grapher 1.16	BoFLC2	36.8 8.7 4.3 3.9 5.8 7.8	O2, BRMS215-F2-R4b, (62.1); O6, E13M1-02-F4-R3b, (9.3); O2, FLC1-pW131, (9.7); O8, BRMS023-F3-R5d, (30.5); O9, BRMS085-pW240, (55.6); O3, BRMS008-F3-R4d, (18.9)	Okazaki et al. (2007)
Resistance to *Xanthomonas campestris* pv. *campestris*	114 F_2 *B. rapa* R-o-18 x B162	glasshouse	6-weeks-stage	AFLP	Interval mapping, Multiple-QTL mapping, (MQM) Correlation coefficients	JoinMap 3.0, MapChart, MapQTL 4.0, PROC CORR of SAS	*XccR1d-1* (race1 symptoms score), *XccR1i-1* (% succesful infection), *XccR4d-1* (race4 symptoms score), *XccR4i-1* (% succesful infection), *XccR4i-2* (% succesful infection), *XccR4i-3* (% succesful infection) infection	63.9 51.0 58.0 24.0 12.6 6.5	A06, E11M50_280b, (20.5); A06, E12M48_171r, (15.7); A06, E12M61_215b, (33.5); A06, E12M61_215b, (33.7); A02, E11M59_178r, (27.6); A09, E12M48_1>330b, (27.2)	Soengas et al. (2007)

Table 5-1 contd...

Table 5-1 contd...

Trait	Plant Material	Spatial Conditions	Temporal Conditions	Marker	Statistical approach	Software	QTL	Variation explained (%)	QTL position LG/Flanking markers/ distances (cM)	Authors
Leafy trait: Leaf edge shape (LES) Leaf trichomes (LT) Leaf number (LN) Flowering time: Flowering time without vernalization (NDF) Flowering time after vernalization (VDF) Phytate: Phytate level in leaves (LPHY) Phosphate level in leaves (LPHO) Phytate level in seeds (SPHO) Phosphate level in seeds (SPHY)	160 *B. rapa* accessions	greenhouse	days to flowering for vernalized plants, days to flowering up to 120 for vernalized plants, 5-week-old plants for LPHY and LPHOL 4 weeks after sowing for LN, LES and LT	AFLP	One-way analysis of variance (ANOVA), correlation analysis, association mapping: Bayesian approach, Markow Chain, Monte Carlo analyses, Fisher's exact test	GenStat 8.1, *structure* 2.1, SAS® software	VDF˘ SPHY VDF VDF LT LPHY LES	0.035ˆ 0.017 0.050 0.000 0.000 0.001 0.014	R02,pAG/mCAC0154.7, (63.4); R02,pTA/mCAT0230.2, (84.5); R03,pAT/ mCCA0135.0,((15.8); R05,pGG/mCAA0339.8, (50.1); R05,pGG/mCAA0339.8, (50.1); R07,pAG/mCAC316.7, (34); R08,pAG/mCAC090.5, (96.5)	Zhao et al. (2007)
Shoot calcium and magnesium content	376 *B. oleracea* accessions; 74 F_1 hybrid cultivars 90 DH lines from AGDH population of *B. oleracea*, 20 SLs from AGSL popualtion,	glasshouse field	Shoots were sampled after 39 and 105 days of growth	RFLP AFLP	Composite interval, mapping (CIM), marker regression analysis	QTL Carto-grapher 2.0, QTL CAFÉ, GenStat	Mg-f [e] Ca-m-mr Mg-m-mr Ca-g Ca-g Ca-f		C2, pN63E1-pW141E1, (70–100); C6, pW197E2-pO104E2, (25–47); C6, pO10E1-pO104E2, (20–55); C7, pO87E2-pN86E1, (0–24); C7, pN86E1-pN20e2, (20–33); C7, pN64E2-pCenE3, (45–75);	Broadley et al. (2008)

	selected AGSLs were backcrossed and selfed to $BC_3S_{2/3}$						Ca-m		C7, pN97J2-pCene3, (62–73);	
							Ca-m		C8, pW138J1-pR97J1, (20–52);	
							Ca-g-mr		C8, pW188J1-AC-CAAE05, (32–56);	
							Ca-m-mr		C8, pW188J1-AC-CAAE05, (34–60);	
							Ca-f-mr		C8, pW188J1- pO143E2, (40–62);	
							Mg-g-mr		C8, pW188J1-AC-CAAE05, (38–58);	
							Mg-m-mr		C8, pW188J1-pN173E1, (40–74);	
							Mg-f-mr		C8, pW188J1-pO143E2, (38–64);	
							Mg-f	33	C9, pN52E2-pO125E1N, (0–10);	
							Mg-f		C9, pO160E1-pW200J1, (70–100);	
							Ca-f		C9, pW137J1-pO119J1, (23–42);	
							Ca-f	55	C9, pO119J1-pW233J1, (42–55);	
							Ca-f-mr		C9, pW233J1-pO160E1, (54–72);	
							Ca-f		C9, pO7E1-pN47E4NM, (73–90);	
							Ca-g	31	C9, pW114E2-pW233J1, (35–55);	
							Ca-g-mr		C9, pO119J1-pW233J1, (42–57);	

Table 5-1 contd...

Table 5-1 contd...

Trait	Plant Material	Spatial Conditions	Temporal Conditions	Marker	Statistical approach	Software	QTL	Variation explained (%)	QTL position LG/Flanking markers/ distances (cM)	Authors
							Ca-m Ca-m-mr Mg-g Mg-g-mr Mg-m-mr Mg-m	14	C9, pO119J1-pW233J1, (42–60); C9, pW233J1-pO160E1, (52–63); C9, pO119J1-pO160E1, (42–69); C9, pW233J1-pO160E1, (44–63); C9, pW233J1-pO160E1,(52–64); C9,pO160E1-pW200J1, (70–105)	
Glucosinolate accumulation	71 lines from DH38 *B. rapa* yellow sarson R500 x pak choi HK Naibaicai, 64 lines from DH30 *B. rapa* yellow sarson R500 x turnip Kairoku Hakata	glasshouse, growth cabinet	leaves of 40- or 50- days-old plants	AFLP SSR	Interval mapping, MQM mapping, ANOVA	JoinMap 4.0, MapChart 2.1, MAPQTL 5.0	Ali-QTL1.1[f] Ali-QTL1.2 Ali-QTL1.3 Ali-QTL3.1 Ali-QTL3.2 Ali-QTL4.1 Ali-QTL5.1 Ali-QTL5.2 Ali-QTL6.1 Ali-QTL7.1 Ali-QTL8.1 Ali-QTL9.1 Ali-QTL9.2 Ali-QTL9.3 Ali-QTL10.1 Ali-QTL10.2 Ind-QTL2.1 Ind-QTL3.1 Ind-QTL8.1	40–56 for DH38 75–47 for DH30 12.7 (DH38) 18 (DH30) 16.4 (DH38)	A1-top; A1-Mid; A1-Bot; A3-Mid; A3-Bot; A4-Mid; A5-Mid; A5-mid; A6-Mid; A7-Mid; A8-Bot; A9-Mid; A9-Mid; A9-Bot; A10-Mid; A10-Bot; A2-mid; A3-Bot; A8-Bot	Lou et al. (2008)

							Aro-QTL4.1	16.2 (DH38)	A4-Top	
							Aro-QTL4.2	24.7 (DH30)	A4-Mid	
							Aro-QTL7.1	24.9 (DH30)	A7-Top	
Taproot thickness (TRT)	144 F_2 lines *B. rapa* ssp. *rapifera* QSH97-24 x *B. rapa* ssp. *chinensis* AJH97-2 144 F_3 lines	field	55 days after sowing, 89 days after sowing	AFLP RAPD	Analysis of variance, trait correlation, Shapiro-Wilk test	SAS software, Mapmaker/ QTL 1.1	*qTRT1*	11.6	LG1,H14-300~E41M49-193;	Lu et al. (2008)
							qTRT2	16.8	LG2, X03-600~F15-350;	
							qTRT3	8.4	LG3, F07-1300~B02-420;	
							qTRT4a	17.8	LG4, A16-250~F08-680;	
							qTRT4b	27.4	LG4, E41M48-180~E36M49-220;	
							qTRT6	11.9	LG6, E38M62-100~E36M61-142;	
Taproot length(TRL)							*qTRT9*	15.1	LG9, W07-920~A09-330;	
							qTRL2	9.2	LG2, S16-1000~T09-300;	
							qTRL3	17.3	LG3, A04-1400~E38M62-315;	
							qTRL4	11.8	LG4, E41M48-180~E36M49-220;	
							qTRL7	19.3	LG7, J04-500~G03-450;	
Taproot weight (TRW)							*qTRL9*	14.5	LG9, A09-330~V18-400;	
							qTRW1	14.0	LG1, H14-300~E41M49-193;	
							qTRW2	12.2	LG2, A09-300~E32M49-590;	
							qTRW4	24.8	LG4, E41M48-180~E36M49-220;	
							qTRW5a	18.6	LG5, S18-850~E36M49-300;	
							qTRW5b	17.1	LG5, E36M49-300~E30M50-405;	
							qTRW9	15.4	LG9, W07-920~A09-330	

Table 5-1 contd...

Table 5-1 contd...

Trait	Plant Material	Spatial Conditions	Temporal Conditions	Marker	Statistical approach	Software	QTL	Variation explained (%)	QTL position LG/Flanking markers/ distances (cM)	Authors
Resistance to *Plasmodiophora brassiceae*	*B. rapa* ssp. *pekinensis* $F_{2/3}$ population K10 x Q5; $F_{2/3}$ population C9 x 6R	glasshouse	30 days after sowing	RFLP RAPD AFLP SSR STS	Composite interval mapping (CIM)	JoinMap 3.0, QTL Cartographer 2.5	M85 resistance K04 resistance	-	R3,HC688, (37.1); R3, HC186, (39.3); R3, E11M2-03, (42.6); R3, OPC11-2S, (46.2); R3, HC688, (37.1); R3, HC186, (39.3); R3, E11M2-03, (42.6); R3, OPC11-2S, (46.2); R2, E14M3-02, (22.5); R2, m6R, (25.5); R2, E15M4-06, (27.6)	Sakamoto et al. (2008)

*named according to their location on linkage groups

[a] located on Sebastian et al. (2000) linkage group

[b]Located on Saal et al. (2001) linkage group

[c]Block 1

[d]Block 2

†Abbreviations used by Sebastian et al. (2002): lamina width (LW), lamina petiole length (LPL), bare petiole length (BPL), auricle petiole length (APL), wing petiole length (WPL), leaf apex shape (LAS), midrib width (MW), lobe number (LN), wing number (WN), leaf type (LT), leaf transition zone (LTZ), vernalisation (VN), axillary buds (AB), node number (NN), leaf number (NL), plant high (PH), stem length (SL);

ˇonly associations to markers with known map position were described (Zhao et al. 2007);

^ significant *p* value in association between measured traits and AFLP markers using multi-QTL model (Zhao et al. 2007)

[1]adventitious root production in the absence of *Agrobacterium*, Oldacres et al. (2005),

[2]adventitious root production in the presence of A. rhizogenes, Oldacres et al. (2005),

[3]transgenic root production, Oldacres et al. (2005);

[e]QTL detected by CIM: Ca-f: shoot Ca in field; Ca-g: shoot Ca in glasshouse, Ca-m: mean of shoot Ca in field and glasshouse Broadley et al. (2008);

[e]QTL detected by multiple marker regression have –mr suffix (e.g. Mg-m-mr) Broadley et al. (2008);

[f]Abbreviations used in QTL description by Lou et al. (2008): Ali (aliphatic), Ind (indole), or Aro (aromatic).

previous reports, evidence for genetic interaction was found between two regions affecting flowering time, on LG 6 and LG 8 (Sachan and Singh 1987; Kennard et al. 1994). Homology relationship between genes related to flowering time in *B. napus* (Ferreira et al. 1995) and *B. rapa* (Teutonico and Osborn 1995) was not detected. Comparative mapping analysis found three segments in *B. nigra* genome to be homologous to the top of *A. thaliana* chromosome 5 (Lagercrantz et al. 1996). This is a region where several well characterized flowering time genes such as *FLOWERING LOCUS C (FLC), FY* and *CONSTANS* (*CO*) are located. In two of those three segments in *B. nigra* flowering time affecting loci were found in close proximity to mapped homologues of the *A. thaliana CO* gene. These findings indicated that *B. nigra CO* homologs could be the candidate gene for flowering time QTL. Such events have also been identified in *B. oleracea* genome (Bohuon et al. 1998; Table 5-1). The objective of this work was an analysis of an extensive genetic variation in *Brassica* crop genus, exploiting the synteny between and within *Brassica* genomes and *A. thaliana* in order to ease the mapping and identification of latent candidate loci. For this purpose, a population of 150 DH lines of rapid cycling *B. oleracea* derived from a cross between *B. oleracea* var. *alboglabra* x *B. oleracea* var. *italica* was developed, and subsequently scored twice for flowering time. QTL mapping identified six QTL in the following linkage groups: one each on O2 and O3 and two each on O5 and O9. The noteworthy difference in genetic variation was observed, as it was explained in 58% and 93% for the two trials studied, respectively. The observation of homology of regions on O2, O3 and O9, marked with three out of six QTLs, both with each other and in respect to chromosome regions of *B. nigra* (Lagercrantz et al. 1996), and to a single region on *A. thaliana* chromosome 5, pointed at the possibility that one or more of these genes may be candidates for the QTL found in *Brassica*.

The development of backcross substitution lines using the same cross as reported above, and performing longer field trials enabled Rae et al. (1999; Table 5.1) the localization of novel QTLs on chromosomes O1, O2, O3, O5 and O9 of *B. oleracea*. These findings coincided with the results of Bohuon et al. (1998; Table 5-1), in respect to the location of five of QTLs.

Lan and Paterson (2000; Table 5-1) continued the work on localization of flowering time affecting loci (Osborn et al. 1997; Bohuon et al. 1998) and identified at least six additional *Brassica* QTLs on the tops of *B. oleracea* linkage groups 1, 4 and 7, regions homologous to a region of *A. thaliana* chromosome 5.

The collection of a great amount of data on the localization within *Brassica* genomes, chromosomal segments equivalent to the top of the fifth *A. thaliana* chromosome, to which the QTL for flowering time has been mapped (Lagercrantz et al. 1996; Bohuon et al. 1998), raised two important questions. First, how strongly do duplicated genes in polyploid *Brassica* genomes influence the variation in flowering time traits? Second, are multiple QTLs for a particular trait often controlled by homologous genes? The purpose of the analysis performed by Axelsson et al. (2001; Table 5-1) was to answer these questions. This work was extended and flowering time related QTLs were mapped in *B. juncea*, *B. oleracea* and *B. rapa*, using rapid-cycling lines of each species as the early flowering parent. This study targeted chromosomal regions of *Brassica* that are homologous to chromosomal regions around the *CO* gene in *A. thaliana*. Using molecular markers they found that each diploid *Brassica* species had three regions homologous to the *CO* region of *A. thaliana* and that the amphidiploid *B. juncea* had six copies. Flowering time QTLs were detected in two of the three replicated segments of each diploid (A, B and C) genome, and in three of six replicated segments in the amphidiploid *B. juncea* (two QTLs in the A-genome and one QTL in the B-genome). These findings provided evidence that at least some of the QTLs in the different species are controlled by orthologous and (or) paralogous genes. It is worth mentioning that the QTLs detected on O3 and O9 of *B. oleracea* genome in this study may be correlative to those previously identified by Bohuon et al. (1998).

In contrast to Axelsson et al. (2001; Table 5-1), who reported that the QTL detected correspond to homologues of *CO*, Kole et al. (2001) and Schranz et al. (2002; Table 5.1) provided strong evidence based on confidence intervals for QTL and map positions for *CO* and *FLC*, that the flowering time loci in *B. rapa* are controlled by *FLC* rather than *CO* homologs. This discrepancy may be due to differences in the parentage of populations used to construct linkage maps: Axelsson et al. (2001; Table 5-1) crossed annual types (no vernalization requirement, controlled by *FLC* in *A. thaliana*) to develop segregating populations, whereas Kole et al. (2001) and Schranz et al. (2002; Table 5-1) used segregating populations derived from a cross between biennial and annual form of *B. rapa* (with vernalization requirement).

Additional information came with a study by Sebastian et al. (2002; Table 5.1), described in the previous section, who localized two QTLs affecting vernalization, on linkage groups O7 and O8, respectively. The latter appears to be equivalent to the QTL identified on linkage group 9 of *B. napus* (Ferreira et al. 1995). Other studies have shown the same region in *B. napus* to have a homology with a region in *B. rapa* that also has a putative QTL controlling vernalization (Teutonico and Osborn 1994; Osborn et al. 1997).

Although multiple QTLs affecting various traits related to flowering time have been localized, it was only the first step toward identification of the nature of genetic variation for those traits. In order to achieve this goal, the analysis of candidate genes, as a following step is necessary (Razi et al. 2008). The candidate gene approach here takes advantage of information from many studies, which have identified flowering time

QTLs in different populations and environmental conditions, and from the detailed knowledge of the key genes involved in controlling flowering time that have mainly used *A. thaliana* as a model system (Michaels and Amasino 1999; Sheldon et al. 1999, 2000). Many genes controlling flowering time have been identified in *A. thaliana* but these are considered as major candidate genes *FRI* and *FLC*, which are a key vernalization requirement and response genes, and *CO*, which plays a key role in a photoperiodic regulation of flowering (for a review, see Razi et al. 2008). Keeping in mind, that due to a whole-genome triplication followed by chromosome rearrangements (Lagercrantz and Lydiate 1996; Lysak et al. 2005; Nelson and Lydiate 2006; Ziolkowski et al. 2006), which led to the formation of *Brassica* ancestors, a single locus in the *A. thaliana* genome is represented by multiple copies in *Brassica* genomes. Hence, in *B. oleracea* four *FLC* copies were isolated (Schranz et al. 2002; Lin et al. 2005; Okazaki et al. 2007). A recent study by Okazaki et al. (2007; Table 5-1) reported that among those *FLC* copies (*BoFLC1, BoFLC2, BoFLC3,* and *BoFLC5*) only *BoFLC2* affected flowering time through vernalization. In amphidiploid *B. napus* in the first instance only five *FLC* genes were identified (Tadege et al. 2001), but more recently Udall et al. (2005) reported the mapping of eight loci, four in each A and C genomes.

Three *B. oleracea FLC* homologs (*BoFLC1, BoFLC3,* and *BoFLC5*) were sequenced and their genomic organization characterized (Razi et al. 2008) to infer their identity as candidate genes. Phylogenetic analysis confirmed that *Brassica FLC* copies are *A. thaliana FLC* orthologs, and supported the occurrence of *FLC* gene duplications before the divergence of *B. oleracea* and *B. rapa* about 4 Mya (Rana et al. 2004). Mapping of four *BoFLC* loci onto the genetic map (A12/GD33 population) confirmed that three copies (*BoFLC1, BoFLC3,* and *BoFLC5*) are located within the confidence interval of their respective flowering time QTLs, while *BoFLC4* is located on linkage group O2, where no flowering time QTL was detected. Although the fragment from GD33 parent was considered as a donor of QTL affecting flowering time in backcross progeny obtained, there was no correlation found between late flowering and segregation of GD alleles for *BoFLC1* and *BoFLC3*. These findings together with results indicating that *BoFLC5* and *BoFLC4* were pseudogenes, strongly contradicted the hypothesis that any of them could be candidate for the flowering time QTL in this population.

In *B. rapa*, several QTLs (*VFR1, VFR2,* and *VFR3*; *FR1, FR2,* and *FR3*) for flowering time were identified in F_2 and RIL populations derived from a cross between annual and biennial oil types (Teutonico and Osborn 1995; Osborn et al. 1997). *VFR2* was considered to have a large effect, and was suggested to be homologous to *FLC* of *A. thaliana*. Further studies have confirmed the co-localization of several QTLs and loci for *B. rapa FLC* homologs *BrFLC2, BrFLC3, BrFLC5,* and *BrFLC1*, which were then assigned to linkage groups R02, R03, R03, and R10, respectively (Kole et al. 2001; Schranz et al. 2002; Kim et al. 2006; Lou et al. 2007). Three QTLs for flowering time traits detected by Lou et al. (2007) (Fig. 5-1; Table 5-1) co-localized with *LFY* (*LEAFY*), *FT* (*FLOWERING LOCUS T*) and *VRN2* (*VERNALIZATION2*) homologs. Among 11 AFLP markers associated with days to flowering, identified in *B. rapa* through association mapping, three were localized in R02, R03, and R05 (Zhao et al. 2007; Table 5.1). Studies on effects of *FLC* regions on flowering time have revealed that in *B. rapa FR1* (*BrFLC2*) and *VFR2* (*BrFLC1*) act in a dosage-dependent manner to control the flowering response and respond to vernalization treatment (Schranz et al. 2002; Table 5-1).

5.2.3 Pathogen Resistance

5.2.3.1 Plasmodiophora brassicae

Clubroot, the most important disease of vegetable *Brassica* crops, is caused by the obligate, biotrophic soil-borne pathogen *Plasmodiophora brassicae.* The infection leads to gall formation on the root system, which inhibits nutrient and water transport, delays plant growth and increases the susceptibility to wilting. As the protection of these crops by using cultural practices or chemical treatments has proven problematic, the development of resistant cultivars is the most promising way to control clubroot for all *Brassica* crops (see Chapter 2).

The reported resistance to clubroot concerns different cruciferous species, including the three most extensively cultivated: *B. napus*, *B. rapa* and *B. oleracea* (Crute et al. 1980). In spite of identification of several sources of resistance in *B. oleracea* (Crute et al. 1980; Crisp et al. 1989; Monteiro and Williams 1989; Dias et al. 1993; Manzanares-Dauleux et al. 2000), only in a few cases has breeding for resistance been successful. This was mainly due to incomplete resistance to clubroot in *B. oleracea*, which was also mainly found in improper, from a market point of view, types (Crute et al. 1983). Variable expression of clubroot resistance and lack of exact information on genetic control have together limited progress towards developing highly resistant varieties. Some authors have presented evidence, that in *B. oleracea* this character is determined by several genes with either recessive (Weisaeth 1974; Crute et al. 1983; Voorips and Visser 1993) or dominant alleles (Laurens and Thomas 1993).

Genes for clubroot resistance were for the first time mapped by Landry et al. (1992; Table 5-1). They have identified two QTLs, which explained 61% of the total variation for resistance to race 2 of *P. brassicae*. The results indicated, that in the case of a cabbage resistant line (*B. oleracea* var. *capitata*), alleles for resistance of two QTLs appeared to be dominant over susceptibility.

Figdore et al. (1993; Table 5.1) have detected three QTLs in an F_2 population obtained from a cross between broccoli (*B. oleracea* var. *italica*), resistant to race 7 of *P. brassicae* and cauliflower (*B. oleracea* var. *botrytis*) susceptible to *P. brassicae*. They reported one major dominant QTL that came from the broccoli line, one QTL—from the cauliflower line and the third—expressed in the heterozygous state.

Grandclement and Thomas (1996; Table 5-1) investigated resistance to *P. brassicae* Woron in an F_2 population of a cross between a clubroot-resistant kale (*B. oleracea* var. *acephala*) and a susceptible cauliflower (*B. oleracea* var. *botrytis*). RAPD markers were used for QTL detection. The disease symptoms were observed twice, at different times after inoculation. Three markers were associated with these two observations and three were specific to only one. QTL analysis demonstrated the existence of at least two genetic mechanisms for resistance to *P. brassicae*.

A significant improvement in QTL mapping and estimating their effects was achieved by Ramsey et al. (1996). They increased the resolution of QTL identification by using molecular markers that limit regions from chromosome of one donor line to be introgressed into recurrent recipient line through subsequent backcrossing and inbreeding. The construction of a library containing *B. oleracea* recombinant backcross lines with single fragment in the BC_2S_1 or BC_3S_1 generations has been proved possible.

QTL analysis for clubroot resistance was performed in a DH population developed from a cross between DH parents of cabbage (*B. oleracea* var. *capitata*) and broccoli (*B. oleracea* var. *italica*) by Voorrips et al. (1997). They tested each plant by determining symptom grades, and by measuring the fresh weights of the healthy and affected parts of the root system. The analysis performed with multiple QTL mapping approach indicated that at least three QTLs responsible for clubroot resistance were present; two that were found, namely *pb-3* and *pb-4*, that explained only 68% of the difference between the parents and 60% of the genetic variance among DH line means. Contrasting results were obtained by Moriguchi et al. (1999; Table 5-1), who found a single QTL affecting clubroot resistance, performing field experiments on plants naturally infected.

Rocherieux et al. (2004; Table 5-1) analyzed the genetic basis of resistance to clubroot from C10, an inbred line selected from a French kale (*B. oleracea* var. *acephala*) landrace, previously reported as highly resistant to several pathotypes of *P. brassicae* (Manzanares-Dauleux et al. 2000). This line was crossed to a highly susceptible broccoli (HDEM), and genetic control of resistance to five different *P. brassicae* isolates (four single spore isolates and one field isolate) was investigated in constructed $F_{2/3}$ population. Eight out of nine QTLs identified were isolate-specific and one conferred resistance to all the isolates tested. The analyses showed that the effect of all QTLs was flexible, and relative to the isolates.

The identification of QTL for clubroot resistance has also been conducted in *B. rapa* (Kuginuki et al. 1997; Suwabe et al. 2002). The genetic origin of clubroot resistance was investigated in an F_3 population derived from a cross between resistant and susceptible DH lines of Chinese cabbage (*B. rapa* ssp. *pekinensis*) (Suwabe et al. 2003, 2006; Table 5.1). The comparative analysis of three QTL (*Crr1*, *Crr2*, and *Crr4*) identified in the latter study, suggested the co-localization of two major QTL regions, *Crr1* and *Crr2* in a region of *A. thaliana* chromosome 4 (Fig. 5-2). These results together with previous analysis, which showed the presence of disease-resistance gene cluster in the above mentioned *A. thaliana* region, indicated their common ancestry. Additionally, in *B. oleracea* also two clubroot resistance related QTLs have been reported (Landry et al. 1992; Voorips et al. 1997; Table 5-1). These similarities, together with the hypothesis that both species evolved from a common ancestor, raised the question whether clubroot resistance in *B. rapa* and *B. oleracea* might share the same evolutionary origin and resistance mechanisms.

Recently, a new locus for clubroot resistance was identified by QTL mapping in an F_2 population derived from a cross between C9 Chinese cabbage clubroot-resistant DH line and susceptible DH line, 6R (Sakamoto et al. 2008). Locus *CRc* was localized on linkage group R2. This study also reported the identification of a locus *CRk* in an F_2 population derived from a cross between another resistant line, K10 and susceptible Q5. This locus was located close to *Crr3* in linkage group R3. These two clubroot resistant Chinese cabbage DH lines, together with a third one, T136-8, have been developed using resistant genes from CR European fodder turnips (*B. rapa* ssp. *rapifera*), and evaluated on resistance against two isolates of *P. brassiceae*, M85 and K04. Additionally, the novel *CR* loci have been tagged by newly developed sequence tagged site (STS) markers.

5.2.3.2 Albugo candida

Another pathogen, which causes the disease known as white rust on cultivated *Brassica* species and several wild crucifers, is a biotrophic fungus, *Albugo candida* (Pers.). The symptoms are: appearance of white, blister-like pustules on the lower surface of leaves and on stems, followed with severe distortions in development of the affected organs, especially flowers and siliqua. The consequence of this disease is severe yield losses in turnip rape (*B. rapa*), mustard (*B. juncea*), and rapeseed (*B. napus*) (Kumari et al. 1970; Petrie 1973; Harper and Pittman 1974; Fan et al. 1983). On leafy types of *B. rapa*, such as pak choi (*B. rapa* ssp. *chinensis* L.) and Chinese cabbages (*B. rapa* ssp. *pekinensis* L.), white rust damage can make all the production unsuitable for markets (Santos et al. 2006). Breeding for resistant cultivars

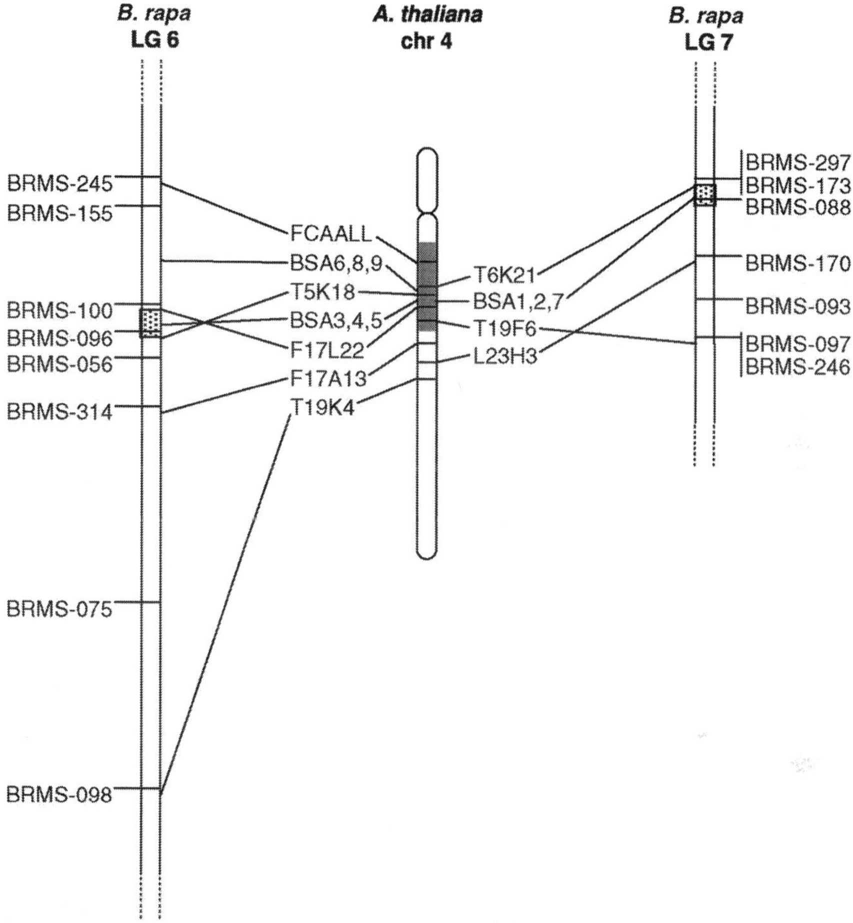

Figure 5-2 Synteny map of the *Crr1* and *Crr2* regions (stippled area) with *B. rapa* and *A. thaliana*. The linkage groups (LG) of *B. rapa* and the chromosome (chr) of *A. thaliana* are presented as vertical bars. The *Arabidopsis* BAC clones that correspond to the SSR loci of *B. rapa* are connected by lines. The shaded region indicates the cluster of disease resistance genes (MRC-H) in the *Arabidopsis* chromosome. (reprinted from Suwabe et al. 2006 with permission of Genetics Society of America).

is the most successive way to control the disease, besides crop rotation and pesticide spraying.

The genetic basis of resistance to white rust in *B. rapa* was investigated by mapping in RIL population derived from a cross between a biennial turnip rape, resistant to AC2 and AC7 races, and an annual sarson, susceptible to both races of *A. candida* (Kole et al. 2002; Table 5-1). The response to both races of *A. candida* was analyzed in young seedlings, in which cotyledons were inoculated and subsequently scored for interactions phenotypes.

The test found that a single major gene (or tightly linked genes) controlled resistance to the two races. Major resistance locus for both races co-localized on linkage group 4 with previously identified QTL for resistance to race 2 (Kole et al. 1996; Kole et al. 2002). A second minor QTL for AC2 was mapped on linkage group 2. The comparative analysis of map positions of white rust resistance genes in *B. rapa* and *B. napus*, based on common loci in both maps (Parkin et al. 1995; Osborn et al. 1997), pointed to the possibility of location of additional loci, that have not been localized. The synteny of a region with resistance genes mapped in this study to the top or the bottom of *A. thaliana* chromosome 5 was also suggested.

Resistance to white rust was also evaluated in 43 accessions of *B. rapa* ssp. *chinensis* and 19 accessions of *B. rapa* ssp. *pekinensis* at the cotyledon stage with a Portuguese *A. candida* isolate Ac506 collected from turnip leaves (Santos et al. 2006). The two crosses between pak choi BRA 117 (the most resistant accession) and the rapid cycling *B. rapa* line CrC 1.19 were used to analyze the genetic basis of resistance. As a result, dominant/recessive epistatic interaction between two genes affecting the resistance to Ac506 in pak choi BRA 117 was suggested.

5.2.3.3 *Xanthomonas campestris pv. campestris*

The disease that is considered the most serious to some crucifer crops, decreasing both yield and quality (Williams et al. 1980), is a black rot, caused by bacteria *Xanthomonas campestris* pv. *campestris* (Pammel) Dawson. First contact with plant tissues occurs through xylem and mesophyll. The symptoms are marginal leaf chlorosis (usually V-shaped), necrosis and darkening of leaf veins and vascular tissue within the stem, as an effect of an extracellular polysaccharide production by bacteria that stops normal water flow, followed with wilting and necrosis.

In *B. rapa*, the most severe damage is done to turnip and turnip greens (Vicente 2004). The presence of at least six races of the pathogen (Vicente et al. 2001) makes breeding for black rot resistant cultivars a complex approach. Although, races 1 and 4 are thought as the main factors causing disease in *B. oleracea* crops, the extensive screening of *B. oleracea* accessions led to a conclusion that either resistance to these two races did not exist or was very rare (Taylor et al. 2002), in contrast to common resistance to less important races (2, 3, and 6). Because race specific resistance to races 1 and/or 4 is frequently found in other species of the genus *Brassica* (Taylor et al. 2002), which could then serve as a source of this resistance, the inheritance of resistance of *B. rapa* Chinese cabbage accession B162 to the races 1 and 4 of *X. campestris* pv. *campestris* was studied (Soengas et al. 2007; Table 5-1). QTL mapping resulted in identification of four highly significant QTL for resistance to both races of *X. campestris* on A06 of linkage map R-o-18 x B162, with two additional QTL for percentage of successful infection by race 4 on A02 and A09. Another piece of evidence indicating the linkage between the genetic factors for resistance to the two races came from correlation of resistance to races 1 and 4 observed in the F_2 population. Although, the QTL localization is an initial step to the proposed transfer of resistance from genome A of *B. rapa* to the genome C of *B. oleracea*, promising results of an attempt to introduce resistance to races 1 and 4 from *B. juncea* into *B. oleracea* (Tonguç and Griffiths 2004) suggest that it can be an alternative way to develop resistant cultivars that help to control the disease.

5.2.4 Transformation and Plant Regeneration

5.2.4.1 Agrobacterium-mediated Transformation

Transformation of *Brassica* species with the use of *Agrobacterium*-based system is a commonly used method for gene introduction (reviewed by Poulsen 1996; Puddephat et al. 1996; Sparrow et al. 2006). Efficiency of transformation is limited by genetic factors dependent either on the plant or bacterial genome.

Cogan et al.'s (2002; Table 5-1) paper is the first report on identification of genetic regions that co-regulate this process, apart from the observation indicating the plant genome as a source of genotypic basis to transformation (Gelvin 2000). Using a *B. oleracea* DH mapping population (AG population), they identified seven QTLs affecting transgenic (GFP-green fluorescent protein inserted into the plant using *Agrobacterium rhizogenes*) and adventitious (non-transgenic) root production. Together, the QTLs account for 26% and 32% of the additive genetic variation in the population, respectively. Two of the QTL regions mapped to the linkage groups O3 and O7, controlled both traits. Both parents were the source of regions of the genome carrying QTLs associated with transgenic root production. QTL analysis was performed with two different methods (marker regression and interval mapping), which detected the same regions, except for the region on linkage group O7 for transgenic root production. This work is significant for two reasons: for the first time, QTL mapping approach was used to identify regions associated with transgenic and adventitious root production and also for the first time identified regions that seemed to co-regulate these two processes.

Further analysis of the above mentioned QTL was carried out using a set of A12/GD33 substitution lines resulted in increased precision of localization (Cogan et al. 2004; Table 5-1). This study confirmed co-localization of QTL regions influencing transgenic and adventitious (non-transgenic) root production.

Oldacres et al. (2005; Table 5-1) tested the hypothesis that performance of transgenic root production is associated with performance of adventitious (non-transgenic) root production in *B. oleracea*. In their investigation a second population derived from a cross between cauliflower line Nedcha (*B. oleracea* var. *botrytis*) and the DH Brussels sprout line Gower (*B. oleracea* var. *gemmifera*) (NG population) was used. The regression analysis between the performance of the two traits and the mapping of a QTL controlling the two traits to the same position (LG1, LG8, and LG9) supported the above mentioned hypothesis. Cogan et al. (2002; Table 5-1) also showed two QTLs (LG3 and LG7) to influence both adventitious and transgenic root production in the AG population, with an additional QTL (on LG1) for transgenic root production alone, map on the same linkage group as reported in Oldacres et al. (2005), but in a different position. Hence, these findings provide independent confirmation for the ability of individual QTLs to control both adventitious and transgenic root production.

The genetic loci controlling susceptibility to *A. tumefaciens* in *B. oleracea* were investigated by Sparrow et al. (2004a; Table 5-1), using the same DH mapping population of *B. oleracea* as reported by Cogan et al. (2002; AG population; Table 5-1), and the associated RFLP map. Heritability of a crown gall-formation, as an effect of infection with *A. tumefaciens*, was studied in 8 x 8 diallel arrangement of parental lines that were selected to include a range of susceptibilities to *A. tumefaciens*. The results suggested that although both additive and dominant gene effects control this process, the former was more important. They identified a suggestive QTL correlated with susceptibility to *A. tumefaciens* on linkage group O9, which was further narrowed to a central region, by using substitution lines of the A12 genome with substituted regions of the GD33 genome. The assignment of linkage groups to chromosomes of *B. oleracea* (Howell et al. 2002) revealed the presence of genes correlated with susceptibility to *A. tumefaciens* on chromosome 3 of cytogenetic map of *B. oleracea*.

The next step was the use of both shoot regeneration and susceptibility to *A. tumefaciens* as phenotypic markers and selection of genotypes that could be transformed with high efficiency from the AG population (Sparrow et al. 2004b).

5.2.4.2 Plant Regeneration

The ability to regenerate plants from in vitro cultures is important for exploiting transformation and somatic hybridization techniques. Hansen et al. (1999) investigated the genetic control of *B. oleracea* plant regeneration from protoplast-derived microcalli by analyzing F_2 progenies from a cross between two parents with high and low regeneration ability. They found that two or three genes control this trait.

The same plant material and the corresponding evaluation data were used by Holme et al. (2004; Table 5-1) to identify two major QTLs connected with the regeneration ability from microcalli of *B. oleracea* protoplasts, and to locate them to linkage groups O2 and O9, and C8 and C7, on two previously published maps, Sebastian et al. (2000) and Saal et al. (2001), respectively. Multiple regression analysis revealed that these two QTLs, both with additive effects from the high-responding parent, accounted for 83% of the total genetic variation for the trait analyzed. The regions carrying those QTLs shared homology with a region from the top of chromosome 5 of *A. thaliana*, known for localization of QTL for plant regeneration and for flowering time (Bohuon et al. 1998; Table 5-1). But in the material used in this study, association between flowering time and regeneration ability was not found.

5.2.5 Nutrient Composition

From the agronomic point of view, not only are the loci affecting traits connected with the development of organs suitable for consumers important, but also the nutrient composition of these edible parts including proteins, vitamin C, secondary metabolites such as glucosinolates, and phosphate and other minerals (see Chapters 1.3, 2.2.4.4, 10.4 and 10.5). These traits could be a subject of plant breeding programs, as long as adequate variation can be found among populations and there is sufficient demand from consumers.

Since consumption of plants with insufficient amount of minerals and vitamins is the cause of dietary deficiencies, there are two ways to overcome this limitation. Both relate to plant nutrient enrichment, through agronomic attempts such as the use of fertilizers, or through breeding crops with increased nutrient content. Glucosinolate content is an example of a trait for which the latter is an attractive strategy. Several studies have provided evidence that increased consumption of cruciferous vegetables could significantly reduce the risk of cancer, particularly of the gastro-intestinal tract (Block et al. 1992; Verhoeven et al. 1996). Physiological experiments performed on mammalian cell cultures and rodents, have demonstrated the effectiveness of 3-methylsulfinylpropyl (3-MSP) and 4-methylsulfinylbutyl (4-MSB) isothiocyanates (ITCs), derived from the corresponding glucosinolates in broccoli, as inducers of phase II detoxification enzymes (Zhang et al. 1992; Faulkner et al. 1998). Glucosinolates also have roles in plant defense, flavor and taste (Blau et al. 1978; Kliebenstein et al. 2002).

Mithen et al. (2003) developed ITC-enriched broccoli lines, through introgression of three small segments from *Brassica villosa* genome, a wild relative of *B. oleracea*, into a standard broccoli genetic background. Besides localization of particular introgressed QTL containing regions, nothing was known concerning the nature of gene or genes within those QTLs.

Recently, the mapping of QTL for glucosinolate content in *B. rapa* leaves was reported (Lou et al. 2008; Table 5.1). This analysis was performed on two DH populations: DH38, derived from a cross between yellow sarson R500 and the pak choi variety HK Naibaicai; and DH30, from a cross between R500 and Kairyou Hakata, a Japanese vegetable turnip variety. Altogether, 16 loci controlling aliphatic glucosinolate accumulation were found, three loci controlling total indolic glucosinolate concentration and three loci regulating aromatic glucosinolate concentrations. The results obtained with comparative analysis between *A. thaliana* and *B. rapa*, together with candidate gene mapping in *B. rapa*, indicated the possibility of several genes involved in the glucosinolate biosynthesis pathway to explain the detected QTLs.

Vegetable Brassicas can serve as a remarkable source of sufficient minerals to satisfy human dietary requirements. However, over half of the world's population fails to meet these requirements regarding one or more minerals (Welch and Graham 2004). Among group II elements both in plants and animals the most common are calcium (Ca) and magnesium (Mg). Their concentration in shoots and genetic variation underlying these traits were for the first time extensively analyzed in *B. oleracea* by Broadley et al. (2008; Table 5-1). A set of 376 accessions, including examples of wild relatives from natural population and 74 modern F_1 hybrid cultivars, was included in a constructed diversity base set covering a great amount of species-wide allelic diversity within domesticated *B. oleracea*. Homozygous lines from a segregating *B. oleracea* mapping population (AGDH) was used to investigate the genetic basis of variation in shoot Ca and Mg content. QTLs affecting those traits were located on chromosomes C2, C6, C7, C8, and C9. The QTLs on C9 accounted for 14% to 55% of the total genetic variance. The localization of QTL on C9 was further narrowed to a 4 cM region with the use of recurrent backcross substitution lines, derived from the same parents (AGSL population). This region was shown to be homologous to a region on *A. thaliana* chromosome 5 (Parkin et al. 2005), which could serve as a source of putative candidate genes affecting Ca and Mg transport in plants.

In contrast to glucosinolates or minerals, which are the subject of breeding programs of plant enrichment, there is a group of antinutritional substances (e.g., phytate), for which reduction is desired. The differences in phytate and phosfate content was studied by Zhao et al. (2007; Table 5-1) in a different set of 160 *B. rapa* accessions and association mapping was performed to identify genomic regions affecting these traits. The results of association mapping for phytate and phosfate content were compared with that for other traits, which were a subject of human selection. Based on the analysis of marker-marker associations within subpopulations, the presence of significant LD up to 80 cM was reported. However, LD decay proportional to genetic distance was not observed, though marker density was relatively low. Three models for analysis of association between markers and quantitative traits (simple *t*-test, model correcting for population structure, and multi-QTL procedure) were used in this study. Depending on the model tested, between 27 and 170 marker/trait associations were detected. The analysis of the same traits in a set of mapping populations developed from eight parents (Lou et al. 2007; Fig. 5-1; Table 5-1) identified a number of QTL related to different traits. Together, these studies demonstrated that a combined association and linkage mapping approach is effective in identifying regions of the genome controlling important traits in *B. rapa*.

5.3 Future Prospects

5.3.1 Array-based, High-Throughput DNA Marker Systems for **Brassica** *Species*

5.3.1.1 Single Nucleotide Polymorphism Arrays

The most common form of polymorphism in any genome is nucleotide substitution, also known as single nucleotide polymorphism (SNP; reviewed by Edwards et al. 2007) (see Chapters 4.2.8 and 11). SNP markers are the markers of choice for whole-genome profiling in well-characterized genomes because of their codominance, simplicity (generally biallelic), ease of automation, high reliability and sheer numerical quantity. There are now several automated platforms whereby many thousands of SNPs can be assayed in parallel array (reviewed by Chagné et al. 2007). Agricultural species with large numbers of validated SNP markers include cattle (Khatkar et al. 2007), rice (Shen et al. 2004) and barley (Rostoks et al. 2006). Currently, there are few validated SNPs and no publicly available SNP array available for vegetable *Brassica* species (see Chapter 11). However, the prospects are good that these deficiencies will be redressed in the near future.

Two approaches for large-scale SNP discovery in *Brassica* have recently been reported. The first of these was reported by Duran et al. (2009) and makes use of the large set of *Brassica* EST accessions listed in the GeneBank (907,656 at the time of writing, March 2010) (see Chapter 11). The second takes advantage of "next generation" sequencing technology for deep-sequencing of transcribed genes from European and Chinese parents of a *B. napus* reference mapping population (Trick et al. 2009). The next step —SNP validation—has yet to be completed but initial promising results suggest that a large proportion of SNPs identified by Trick et al. (2009) appear to be heritable in a small sample of the *B. napus* reference mapping population. Heritability is one aspect of SNP validation; another aspect is informativeness. For a marker to be informative, it must be polymorphic in the target population (breeding or wild), and for measurement of linkage disequilibrium each allele should be present at a frequency greater than 0.1 (Rostoks et al. 2006).

5.3.1.2 Expression Microarrays

Expression microarrays are available for many species, including those for which there are relatively few validated SNP markers (see Chapters 9 and 11). Microarrays consist of many thousands of cloned DNA molecules or oligonucleotides spotted onto a solid matrix. Gene expression analysis involves taking expressed mRNA from a test subject tissue, synthesizing cDNA and then hybridizing to the spotted microarray. More recently, expression microarrays have been used to identify allelic variation—termed single feature polymorphisms (SFPs)—which can be used to measure gene expression and simultaneously genotype test populations (Borevitz et al. 2007). SFPs are based on a range of sequence polymorphisms including SNPs, insertions/deletions and repeat number variation. Several expression microarrays have been developed for *Brassica*: 95K Agilent array for *Brassica* species (*http://www.cogenics.com/geneexpression/service/brassica.cfm*), 94K CombiMatrix array for *Brassica* species (*http://www.intl-pag.org/16/abstracts/PAG16_P07a_708.html*) and 24K NimbleGen array for *B. rapa* (*http://www.brassica-rapa.org/BrEMD/*). It remains to be seen how effective these arrays will be in detecting SFPs in *Brassica* species, which have undergone multiple rounds of polyploidization (Nelson and Lydiate 2006) (see Chapter 8).

5.3.1.3 Diversity Arrays Technology

Another option for large-scale, array-based whole-genome profiling in *Brassica* is diversity arrays technology (DArT, Jaccoud et al. 2001). The advantage of this method over large-scale SNP assay is that no prior knowledge of nucleotide sequence is needed and is therefore an option for any species for which there is little sequence information available. While the DArT marker system is dominant in nature (i.e. presence or absence of marker allele), the hybridization-based assay is highly quantitative as demonstrated in quantitative bulked segregant analysis applications (Wenzl et al. 2007). The dominant scoring of individual markers is somewhat mitigated by the high density of markers (hundreds to thousands per assay) allowing the use of haplotypes for closely linked markers to determine zygosity status for any region. If required, the sequence of the markers can be easily determined from cloned fragments used to spot the array and this sequence can be used to design locus-specific markers. In contrast to SNP markers where polyploidy makes marker development and validation difficult, the DArT marker system has demonstrated its efficacy in genotyping applications in hexaploid wheat (White et al. 2008) and in hexaploid oats (Tinker et al. 2009). Since *Brassica* diploid species show clear evidence of ancient polyploidy (Lagercrantz and Lydiate 1996), the DArT marker approach may be easier to implement than SNP markers in *Brassica* species. As part of an industry-supported consortium, a DArT array has recently been developed for oilseed *B. napus* and *B. rapa* and currently comprises 2000 polymorphic markers (Andrzej Kilian, pers. comm.). To our knowledge, the array has not yet been tested on vegetable *Brassica* types but is likely to be suitable for the vegetable forms of *B. oleracea* and *B. rapa*.

5.3.2 *Gene Expression Analysis of Agronomic Traits*

QTL analysis has been used extensively for analyzing the genetic control of agriculturally important traits (see section 2. above). A variation on this approach is expression QTL (eQTL) in which global gene expression is analyzed in a mapping population (for details on global gene expression see Chapter 9). Gene transcript levels, when assayed in a mapping population, can be considered as quantitative traits and their variation used to map eQTL. The highest resolution studies employ whole-genome microarrays in reasonably large mapping populations of species with completely sequenced genomes. For example, West et al. (2007) used > 1,000 Affymetrix ATH1 GeneChips (each containing 22,810 genes) to analyze global expression in 211 RILs of Arabidopsis. In that study, 36,871 eQTL were detected: about one-third of which were *cis*-eQTLs (where gene expression is controlled by the gene itself) and two-thirds of which were *trans*-eQTLs (where gene expression is controlled by another locus or loci). The incredible complexity of gene interactions observed in this controlled and relatively simple population is an indicator of how challenging the eQTL approach will be to understand genetic control of agronomic traits in diverse populations of more complex crop species genomes. Comprehensive expression microarrays for *Brassica* species are now available but so far they have not been used for eQTL analysis.

One potential application of gene expression microarrays in *Brassica* species is investigating the nature of heterosis (see Chapter 2.3.3). This approach was show-cased by an elegant study by Swanson-Wagner et al. (2006) in maize, the archetypal hybrid crop. In that study, global gene expression analysis revealed all possible modes of gene action in maize hybrids compared to inbred parents including overdominance. The challenge with any such analysis will be to tease out the causes of heterosis versus the effects of heterosis.

5.3.3 Elucidating the Genetic Basis of Crop Domestication and Adaptation

Domestication has in every case involved one or several "population bottlenecks" that resulted in reduced genetic diversity in domesticates compared to their wild progenitors (Doebley et al. 2006). This is a problem for plant breeders because availability of adequate genetic diversity is a prerequisite for generating new and better allele combinations. Crop domestication can also be considered a microcosm of species evolution, as Darwin expounded in "The Origin of Species" (1859). Therefore, understanding the genetic basis of crop domestication and adaptation has both basic and applied significance.

There has been some success in identifying (via linkage or association mapping) major domestication genes such as curd development in cauliflower (Purugganan et al. 2000; Smith and King 2000), naked grains in maize (Wang et al. 2005) and large fruit in tomato (Frary et al. 2000). Ross-Ibarra et al. (2007) support an alternative reverse genetic approach that first identifies "signatures of adaptation" in the genomes of domesticated crops and then seeks to identify phenotypic changes associated with those genomic changes. These signatures of adaptation are manifested as lower genetic diversity in genes that have been subjected to selective pressure compared to adaptively neutral genes that have not. With continually improving genotyping tools and the wealth of phenotypic variation present in vegetable Brassicas, both forward and reverse genetic approaches are likely to uncover many of the key genes governing morphological variation and adaptation to domestication.

5.3.4 Epigenetic Regulation of Crop Development

A rapidly growing area of genetics is epigenetics, which can be defined as the study of heritable changes in gene function that occur without a change in the DNA sequence. For an excellent general review of epigenetics in plants, see Grant-Downton and Dickinson (2005). Epigenetics provides a unifying theme that connects an array of what originally seemed to be disparate exceptions to normal Mendelian principles that have accrued over several decades, originally described as paramutation, somaclonal variation, nucleolar dominance, genomic imprinting and transgene silencing. Epigenetic modification is mediated by what has been referred to as the "Three Pillars of Epigenetics": (1) cytosine DNA methylation; (2) changes in chromatin structure by histone modification; and (3) post-transcriptional gene silencing by short interfering RNAs (Grant-Downton and Dickinson 2005).

The role of epigenetic modification in crop development and performance is not well understood. Indeed, it appears that epigenetic mechanisms are more complex in plants than in other eukaryotes (Grant-Downton and Dickinson 2005). One well known example of an agriculturally significant trait, which is epigenetically controlled, is vernalization—the requirement of a cold period to initiate floral development in some winter annual and perennial plants. Vernalization in *Arabidopsis* has been shown to be mediated by changes in histone methylation and acetylation of chromatin at the *FLC* gene (reviewed by Amasino 2005). It may be that many other agronomic traits also have significant epigenetic components.

There have been a few studies of genome-wide DNA methylation and changes in chromatin structure in *Brassica* species and these have mainly used the low-resolution approach of methylation-sensitive RFLP marker analysis (e.g., Lukens et al. 2006). Until recently, high-resolution methylation analysis was not feasible except in model species. However, the advent of next generation DNA sequencing technology now makes high-resolution methylation analysis (by bisulphite sequencing, Cokus et al. 2008) feasible even for non-model species and is likely to shed further insight into the role of epigenetic mechanisms in crop development.

5.4 Conclusions

The large number of plant QTL that have been mapped to a high resolution will evolve in the coming years to associations of complex phenotypes with their underlying factors. This will create a framework in which to examine how the elusive biological networks interact to create a phenotype (Paran and Zamir 2003).

References

Amasino RM (2005) Vernalization and flowering time. Curr Opp Plant Biotechnol 16: 154–158.

Axelsson T, Shavorskaya O, Lagercrantz U (2001) Multiple flowering time QTLs within several *Brassica* species could be the result of duplicated copies of one ancestral gene. Genome 44: 856–864.

Baggett JR, Wahlert W (1975) Annual flowering and growth habit in cabbage-broccoli crosses. HortScience 10: 170–172.

Baggett JR, Kean D (1989) Inheritance of annual flowering in *Brassica oleracea*. HortScience 24: 662–664.

Barton NH, Keightley PD (2002) Understanding quantitative genetic variation. Nat Rev Genet 3: 11–21.

Bettey M, Finch-Savage WE, King GJ, Lynn JR (2000) Quantitative genetic analysis of seed vigour and pre-emergence seedling growth traits in *Brassica oleracea*. New Phytol 148: 277–286.

Blau PA, Feenny P, Contrado L (1978) Allyglucosinolate and herbivorous caterpillars: a contrast in toxicity and tolerance. Science 200: 1296–1298.

Block G, Patterson B, Subar A (1992) Fruit, vegetables and cancer prevention: a review of the epidemiological evidence. Nutr Cancer 18: 1–29.

Bohuon EJR, Ramsay LD, Craft JA, Arthur AE, Marshall DF, Lydiate DJ, Kearsey MJ (1998) The association of flowering time quantitative trait loci with duplicated regions and candidate loci in *Brassica oleracea*. Genetics 150: 393–401.

Bonierbale MW, Plaisted RL, Tanksley SD (1988) RFLP maps based on a common set of clones reveal modes of chromosomal evolution in potato and tomato. Genetics 120: 1095–1103.

Borevitz JO, Hazen SP, Michael TP, Morris GP, Baxter IR, Hu TT, Chen H, Werner JD, Nordborg M, Salt DE, Kay SA, Chory J, Weigel D, Jones JDG, Ecker JR (2007) Genome-wide patterns of single-feature polymorphism in *Arabidopsis thaliana*. Proc Natl Acad Sci USA 104: 12057–12062.

Broadley MR, Hammond JP, King GJ, Astley D, Bowen HC, Meacham MC, Mead A, Pink DAC, Teakle GR, Hayden RM, Spracklen WP, White PJ (2008) Shoot calcium and magnesium concentrations differ between subtaxa, are highly heritable, and associate with potentially pleiotropic loci in *Brassica oleracea*. Plant Physiol 146: 1707–1720.

Camargo LEA, Osborn TC (1996) Mapping loci controlling flowering time in *Brassica oleracea*. Theor Appl Genet 92: 610–616.

Camargo LEA, Williams PH, Osborn TC (1995) Maping of quantitative trait loci controlling resistance to *Xanthomonas campestris* pv *campestris* in the field and glasshouse. Phytopathology 85: 1296–1300.

Chagné D, Batley J, Edwards D, Forster J (2007) Single nucleotide polymorphisms genotyping in plants. In: NC Oraguzie , EHA Rikkerink ,SE Gardiner ,HN De Silva (eds) Association Mapping in Plants. Springer, New York, USA, pp 77–94.

Cogan NOI, Lynn JR, King GJ, Kearsey MJ, Newbury HJ, Puddephat IJ (2002) Identification of genetic factors controlling the efficiency of *Agrobacterium rhizogenes*-mediated transformation in *Brassica oleracea* by QTL analysis. Theor Appl Genet 105: 568–576.

Cogan NOI, Newbury HJ, Oldacres AM, Lynn JR, King GJ, Kearsey MJ, Puddephat IJ (2004) Identification and characterization of QTL controlling *Agrobacterium*-mediated transient and stable transformation of *Brassica oleracea*. Plant Biotechnol 2: 59–69.

Cokus SJ, Feng S, Zhang X, Chen Z, Merriman B, Haudenschild CD, Pradhan S, Nelson SF, Pellegrini M, Jacobsen SE (2008) Shotgun bisulphite sequencing of the Arabidopsis genome reveals DNA methylation patterning. Nature 452: 215–219.

Cong B, Liu J, Tanksley SD (2002) Natural alleles at a tomato fruit size quantitative trait locus differ by heterochronic regulatory mutations. Proc Natl Acad Sci USA 99: 13606–13611.

Crisp P, Crute IR, Sutherland RA, Angeli SM, Bloor K, Burgess H, Gordon P (1989) The exploitation of genetic resources of *Brassica oleracea* in breeding for resistance in clubroot (*Plasmodiophora brassicae*). Euphytica 42: 215–226.

Crute IR, Gray AR, Crisp P, Buczacki ST (1980) Variation in *Plasmodiophora brassicae* and resistance to clubroot disease in *Brassica* and allied crops—a critical review. Plant Breed Abstr 50: 91–104.

Crute IR, Phelps K, Barnes A, Buczacki ST, Crisp P (1983) The relationship between genotypes of three *Brassica* species and collections of *Plasmodiophora brassicae*. Plant Pathol 32: 405–420.

Darvasi A, Soller M (1995) Advanced intercrossed lines, an experimental population for fine genetic mapping. Genetics 141: 1199–1207.

Darwin CR (1859) On the Origin of Species by Means of Natural Selection, or the Preservation of Favoured Races in the Struggle for Life. Murray, London, UK.

Detjen LR (1926) A preliminary report on cabbage breeding. Proc Am Soc Hort Sci 23: 325–332.

Dias JS, Ferreira ME, Williams PH (1993) Screening of Portuguese cole landraces (*Brassica oleracea* L.) with *Peronospora parasitica* and *Plasmodiophora brassicae*. Euphytica 67: 135–141.

Doebley J, Stec A, Hibbard L (1997) The evolution of apical dominance in maize. Nature 386: 485–488.

Doebley JF, Gaut BS, Smith BD (2006) The molecular genetics of crop domestication.Cell 127: 1309–1321.

Duran C, Appleby N, Clark T, Wood D, Imelfort M, Batley J, Edwards D (2009) AutoSNPdb: an annotated single nucleotide polymorphism database for crop plants. Nucl Acids Res 37: D951–953.

Edwards D, Forster J, Cogan N, Batley J, Chagné D (2007) Single nucleotide polymorphism discovery. In: NC Oraguzie , EHA Rikkerink , SE Gardiner , HN De Silva (eds) Association Mapping in Plants. Springer, New York, USA pp 53–76.

Falconer DS, Mackay TFC (1996) Introduction to Quantitative Genetics. Addison Wesley Longman, Harlow, Essex, UK.

Fan Z, Rimmer SR, Stefansson BR (1983) Inheritance of resistance to *Albugo candida* in rape (*Brassica napus* L.). Can J Genet Cytol 25: 420–424.

Farrall M (2004) Quantitative genetic variation: a post-modern view. Hum Mol Genet 13: R1–R7. doi: 10.1093/hmg/ddh084.

Faulkner K, Mithen R, Williamson G (1998) Selective increase of the potential anticarcinogen 4-methylsulphinylbutyl glucosinolate in broccoli. Carcinogenesis 19: 605–609.

Ferreira ME, Satagopan J, Yandell BS, Williams PH, Osborn TC (1995) Mapping loci controlling vernalisation requirement and flowering time in *Brassica napus*. Theor Appl Genet 90: 727–732.

Figdore SS, Fereira ME, Slocum MK, Williams PH (1993) Association of RFLP markers with trait loci affecting clubroot resistance and morphological characters in *Brassica oleracea* L. Euphytica 69: 33–44.

Fisher RA (1930) The Genetics of Natural Selection. Oxford Univ Press, Oxford, UK.

Flint-Garcia SA, Thornsberry JM, Buckler ES (2003) Structure of linkage disequilibrium in plants. Annu Rev Plant Biol 54: 357–374.

Frary A, Nesbitt TC, Frary A, Grandillo S, van der Knapp E, et al. (2000) *fw2.2*: a quantitative trait locus key to the evolution of tomato fruit size. Science 289: 85–88.

Fridman E, Pleban T, Zamir D (2000) A recombination hotspot delimits a wild species quantitative trait locus for tomato sugar content to 484 bp within an invertase gene. Proc Natl Acad Sci USA 97: 4718–4723.

Fusari CM, Lia VV, Hopp HE, Heinz RA, Paniego NB (2008) Identification of single nucleotide polymorphisms and analysis of linkage disequilibrium in sunflower elite inbred lines using the candidate gene approach. BMC Plant Biol 8: 7.

Gelvin SB (2000) *Agrobacterium* and plant genes involved in T-DNA transfer and integration. Annu Rev Plant Physiol Plant Mol Biol 51: 223–256.

Glazier AM, Nadeau JH, Aitman TJ (2002) Finding genes that underlie complex traits. Science 298: 2345–2349.

Grandclement C, Thomas G (1996) Detection and analysis of QTLs based on RAPD markers for plygenic resistance to *Plasmodiophora brassicae* Woron in *Brassica oleracea* L. Theor Appl Genet 93: 86–90

Grant-Downton RT, Dickinson HG (2005) Epigenetics and its implications for plant biology. 1. The epigenetic network in plants. Ann Bot 96: 1143–1164.

Gupta PK, Rustgi S, Kulwai PI (2005) Linkage disequilibrium and association studies in higher plants: present status and future prospects. Plant Mol Biol 57: 461–485.

Gupta PK, Rustgi S, Mir RR (2008) Array-based high-throughput DNA markers for crop improvement. Heredity 101: 5–18.

Hansen LN, Rodomiro O, Andersen SB (1999) Genetic analysis of protoplast regeneration ability in *Brassica oleracea*. Plant Cell Tiss Org Cult 58: 127–132.

Harper FR, Pittman UJ (1974) Yield loss by *Brassica campestris* and *Brassica napus* from systemic stem infection by *Albugo cruciferatum*. Phytopathology 64: 408–410.

Hartl DS, Clark AG (1997) Principles of Population Genetics. 3rd edn. Sinauer Assoc, Sunderland, MA, USA.

Hill WG (1982) Rates of change in quantitative traits from fixation of new mutations. Proc Natl Acad Sci USA 79: 142–145.

Hill WG, Robertson A (1968) Linkage disequilibrium in finite populations. Theor Appl Genet 38: 226–231.

Holland JB (2007) Genetic architecture of complex traits in plants. Curr Opp Plant Biol 10: 156–161.

Holme IB, Trop AM, Hansen LN, Andersen SB (2004) Quantitative trait loci affecting plant regeneration from protoplasts of *Brassica oleracea*. Theor Appl Genet 108: 1513–1520.

Horovitz S, Perlasca G (1954) Genetics of normal flowering of cabbage in the torrid tropics. Agron Trop 4: 81–93.

Howell EC, Barker GC, Jones GH, Kearsey MJ, King GJ, Kop EP, Ruder CD, Teakle GR, Vicente JG, Armstrong SJ (2002) Integration of the cytogenetic and genetic linkage maps of *Brassica oleracea*. Genetics 161: 1225–1234.

Hubbard L, McSteen P, Doebley J, Hake S (2002) Expression patterns and mutant phenotype of *teosinte branched1* correlate with growth suppression in maize and teosinte. Genetics 162: 1927–1935.

Jaccoud D, Peng K, Feinstein D, Kilian A (2001) Diversity Arrays: a solid state technology for sequence information independent genotyping. Nucl Acids Res 29: e25.

Jansen RC, Stam P (1994) High resolution of quantitative traits into multiple loci via interval mapping. Genetics 136: 1447–1455.

Kalloo G, Bergh BO (1993) Genetic Improvement of Vegetable Crops. Pergamon Press, Oxford, UK.

Kao C-H, Zeng Z-B, Teasdale R (1999) Multiple interval mapping for quantitative trait loci. Genetics 152: 1203–1216.

Kashkush K, Feldmen M, Levy AA (2003) Transcriptional activation of retrotransposons alters the expression of adjacent genes in wheat. Nat Genet 33: 102–106.

Kennard WC, Slocum MK, Figdore SS, Osborn TC (1994) Genetic analysis of morphological variation in *Brassica oleracea* using molecular markers. Theor Appl Genet 87: 721–732.

Khatkar MS, Zenger KR, Hobbs M, Hawken RJ, Cavanagh JAL, Barris W, McClintock AE, McClintock S, Thomson PC, Tier B, Nicholas FW, Raadsma HW (2007) A primary assembly of a bovine haplotype block map based on a 15,036-single-nucleotide polymorphism panel genotyped in Holstein-Friesian cattle. Genetics 176: 763–772.

Kim JS, Chung TY, King GJ, Jin M, Yang TJ, Jin YM, Kim HI, Park BS (2006) A sequence-tagged linkage map of *Brassica rapa*. Genetics 174: 29–39.

Kim S, Plagnol V, Hu TT, Toomajian C, Clark RM, Ossowski S, Ecker JR, Weigel D, Nordborg M (2007) Recombination and linkage disequilibrium in *Arabidopsis thaliana*. Nat Genet 39: 1151–1155.

Kimura M (1983) The Neutral Theory of Molecular Evolution. Cambridge Univ Press, Cambridge, UK.

Kliebenstein DJ, Pedersen D, Barker B, Mitchell-Olds T (2002) Comparative analysis of quantitative trait loci controlling glucosinolates, myrosinase and insect resistance in *Arabidopsis thaliana*. Genetics 161: 325–332.

Knapp SJ, Bridges WC, Birkes D (1990) Mapping quantitative trait loci using molecular linkage maps. Theor Appl Genet 79: 583–592.

Kole C, Teutonico R, Mengistu A, Williams PH, Osborn TC (1996) Molecular mapping of a locus controlling resistance to *Albugo candida* in *Brassica rapa*. Phytopathology, 86: 367–369.

Kole C, Quijada P, Michaels SD, Amasino RM, Osborn TC (2001) Evidence for homology of flowering-time genes *VFR2* from *Brassica rapa* and *FLC* from *Arabidopsis thaliana*. Theor Appl Genet 102: 425–430.

Kole C, Williams PH, Rimmer SR, Osborn TC (2002) Linkage mapping of genes controlling resistance to white rust (*Albugo candida*) in *Brassica rapa* (syn. *campestris*) and comparative mapping to *Brassica napus* and *Arabidopsis thaliana*. Genome 45: 22–27.

Kristensen CN, Kelefiotis D, Kriestensen T, Borrensen-Dale A-L (2001) High-throughput methods for detection of genetic variation. Biotechniques 30: 318–332.

Kuginuki Y, Ajisaka H, Yui M, Yoshikawa H, Hida KM, Hirai (1997) RAPD markers linked to a clubroot-resistance locus in *Brassica rapa* L. Euphytica 98: 149–154.

Kumari K, Varghese TM, Suryanarayana D (1970) Qualitative changes in the amino acid contents of hypertrophied organs in mustard due to *Albugo candida*. Curr Sci 39: 240–241.
Lagercrantz U, Lydiate DJ (1996) Comparative genome mapping in *Brassica*. Genetics 144: 1903–1910.
Lagercrantz U, Putterill J, Coupland G, Lydiate D (1996) Comparative mapping in *Arabidopsis* and *Brassica*, fine scale genome collinearity and congruence of genes controlling flowering time. Plant J 9: 13–20.
Lan T-H, Paterson AH (2000) Comparative mapping of quantitative trait loci sculpting the curd of *Brassica oleracea*. Genetics 155: 1927–1954.
Lan T-H, Paterson AH (2001) Comparative mapping of QTLs determining the plant size of *Brassica oleracea*. Theor Appl Genet 103: 383–397.
Lander ES, Botstein D (1989) Mapping Mendelian factors underlying quantitative traits using RFLP linkage maps. Genetics 121: 185–200.
Landry BS, Hubert N, Crete R, Chiang MS, Lincoln SE, Etoh T (1992) A genetic map for *Brassica oleracea* based on RFLP markers detected with expressed DNA sequences and mapping of resistance genes to race 2 of *Plasmodiophora brassicae* (Woronin). Genome 35: 409–420.
Laurens F, Thomas G (1993) Inheritance of resistance to clubroot (*Plasmodiophora brassicae* Wor.) in kale (*Brassica oleracea* ssp. *acephala*). Hereditas 119: 253–262.
Lijavetzky D, Cabezas JA, Ibáñez A, Rodríguez V, Martínez-Zapater JM (2007) High throughput SNP discovery and genotyping in grapevine (Vitis vinifera L.) by combining a re-sequencing approach and SNPlex technology. BMC Genomics 8: 424.
Lin S, Wang J, Poon S, Su C, Wang S, Chiou T (2005) Differential regulation of FLOWERING LOCUS C expression by vernalization in cabbage and *Arabidopsis*. Plant Physiol 137: 1037–1048.
Long A.D, Mullaney SL, Mackay TFC, Langley CH (1996) Genetic interactions between naturally occurring alleles at quantitative trait loci and mutant alleles at candidate loci affecting bristle number in *Drosophila melanogaster*. Genetics 114: 1497–1510.
Long AD, Lyman RF, Langley CH, Mackay TFC (1998) Two sites in the Delta gene region contribute to naturally occurring variation in bristle number in *Drosophila melanogaster*. Genetics 149: 999–1017.
Long AD, Langley CH (1999) Power of association studies to detect the contribution of candidate genetic loci to complexy inherited phenotypes. Genome Res 9: 720–731.
Lou P, Zhao J, Kim JS, Shen S, Pino Del Carpio D, Song X, Jin M, Vreugdenhil D, Wang X, Koornneef M, Bonnema G (2007) Quantitative trait loci for flowering time and morphological traits in multiple populations of *Brassica rapa*. J Exp Bot 58: 4005–4016 doi:10.1093/jxb/erm255.
Lou P, Zhao J, He H, Hanhart C, Pino del Carpio D, Verkerk R, Custers J, Koornneef M, Bonnema G (2008) Quantitative trait loci for glucosinolate accumulation in *Brassica rapa* leaves. New Phytol 179: 1017–1032 doi: 10.1111/j.1469–8137.2008.02530.x.
Lu G, Cao J, Yu X, Xiang X, Chen H (2008) Mapping QTLs for root morphological traits in *Brassica rapa* L. based on AFLP and RAPD markers. J Appl Genet 49: 23–31.
Lukens L, Doebley J (2001) Molecular evolution of the *teosinte* branched gene among maize and related grasses. Mol Biol Evol 18: 627–638.
Lukens LN, Pires JC, Leon E, Vogelzang R, Oslach L, Osborn T (2006) Patterns of sequence loss and cytosine methylation within a population of newly resynthesized *Brassica napus* allopolyploids. Plant Physiol 140: 336–348.
Lysak MA, Koch MA, Pecinka A, Schubert I (2005) Chromosome triplication found across the tribe Brassiceae. Genome Res 15(4): 516–25.
Mackay I, Powell W (2007) Methods for linkage disequilibrium mapping in crops. Trends Plant Sci 12: 57–63.
Mackay TFC (1985) Transposable element-induced response to artificial selection in *Drosophila melanogaster*. Genetics 111: 351–374.
Mackay TFC (2001) The genetic architecture of quantitative traits. Annu Rev Genet 35: 303–339.

Manzanares-Dauleux MJ, Divaret I, Baron F, Thomas G (2000) Evaluation of French *Brassica oleracea* landraces for resistance to *Plasmodiophora brassicae*. Euphytica 113: 211–218.

Mather K (1941) Variation and selection of polygenic characters. J Genet 41: 159–193.

Mather K, Harrison BJ (1949) The manifold effect of selection. Heredity 3: 131–162.

Mather KA, Caicedo AL, Polato NR, Olsen KM, McCouch S, Purugganan MD (2007) The extent of linkage disequilibrium in rice (Oryza sativa L.). Genetics 177(4): 2223–32.

Michaels SD, Amasino RM (1999) *FLOWERING LOCUS C* encodes a novel MADS domain protein that acts as a repressor of flowering. Plant Cell 11: 949–956.

Mithen R, Faulkner K, Magrath R, Rose P, Williamson G, Marquez J (2003) Development of isothiocyanate-enriched broccoli, and its enhanced ability to induce phase 2 detoxification enzymes in mammalian cells. Theor Appl Genet 106: 727–734 doi 10.1007/s00122-002-1123-x.

Monteiro AA, Williams PH (1989) The exploration of resources of Portuguese cabbage and kale for resistance to several *Brassica* diseases. Euphytica 41: 215–225.

Morgante M, Salamini F (2003) From plant genomics to breeding practice. Curr Opp Plant Biotechnol 14: 214–219.

Moriguchi K, Kimizuka-Takagi Ch, Ishii K, Nomura K (1999) A genetic map based on RAPD. RFLP. Isozyme. morphological markers and QTL analysis for clubroot resistance in *Brassica oleracea*. Breed Sci 49: 257–265.

Mott R, Talbot CJ, Turri MG, Collins AC, Flint J (2000) A method for fine mapping quantitative trait loci in outbreed animal stocks. Proc Natl Acad Sci USA 97: 12649–12654.

Nadeau JH, Frankel WN (2000) The roads from phenotypic variation to gene discovery: mutagenesis versus QTLs. Nat Genet 25: 381–384.

Nasrallah JB, Kao TH, Goldberg ML, Nasrallah ME (1985) A cDNA clone encoding an S-locus specific glycoprotein from *Brassica oleracea*. Nature 318: 262–267.

Nelson MN, Lydiate D (2006) New evidence from *Sinapis alba* L. for ancestral triplication in a crucifer genome. Genome 49: 230–238.

Nesbitt TC, Tanksley SD (2002) Comparative sequencing in the genus *Lycopersicon*. Implications for the evolution of fruit size in the domestication of cultivated tomatoes. Genetics 162: 365–379.

Nuzhdin SV, Dilda CL, Mackay TFC (1999) The genetic architecture of selection response: inferences from fine- scale mapping of bristle number quantitative trait loci in *Drosphila melanogaster*. Genetics 153: 1317–1331.

Okazaki K, Sakamoto K, Kikuchi R, Saito A, Togashi E, Kuginuki Y, Matsumoto S, Hirai M (2007) Mapping and characterization of *FLC* homologs and QTL analysis of flowering time in *Brassica oleracea*. Theor Appl Genet 114: 595–608.

Oldacres AM, Newbury HJ, Puddephat IJ (2005) QTLs controlling the production of transgenic and adventiotious roots in *Brassica oleracea* following treatment with *Agrobacterium rhizogenes*. Theor Appl Genet 111: 479–488.

Orr HA (1998) The population genetics of adaptation: the distribution of factors fixed during adaptive evolution. Evolution 52: 935–949.

Osborn TC, Kole C, Parkin IAP, Sharpe AG, Kuiper M, Lydiate DJ, Trick M (1997) Comparison of flowering time genes in *Brassica rapa, B. napus* and *Arabidopsis thaliana*. Genetics 146: 1123–1129.

Paran I, Zamir D (2003) Quantitative traits in plants: beyond the QTL. Trends Genet 19: 303–305.

Parkin IAP, Sharpe AG, Keith DJ, Lydiate DJ (1995) Identification of the A and C genomes of amphidiploid *Brassica napus* (oilseed rape). Genome, 38: 1122–1133.

Parkin IAP, Gulden SM, Sharpe AG, Lukens L, Trick M, Osborn TC, Lydiate DJ (2005) Segmental structure of the *Brassica napus* genome based on comparative analysis with *Arabidopsis thaliana*. Genetics 171: 765–781.

Pease MA (1926) Genetic situation in *Brassica oleracea*. J Genetics 16: 363–385.

Pelofske PJ, Baggett JR (1979) Inheritance of internode length, plant form and annual habit in a cross of cabbage and broccoli (*Brassica oleracea* var. *capitata* and var. *italica*). Euphytica 28: 189–197.

Petrie GA (1973) Diseases of *Brassica* species in Saskatchewan 1970–72. I. Staghead and aster yellows. Can Plant Dis Surv 53: 19–25.

Poulsen GB (1996) Genetic transformation of *Brassica*. Plant Breed 115: 209–225.

Puddephat IJ, Riggs TJ, Fenning TM (1996) Transformation of *Brassica oleracea* L.: a critical review. Mol Breed 2: 185–210.

Purugganan MD, Boyles AL, Suddith JI (2000) Variation and selection at the *CAULIFLOWER* floral homeotic gene accompanying the evolution of domesticated *Brassica oleracea*. Genetics 155: 855–862.

Rae AM, Howell EC, Kearsey MJ (1999) More QTL for flowering time revealed by substitution lines in *Brassica oleracea*. Heredity 83: 586–596.

Ramsay LD, Jennings DE, Bohuon EJR, Arthur AE, Lydiate DJ, Kearsey MJ, Marshall DF (1996) The construction of a substitution library of recombinant backcross lines in *Brassica oleracea* for the precision mapping of quantitative trait loci. Genome 39: 558–567.

Rana D, Boogaart TV, O'Neill CM, Hynes L, Bent E, Macpherson L, Park JY, Lim YP, Bancroft I (2004) Conservation of the microstructure of genome segments in *Brassica napus* and its diploid relatives. Plant J 40: 725–733.

Razi H, Howell EC, Newbury HJ, Kearsey MJ (2008) Does sequence polymorphism of *FLC* paralogues underlie flowering time QTL in *Brassica oleracea*? Theor Appl Genet 116: 179–192 doi 10.1007/s00122-007-0657-3.

Robertson DS (1985) A possible technique for isolating genic DNA for quantitative traits in plants. J Theor Biol 117: 1–10.

Robertson FW, Reeve ERC (1952) Studies in quantitative inheritance. I. The effects of selection of wing and thorax length in *Drosophila melanogaster*. J Genet 50: 414–418.

Rocherieux J, Glory P, Giboulot A, Boury S, Barberyon G, Thomas G, Manzanares-Dauleux MJ (2004) Isolate-specific and broad-spectrum QTLs are involved in the control of clubroot in *Brassica oleracea*. Theor Appl Genet 108: 1555–1563.

Ross-Ibarra J, Morrell PL, Gaut BS (2007) Plant domestication, a unique opportunity to identify the genetic basis of adaptation. Proc Natl Acad Sci USA 104: 8641–8648.

Rostoks N, Ramsay L, MacKenzie K, Cardle L, Bhat PR, Roose ML, Svensson JT, Stein N, Varshney RK, Marshall DF, Graner A, Close TJ, Waugh R (2006) Recent history of artificial outcrossing facilitates whole-genome association mapping in elite inbred crop varieties. Proc Nat Acad Sci USA 103: 18656–18661.

Saal B, Plieske J, Hu J, Quiros CF, Struss D (2001) Microsatellite markers for genome analysis in *Brassica*. II. Assignment of rapeseed microsattelites to the A and C genomes and genetic mapping in *Brassica oleracea* L. Theor Appl Genet 102: 695–699.

Sachan JS, Singh B (1987) Generation mean analysis for flowering and maturity in Indian mustard [*Brassica juncea* (L.) Czern & Cross]. Theor Appl Genet 73: 571–574.

Sakamoto K, Saito A, Hayashida N, Taguchi G, Matsumoto E (2008) Mapping of isolate-specific QTLs for clubroot resistance in Chinese cabbage (*Brassica rapa* L. ssp. *pekinensis*). Theor Appl Genet 117: 759–767 doi 10.1007/s00122-008-0817-0.

Salvi S, Tuberosa R (2005) To clone or not to clone plant QTLs: present and future challenges. Trends Plant Sci 10: 297–304.

Santos MR, Dias JS, Silva MJ, Ferreira-Pinto MM (2006) Resistance to white rust in pak choi and Chinese cabbage at the cotyledon stage. Commun Agri Appl Biol Sci 71: 963–971.

Sax K (1923) The association of size differences with seed-coat pattern and pigmentation in *Phaseolus vulgaris*. Genetics 8: 552–560.

Schranz M, Quijada EP, Sung SB, Lukens L, Amasino R, Osborn TC (2002) Characterization and effects of the replicated flowering time gene *FLC* in *Brassica rapa*. Genetics 162: 1457–1468.

Sebastian RL, Howell EC, King GJ, Marshall DF, Kearsey MJ (2000) An integrated AFLP and RFLP *Brassica oleracea* linkage map from two morphologically distinct doubled-haploid mapping populations. Theor Appl Genet 100: 75–81.

Sebastian RL, Kearsey MJ, King GJ (2002) Identification of quantitative trait loci controlling developmental characteristics of *Brassica oleracea* L. Theor Appl Genet 104: 601–609.

Sheldon CC, Burn JE, Perez PP, Metzger J, Edwards JA, Peacock WJ, Dennis ES (1999) The *FLF* MADS Box gene: a repressor of flowering in *Arabidopsis* regulated by vernalization and methylation. Plant Cell 11: 445– 458.

Sheldon CC, Finnegan EJ, Rouse DT, Tadege M, Bagnall DJ, Helliwell CA, Peacock WJ, Dennis ES (2000) The control of flowering by vernalization. Curr Opin Plant Biol 3: 418–422.

Shen Y-J, Jiang H, Jin J-P, Zhang Z-B, Xi B, He Y-Y, Wang G, Wang C, Qian L, Li X, Yu Q-B, Liu H-J, Chen D-H, Gao J-H, Huang H, Shi T-L, Yang Z-N (2004) Development of genome-wide DNA polymorphism database for map-based cloning of rice genes. Plant Physiol 135: 1198–1205.

Smith LB, King GJ (2000) The distribution of *BoCAL-a* alleles in *Brassica oleracea* is consistent with a genetic model for curd development and domestication of the cauliflower. Mol Breed 6: 603–613.

Soengas P, Hand P, Vicente JG, Pole JM, Pink DAC (2007) Identification of quantitative trait loci for resistance to *Xanthomonas campestris* pv. *campestris* in *Brassica rapa*. Theor Appl Genet 114: 637–645 doi 10.1007/s00122-006-0464-2.

Soller M, Brody T, Genezi A (1976) On the power of experimental designs for the detection of linkage between marker loci and quantitative loci in crosses between inbred lines. Theor Appl Genet 47: 35–39.

Sparrow PAC, Townsend TM, Arthur AE, Dale PJ, Irwin JA (2004a) Genetic analysis of *Agrobacterium tumefaciens* susceptibility in *Brassica oleracea*. Theor Appl Genet 108: 644–650.

Sparrow PAC, Dale PJ, Irwin JA (2004b) The use of phenotypic markers to identify *Brassica oleracea* genotypes for routine high-throughput *Agrobacterium*-mediated transformation. Plant Cell Rep 23: 64–70.

Sparrow PA, Dale PJ, Irwin JA (2006) *Brassica oleracea*. Meth Mol Biol 343: 417–426.

Stam M, Belele C, Ramakrishna W, Dorweller JE, Bennetzen JL, Chandler VL (2002) The regulatory regions required for B′ paramutation and expression are located far upstream of the maize b1 transcribed sequences. Genetics 162: 917–930.

Suwabe K, Ikehashi H, Nunome T, Kage T, Hirai M (2002) Isolation and characterization of microsatellites in *Brassica rapa* L. Theor Appl Genet 104: 1092–1098.

Suwabe K, Tsukazaki H, Iketani H, Hatakeyama K, Fujimura M, Nunome T, Fukuoka H, Matsumoto S, Hirai M (2003) Identification of two loci for resistance to clubroot (*Plasmodiophora brassicae* Woronin) in *Brassica rapa* L. Theor Appl Genet 107: 997–1002.

Suwabe K, Tsukazaki H, Iketani H, Hatakeyama K, Kondo M, Fujimura M, Nunome T, Fukuoka H, Hirai M, Matsumoto S (2006) Simple sequence repeat-based comparative genomics between *Brassica rapa* and *Arabidopsis thaliana*: the genetic origin of clubroot resistance. Genetics 173: 309–319.

Swanson-Wagner RA, Jia Y, DeCook R, Borsuk LA, Nettleton D, Schnable PS (2006) All possible modes of gene action are observed in a global comparison of gene expression in a maize F1 hybrid and its inbred parents. Proc Natl Acad Sci USA 103: 6805–6810.

Tadege M, Sheldon CC, Helliwell CA, Stoutjesdijk P, Dennis ES, Peacock WJ (2001) Control of flowering time by *FLC* orthologues in *Brassica napus*. Plant J 28: 545–553.

Talbot CJ, Nicod A, Cherny SS, Fulker DW, Collins AC et al. (1999) High resolution mapping of quantitative trait loci in outbreed mice. Nat Genet 21: 305–308.

Taylor JD, Conway J, Roberts SJ, Astley D, Vicente JG (2002) Sources and origin of resistance to *Xanthomonas campestris* pv. *campestris* in *Brassica* genomes. Phytopathology 92: 105–111.

Teutonico RA, Osborn TC (1994) Mapping of RFLP and qualitative traits loci in *Brassica rapa* and comparison to the linkage maps of *B. napus*, *B.oleracea* and *Arabidopsis thaliana*. Theor Appl Genet 89: 885–894.

Teutonico RA, Osborn TC (1995) Mapping loci controlling the vernalization requirement in *Brassica rapa*. Theor Appl Genet 91: 1279–1283.

Tinker N, Kilian A, Wight C, Heller-Uszynska K, Wenzl P, Rines H, Bjornstad A, Howarth C, Jannink J-L, Anderson J, Rossnagel B, Stuthman D, Sorrells M, Jackson E, Tuvesson S, Kolb F, Olsson O, Federizzi L, Carson M, Ohm H, Molnar S, Scoles G, Eckstein P, Bonman J, Ceplitis A, Langdon T (2009) New DArT markers for oat provide enhanced map coverage and global germplasm characterization. BMC Genom 10: 39.

Tonguç M, Griffths PD (2004) Development of black rot resistant interspecific hybrids between *Brassica oleracea* L. cultivars and *Brassica* accession A 19482, using embryo rescue. Euphytica 136: 313–318.

Trick M, Long Y, Meng J, Bancroft I (2009) Single nucleotide polymorphism (SNP) discovery in the polyploid *Brassica napus* using Solexa transcriptome sequencing. Plant Biotechnol J 7: 334–346.

Udall JA, Quijada P, Osborn TC (2005) Detection of chromosomal rearrangements derived from homologous recombination. Genetics 169: 967–979.

Verhoeven DTH, Goldbohmm RA van Poppelm G, Verhagenm H, van den Brandt PA (1996) Epidemiological studies on *Brassica* vegetables and cancer risk. Cancer Epidemiol, Biomark Prevent 5: 733–748.

Vicente JG (2004) A podridão negra das crucíferas. In: G Lopes (ed) COTHN Centro Operativo e Tecnologico Hortofruticola Nacional. Alcobaça, Portugal, p 102.

Vicente JG, Conway J, Roberts SJ, Taylor JD (2001) Identification and origin of *Xanthomonas campestris* pv. *campestris* races and related pathovars. Phytopathology 91: 492–499.

Volpe TA, Kidner C Hall IM, Teng G, Grewal SI, Martienssen RA (2002) Regulation of heterochromatic silencing and histone H3 lysine-9 methylation by RNAi. Science 297: 1833–1837.

Voorips RE, Visser DL (1993) Examination of resistance to clubroot in accessions of *Brassica oleracea* using a glasshouse seedling test. Neth J Plant Pathol 99: 269–276.

Voorips RE, Jongerius,MC, Kanne HJ (1997) Mapping of two genes for resistance to clubroot (*Plasmodiophora brassicae*) in a population of double haploid lines of *Brassica oleracea* by means of RFLP and AFLP markers. Theor Appl Genet 94: 75–82.

Walkof C (1963) A mutant annual cabbage. Euphytica 12: 77–80.

Wang H, Nussbaum-Wagler T, Li B, Zhao Q, Vigouroux Y, Faller M, Bomblies K, Lukens L, Doebley JF (2005) The origin of the naked grains of maize. Nature 436: 714–719.

Watts LE (1964) Studies of maturity in F_1 and F_2 generation of cauliflowers from crosses between summer, autumn and winter types. J Hort Sci 39: 84–91.

Welch RM, Graham RD (2004) Breeding for micronutrients in staple food crops from a human nutrition perspective. J Exp Bot 55: 353–364.

Wenzl P, Raman H, Wang J, Zhou M, Huttner E, Kilian A (2007) A DArT platform for quantitative bulked segregant analysis. BMC Genomics 8: 196.

Weisaeth G (1974) Some problems and results in breeding for clubroot resistance in cole crops. Proc Eucarpia Meetings—Cruciferae, 1974, 101–107. Scott. Hort. Res. Inst., Dundee, Scotland.

West MAL, Kim K, Kliebenstein DJ, van Leeuwen H, Michelmore RW, Doerge RW, St.Clair DA (2007) Global eQTL Mapping reveals the complex genetic architecture of transcript level variation in Arabidopsis. Genetics 175: 1441–1450.

White J, Law JR, MacKay I, Chalmers KJ, Smith JSC, Kilian A, Powell W (2008) The genetic diversity of UK, US and Australian cultivars of *Triticum aestivum* measured by DArT markers and considered by genome. Theor Appl Genet 116: 439–453.

Williams PH (1980) Black rot: a continuing threat to world crucifers. Plant Dis 64: 736–742.

Xie C, Gessler DDG, Xu S (1998) Combining different line crosses for mapping quantitative trait loci using the identical by descent-based variance component method. Genetics 149: 1139–1146.

Xing Y, Frei U, Schejbel B, Asp T, Lübberstedt T (2007) Nucleotide diversity and linkage disequilibrium in 11 expressed resistance candidate genes in Lolium perenne. BMC Plant Biol 7: 43.

Xu S (1996) Mapping quantitative trait loci using four-way crosses. Genet Res 68: 175–181.

Yan J, Shah T, Warburton ML, Buckler ES, McMullen MD, Crouch J (2009) Genetic characterization and linkage disequilibrium estimation of a global maize collection using SNP markers. PLoS One 4(12): e8451.

Yedid G, Bell G (2002) Macroevolution simulated with autonomously replicating computer programs. Nature 420: 810–812.

Zeng Z-B (1994) Precision mapping of quantitative trait loci. Genet Res 68: 1457–1468.

Ziolkowski PA, Kaczmarek M, Babula D, Sadowski J (2006) Genome evolution in Arabidopsis/ Brassica: conservation and divergence of ancient rearranged segments and their breakpoints. Plant J 47: 63–74.

Zhang Y, Talalay P, Cho C-G, Posner GH (1992) A major inducer of anticarcinogenic protective enzymes from broccoli: isolation and elucidation of structure. Proc Natl Acad Sci USA 89: 2399–2403.

Zhao J, Paulo M-A, Jamar D, Lou P, van Eeuwijk F, Bonnema G, Vreugdenhil D, Koornneef M (2007) Association mapping of leaf traits, flowering time, and phytate content in *Brassica rapa*. Genome 50: 963–973 doi:10.1139/G07–078.

Zilberman D, Cao X, Jacobsen SE (2003) ARGONAUTE4 control of locus-specific siRNA accumulation and DNA and histone methylation. Science 299: 716–719.

6

Cytogenetic Studies in Vegetable Brassicas

Piotr A. Ziolkowski,[1]* *Martin A. Lysak,*[2] and *Waheeb K. Heneen*[3]

ABSTRACT

Cytogenetics of the crop *Brassica* species will soon celebrate 100 years of history. Although the objectives of cytogenetic studies have changed in the course of time, the field is experiencing a renaissance with the advent of new techniques (multicolor fluorescence *in situ* hybridization, chromosome painting) and due to expanding genomic resources (large-insert DNA libraries, sequence data). Moreover, cytogenomics of Brassicas benefits to a great extent from a variety of genomic tools, resources, and the outstanding genome annotation developed for *Arabidopsis thaliana*. In this chapter we present a brief account of the most significant achievements in *Brassica* cytogenetics and review the latest findings of the genome evolution in *Brassica* crop species. We summarize fundamental facts of the genome composition, homeologous chromosome pairing, chromosome identification, chromosome characterization, and evolutionary dynamics of dispersed and satellite repeats in diploid and allopolyploid *Brassica* genomes. Furthermore, the recent progress in comparative cytogenetic studies by chromosome painting mapping with bacterial artificial chromosomes (BAC) and their impact on the reconstruction of genome evolution in the genus *Brassica* are discussed.

Keywords: cytogenetics, fluorescence *in situ* hybridization (FISH), comparative chromosome painting (CCP), chromosomal rearrangements, Ancestral Crucifer Karyotype (ACK), hexaploidization, *Brassica* genome evolution

[1]Department of Biotechnology, Adam Mickiewicz University, Poznań, Poland.
[2]Department of Functional Genomics and Proteomics, Masaryk University, Brno, Czech Republic.
[3]Department of Plant Breeding and Biotechnology, Swedish University of Agricultural Sciences, Alnarp, Sweden.
*Corresponding author: *pzio@amu.edu.pl*

6.1 Classical Cytogenetic Approaches and the Genetic Background of the *Brassica* Genus

Despite relatively tiny and numerous chromosomes, cytogenetic studies on *Brassica* species have a very long history. The first chromosome analyses in the genus are from the beginning of the 20th century (Takamine 1916; Karpechenko 1922; Morinaga 1928, 1933, 1934; U 1935). Takamine (1916), Karpechenko, (1922) and Morinaga (1933, 1934) reported the chromosome numbers of the economically important species within the *Brassica* genus. Since that time, chromosome numbers have been determined for more than 1,500 different Brassicaceae taxa (summarized in Warwick and Al-Shehbaz 2006). Chromosome numbers for vegetable Brassicas are shown in Table 6-1.

Meiotic analysis of chromosome pairing and segregation in *Brassica* interspecific crosses elucidated evolutionary relationships among the genomes of diploid and allotetraploid species within the genus (Morinaga 1933, 1934; U 1935). It was suggested that the natural amphidiploid (an allotetraploid formed by hybridization of two species followed by chromosome doubling; in consequence it appears to be a functional diploid) *Brassica* species originated from interspecific crosses between diploid species (Fig. 6-1; U 1935). According to this hypothesis, the diploid species *Brassica rapa* (the so called genome A), and diploid *B. nigra* (genome B) are parents of natural amphidiploid *B. juncea* (genomes A and B), *B. rapa* and *B. oleracea* (genome C) are parents of the amphidiploid *B. napus* (genomes A and C), whereas *B. nigra* and *B. oleracea* formed the amphidiploid *B. carinata* (genomes B and C). These relationships are known as the "U-triangle" (U 1935) and have been widely accepted, since they were confirmed by many subsequent genetic and biochemical analyses.

6.2 Addition and Substitution Lines

Access to a natural or synthetic amphidiploid is a prerequisite for the development of mono- and disomic addition lines of diploid species. Crossing the amphidiploid to one of the parental species results in a sesquidiploid hybrid, with two sets of one genome and one set of the other genome. Selfing or recurrent backcrossing generates aneuploids comprising monosomic addition lines. Disomic addition lines arise in the progeny of monosomics. Characterization of the addition lines spans over the documentation of plant morphology, cytology, isozyme patterns, and genome-specific DNA markers that distinguish the various alien chromosomes. Monosomic addition lines offer unique opportunities for cytogenetic analysis (Chang and de Jong 2005). They are well suited for localization of genes and identification of linkage groups, definition

Table 6-1 Chromosome numbers for vegetable Brassicas (data from Warwick and Al-Shehbaz 2006). Some other important species of the Brassicaceae family are included for comparison.

	Taxon	**Common name**	**Genome structure**	**Ploidy level and chromosome number**
Leaf Brassicas	*B. oleracea* var. *capitata*	cabbage	CC	$2n = 2x = 18$
	B. oleracea var. *gemmifera*	Brussels sprouts		
	B. oleracea var. *acephala*	kale, collard greens		
	B. oleracea var. *alboglabra*	Chinese broccoli (kai-lan)		
	B. rapa subsp. *pekinensis*	Chinese cabbage	AA	$2n=2x=20$
	B. rapa subsp. *chinensis*	bok-choy (pak-choi)		
	B. rapa var. *pervidis*	komatsuna		
	B. rapa subsp. *narinosa*	tatsoi		
	B. juncea	mustard greens	AABB	$2n=4x=36$
Stem Brassicas	*B. olearacea* var. *gongylodes*	kohlrabi	CC	$2n=2x=18$
Flowerhead Brassicas	*B. oleracea* var. *italica*	calabrese, broccoli		
	B. oleracea var. *botrytis*	cauliflower, Romanesco broccoli		
Root Brassicas	*B. rapa*	turnip	AA	$2n=2x=20$
	B. napus var. *napobrassica*	swede (rutabaga)	AACC	$2n=4x=38$
Other Brassicas	*B. nigra*	black mustard	BB	$2n=2x=16$
	B. napus subsp. *napus*	rapeseed	AACC	$2n=4x=38$
	B. carinata	Ethiopian mustard	BBCC	$2n=4x=34$
Other Brassicaceae	*Raphanus sativus*	radish		$2n=2x=18$
	Sinapis alba	white mustard		$2n=2x=24$
	Arabidopsis thaliana	thale cress		$2n=2x=10$
	A. lyrata	lyrate rockcress		$2n=2x, 4x=16, 32$
	Capsella rubella	pink shepherd's purse		$2n=2x=16$

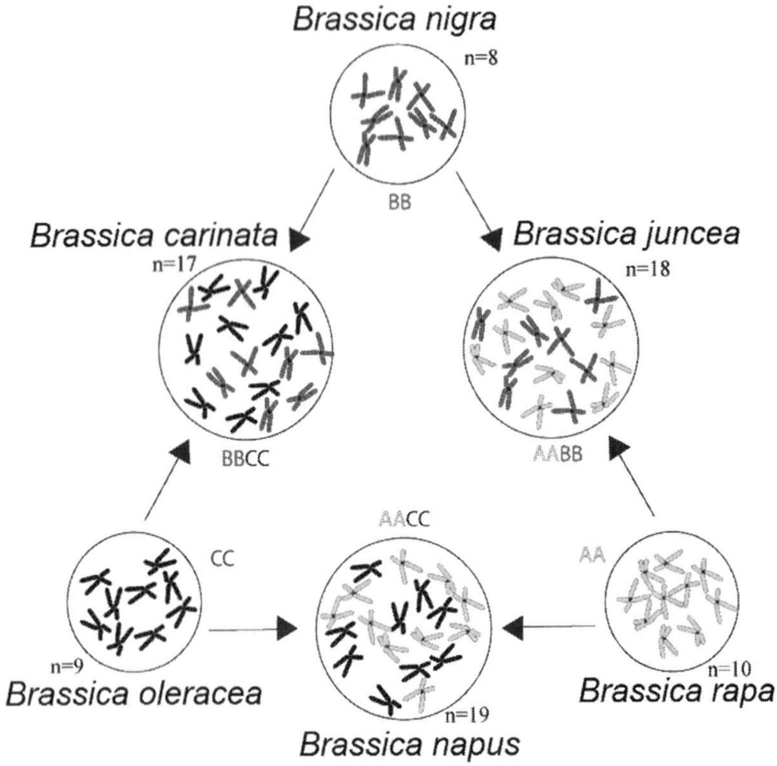

Figure 6-1 Overview of genetic relationships between *Brassica* species, as described by Nagaharu U in 1935 [picture modified from Mike Jones, Wikipedia (*http://en.wikipedia.org/wiki/Triangle_of_U*)].

and physical mapping of chromosome-specific DNA markers, analysis of homeologous pairing and monitoring of intergenomic gene transfer. A nullisomic condition or the substitution of a certain chromosome pair by its homeologues, or by another pair, from a related species, has been also informative regarding gene linkage and the genetic control of certain characters.

Addition and substitution lines, as well as a nullisomic, portraying the dissection of the genomes of the vegetable Brassicas (*B. oleracea, B. rapa, B. juncea* and *B. napus*) are presented in Table 6-2. Most exploited and informative are the monosomic addition lines resulting from the dissection of *B. oleracea*. Thus up to seven or eight of the nine chromosome types of *B. oleracea* have been incorporated in the background of *B. rapa* (McGrath and Quiros 1990; McGrath et al. 1990; Heneen and Jørgensen 2001) and *Raphanus sativus* (Kaneko et al. 1987, 1991). The addition lines were morphologically distinguishable when in the *R. sativus* background, and to a much less extent

Table 6-2 Dissected and background genomes of chromosome addition and substitution lines, and a nullisomic, involving *B. oleracea*, *B. rapa*, *B. juncea* and *B. napus*.

Background genome	Dissected genome	Reference
Additions		
B. rapa	*B. oleracea*	Quiros et al. 1987
		McGrath and Quiros 1990
		McGrath et al. 1990
		Hu and Quiros 1991
		Lee and Namai 1992, 1994
	B. oleracea var. *alboglabra*	Chen et al. 1992,1997a,b
		Cheng et al. 1994a,b, 1995a
		Jørgensen et al. 1996
		Heneen and Brismar 2001
		Heneen and Jørgensen 2001
		Hasterok et al. 2005b
	B. napus	Fantes and MacKay 1979
Raphanus sativus	*B. oleracea*	Kaneko et al. 1987, 1991, 2000, 2002
	B. rapa	Kaneko et al. 2001, 2003
B. napus	*B. juncea*	Struss et al. 1991, 1992
Substitutions		
B. rapa	*B. oleracea*	McGrath et al. 1990
		Chen et al. 1997a, 1997b
B. oleracea	*B. napus*	Chiang et al. 1980
B. juncea	*B. napus*	Banga 1988
Nullisomic		
B. juncea	*B. juncea*	Cheng et al. 2001

in the *B. rapa* background, apparently due to the fact that *B. oleracea* and *B. rapa* are more closely related. Substitution of a chromosome pair of *B. rapa* by another pair of *B. oleracea* has been also encountered in these materials (McGrath et al. 1990; Chen et al. 1997a).

In addition to plant morphology, analysis of the *B. rapa-B. oleracea* monosomic addition lines highlighted further features (references in Table 6-2). Thus, the molecular characterization of these lines characterized synteny groups and disclosed the prevalence of duplicated gene loci, in agreement with reported gene linkage and physical localization maps, and supporting the hypothesis that these diploid species are actually secondary polyploids. The ability to differentiate between the chromosomes of the two species during diakinesis, based on chromatin condensation patterns, made it possible to accurately define if univalents are alien or background chromosomes, and to monitor homeologous pairing in bivalents and multivalents (Fig. 6-2). That the short and long arms of the alien *B. oleracea* chromosome can pair with arms that belong to different *B. rapa* chromosomes is of relevance as to the relatedness and evolution of these genomes. The use of rDNA probes in combination with fluorescence in situ hybridization (FISH) was helpful in defining the chromosomes with rDNA

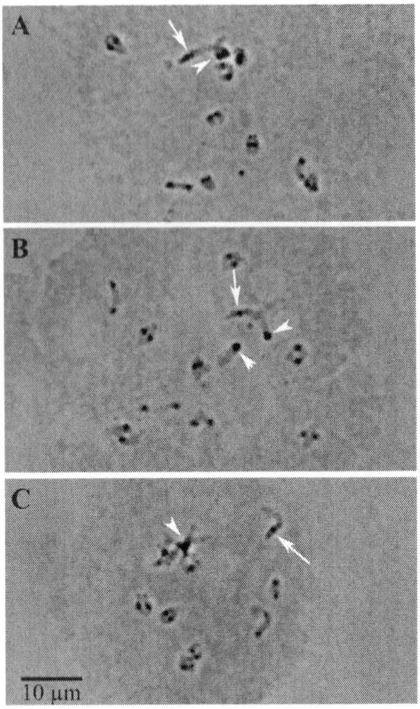

Figure 6-2 Diakinesis of *Brassica rapa* – *B. oleracea* var. *alboglabra* monosomic addition line for chromosome 4 (2n=21, chromosome designation according to Cheng et al. 1995a). The background chromosomes of *B. rapa* exhibit differential chromatin condensation and staining, expressed in the characteristic densely condensed and darkly stained pericentromeric regions that incorporated the rest of the satellited short arm in the case of nucleolar organizer region (NOR)-bearing chromosomes (arrowheads). The alien chromosome of *B. oleracea* var. *alboglabra* has a specific different condensation pattern that enables its identification (arrow). A. 1III + 9II, the trivalent is composed of the alien chromosome and NOR-bearing chromosomes of *B. rapa*. B. 10II + 1 I, the alien chromosome is paired with a NOR-bearing chromosome of *B. rapa*, composing a heterobivalent, while the other NOR-bearing chromosome appears as a univalent. C. 10II + 1I, the alien chromosome appears as a univalent.

sites and the meiotic behavior of these chromosomes. The occurrence of intra- and intergenomic gene recombination and chromosomal structural changes, like deletions and translocations, has been documented during the development of the addition lines and in their progenies. Obtained double monosomics are useful to reveal intragenomic homeology. It was possible to define which chromosomes, and in some cases to infer which regions on chromosomes, that harbor genes of agronomic importance, like those controlling erucic acid content and seed color. At least three chromosomes carried genes that controlled seed color either maternally or biparently. The transmission of the alien chromosome in the addition lines occurred at different frequencies, and seed setting was variable, but enough for maintenance of these lines.

The potential of developing new cultivars from disomic, double disomic or multiple disomic addition lines of *B. rapa-B. oleracea* has been pointed out (Lee and Namai 1992, 1994). Further use of addition and substitution lines in plant breeding relate to the inheritance of the boltic trait (Banga 1988; Kaneko et al. 2000). Of the 10 chromosome types of *B. rapa*, eight were represented in a *R. sativus* background, and similar to the situation with *B. oleracea*, they gave rise to morphologically differentiable addition lines (Kaneko et al. 1991, 2003). A comparison of a double monosomic *R. sativus* line, harboring one chromosome from each of *B. oleracea* and *B. rapa*, with its parental monosomic addition lines elucidated chromosome homeological relationships among the three species (Kaneko et al. 2002). The dissection of *B. napus* and *B. juncea* in addition and substitution lines, and in a nullisomic, have been restricted to a limited number of chromosome types (Chiang et al. 1980; Banga 1988; Struss et al 1991, 1992; Cheng et al. 2001).

In summary, chromosome addition and substitution lines and nullisomics are valuable tools for the elucidation of genome composition and the assignment of genes, DNA markers and synteny groups to defined chromosomes. They are also suitable for chromosome morphological and structural studies, physical mapping of genes and specific DNA sequences, monitoring of phylogenetical relationships and intergenomic gene transfer, as well as providing bridging materials for plant breeding.

6.3 Interspecific Hybridization and Homeologous Chromosome Pairing

A large number of interspecific and intergeneric sexual and somatic hybrids, involving the Brassicas, has been obtained (see reviews by Warwick et al. 2000; FitzJohn et al. 2007; Prakash et al. 2009). In addition to the goal of expanding the genetic variability for breeding purposes, hybridization has contributed to the elucidation of interrelationships between genomes, expressed by homeologous chromosome pairing (Snowdon 2007). The high crossability among *Brassica* species and close relatives, and the occurrence of homeologous pairing and genetic recombination in many of these hybrids is a suitable means of transferring desirable characters from one species to the other. Only hybrids denoting intergenomic chromosome homeology that pertain to the vegetable Brassicas (*B. oleracea, B. rapa, B. napus, B. juncea* and *B. carinata*), will be considered here.

Intergenomic relationships inferred from meiotic studies are usually based on pairing formulas expressing numbers of univalents, bivalents and multivalents, reflecting the affinities between chromosomes. Most of the reported meiotic pairing formulae are not based on actual chromosome identification of parental genomes, due to lack of differentiating morphological features, or not applying discriminating techniques, like genomic in situ hybridization (GISH) or FISH and chromosome painting using chromosome-specific DNA probes. Thus, caution is needed when inferring auto- and allosyndesis of paired chromosomes, and parental origin of univalents, if chromosome identification has not been ascertained. The frequent occurrence of duplications in the *Brassica* genomes contributes to the often documented intragenomic chromosome pairing (Li et al. 2005; Leflon et al. 2006; Ge and Li 2007).

Differences in chromatin condensation patterns of parental chromosomes are useful for the detection of homeologous pairing in bivalents and multivalens, and determination of the parental origin of univalents (Heneen et al. 1995). Discrimination between genomes in hybrids and allopolyploids, intergenomic recombination and introgression have been more accurately determined by the application of FISH and GISH (Li et al. 2005; Wang et al. 2005; Leflon et al. 2006; Ge and Li 2007; Snowdon 2007; Nicolas et al. 2008).

The earliest *Brassica* hybridizations were among the three diploids *B. rapa*, *B. oleracea* and *B. nigra* and their three amphidiploids *B. napus*, *B. juncea* and *B. carinata*. All three diploid basic genomes have been further brought together in trigenomic allohexaploids. Most information on the relationship between homeologous chromosomes has been inferred from hybrids, synthetic allopolyploids, digenomic triploids, and monosomic addition lines. While multivalents occurred in synthetic amphidiploids, diploid-like meiosis achieved during evolution, prevailed in natural polyploids. A genetic control of the diploid-like meiosis in natural *Brassica* allopolyploids has been advocated (Jenczewski et al. 2003 and references therein). Apparently, crossover formation and frequency are not only determined by the level of intra- and intergenomic relatedness, and genetic control, but also by the composition of the chromosome complement, whether constituted of different genomes or a mixture of similar and different genomes (Nagpal et al. 1996; Nicolas et al. 2008).

Analysis of hybrids, synthetic allopolyploids, backcrosses and addition lines involving the six *Brassica* species of the U-triangle led to the conclusion that the A and C genomes are phylogenetically closer to each other than to the B genome. This has been corroborated by molecular DNA analysis (Truco et al. 1996 and references therein). In addition to the established chromosome homeologies between the A, B and C genomes documented in the six cultivated species, homeologies were inferred between the vegetable Brassicas and *B. balearica*, *B. barrelieri*, *B. bourgeaui*, *B. cossoneana*, *B. cretica*, *B. fruticulosa*, *B. gravinae*, *B. incana*, *B. insularis*, *B. macrocarpa*, *B. maurorum*, *B. montana*, *B. oxyrrhina*, *B. rupestris*, *B. tournefortii* and *B. villosa* (Inomata 1993; Bothmer et al. 1995; Prakash et al. 2009). Intergeneric hybrids, exhibiting homeologous chromosome pairing, have been obtained between vegetable Brassicas and various other genera including *Coicya*, *Diplotaxis*,

Enarthrocarpus, Eruca, Erucastrum, Hirschfeldia, Moricandia, Orychophragmus, Raphanus, Sinapidendron and *Synapsis* (Du et al. 2008; Prakash et al. 2009). In hybrids involving *Orychophragmus violaceus*, genome separation and elimination of the *O. violaceus* genome prevailed (Hua et al. 2006; Li and Ge 2007).

Hybridization and backcrossing have been employed for the introgression of desirable characters, such as disease resistance and characters relating to seed quality, into *Brassica* genomes. Monitoring the incidence of hybridization between crop plants and related wild species, homeologous chromosome pairing and consequent gene transfer, are pertinent issues before marketing of transgenic crops.

6.4 Genome Organization at Chromosome Level—Euchromatin and Heterochromatin

One of the molecular mechanisms modulating the gene function in higher eukaryotes is the make-up of chromatin in their immediate vicinity. At the same time, it is also clear that genes constitute only a fraction of genomes, and that usually a large part of the DNA has a noncoding nature. Chromatin incorporates several classes of protein complexes, which are responsible for a level of its compaction. From a cytogenetic point of view, there are two types of chromatin which can be easily identified under a microscope: euchromatin, a less condensed fraction in which most of transcriptionally active genes are found, and heterochromatin, a fraction that remains highly condensed throughout the cell cycle, occupied mainly by repetitive sequences (reviewed by Richards and Elgin 2002).

In higher eukaryotes heterochromatin is located at telomeres, centromeres and pericentromeric regions; in plants, in addition to these regions, heterochromatin is located at the nucleolar organizer regions (NORs), at knobs and in some cases along B chromosomes (Alfenito and Birchler 1993; Copenhaver et al. 1999; Fransz et al. 2000; McCombie et al. 2000).

Under phase-contrast microscopy, heterochromatin in interphase and prophase nuclei may be clearly recognized as dark regions in a light background of euchromatin; conversely, DAPI (4′,6-diamidino-2-phenylindole) staining reveals bright heterochromatin foci under an epifluorescence microscope. In *Brassica* species, heterochromatin and euchromatin can be easily distinguished throughout the meiotic prophase I, and in interphase nuclei (see Fig. 6-3). At the mitotic metaphase, chromosomes are usually too compact to discriminate both types of chromatin.

Plants often possess much larger genomes than animals (Bennett and Leitch 2005), mainly due to a higher frequency of the transposable elements

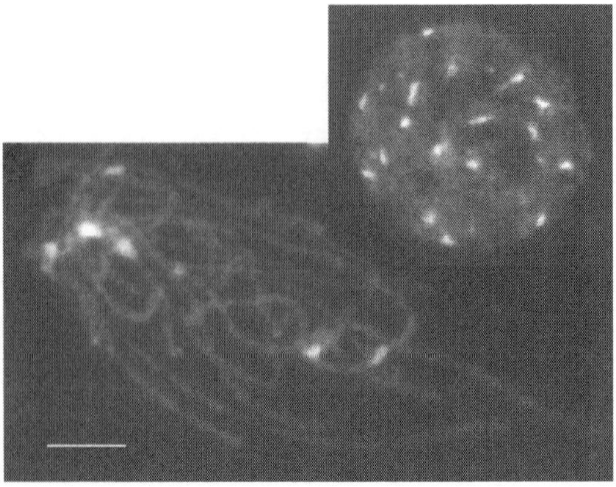

Figure 6-3 Euchromatin and heterochromatin (bright areas) in *B. oleracea* pachytene bivalents (lower panel) and an interphase nucleus (upper panel), as observed after DAPI staining. Heterochromatic sites correspond to (peri)centromeric regions and NORs. At the pachytene stage, bivalents usually coalesce at centromeric regions. Bar, 10 µm.

(TEs) amplification. In consequence, large regions of non-genic DNA are formed, which are subsequently methylated to stop further expansion of TEs (Bennetzen 2005). At the same time, chromatin remodeling converts initially relaxed chromatin of the chromosome region into the compact state, mainly by histone methylation (Lippman et al. 2004; Fransz et al. 2006). This can cause formation of heterochromatin knobs, which in an extreme situation can be the major constituent of all chromosomes in the genome, as in the case of onion *Allium cepa* (de Jong et al. 1999). However, *Brassica* genomes are relatively small and compact as compared to many other plant genomes, with chromosome architecture resembling that of *Arabidopsis thaliana*. Thus, heterochromatin blocks can be found preferentially around centromeres, at NORs, and in a few distal and/or interstitial sites (usually described as knobs or chromomeres). In *B. oleracea*, only one non-centromeric knob was found, which was further identified as the major 5S ribosomal DNA (rDNA) locus (Ziolkowski and Sadowski 2002). In *B. rapa* five non-centromeric heterochromatic knobs were reported, one of them bearing a 5S rDNA locus (Koo et al. 2004). Up to now, such an analysis has not been performed in other *Brassica* species.

6.5 Towards Identification of *Brassica* Chromosomes

One of the main goals of cytogenetic analysis is to identify and efficiently distinguish chromosomes within the genome(s) under study. The reasons to do so are numerous: at the cytological level, it enables the characterization of chromosomal aberrations, and to assign chromosomes to parental species in allopolyploids, and introgression lines. This is of major importance in plant breeding strategies, when the identification of genetic material of donor species may significantly accelerate production of new cultivars. On the other hand, identification of alien chromosomes in addition or substitution lines can be beneficial for mapping of quantitative trait loci (QTL), and for interspecies gene transfer in breeding programs. On the molecular level, identification of chromosomes is needed for integration of genetic maps with physical maps produced by karyotype analysis and cytogenetic determination of the position of DNA sequences on chromosomes. This is an important step in advanced genome sequencing programs, where particular bacterial artificial chromosome (BAC) or shotgun-derived contig could be assigned to a chromosome.

Chromosome identification is relatively simple for species with a low number of large chromosomes, which have different morphology and different condensation patterns/distribution of repetitive sequences. In such a case, constructing a karyotype is then possible through classical cytogenetic analyses. Unfortunately, *Brassica* species possess relatively small genome sizes (~ 650 Mb for *B. oleracea*; for more details see Table 6-3), and thus their chromosomes are very small. As they measure only 2–5 µm at the mitotic metaphase stage (Cheng et al. 1995a), karyotype construction is a hard task. Because of these difficulties, researchers have been looking for some other possibilities to analyze *Brassica* chromosomes. One of them is to use pachytene stage of meiosis, where incomplete condensed bivalents are much longer than their counterparts at the mitotic metaphase. The difference in chromosome length between mitotic metaphase and meiotic pachytene chromosomes is 10- to 40-fold, depending on a species (de Jong et al. 1999). For *Brassica,* the length of pachytene bivalents is 15- to 20-fold the length of mitotic metaphase chromosomes (Ziolkowski and Sadowski 2002). Thus, the resolution can be greatly improved by analysis of pachytene chromosomes.

The first trial to construct a *Brassica* karyotype dates to 1960. Röbbelen (1960) used pachytene bivalents for constructing karyotypes of three diploid *Brassica* species on the basis of structural characteristics of total chromosome length, symmetry of the arms, and the shape of the heterochromatic pericentromeric regions. Based on their absolute length, chromosomes were classified into five different types: very short (up to 20 µm), short (20–25 µm), medium (25–30 µm), long (30–40 µm), and very long (more than 40 µm). Based on the centromere position, they were classified as chromosomes with a median or submedian centromere, respectively, and a chromosome with subterminal centromere and a satellite. The work reports the only complete pachytene karyotype for *B. oleracea* and *B. nigra* published to date. Recently, a high-resolution karyotype of *B. rapa* subsp. *pekinensis* based on

Table 6-3 Genome sizes in *Brassica* species of the U triangle (Table redrawn from *http://www.brassica.info/info/reference/genome-sizes.php* with permission).

Species (genome)	Genome size (Mb)	Variety/ common name	Method	Reference
B. rapa (A)	468–516		FCM	Arumuganathan and Earle 1991
	507	var. *chinensis* Pak choi		Arumuganathan and Earle 1991
	511	var. *rapifera* turnip		Arumuganathan and Earle 1991
	529			Johnston et al. 2005
	564			Bennett and Smith 1991
	784			Bennett et al. 1982
B. nigra (B)	468	black mustard	FCM	Arumuganathan and Earle 1991
	632		FCM	Johnston et al. 2005
	760		FD	Bennett and Smith 1976
B. oleracea (C)	599–618	var. *italica* broccoli	FCM	Arumuganathan and Earle 1991
	603	var. *capitata* cabbage	FCM	Arumuganathan and Earle 1991
	628	var. *gemmifera* Brussels sprout	FCM	Arumuganathan and Earle 1991
	628–662	var. *botrytis* cauliflower	FCM	Arumuganathan and Earle 1991
	696		FCM	Johnston et al. 2005
	752		FD	Bennett and Smith 1991
	868		FD	Bennett and Smith 1976
B. juncea (AB)	1068	brown mustard	FCM	Johnston et al. 2005
	1105		FCM	Arumuganathan and Earle 1991
	1495		FD	Bennett and Smith 1976
B. napus (AC)	1129–1235	Rapeseed	FCM	Arumuganathan and Earle 1991
	1127		FD	Bennett and Smith 1991
	1132		FCM	Johnston et al. 2005
B. carinata (BC)	1284	Ethiopian mustard	FCM	Johnston et al. 2005
	1544		FD	Bennett and Smith 1976

FCM, flow cytometry; FD, Feulgen densitometry.

multicolor FISH using pachytene and mitotic metaphase chromosomes has been published (Koo et al. 2004).

Chromosome banding techniques have been applied to *Brassica* species starting from the late 1980s (Wang and Luo 1987; Wang et al. 1989). Both C-banding and G-banding techniques revealed that *Brassica* chromosomes are not enough structurally differentiated to be fully distinguished based on these methods (Nishibayasahi 1992). The main problem was still the length of mitotic metaphase chromosomes. One possible solution was to use late prophase preparations for karyotyping. Olin-Fatih and Heneen (1992) described karyotypes of *B. oleracea*, *B. rapa* and *B. napus* studied using C-banding. The chromosomes were grouped as with a median, submedian, subterminal, and terminal centromere, respectively. According to the authors, all chromosome pairs were morphologically distinguishable. This karyotyping technique was further utilized by Olin-Fatih in a subsequent work conducted on protoplasts (Olin-Fatih 1996). A study on root tips was also conducted by Cheng et al. (1995a), who published a revised karyotype of *B. oleracea* var. *alboglabra.* However, it should be noted that a relatively weak confidence of chromosome identification precludes other researchers from adopting the banding methods in their research as a standard protocol for chromosome discrimination in *Brassica.* The karyotyping based on G-banding applied by Olin-Fatih (1994) also cannot be accepted as a routine way for this purpose. All the studies mentioned above show that in *Brassica,* morphometric uniformity of chromosomes and their relatively high number are the main limiting factors for efficient karyotype construction.

Like in the pachytene stage of meiosis, mitotic prometaphase chromosomes exhibit an uneven pattern of chromatin condensation. This fact has been utilized in measuring chromatin density along prometaphase chromosomes (CP, condensation pattern = density profile) (Nakayama and Fukui 1997). This method, together with 45S rDNA FISH-based mapping, were further used for constructing karyotypes of three diploid *Brassica* species (Fukui et al. 1998), and rapeseed (Kamisugi et al. 1998). Nevertheless, "quantitative karyotyping" has not been widely adopted in *Brassica* research, as it is relatively time-consuming and needs special software equipment.

Later trials of karyotype construction utilized both chromosome length/ morphology, as well as 45S and 5S rDNA FISH analysis for chromosome identification (Kulak et al. 2002; Snowdon et al. 2002). These attempts will be further described in Section 6. In *B. oleracea,* the problem lies in a lower number of chromosomes with 5S and 45S rDNA loci, which permits the identification of only two out of nine chromosome pairs. More reliable chromosome identification is achieved by using low-copy sequences, especially derived from *Brassica* BAC libraries currently developed. A more promising situation exists in *B. rapa,* where six out of 10 chromosome pairs can be identified using 5S and 45S rDNA probes (Snowdon et al. 2002). Koo et al. (2004) used 5S and 45S rDNA in combination with a 350-bp centromeric tandem repeat, to distinguish both mitotic metaphase and pachytene *B. rapa* chromosomes. Due to the specific hybridization pattern of the centromeric repeat hybridizing to eight out of 10 chromosome pairs, the construction of the *B. rapa* karyotype was greatly facilitated. The pachytene karyotype described by Koo et al. (2004) comprises one acrocentric, two metacentric, five submetacentric, and two subtelocentric chromosomes. Moreover, the study gives us the opportunity to compare sizes of mitotic metaphase *B. rapa* chromosomes with their pachytene counterparts (Table 6-4). All *B. rapa* chromosomes are 15–20 times longer at the pachytene stage as compared with mitotic metaphase chromosomes; similar to what was described for *B. oleracea* (Ziolkowski and Sadowski 2002).

Table 6-4 A comparison of *B. rapa* subsp. *pekinensis* mitotic metaphase chromosomes with their pachytene counterparts (according to Koo et al. 2004).

Chromosome no.[a]	Chromosome length (μm ± SD)		Centromeric
	mitotic metaphase	pachytene	index (%±SD)[b]
1	3.30±0.21	51.3±7.45	38.8±1.59
2	2.89±0.25	48.0±3.05	9.38±0.91
3	2.53±0.17	47.1±6.20	29.5±2.78
4	2.35±0.16	41.4±3.64	30.4±3.26
5	2.20±0.21	40.8±6.16	32.3±2.18
6	2.00±0.20	36.8±2.49	41.0±4.66
7	1.84±0.12	36.8±4.41	17.4±4.73
8	1.73±0.16	32.7±3.18	20.2±2.08
9	1.62±0.15	27.0±2.62	36.7±1.94
10	1.46±0.14	23.7±2.97	32.9±2.43

[a]Chromosomes arranged according the decreasing length.
[b]Centromeric index is the percentage of short arm by total chromosome length (Levan et al. 1964).

Armstrong et al. (1998) published an idiogram of *B. oleracea* var. *alboglabra* short-cycle line A12DHd, which was a parental genotype of one of the current *B. oleracea* mapping populations (Bohuon et al. 1996; Sebastian et al. 2000), and has been used in genome sequencing projects. Idiograms of mitotic prometa phase chromosomes were constructed and further used as a basis for more detailed studies, in which chromosome-specific probes were developed (Howell et al. 2002). Howell et al. (2002) used a number of genomic clones as FISH landmarks for individual chromosomes. These clones included a genomic restriction fragment length polymorphism (RFLP) clone, cDNA plasmid clone, cosmid clone, as well as 17 *B. oleracea* BACs. Selection of the BAC clones was based on BAC-filter hybridization with a number of known genomic sequences, and their assignment to specific loci was further confirmed by BAC clone fingerprinting, Southern hybridization, and locus-specific PCR (Howell et al. 2002). Most BACs contained, beside unique sequences, also DNA repeats (mainly TEs), which showed unspecific hybridization signals. Therefore, the inclusion of C_0t-1 blocking DNA was required to suppress the hybridization of repetitive sequences. Once having chromosome-specific landmarks at hand, emphasis was put on assignment of linkage groups to chromosomes. This assignment was done by applying a number of mapping strategies, including RFLP, cleaved amplified polymorphic sequences (CAPS) and single nucleotide polymorphism (SNP). Since there is more than one probe specific for each of the nine chromosomes of *B. oleracea*, these chromosomes can be oriented with respect to the linkage groups. The integrated cytogenetic and genetic maps of *B. oleracea* may be useful for estimation of the physical distance between markers applied in positional cloning. It should also be possible to estimate the physical length of a DNA region from a donor parent in a substitution line using FISH probes associated with markers at each end of the substitution chromosome. But the most significant application of the integrated map lies in assignment of BACs of an overlapping set of clones (i.e. a BAC contig) to linkage groups for the construction of physical *Brassica* maps. Integrating genetic, cytogenetic and physical information coming from different research groups will lead to a much more comprehensive understanding of the *Brassica* genome architecture and its structural evolution.

6.6 Chromosomal Localization of rDNA in *Brassica*

The two types of ribosomal RNA genes, 5S rDNA and 45S rDNA encoding for 5S and 18S-5.8S-25S rRNAs, respectively, are widely used in FISH studies, providing convenient chromosome markers. As these two gene clusters are organized in head-to-tail tandem arrays localized on a few chromosomes in all known eukaryotic species, with each locus built of several hundred to a few thousand gene copies (Copenhaver and Pikaard 1996), they are easy to detect by the FISH technique.

Ribosomal RNAs, together with some protein complexes, build up ribosomes, the translation factory for protein synthesis. There are four classes of rRNA molecules in plant ribosomes, named 5S, 5.8S, 18S and 25S according to their Svedberg coefficient (see Pederson and Politz 2000, for review). Of these, the 5.8, 18 and 25S rRNAs result from splicing of a single 45S transcript encoded by rDNA repetitive units clustered at particular chromosomal loci. The fourth rRNA class, 5S rRNA, is transcribed from different gene arrays located usually in distinct chromosomal loci (Heslop-Harrison 2000). Transcription of the 45S rDNA gives rise to the most conspicuous morphological element within the nucleus, the nucleolus; hence, the loci where 5.8-18-25S rRNA genes are localized are termed nucleolar organizer regions (NORs) (McClintock 1934; Fig. 6-4). rRNA gene transcription is suspended during meiosis, which results in the disappearance of the nucleolus in the late meiotic prophase I in higher eukaryotes. The activity of the 45S rDNA loci at mitotic interphase can often be visualized at the following metaphase as a secondary constriction (McClintock 1934), thought to reflect the necessary DNA decondensation that allows the access of transcription machinery at interphase. At the loci of 45S rDNA, which were involved in nucleolus formation during interphase, some of the rRNA gene transcription proteins are retained at metaphase (Roussel et al. 1996). These proteins possess the ability to reduce silver under acid conditions, and this enables a differential staining of mitotic NORs (so called Ag-NORs) that were transcribed during the preceding interphase, as well as the staining of nucleoli in interphase nuclei (Goodpasture and Bloom 1975; Robert-Fortel et al. 1993). For the first time, this procedure was applied to *Brassica* chromosomes by Cheng and Heneen (1995), which resulted in a much more precise description of NORs in dividing cells.

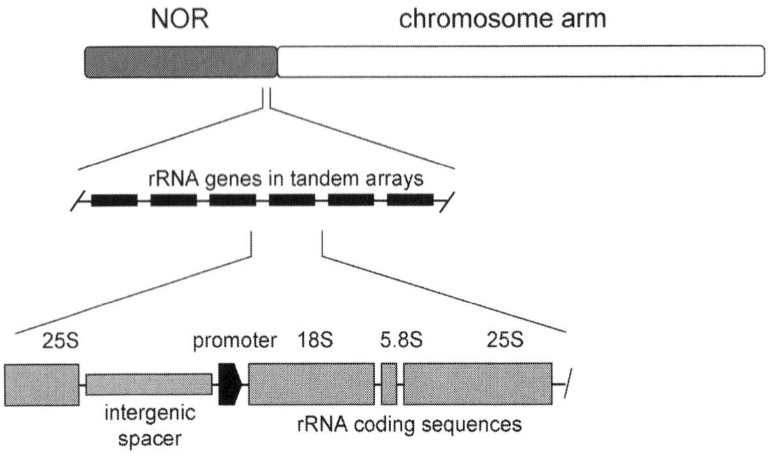

Figure 6-4 Organization of 45S rDNA locus in plants. Genes are organized as head-to-tail repeats, each block consists of 18S-5.8S-25S rRNA gene units. Similar but simpler organization exists for the 5S rRNA gene cluster.

In *Brassica* cytogenetics, NORs were used as convenient markers distinguishing chromosomes (e.g., Röbbelen 1960; Olin-Fatih and Heneen 1992; Cheng et al. 1995a; Olin-Fatih 1996). Adaptation of the silver staining technique also contributed to the use of rDNA loci as convenient chromosomal markers (Cheng and Heneen 1995). On the other hand, by these means not all 45S rDNA loci can be identified, because only transcriptionally active genes contribute to the nucleolus formation. The progress in FISH techniques provided a much needed tool for the chromosome localization of rRNA genes.

Maluszynska and Heslop-Harrison (1993) were the first to successfully apply 45S rDNA as a molecular marker in FISH to *Brassica* chromosomes. The authors analyzed the presence of the 45S rDNA in diploid species (*B. nigra*, *B. oleracea*, and *B. rapa*), as well as in their allotetraploid derivates (*B. carinata*, *B. juncea*, and *B. napus*). They found two major pairs of rDNA loci in *B. nigra*, two major pairs and one minor pair in *B. oleracea*, and five pairs of loci in *B. rapa*. Among the three tetraploid species, *B. carinata* had four loci, *B. juncea* had five major pairs and one minor pair, and *B. napus* had six pairs of loci. These results indicated that the number of 45S rDNA loci has been reduced during the evolution of allotetraploid species. Later, Snowdon et al. (1997a) confirmed these results for *B. juncea* and *B. napus*. However, the authors found the number of rDNA sites in *B. carinata* equaled the sum of loci observed in the parental species. Furthermore, based on the morphology of prometaphase chromosomes and rDNA hybridization patterns, Snowdon et al. (1997a) were able to assign individual *B. napus* rDNA loci to the parental A and C genomes, respectively. They proposed the reduction of rDNA loci number in the amphidiploid species resulting from the loss of the smallest site of the A genome (Snowdon et al. 1997a). The study of Fukui et al. (1998), based on the analysis of prometaphase chromosomes, was only partially consistent with the results obtained by Maluszynska and Heslop-Harrison (1993) and Snowdon et al. (1997a). The FISH experiments revealed five, three and two 45S rDNA loci in *B. rapa*, *B. nigra* and *B. oleracea*, respectively. The authors suggested that these differences may reflect an intraspecific polymorphism of rDNA loci.

In *B. oleracea*, the integration of genetic and physical chromosome maps started with the study of Armstrong et al. (1998), as described in Section 5. The authors used 5S and 45S rDNA as FISH probes, confirming the presence of three loci of 45S rDNA loci, previously reported for the species (Maluszynska and Heslop-Harrison 1993; Snowdon et al. 1997a). Using both mitotic and pachytene chromosomes, they found two 5S rDNA sites located on the long arm of chromosome O4 (chromosome numbering according to Howell et al. 2002). Later analysis, also performed on pachytene chromosomes, revealed that in fact there are three 5S rDNA loci on this chromosome arm (Ziolkowski and Sadowski 2002; Fig. 6-5).

Other studies used 5S and 45S rDNA probes for karyotype construction of the species of the U-triangle (Hasterok et al. 2001; Kulak et al. 2002; Snowdon et al. 2002). Although multicolour FISH (5S and 45S rDNA probes applied simultaneously) in combination with more extended prometaphase chromosomes results in more confident chromosome identification than with metaphase preparations, chromosome size and morphology still remain the main limiting factors in such approaches. Although a polymorphism in rDNA loci has been reported (see below), eight types of rDNA-bearing chromosomes in *Brassica* species of the U-triangle could be distinguished according to Hasterok et al. (2001) and Kulak et al. (2002) (Fig. 6-5).

6.6.1 Nucleolar Dominance in Brassica

An interesting epigenetic phenomenon observed for genes encoding for 18S-5.8S-25S rRNA is nucleolar dominance. It is based on transcriptional dominance of the ribosomal genes from one species over the genes from the second species in interspecific hybrids.

Both molecular and cytogenetic approaches were used to analyze the nucleolar dominance in *Brassica* species. For cytogenetic studies, the silver staining method (AgNOR; for principles of the technique, see above) could be applied, as this technique enables detection of only transcriptionally active 45S rRNA genes (Goodpasture and Bloom 1975). These studies were

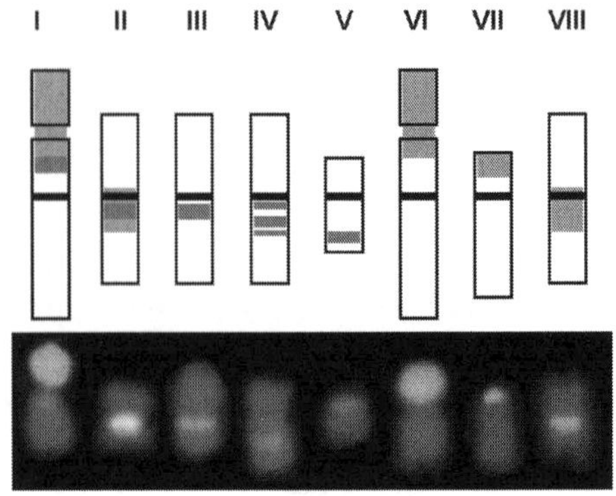

Species	Genome	A	A	B	C	A	B,C	B,C	A
B. nigra	BB	-	-	1	-	-	2	1	-
B. oleracea	CC	-	-	-	1	-	1	1	-
B. rapa	AA	1	1	-	-	1	-	-	3
B. carinata	BBCC	-	-	1	1	-	2	2	-
B. juncea	AABB	1	2	1	-	1	2	1	2
B. napus	AACC	1	3	-	1	1	1	1	1

Figure 6-5 Types of mitotic chromosomes with 5S rDNA (red) and 45S rDNA (green) and their occurrence within the genomes of six *Brassica* species of the U triangle. Chromosome types (Roman numerals) are numbered according to Hasterok et al. (2001). Modified from Hasterok et al. (2001, 2006) with permission.

Color image of this figure appears in the color plate section at the end of the book.

conducted in Brassicas from the 1990s, when the nucleolar activity of 45S rDNA loci was described for three diploid *Brassica* species (Cheng and Heneen 1995; Cheng et al. 1995b). Combining FISH with AgNOR facilitated discrimination of active and inactive loci in *Brassica* amphidiploids (Hasterok and Maluszynska 2000b). This study showed that although the number of loci in amphidiploids is reduced, the number of active loci equals to the number of AgNORs in their diploid ancestors. Active rDNA loci are located at the secondary constrictions, while loci at pericentromeric or terminal locations are inactive or their activity is below detection. More interestingly, these findings indicate that there is no nucleolar dominance in root meristematic cells of *Brassica* species. However, the result seems to be contradictory to the molecular studies of Chen and Pikaard (1997b), who presented convincing evidence for nucleolar dominance in young leaves of allotetraploid *Brassica* species. The discrepancy can be explained by different patterns of rRNA gene expression in various organs and developmental stages, which was partially confirmed by other molecular studies (Chen and Pikaard 1997a).

Molecular studies elucidated the epigenetic nature of nucleolar dominance in amphidiploid *Brassica* species (Chen and Pikaard 1997a). It was shown that nucleolar dominance can be reversed by chemical inhibitors of DNA methylation or histone deacetylation. These findings led to the conclusion that nucleolar dominance is a consequence of the selective silencing of rRNA genes from one progenitor rather than the selective activation of the other (Chen and Pikaard 1997a). Further studies carried out on *Arabidopsis* plants have shown that small interfering RNA (siRNA)-directed DNA methylation and changes in posttranslational histone modifications (mainly acetylation and methylation) are responsible for the "on-off switch mechanism" controlling the number of active rRNA genes (Lawrence et al. 2004; Preuss et al. 2008). Moreover, the mechanism seems to be the same both for interspecific hybrids exhibiting nucleolar dominance and for non-hybrid, diploid plants, where it controls the effective dosage of rRNA in response to the particular physiological demands of the cell (Lawrence et al. 2004).

6.6.2 Evolutionary Trends in rDNA Loci Number Change

One of the most intriguing aspects of the evolution of rDNA in *Brassica* species is connected with the differences in the number of rDNA loci found among closely related species, their intraspecific variation, and increase/decrease of their number in *Brassica* amphidiploids.

First clues of rDNA copy number variation came from molecular analyses. Restriction fragment length polymorphism (RFLP) applied to *B. rapa-B.oleracea* addition lines revealed two loci of 45S rDNA (Delseny et al. 1990). A similar approach applied on dissected *B. oleracea* chromosomes (using monosomic addition lines) revealed strong rDNA hybridization signals linked to two linkage groups, and a weak signal linked to a third linkage group (McGrath et al. 1990). Further studies on the segregation of rDNA RFLP markers in cross progenies have indicated the presence of three pairs of rRNA gene loci in one of the four *B. oleracea* populations investigated (Kianian and Quiros 1992a, b).

These initial investigations conducted in *B. oleracea* only indicated the phenomenon of a rapid rDNA evolution in the *Brassica* genus. Much more data were collected based on FISH analyses. One of the first unexpected results was a finding that the number of 45S rDNA loci in allotetraploids is reduced in comparison with their diploid ancestors (Maluszynska and Heslop-Harrison 1993; Snowdon et al. 1997a). Reduction of 45S rDNA loci number was also observed in some other polyploid species and genera, including *Nicotiana tabacum* (Volkov et al. 1999), and rice (Shishido et al. 2000).

It is still not clear what genetic mechanisms triger rapid evolution of 45S rDNA loci. However, comparative mapping in the Brassicaceae family shows that most of the chromosome breakages have occurred close to centromeres or telomeres, where long stretches of tandem repeats exist

(Boivin et al. 2004; Kuittinen et al. 2004; Yogeeswaran et al. 2005; Henry et al. 2006; Lysak et al. 2006). As 45S rDNA loci are found usually at chromosome ends or within pericentromeric regions, it may imply that their extremely high mobility within the genome and frequent loss in hybrids is a result of their breakage-prone location (Schubert 2007). In addition, 45S rDNA tandem repeats are prone to recombination and rearrangements.

The reduction of loci number after polyploid formation is also observed in the case of 5S rRNA genes. A loss of rDNA loci has apparently occurred in some polyploid species within the Brassicaceae family, as in polyploid species *Moricandia arvensis* ($2n = 28$), and *Alyssum spinosum* ($2n = 32$), only one 5S rDNA locus was reported (Ali et al. 2005). Reduction of loci number has been also found in *B. carinata* (both parental species have six 5S rDNA loci, while the polyploid only four; see Table 6-5). Because of the documented large numbers of 5S rDNA loci and often contradictory reports, the question of 5S rDNA loci decrease/increase in *B. napus* and *B. juncea* could not be resolved as yet (see Table 6.5 for references).

The phenomenon of reduction of rDNA loci in polyploids is usually explained as one of the first steps in genome diploidization, the process by which a polyploid organism attains the diploid state (Wendel 2000). After hybridization of genomes derived from two different species, an imbalance in gene dosage can prevail, which leads to inadequate amounts of different gene products in cells. In the case of rDNA genes, an excess of ribosomal RNA produced by multiplied rDNA loci can further cause disturbances

Table 6-5 Chromosome localization of 5S and 45S rRNA genes in *Brassica* species.

Species	Genome(s)	No. of chromosomes with 5S rDNA loci	No. of chromosomes with 45S rDNA loci	No. of chromosomes with 5S and 45S rDNA loci	References
B. nigra	BB	2 (t)	6		1,4,6,9,13,14
B. oleracea	CC	4* (4,6)*** (i)	4+2 minor (i)		1,2,3,4,6,9,10,12,14
B. rapa	AA	6 (6-10)**	10 (i) (6,10)**	4 (4-8)**	1,2,4,6,9,12,14
B. carinata	BBCC	4 (i)	8 (8-10)**		1,6,9,11,14
B. juncea	AABB	10 (8,10)**	12-16 (i)	6 (4-6)**	1,6,9,11,14
B. napus	AACC	14* (10,12)**	12-14 (i)	6-8	1,2,5,6,7,8,9,10,11,12,14

1, Maluszynska and Heslop-Harrison 1993 (only 45S rDNA), 2, Snowdon et al. 1997a, 3, Armstrong et al. 1998, 4, Fukui et al. 1998 (only 45S rDNA), 5, Kamisugi et al. 1998, 6, Hasterok and Maluszynska 2000a (only 45S rDNA), 7, Schrader et al. 2000, 8, Snowdon et al. 2000a, 9 Hasterok et al. 2001, 10, Ziolkowski and Sadowski 2002, 11, Kulak et al. 2002, 12, Snowdon et al. 2002, 13, Ali et al. 2005, 14, Hasterok et al. 2006.

*Two 5S rDNA sites on the same chromosome.

** Polymorphism of loci number reported by Hasterok et al. (2006).

*** Polymorphism of loci number reported by Ziolkowski and Sadowski (2002).

(i), interstitial positions of rDNA exclusively; (t), terminal/subterminal positions of rDNA exclusively.

in the ribosome function. For this reason, reduction of unnecessary 5S or 45S rDNA can be beneficial for the newly formed polyploid (Dvorak 1990; Wendel 2000). However, opposite trends during polyploid formation are sometimes observed, as shown in cotton (Hanson et al. 1996) and *Sanguisorba* species (Mishima et al. 2002). In those species the trend towards the increase of rDNA loci number in newly formed polyploids can be explained in an analogous way, by a requirement to compensate a deficiency in rRNA production. This could be true especially in the cases, in which a polyploid plant tends to be larger and demonstrates a more rapid growth than its progenitor(s). However, spontaneous chromosomal rearrangements inside and between the rRNA gene loci cannot be also ruled out.

Comparative studies within the *Brassica* genus revealed that polymorphism in the number and chromosomal position of rDNA occurs among various subspecies, varieties, and even within a population of one cultivar or variety (Hasterok et al. 2006). A high level of polymorphism was especially found in the species containing the A genome, i.e. *B. rapa*, *B. juncea*, and *B. napus*, and it is supposed that this phenomenon is linked to extraordinary high numbers of both 5S and 45S rDNA loci in the A genome (Hasterok et al. 2006). The polymorphism observed may reflect differences in transcriptional activity of a particular rRNA gene loci among different forms of the species. These differences can be indirectly inferred from variation in length of secondary constrictions. Other examples of polymorphism are due to the significant variation in copy number of tandemly arrayed rRNA genes in a locus, e.g., one to three copies reported for 5S rDNA loci in *B. oleracea* (see Hasterok et al. 2006 for a review).

Interesting conclusions about 5S rDNA loci spreading in *Brassica* can be drawn from analysis of triplicated chromosomal block N (Ziolkowski et al. 2006; see also below). This chromosomal region is unique in the *A. thaliana* genome, but occurs in three copies in diploid *Brassica* genomes as a consequence of a whole-genome triplication (Lysak et al. 2005; Parkin et al. 2005; Ziolkowski et al. 2006). One of the three genomic copies in the *B. oleracea* genome contains two loci of 5S rRNA genes, at interstitial location (Ziolkowski et al. 2006). The whole block is close to another pericentromeric 5S rDNA locus, which is considered to be ancestral, based on the interspecies comparison (Hasterok et al. 2006). A most probable explanation of such a pattern is a small-scale transposition of the 5S rDNA locus from its original pericentromeric location to two sites within the adjacent block N by an unknown mechanism, possibly by rDNA-containing extrachromosomal circular DNA (eccDNA; Cohen et al. 2008). Further expansion of copy number was probably the result of unequal crossing-over (Fig. 6-6).

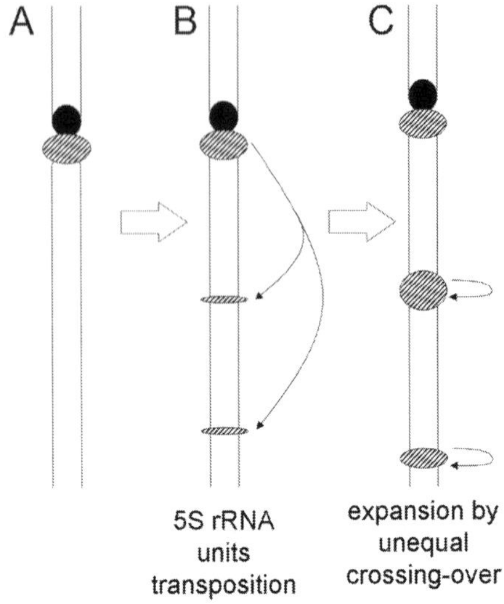

Figure 6-6 Presumed scenario of 5S rDNA amplification in the *B. oleracea* genome. (A) The original 5S rRNA gene locus (hatched oval) is adjacent to the centromere (black circle). (B) Transposition events form new loci within the adjacent chromosomal region, breaking its ancestral gene order. (C) Number of repeats within the new rDNA loci may increase due to unequal recombination events. The same mechanism may exist in other plant species (P.A. Ziolkowski and J. Sadowski).

6.7 Genome-Specific Repeats and Genomic in Situ Hybridization

That both tandem and dispersed TE repetitive sequences can exhibit a very variable level of the sequence similarity among inter-related taxa. DNA repeats shared among a group of species usually underline their close relatedness and common ancestry. On the contrary, the dynamic evolution and/or long-term genetic isolation of repetitive elements results in their sequence divergence, and the origin of genome- and taxon-specific repeats. Both opposed phenomena mark the evolution of repetitive elements in *Brassica* and are being used in cytogenetic and phylogenomic analyses of this genus (Lysak and Lexer 2006).

Genome-specific dispersed repetitive elements allow for a specific discrimination of individual genomes or chromosomes within composite genomes of allopolyploid species, interspecific hybrids, and introgression lines. In GISH experiments, typically genomic DNA (gDNA) of one genome/species is fluorescently labeled and hybridized to chromosomes of the composite genome in an excess of unlabeled gDNA or repeat-enriched (C_0t-1) DNA fraction of the second parental genome. The unlabeled gDNA hybridizes with common repeats in the probe gDNA, minimizing the potential intergenomic cross-hybridization. Alternatively, both parental gDNAs can be differentially labeled and hybridized simultaneously without the addition of blocking DNA.

As three important *Brassica* species (*B. carinata, B. juncea, B. napus*) have an allopolyploid origin (U 1935), GISH has been used frequently to discern parental genomes or individual chromosomes in these species, their derivates, and intergeneric hybrids between *B. napus* and other crucifer species. Parental genomes (A, B, and C) can be identified using GISH in the allotetraploid genomes of *B. carinata* (BBCC) and *B. juncea* (AABB) (Snowdon et al. 1997b; Hasterok et al. 2005a; Maluszynska and Hasterok 2005). However, donor genomes of *B. napus* (AACC) could not be clearly distinguished (Snowdon et al. 1997b; Hasterok et al. 2005a) due to the close phylogenetic relationship between *B. oleracea* (CC) and *B. rapa* (AA) (e.g., Lysak et al. 2005; Warwick and Sauder 2005). Contrary to these results, recently the C genome has been unequivocally identified in *B. napus* using labeled *B. oleracea* as a probe and approximately a 10-fold excess of unlabeled blocking DNA of *B. rapa* (Howell et al. 2008; Fig. 6-7). The reverse combination did not yield convincing results as the signals of labeled *B. rapa* DNA were overall weaker and cross-hybridization to pericentromic regions of *B. oleracea*-derived chromosomes has been observed. As crucifer species including *Brassica* species possess generally small genomes and correspondingly low amount of genome-specific repetitive elements (Lysak et al. 2009), GISH has a limited application in *Brassica* cytogenetics. Genome-

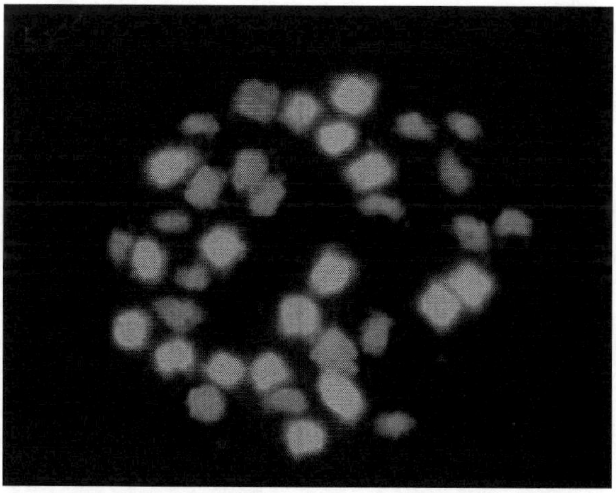

Figure 6-7 GISH in *Brassica napus*. GISH with labeled genomic DNA of *B. oleracea* (red) revealed 18 mitotic chromosomes corresponding to the C genome. Unlabeled DNA of *B. rapa* (A genome) and the intergenic spacer region of *B. oleracea* 45S rDNA were used as a blocking DNA. Chromosomes counterstained with DAPI. Courtesy of S. J. Armstrong (University of Birmingham).

Color image of this figure appears in the color plate section at the end of the book.

specific repeats are primarily clustered within pericentromeric regions, and therefore chromosome arms remain frequently unlabeled and potential intergenomic translocations undetected (e.g., Snowdon et al. 1997b; Wang et al. 2004). This spatially limited labeling was also observed by Howell et al. (2008) on A genome chromosomes in *B. napus* when gDNA of *B. rapa* has been used as a probe. Conversely, gDNA of *B. oleracea* labeled corresponding C-genome chromosomes homogeneously (see Fig. 6-7) to such an extent that a reciprocal translocation between A genome and C genome chromosomes could be detected (Howell et al. 2008). Different labeling efficiency of the two gDNA probes also suggests that *B. oleracea* chromosomes contain more and equally distributed genome-specific repetitive sequences than found in the chromosomes of *B. rapa*.

GISH was also extensively used to identify parental chromosomes in intergeneric hybrids and addition lines of *B. napus* and a number of other crucifer species including *A. thaliana* (Leino et al. 2004), *Crambe abyssinica* (Wang et al. 2004), *Eruca sativa* (Fahleson et al. 1997), *Lesquerella fendleri* (Skarzhinskaya et al. 1998), *Orychophragmus violaceus* (Hua et al. 2006; Zhao et al. 2008 and references therein), *Raphanus sativus* (Snowdon et al. 1997b), *R. raphanistrum* (Benabdelmouna et al. 2003), *Sinapis arvensis* (Snowdon et al. 2000a), and *S. alba* (Wang et al. 2005).

6.8 FISH with Repetitive Sequences Other Than rDNA

The chromosome organization of repetitive elements does not have only an important evolutionary significance but diverse classes of *Brassica* repeats serve genome and chromosome identification. In *Brassica* genomes, heterochromatic pericentromeric and centromeric regions are dominated by tandem (satellite) repeats, whereas transposable elements reside in these regions as well as along largely euchromatic chromosome arms.

Similar to other crucifer species (Lysak and Lexer 2006), satellite repeats are an abundant component of *Brassica* centromeres. In *B. rapa*, two 176-bp centromeric satellite repeats sharing ca. 80% sequence homology have been identified (Harrison and Heslop-Harrison 1995; Koo et al. 2004; Lim et al. 2005). The CentBr1 and CentBr2 satellites reside at eight and two centromeres, respectively (Lim et al. 2005). Both CentBr1 and CentBr2 were also found in *B. oleracea*, with the former and latter repeat localized to all and at least five centromeres, respectively, but not detected in the *B. nigra* genome (Lim et al. 2007). The expected number of chromosomes with CentBr1 and CentBr2 repeats corresponding to the donor (A and C) genomes was identified in *B. carinata*, *B. juncea*, and *B. napus* (Lim et al. 2007). In *B. nigra*, a 329-bp tandem repeat was shown to be specific for the B genome, and confirmed also on 16 B-genome chromosomes in the allotetraploid *B. juncea* (Schelfhout et al. 2004). The genome specificity of centromeric satellites further underlines closer phylogenetic relationship between A and C genomes (*B. rapa* and *B. oleracea*) as compared to the B genome (*B. nigra*) (Lysak et al. 2005; Warwick and Sauder 2005). Distal parts of *Brassica* chromosomes were explored much less than (peri)centromeric regions, and except the *Arabidopsis*-type telomeric sequence (Hasterok et al. 2005a), only a single telomere-like repeat has been in situ localized on *Brassica* chromosomes (dos Santos et al. 2007). The clone pBo1.6 isolated from *B. oleracea* was localized at chromosome termini as well as presumably interstitial positions in *B. oleracea*, whereas almost no hybridization to *B. rapa* chromosomes has been observed. In *B. napus*, the repeat labeled all C genome chromosomes and interestingly up to six chromosomes of the A genome.

Several TEs were identified in *Brassica* species and localized on chromosomes. Alix and Heslop-Harrison (2004) isolated a number of retrotransposons from the six species of the U's triangle, and analyzed chromosomal distribution of two Ty1-*copia*-like and one Ty3-*gypsy*-like retrotransposons in *B. oleracea* and *B. rapa* (Alix et al. 2005). One Ty1-*copia* probe hybridized along the full chromosome length with some interstitial clustering, whereas another *copia* element, one Ty3-*gypsy* probe, and an Athila-like *gypsy* element hybridized along entire chromosomes with stronger signals at (peri)centromeric regions. Centromere- and pericentromere-specific LTR retrotransposons of *B. rapa* were analyzed by Lim et al. (2007). The CRBs (centromeric retrotransposons of *Brassica*) were localized as nested within the centromeric CentBr satellite repeat in all species of the U's triangle. Peri-centromeric retrotransposons of *B. rapa* (PCRBr) are Ty3-*gypsy*-like elements specific for the A genome (localized on three *B. rapa* chromosomes) and absent in the B and C genomes. Surprisingly, the PCRBr probe hybridized to B and C genome chromosomes in *Brassica* allotetraploids (Lim et al. 2007). Alix et al. (2008) isolated and characterized a CACTA transposon *Bot1* (*B. oleracea* transposon 1) comprising ca. 2.3% of the *B. oleracea* genome. The *Bot1* is a C genome-specific transposon amplified in the genome of *B. oleracea* but not in *B. rapa*. In *B. napus*, the *Bot1* used as a FISH probe specifically labels 18 chromosomes of the C genome with only weak hybridization to *B. rapa*-derived chromosomes (Fig. 6-8). The C genome specificity and homogeneous chromosome distribution of this transposon can be probably linked with uniform GISH labeling pattern of the *B. oleracea* gDNA probe in *B. napus* (Howell et al. 2008; Fig. 6-6). *Bot1* has been found to be useful for cytogenetic identification of A and C genome chromosomes, respectively. A *B. oleracea* BAC clone containing *Bot1* was used to identify C genome chromosomes in inter-species AAC triploid hybrids (Leflon et al. 2006) and in *B. napus* haploids (Nicolas et al. 2007). Similarly, a *B. rapa* BAC containing an unknown repetitive sequences labeled specifically three pairs of A genome chromosomes in AAC hybrids (Leflon et al. 2006).

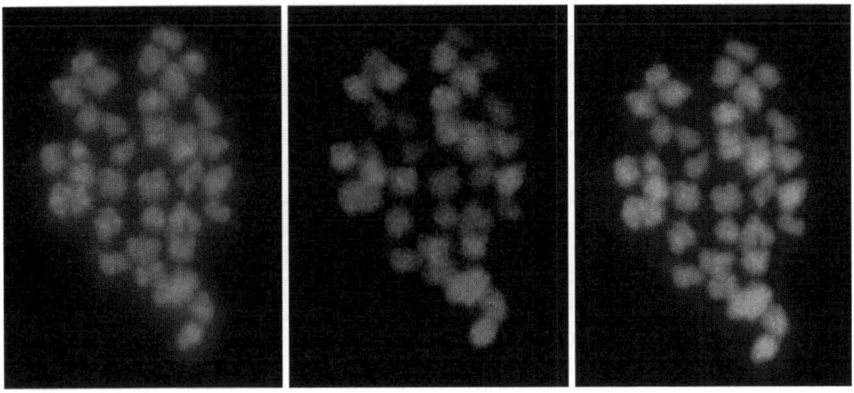

Figure 6-8 Fluorescence in situ hybridization of a CACTA transposon *Bot1* to metaphase chromosomes of *B. napus* cv. "Yudal". The *Bot1* probe (red) labels 18 chromosomes corresponding to the C genome; only weak hybridization to 20 chromosomes of the A genome is apparent. Chromosomes counterstained with DAPI and shown in blue (left) or gray (right). Adapted from Alix et al. (2008) with permission.

Color image of this figure appears in the color plate section at the end of the book.

6.9 BAC FISH and Chromosome Painting

The specific structure of crucifer genomes with most repeats clustered in discrete heterochromatic arrays and well developed genomic resources including chromosome-specific libraries of high-capacity DNA vectors were the main essentials for establishing BAC FISH and chromosome painting in Brassicaceae. There are several types of large-insert DNA vectors available [BACs, yeast artificial chromosomes (YACs), P1-derived artificial chromosomes (PACs), and transformation-competent artificial chromosomes (TACs)], however, only bacterial artificial chromosomes (BACs) became widely used as DNA probes in plant cytogenetics. Without making a clear-cut distinction, BAC FISH generally refers to in situ localization of individual or several BAC clones, whereas chromosome painting applies to visualization of large chromosome regions such as chromosome arms or whole chromosomes using BAC contigs, i.e., continuous sequences of overlapping or non-overlapping BACs.

In *Brassica* cytogenetics, BAC FISH has been mainly used for chromosome identification and integration of genetic and cytogenetic maps. Several BAC libraries of three *Brassica* species (*B. oleracea*, *B. rapa*, and *B. napus*) are available (see The Multinational *Brassica* Genome Project, *http://www.brassica.info*). BAC clones of the two diploid species have been applied most frequently as cytogenetic markers. Howell et al. (2002) assigned nine genetic linkage groups of *B. oleracea* to corresponding mitotic chromosomes using chromosome-specific BAC clones (see Section 5 for details). More recently, FISH of three chromosome-specific BACs followed by GISH was used to characterize a reciprocal translocation in *B. oleracea* (Howell et al. 2008). BAC FISH has been extensively applied in the integration of genetic and cytogenetic maps within the *Brassica rapa* Genome Project (*http://www.brassica-rapa.org*). Chromosome/linkage-specific *B. rapa* BACs have been identified by unique expressed sequence tag (EST) clones and other genetic markers, and mapped to pachytene chromosomes by FISH (Yang et al. 2005; Lim et al. 2006). *B. rapa* BACs can also be localized on mitotic metaphase chromosomes as shown for BACs containing paralogs of the *FLOWERING LOCUS C* gene (Yang et al. 2006). In on-going *Brassica* sequencing projects, BAC FISH will be used as an important tool anchoring the sequenced clones and contigs on chromosomes (see also Chapter 7).

Comparative cytogenetic studies of non-repetitive chromosome regions are based on cross-species FISH of BACs derived from one species to chromosomes of another related species. In cytogenetics of Brassicas, *Arabidopsis* BACs and BAC contigs were exclusively used to reveal cross-species chromosome homeology. Comparative chromosome painting (CCP) and BAC FISH in *Brassica* species serve practical as well as more theoretical objectives. FISH with *Arabidopsis* BAC clones revealing homeologous chromosome regions in *Brassica* genomes can expedite the chromosome landing of *Brassica* BACs and the construction of cytogenetic maps. As large-scale CCP discerns the extent of interspecies chromosome homeology, valuable insights into karyotype and whole-genome evolution can be gained from such analyses. The first comparative BAC FISH in *Brassica* has been shown by Jackson et al. (2000) when a short contig of six *Arabidopsis* BACs was localized on DNA fibers and four to six mitotic chromosomes of *B. rapa*. Hybridization of *Arabidopsis* BAC clones to multiple chromosomes was also observed in *B. oleracea* as two partly overlapping *Arabidopsis* BACs hybridized to two different bivalents (Ziolkowski and Sadowski 2002), and 12 BAC clones were found segmentally duplicated on chromosome O6 (Howell et al. 2005). All three studies suggested that genomes of both the *Brassica* species were duplicated as compared to the genome of *A. thaliana*. Nevertheless, only three other large-scale CCP studies demonstrated the nature of the whole-genome duplication characterizing *Brassica* species. Lysak et al. (2005) analyzed the structure of a ~ 8-Mb *Arabidopsis* BAC contig in 21 species classified as members of the tribe Brassiceae including the six species of the U-triangle. As most species analyzed possessed three or six homeologous copies of the *Arabidopsis* contig, the data indicated that the *Brassica* genome duplication has to be specified as a whole-genome triplication proposed earlier by Lagercrantz (1998) and more recently by Parkin et al. (2005). In agreement with the phylogenetic analysis, revealed chromosome homeologies advocated the existence of a common hexaploid ancestor for the whole tribe Brassiceae. The 8 to 15 million years old whole-genome triplication has been followed by genome diploidization manifested by chromosome number reduction accompanied by various chromosome rearrangements (Lysak et al. 2005). These findings have been corroborated by a CCP analysis of another *Arabidopsis* BAC contig in 10 species traditionally treated as members of Brassiceae (Lysak et al. 2007). Three copies of the analyzed chromosome segment found in the Brassiceae species further supported the hexaploid ancestry and monophyletic origin of the tribe. Ziolkowski et al. (2006) analyzed conservation of ancient rearrangement breakpoints between *A. thaliana* and *B. oleracea* genomes using three *Arabidopsis* BAC contigs spanning a total of 8.25 Mb. Also this work confirmed the triplication event preceding the species divergence in *Brassica* and Brassiceae (Fig. 6-9). Further CCP studies are discussed in Section 6.11.

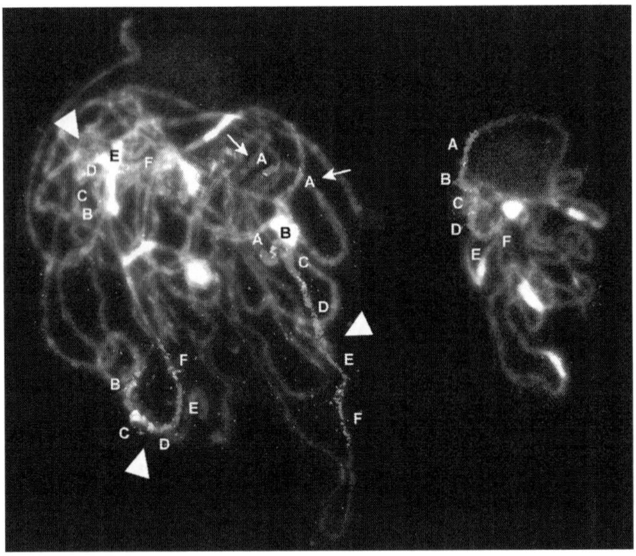

Figure 6-9 An example of comparative chromosome painting in *B. oleracea*. A 5.4 Mb long Arabidopsis BAC contig from chromosome 3 divided into six differentially labeled segments, was hybridized to pachytene chromosomes of *A. thaliana* (right) and *B. oleracea* (left). Whilst the BAC contig reveals a single chromosome region in *A. thaliana*, three copies of the contig were found in *B. oleracea* (arrowheads). Bioinformatic and genetic mapping studies suggest that the arrangement of A-F segments observed in *A. thaliana* has ancestral-like structure [(Henry et al. 2006, also data from *http://chibba.agtec.uga.edu/duplication/index/home* (Tang et al. 2008)]. This is observed in one of the three *B. oleracea* copies, as well. However, segment A was separated from segments B-F in the two other copies (dissociated A segments indicated by arrows). Modified from Ziolkowski et al. (2006).

Color image of this figure appears in the color plate section at the end of the book.

6.10 Whole-Genome Duplications in the Evolution of Crucifer Genomes

In the last decade the major progress in studies on genome evolution of Brassicaceae species was enabled due to the *Arabidopsis* genome sequencing project (AGI 2000). One of the most important reasons for choosing *A. thaliana* as a model plant species was a belief that its small genome (~ 157 Mbp; Bennett et al. 2003) built from five chromosomes, has a typical diploid-like structure. Thus, an intriguing result of genome sequencing was the discovery that the *Arabidopsis* genome is a mosaic of duplicated chromosomal segments (AGI 2000; Blanc et al. 2000; Paterson et al. 2000; Vision et al. 2000). This initial finding has led to an upsurge in bioinformatic studies uncovering the ancient evolutionary history of the *Arabidopsis* genome. At present, new data from genome sequencing projects in *Oryza sativa* (rice), *Populus trichocarpa* (poplar), *Vitis vinifera* (grapevine), *Carica papaya* (papaya) and *Sorghum bicolor* (sorghum) complement the data on the *Arabidopsis* genome. Such studies are based mainly on finding and comparing collinear regions (chromosomal regions with a similar gene content and order) within a genome and/or between different genomes. As it is unlikely that a pair of collinear regions could originate independently, it is assumed that it has a common origin, and was derived from independent segmental duplications and/or whole-genome duplication (WGD) events. In plants, WGDs are much more widespread than large segmental duplications, and played a pivotal role in the course of evolution (Adams and Wendel 2005). Most evolutionary events uncovered by mining the *Arabidopsis* genome are shared also by *Brassica* species, as they predate the divergence of the *Arabidopsis* and *Brassica* lineages.

Currently it is widely accepted that there were three WGDs during the evolution of the *Arabidopsis* genome (named γ, β, and α, or 1R, 2R and 3R, starting from the oldest one; Simillion et al. 2002, Blanc et al. 2003, Bowers et al. 2003). None of these events can be detected cytogenetically (i.e. by BAC FISH or chromosome painting) due to the significant structural divergence of duplicated chromosomal regions (Lysak et al. 2001). Recent comparisons of the *Arabidopsis* and poplar genome sequences with those of grapevine and papaya (Jaillon et al. 2007; Ming et al. 2008) demonstrated that the γ WGD was a hexaploidization event, and took place in a common ancestor of *A. thaliana*, *C. papaya*, *P. trichocarpa*, and *V. vinifera* (Jaillon et al. 2007). A similar comparison of sequence data from rice and sorghum showed that the γ event occurred in a lineage leading to Rosids after the separation from grasses (Jaillon et al. 2007; Paterson et al. 2009; Fig. 6-10).

Comparisons of the *Arabidopsis* genome architecture with those of poplar, papaya and grapevine also shed more light on the rate of chromosome rearrangements within the Brassicaceae lineage. It can be concluded that the extant *Arabidopsis* genome, and hence all the *Brassica* genomes, are much more rearranged, than those of *V. vinifera* and *C. papaya*, while the *P. trichocarpa* genome represents an intermediate situation (Jaillon et al. 2007; Tang et al. 2008). At the level of gene collinearity, only remnants of the ancestral gene content and order could be identified in the *Arabidopsis* genome. Moreover, not only the frequency of rearrangements was significantly higher in the Brassicaceae ancestor (Tang et al. 2008), but also the gene content and order were much more eroded by a frequent gene loss, than in grapevine and papaya (Thomas et al. 2006). Both these phenomena are due to additional rounds of WGD followed by extensive diploidization process in the Brassicaceae family (Adams and Wendel 2005).

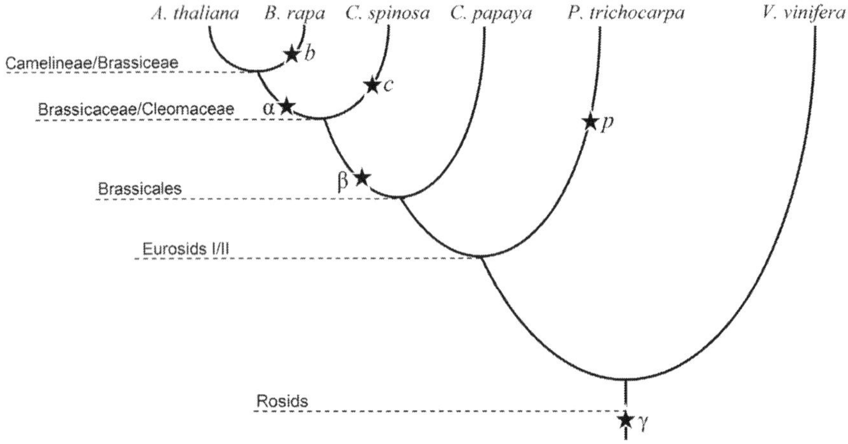

Figure 6-10 Described WGD events positioned on a phylogenetic tree in the evolution of selected dicotyledonous plants. WGDs are represented as stars and refer to the hexaploidization γ shared by all Rosids, WGDs α and β within the Brassicales, the salicoid duplication *p* in the poplar lineage, and hexaploidization events *c* and *b* in Cleomaceae and Brassiceae, respectively. Branch lengths do not correspond to evolutionary distances.

Two more recent WGDs (β and α) described in *Arabidopsis* were presumably tetraploidizations (Blanc et al. 2003; Bowers et al. 2003; Simillion et al. 2003; Fig. 6-10). Recent intergenomic comparisons suggest that both the WGD events took place in Brassicaceae after the separation from the papaya lineage (Ming et al. 2008). Relatively well conserved gene collinearity between segments of β and α duplications suggests that both events occurred within a short interval (Ziolkowski et al. 2003). It has been estimated that β and α duplications cover 52% and 89% of the genome, respectively (Bowers et al. 2003), and only pericentromeric regions were not found to be duplicated, being too eroded by the activity of TEs (Blanc et al. 2003).

Several approaches have been applied for dating β and α WGDs (e.g., Blanc et al. 2003; Bowers et al. 2003; Maere et al. 2005; Schranz and Mitchell-Olds 2006). Most estimates were based on a comparison of the level of synonymous (silent) substitutions between genes duplicated through successive WGDs in *Arabidopsis*, and also with orthologous genes (homologous genes that originated through speciation) from other plant species. Both β and α events took place after the *Arabidopsis*–papaya divergence [~72 million years ago (Mya); Wikström et al. 2001], as there is no evidence for the two events in *C. papaya* (Ming et al. 2008). More research is needed to establish the position of the β WGD within Brassicales. On the other hand, data on the position of the α event is more accurate, as it was established that the event occurred shortly after the divergence of two sister families, Brassicaceae and Cleomaceae (represented by *Cleome spinosa*; Schranz and Mitchell-Olds 2006), some 24 to 40 Mya (Blanc et al. 2003, 2004). Therefore, it has been suggested that the α event is specific for Brassicaceae, however, more effort is needed to determine when the event occurred during the evolution of the family (Schranz and Mitchell-Olds 2006; Fig. 6-10). A detailed investigation of the genus *Aethionema*, which diverged early from the rest of the family (Hall et al. 2002) seems to be of major importance (Schranz and Mitchell-Olds 2006).

Even early cytogenetic studies suggested that present-day diploid *Brassica* genomes exhibit intragenomic chromosome homeology and duplications, and thus represent "balanced secondary polyploids" (e.g., Catcheside 1934; Thompson 1956; Röbbelen 1960). Comparative genetic mapping among *B. nigra*, *B. oleracea* and *B. rapa* (Lagercrantz and Lydiate 1996), and between *A. thaliana* and *B. oleracea* suggested an ancient hexaploidization as a cause of the intragenomic duplications observed in *Brassica* genomes (Lagercrantz 1998; Babula et al. 2003). An alternative hypothesis referred to as "cyclic amphiploidy" proposed that highly duplicated *Brassica* genomes originated from various hybridizations among different species with a genome based on $x = 4$ and/or $x = 5$ (Truco et al. 1996; Quiros 1999). However, a wealth of data coming from different research fields speaks for the hexaploidy hypothesis (i.e., Cavell et al. 1998; O'Neill and Bancroft 2000; Babula et al. 2003; Rana et al. 2004; Lysak et al. 2005, 2007; Parkin et al. 2005; Yang et al. 2005; Town et al. 2006; Ziolkowski et al. 2006). The whole-genome triplication has been also corroborated by comparative chromosome painting with large *Arabidopsis* BAC contigs as discussed in Section 6.9 (Lysak et al. 2005, 2007; Ziolkowski et al. 2006; Fig. 6-9). Moreover, CCP analysis of a number of Brassiceae species revealed that the polyploidization event is common for the whole tribe, and may serve as an apomorphic character defining the tribe (Lysak et al. 2005).

6.11 Chromosome Evolution in Brassiceae and Brassicaceae

To reconstruct the chromosomal evolution across the Brassicaceae, it is crucial to distinguish between derived and ancestral genome organization. Gene-based phylogenies enable us to estimate evolutionary distances and divergence times between individual species and species groups, and in consequence help to position genome alteration events to particular nodes within a phylogenetic tree. The 338 genera of the Brassicaceae family have been recently classified into 25 tribes based on nuclear- and chloroplast-encoded markers (Al-Shehbaz et al. 2006; Bailey et al. 2006; Beilstein et al. 2006, 2008). Sixteen of the 25 tribes can be further grouped into one of three major crucifer lineages (referred to as Lineages I-III; Al-Shehbaz et al. 2006; Beilstein et al. 2006, 2008). This treatment provides the most up to date reference for comparative studies in Brassicaceae, although it is being continuously modified as new data become available (e.g. German et al. 2008).

Chromosomal studies in Brassicaceae are hampered due to the high frequency of polyploidy. Polyploidy has played an important role in the evolution of the Brassicaceae family (Marhold and Lihova 2006), and nearly 37% of the species are polyploid (Warwick and Al-Shehbaz 2006). Up to date chromosome numbers have been reported for more than 1,500 crucifer species (summarized by Warwick and Al-Shehbaz 2006). Although a continuous series of base chromosome numbers from $x = 4$ to $x = 13(17)$ exists, a surprisingly high percentage (37%) of genera appears to be based on $x = 8$ (Warwick and Al-Shehbaz 2006). This finding led to a hypothesis that an ancestral chromosome number in the family, and particularly in Lineage I, was $n = 8$ (Koch et al. 1999; Henry et al. 2006; Lysak et al. 2006; Schranz et al. 2006).

Investigation of the chromosomal evolution within the tribe Brassiceae has progressed significantly due to the comparisons with the *Arabidopsis* genome ($n = 5$). *A. thaliana* is the best studied plant genome furnished with a wealth of genomic resources, and close enough phylogenetically to the Brassiceae. However, it is now well documented that the *Arabidopsis* genome experienced reshuffling coupled with chromosome number reduction from an ancestral $n = 8$ genome (Koch et al. 1999; Henry et al. 2006; Lysak et al. 2006). The reshuffled nature of the *Arabidopsis* genome makes any interspecies comparison difficult as it must account for these species-specific rearrangements (Schranz et al. 2006). To facilitate further intrafamily studies, a tentative Ancestral Crucifer Karyotype (ACK) has been recently proposed (Lysak et al. 2006; Schranz et al. 2006).

The ACK concept has been based on comparative genetic maps of *Arabidopsis lyrata* and *Capsella rubella* (both $n = 8$) constructed using *Arabidopsis* genetic markers (Boivin et al. 2004; Kuittinen et al. 2004; Yogeeswaran et al. 2005). This three-way comparison showed that both *A. lyrata* and

C. rubella possess genomes largely collinear with that of *A. thaliana*, and all three species share large conserved chromosomal blocks (Boivin et al. 2004; Kuittinen et al. 2004; Yogeeswaran et al. 2005). Moreover, genomes of *A. lyrata* and *C. rubella* are almost identical with respect to the gene order (Koch and Kiefer 2005), and thus are believed to resemble an ancestral $n = 8$ genome (Lysak et al. 2006; Schranz et al. 2006). This hypothesis was further tested by CCP approach (Lysak et al. 2006), when *Arabidopsis* BAC contigs were arranged according to the genetic map of *A. lyrata*/*C. rubella*, fluorescently labeled and hybridized to pachytene chromosomes of five Brassicaceae species with $n = 5$-8. The study corroborated results of genetic mapping, provided new insights as to the centromere positions within the ancestral karyotype, and reconstructed species-specific karotype reshuffling of the ancestral karyotype in the five species (Lysak et al. 2006).

Another important step towards the construction of ACK was a comprehensive genetic map of the *B. napus* genome (Parkin et al. 2005). The authors placed 1,327 genetic loci corresponding to *A. thaliana* genes on the 19 linkage groups of *B. napus*. Knowledge about the exact position of the analyzed markers within the *Arabidopsis* genome made it possible to infer corresponding collinear regions within the rapeseed map. This study revealed 21 collinear blocks shared by *A. thaliana* and *B. napus*, which cover 90% of the latter genome. As the amphidiploid *B. napus* genome originated from an interspecific cross between *B. oleracea* ($n = 9$) and *B. rapa* ($n = 10$) (U 1935), structure of the two subgenomes has been reconstructed. Up to date, this is the most comprehensive comparative mapping of *Brassica* genomes (Schranz et al. 2006). The data obtained by Parkin et al. (2005) together with those from comparative mapping in *A. lyrata* and *C. rubella* (Boivin et al. 2004; Kuittinen et al. 2004; Yogeeswaran et al. 2005), and the reconstruction of karyotype evolution in $n = 5$-8 crucifers (Lysak et al. 2006) have been used to propose the model of ancestral karyotype (Schranz et al. 2006). The ACK consisting of eight ancestral chromosomes (AK1-8) and 24 conserved genomic blocks (A-X; see Schranz et al. 2006 for details) has become an important reference genome facilitating the interpretation of cross-species chromosome and genome collinearity across Brassicaceae (see also Chapter 8).

In crucifer cytogenetics, major progress in tracking evolutionary changes of the ACK has been achieved with the use of the CCP approach, whereby *Arabidopsis* BAC contigs are arranged correspondingly to conserved genomic blocks and then used as chromosome painting probes. Although the study of Lysak et al. (2006) can be considered as the first intentional use of the ACK in CCP, the interspecies collinearity of some genomic blocks was explored in some earlier CCP analyses (Lysak et al. 2003, 2005; Ziolkowski et al. 2006). Lysak et al. (2005) analyzed a large chromosome segment (~ 8.7 Mb in *A. thaliana*) corresponding to a major part of block U of the ACK in 21 species assigned to the Brassiceae. This study allowed the time of the polyploidization event to be estimated by comparing CCP results with the phylogeny based on the chloroplast *trnL* intron/*trn*L-F intergenic spacer region. It was shown that members of *Calepina, Conringia* (both genera tentatively assigned to Brassiceae), and *Sisymbrium* had only one copy of block U, while in the remaining Brassiceae taxa of the tribe Brassiceae three or six copies were revealed. As the genus *Sisymbrium* has been separated from the Brassiceae lineage and from the genus *Conringia* ~ 14.6 Mya (Koch et al. 2001), and the *Nigra* and *Rapa/Oleracea* clades within the Brassiceae lineage diverged ~ 7.9 Mya (Koch et al. 2001), it was concluded that the hexaploidization took place between 7.9 and 14.6 Mya (Lysak et al. 2005). Evidence for the triplication was also obtained for block F, which was studied by CCP within 10 Brassiceae species, including *B. oleracea* and *B. rapa* (Lysak et al. 2007), and block N analyzed in *B. oleracea* (Ziolkowski et al. 2006). Triplicated chromosomal blocks differ significantly in the gene content, as gene loss proceeds independently in each paralogous copy (Mayer et al. 2001; Rana et al. 2004). Comparison of triplicated blocks in *Brassica* with their homeologs in *A. thaliana* reveals that the gene loss is much more extensive in *Brassica* (Mayer et al. 2001).

Recently, CCP using *Arabidopsis* BAC contigs arranged according to the 24 ancestral genomic blocks of the ACK was applied to reconstruct karyotype evolution in eight $x = 7$ species belonging to six tribes from the phylogenetic neighborhood of the Brassiceae (Mandáková and Lysak 2008). Patterns of interspecies chromosome collinearity allowed for the reconstruction of an ancestral Proto-Calepineae Karyotype (PCK; $n = 7$), which was most likely shared by all $x = 7$ species analyzed. A detailed analysis of large-scale chromosome rearrangements was carried out, resulting in a scenario of genome evolution leading to the present-day karyotypes (Fig. 6-11). Considering the current phylogenetic relationship among the tribes and the Brassiceae (Al-Shehbaz et al. 2006; Koch and Al-Shehbaz 2009), one can conclude that the PCK karyotype could apply also to the tribe Brassiceae. Comparing the CCP patterns of the $x = 7$ species with the genome structure of *B. napus* (Parkin et al. 2005), the PCK presumably represents an ancestral karyotype of Brassiceae prior to the whole-genome triplication.

Information from ACK is also used to investigate mechanisms responsible for changes in genome structure, especially chromosome fusions, which are frequent in Brassicaceae (Lysak et al. 2006). Based on reconstruction of karyotype evolution in *A. thaliana, A. lyrata, Neslia paniculata, Turritis glabra,* and *Hornungia alpina*, a general model of chromosome number reduction was proposed (Fig. 6-12): First, a pericentric inversion converts a metacentric chromosome into an acrocentric one. Then, a reciprocal translocation occurs between the centric end of the acrocentric chromosome and a telomeric end of another chromosome. In consequence, two novel chromosomes are formed: one large "fusion" chromosome and a minichromosome. As the minichromosome is unstable during meiosis, it can be eliminated without any significant loss of genetic information, and the chromosome number decreases (Fig. 6-12). This rearrangement could have been involved in chromosome number reduction in *Brassica* and other Brassiceae taxa, following the ancient whole-genome triplication.

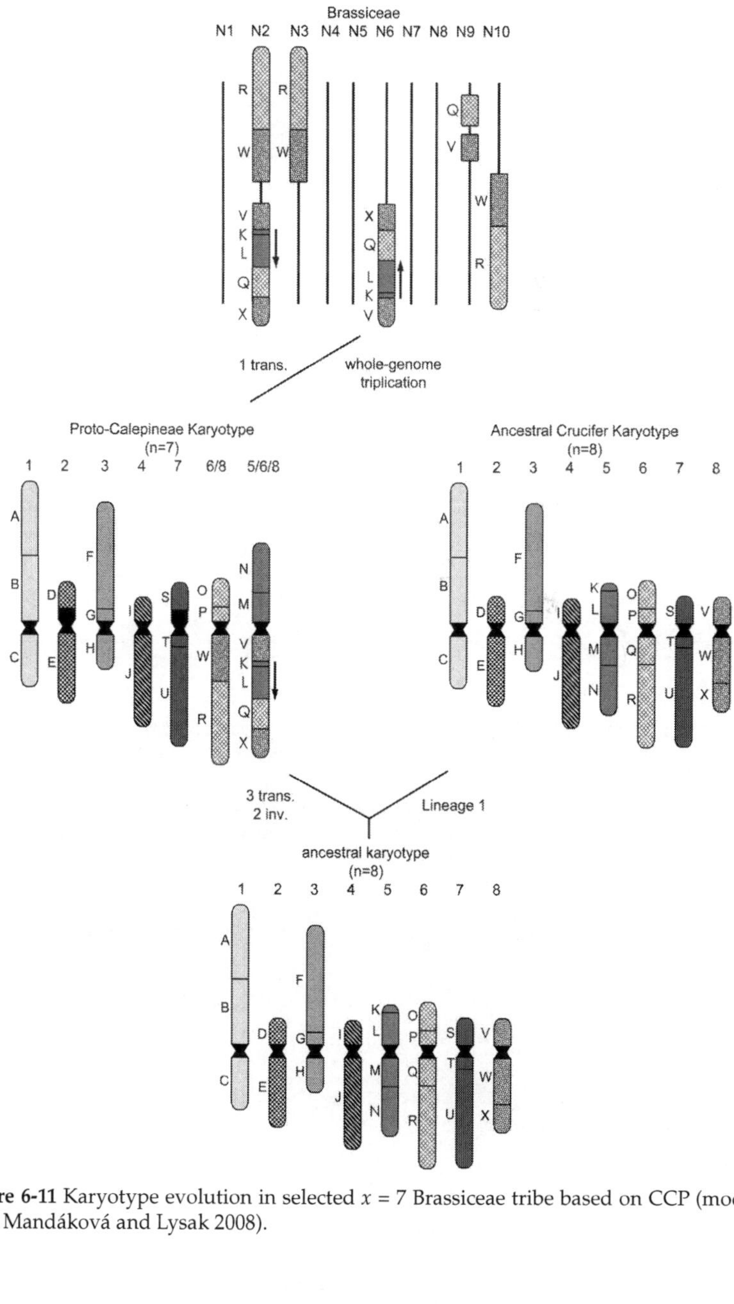

Figure 6-11 Karyotype evolution in selected $x = 7$ Brassiceae tribe based on CCP (modified from Mandáková and Lysak 2008).

Up to date, the fate of the triplicated ancestral genome in the extant *Brassica* species was not resolved using CCP. Some data come from genetic mapping, although high level of internal genome duplication, numerous rearrangements and selective gene loss within duplicated segments hamper certain evaluation of karyotype changes. It is widely accepted that the tribe Brassiceae has monophyletic origin, thus hexaploidization took place as a result of hybridization of closely related ancestral species. Differences in rearrangement breakpoints conservation among the three copies within the *B. oleracea* genome (see Fig. 6-9) may suggest that the polyploidization proceeded by at least two steps, separated in time, and that *Brassica* hexaploid progenitor was formed by hybridization of a tetraploid genome and an ACK-like diploid genome (Ziolkowski et al. 2006; Fig. 6-13). This hypothesis needs however much more experimental research to be confirmed.

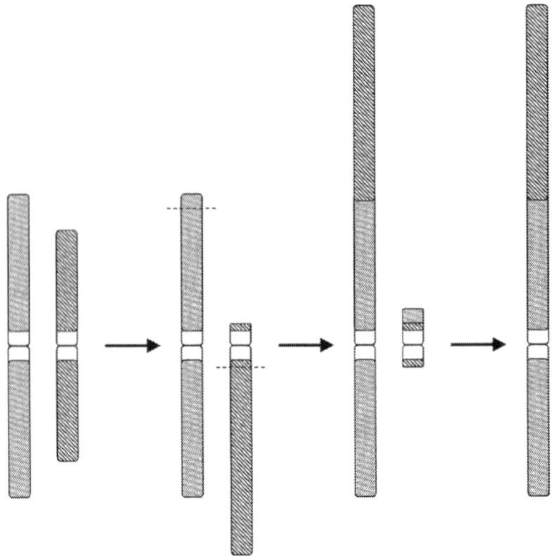

Figure 6-12 The mechanism of chromosome number reduction via a pericentric inversion and reciprocal translocation events as proposed by Lysak et al. (2006). Centromeres are white; dashed lines show breakpoints of chromosome rearrangements. Modified from Schranz et al. (2006) with permission.

Some of the important points describing the *Brassica* genome evolution can be drawn from cytogenetic studies. First of all, it is suggested that *Brassica* species separated within the family due to the whole-genome triplication (Lysak et al. 2005). As the disomic inheritance is more favorable than tetra- or hexasomic, genome diploidization has occurred. This process embraced point mutations, small-scale rearrangements as well as large chromosome aberrations (Song et al. 1995). At the chromosomal level, reciprocal translocations and inversions were the most important rearrangements, as they do not cause any genetic imbalances; these events were extensively documented by both genetic mapping and CCP studies. However, further studies are needed to reveal chromosome rearrangements, which took place during the speciation in *Brassica*. It is likely that the extant amphidiploid *Brassica* species will also evolve through diploidization within the next few millions years, in a similar way as the hexaploid ancestor of the tribe.

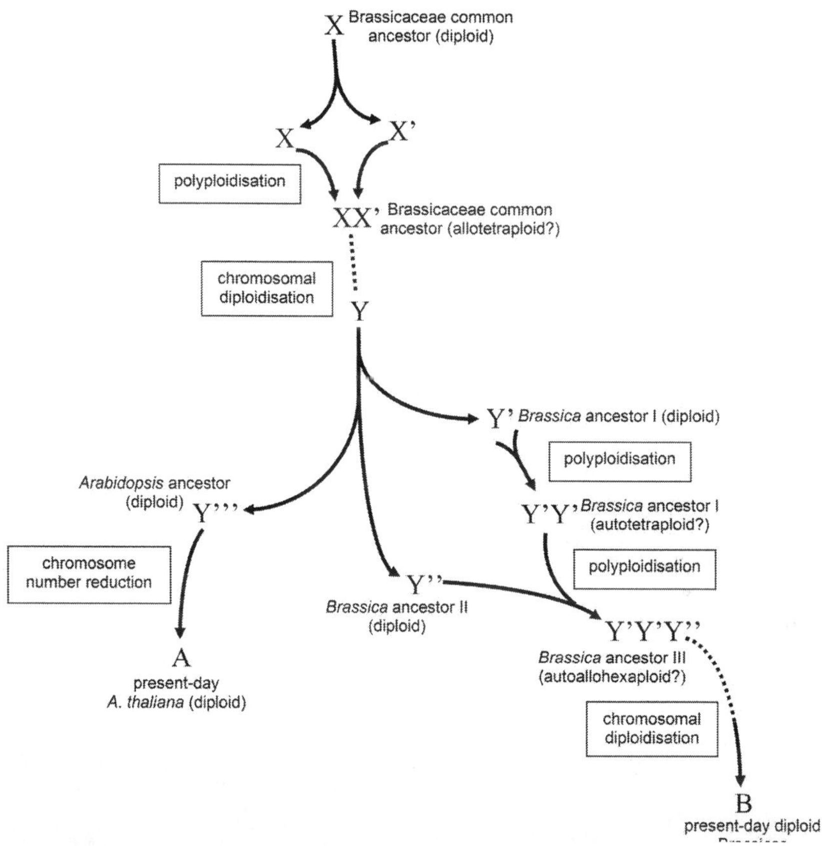

Figure 6-13 A diagram illustrating the hypothetical genome evolution of diploid *Brassicas* via hexaploidization. Tetraploid (ancestor I) and diploid (ancestor II) ancestors are assumed, respectively. Both genomes formed a hexaploid ancestor, which gave rise to the present-day diploid Brassicas. The length of the arrows does not indicate evolutionary distances but only evolutionary relations (modified from Ziolkowski et al. 2006).

Acknowledgements

The authors wish to thank Robert Hasterok for critical reading of the manuscript. This work was partly supported by research grants from the Polish State Committee for Scientific Research (grant nos. PBZ-MNiSW-2/3/2006/19 and N/N303/313437), the Grant Agency of the Czech Academy of Science (grant no. IAA601630902) and the Czech Ministry of Education (grant no. MSM0021622415).

References

Adams KL, Wendel JF (2005) Polyploidy and genome evolution in plants. Curr Opin Plant Biol 8: 135–141.

AGI, the Arabidopsis Genome Initiative (2000) Analysis of the genome sequence of the flowering plant *Arabidopsis thaliana*. Nature 408: 796–815.

Al-Shehbaz IA, Beilstein MA, Kellogg EA (2006) Systematics and phylogeny of the Brassicaceae (Cruciferae): An overview. Plant System Evol 259: 89–120.

Alfenito MR, Birchler JA (1993) Molecular characterization of a maize B chromosome centric sequence. Genetics 135: 589–597.

Ali HB, Lysak MA, Schubert I (2005) Chromosomal localization of rDNA in the Brassicaceae. Genome 48: 341–346.

Alix K, Heslop-Harrison JS (2004) The diversity of retroelements in diploid and allotetraploid *Brassica* species. Plant Mol Biol 54: 895–909.

Alix K, Ryder C, Moore J, King GJ, Heslop-Harrison JS (2005) The genomic organization of retrotransposons in *Brassica oleracea*. Plant Mol Biol 59: 839–851.

Alix K, Joets J, Ryder CD, Moore J, Barker GC, Bailey JP, King GJ, Heslop-Harrison JS (2008) CACTA transposon Bot1 played a major role in *Brassica* genome divergence and gene proliferation. Plant J 56: 1030–1044.

Armstrong SJ, Fransz P, Marshall DF, Jones GH (1998) Physical mapping of DNA repetitive sequences to mitotic and meiotic chromosomes of *Brassica oleracea* var. *alboglabra* by fluorescence *in situ* hybridisation. Heredity 81: 666–673.

Arumuganathan K, Earle ED (1991) Nuclear DNA content of some important plant species. Plant Mol Biol 9: 208–218.

Babula D, Kaczmarek M, Barakat A, Delseny M, Quiros CF, Sadowski J (2003) Chromosomal mapping of *Brassica oleracea* based on ESTs from *Arabidopsis thaliana*: complexity of the comparative map. Mol Genet Genom 268: 656–665.

Bailey CD, Koch MA, Mayer M, Mummenhoff K, O'Kane SL, Warwick SI, Windham MD, Al-Shehbaz IA (2006) Toward a global phylogeny of the Brassicaceae. Mol Biol Evol 23: 2142–2160.

Banga SS (1988) C-genome chromosome substitution lines in *Brassica juncea* (L.) Coss. Genetica 77: 81–84.

Beilstein MA, Al-Shehbaz IA, Kellogg EA (2006) Brassicaceae phylogeny and trichome evolution. Am J Bot 93: 607–619.

Beilstein MA, Al-Shehbaz IA, Mathews S et al. (2008) Brassicaceae phylogeny inferred from phytochrome A and ndhF sequence data: tribes and trichomes revisited. Am J Bot 95: 1307–1327.

Benabdelmouna A, Guéritaine G, Abirached-Darmency M, Darmency H (2003) Genome discrimination in progeny of interspecific hybrids between *Brassica napus* and *Raphanus raphanistrum*. Genome 46: 469–472.

Bennett MD, Smith JB (1976) Nuclear DNA amounts in angiosperms. Phil Trans R Soc Lond B Biol Sci 274: 227–274.

Bennett MD, Smith JB (1991) Nuclear DNA amounts in angiosperms. Phil Trans R Soc Lond B Biol Sci 334: 309–345.

Bennett MD, Leitch IJ. (2005) Genome size evolution in plants. In: TR Gregory (ed) The Evolution of the Genome. Elsevier, San Diego, CA, USA, pp 89–162.

Bennett MD, Smith JB, Heslop-Harrison JS (1982) Nuclear DNA amounts in angiosperms. Phil Trans R Soc Lond B Biol Sci 216: 179–199.

Bennett MD, Leitch IJ, Price HJ, Johnston JS (2003) Comparisons with *Caenorhabditis* (~100 Mb) and *Drosophila* (~ 175 Mb) using flow cytometry show genome size in Arabidopsis to be ~157 Mb and thus ~25 % larger than the Arabidopsis Genome Initiative estimate of ~125 MB. Ann Bot 91: 547–557.

Bennetzen JL (2005) Transposable elements, gene creation and genome rearrangement in flowering plants. Curr Opin Genet Dev 15: 621–627.

Blanc G, Wolfe KH (2004) Widespread paleopolyploidy in model plant species inferred from age distributions of duplicate genes. Plant Cell 16: 1667–1678.

Blanc G, Barakat A, Guyot R, Cooke R, Delseny M (2000) Extensive duplication and reshuffling in the Arabidopsis genome. Plant Cell 12: 1093–1101.

Blanc G, Hokamp K, Wolfe KH (2003) A recent polyploidy superimposed on older large-scale duplications in the Arabidopsis genome. Genome Res 13: 137–144.

Bohuon EJR, Keith DJ, Parkin IAP, Sharpe AG, Lydiate DJ (1996) Alignment of the conserved C genomes of *Brassica oleracea* and *Brassica napus*. Theor Appl Genet 93: 833–839.

Boivin K, Acarkan A, Mbulu RS, Clarenz O, Schmidt R (2004) The Arabidopsis genome sequence as a tool for genome analysis in Brassicaceae. A comparison of the Arabidopsis and *Capsella rubella* genomes. Plant Physiol 135: 735–744.

Bothmer R von, Gustafsson M, Snogerup S (1995) *Brassica* sect. *Brassica* (Brassicaceae). II. Inter- and intraspecific crosses with cultivars of *B. oleracea*. Genet Resour Crop Evol 42: 165–178.

Bowers JE, Chapman BA, Rong J, Paterson AH (2003) Unravelling angiosperm genome evolution by phylogenetic analysis of chromosomal duplication events. Nature 422: 433–438.

Catcheside DG (1934) The chromosomal relationship in the Swede and turnip groups of *Brassica*. Ann Bot Lond 601: 33–56.

Cavell AC, Lydiate DJ, Parkin IA, Dean C, Trick M (1998) Collinearity between a 30-centimorgan segment of *Arabidopsis thaliana* chromosome 4 and duplicated regions within the *Brassica napus* genome. Genome 41: 62–69.

Chang S-B, de Jong H (2005) Production of alien chromosome addition lines and their utility in plant genetics. Cytogenet Genome Res 109: 335–343.

Chen BY, Heneen WK (1991) The basic number of *Brassica* genome: x=3? Crucifer Newsl 14/15: 20–21.

Chen BY, Simonsen V, Lannér-Harrera C, Heneen WK (1992) A *Brassica campestris-alboglabra* addition line and its use for gene mapping, intergenomic gene transfer and generation of trisomics. Theor Appl Genet 84: 592–599.

Chen BY, Cheng BF, Jørgensen RB, Heneen WK (1997a) Production and cytogenetics of *Brassica campestris-alboglabra* chromosome addition lines. Theor Appl Genet 94: 633–640.

Chen BY, Jørgensen RB, Cheng BF, Heneen WK (1997b) Identification and chromosomal assignment of RAPD markers linked with a gene for seed colour in a *Brassica campestris-alboglabra* addition line. Hereditas 126: 133–138.

Chen ZJ, Pikaard CS (1997a) Epigenetic silencing of RNA polymerase I transcription: a role for DNA methylation and histone modification in nucleolar dominance. Genes Dev 11: 2124–2136.

Chen ZJ, Pikaard CS (1997b) Transcriptional analysis of nucleolar dominance in polyploid plants: biased expression/silencing of progenitor rRNA genes is developmentally regulated in *Brassica*. Proc Natl Acad Sci USA 94: 3442–3447.

Cheng BF, Heneen WK (1995) Satellited chromosomes, nucleolus organizer regions and nucleoli of *Brassica campestris* L., *B. nigra* (L.) Koch, and *Sinapis arvensis* L. Hereditas 122: 113–118.

Cheng BF, Chen BY, Heneen WK (1994a) Addition of *Brassica alboglabra* Bailey chromosomes to *B. campestris* L. with special emphasis on seed colour. Heredity 73: 185–189.

Cheng BF, Heneen WK, Chen BY (1994b) Meiotic studies on a *Brassica campestris-alboglabra* monosomic addition line and derived *B. campestris* primary trisomics. Genome 37: 584–589.

Cheng BF, Heneen WK, Chen BY (1995a) Mitotic karyotypes of *Brassica campestris* and *Brassica alboglabra* and identification of the *B. alboglabra* chromosome in an addition line. Genome 38: 313–319.

Cheng BF, Heneen WK, Pedersen C (1995b) Ribosomal RNA gene loci and their nucleolar activity in *Brassica alboglabra* Baily. Hereditas 123: 169–173.

Cheng BF, Séguin-Swartz G, Somers DJ, Rakow G (2001) Low glucosinolate *Brassica juncea* breeding line revealed to be nullisomic. Genome 44: 738–741.

Chiang BY, Chiang MS, Grant WF, Crete R (1980) Transfer of resistance to race 2 of *Plasmodiophora brassicae* from *Brassica napus* to cabbage (*B. oleracea* ssp. *capitata*). IV. A resistant 18-chromosome B1 plant and its B2 progenies. Euphytica 47: 47–55.

Cohen S, Houben A, Segal D (2008) Extrachromosomal circular DNA derived from tandemly repeated genomic sequences in plants. Plant J 53: 1027–1034.

Copenhaver GP, Pikaard CS (1996) Two-dimensional RFLP analyses reveal megabase-sized clusters of rRNA gene variants in *Arabidopsis thaliana*, suggesting local spreading of variants as the mode for gene homogenization during concerted evolution. Plant J 9: 273–282.

Copenhaver GP, Nickel K, Kuromori T, Benito MI, Kaul S, Lin X, Bevan M, Murphy G, Harris B, Parnell LD, et al. (1999) Genetic definition and sequence analysis of Arabidopsis centromeres. Science 286: 2468–2474.

de Jong HJ, Fransz P, Zabel P (1999) High resolution FISH in plants—techniques and applications. Trends Plant Sci 4: 258–263.

Delseny M, McGrath JM, This P, Chevre AM, Quiros CF (1990) Ribosomal RNA genes in diploid and amphidiploid *Brassica* and related species: organization, polymorphism, and evolution. Genome 33: 733–744.

dos Santos KGB, Becker HC, Ecke W, Bellin U (2007) Molecular characterisation and chromosomal localisation of a telomere-like repetitive DNA sequence highly enriched in the C genome of *Brassica*. Cytogenet Genome Res 119: 147–153.

Du X-Z, Ge X-H, Zhao Z-G, Li Z-Y (2008) Chromosome elimination and fragment introgression and recombination producing intertribal partial hybrids from *Brassica napus* × *Lesquerella fendleri* crosses. Plant Cell Rep 27: 261–271.

Dvorak J (1990) Evolution of multigene families: the ribosomal RNA loci of wheat and related species. In: AHD Brown ,MT Clegg , AL Kahler ,BS Weir (eds) Plant Population Genetics, Breeding and Genetic Resources. Sinauer Associates, Sunderland, MA, USA, pp 83–97.

Fahleson J, Lagercrantz U, Mouras A, Glimelius K (1997) Characterization of somatic hybrids between *Brassica napus* and *Eruca sativa* using species-specific repetitive sequences and genomic *in situ* hybridization. Plant Sci 123: 133–142.

Fantes JA, MacKay GR (1979) The production of disomic addition lines of Bras*sica campestris*. Crucifer Newsl 4: 36–37.

FitzJohn RG, Armstrong TT, Newstrom-Lloyd LE, Wilton AD, Cochrane M (2007) Hybridisation within *Brassica* and allied genera: evaluation of potential for transgene escape. Euphytica 158: 209–230.

Fransz PF, Armstrong S, de Jong JH, Parnell LD, van DC, Dean C, Zabel P, Bisseling T, Jones GH (2000) Integrated cytogenetic map of chromosome arm 4S of *A. thaliana*: structural organization of heterochromatic knob and centromere region. Cell 100: 367–376.

Fransz P, ten Hoopen R, Tessadori F (2006) Composition and formation of heterochromatin in *Arabidopsis thaliana*. Chrom Res 14: 71–82.

Fuchs J, Kuchne N, Schubert I (1998) Assignment of linkage groups to pea chromosomes after karyotyping and gene mapping by fluorescent *in situ* hybridization. Chromosoma 107: 272–276.

Fukui K, Nakayama S, Ohmido N, Yoshiaki H, Yamabe M (1998) Quantitive karyotyping of three diploid *Brassica* species by imaging methods and localization of 45S rDNA loci on the identified chromosomes. Theor Appl Genet 96: 325–330.

Ge X-H, Li Z-Y (2007) Intra- and intergenomic homology of B-genome chromosomes in trigenomic combinations of the cultivated *Brassica* species revealed by GISH analysis. Chrom Res. 15: 849–861.

German DA, Al-Shehbaz IA (2008) Five additional tribes (Aphragmeae, Biscutelleae, Calepineae, Conringieae, and Erysimeae) in the Brassicaceae. Harvard Pap Bot 13: 165–170.

Goodpasture C, Bloom SE (1975) Visualization of nucleolar organizer regions in mammalian chromosomes using silver staining. Chromosoma 53: 37–50.

Hall JC, Sytsma KJ, Iltis HH (2002) Phylogeny of Capparaceae and Brassicaceae based on chloroplast sequence data. Am J Bot 89: 1826–1842.

Hanson RE, Islam-Faridi MN, Percival EA, Crane CF, Ji Y, McKnight TD, Stelly DM, Price HJ (1996) Distribution of 5S and 18S–28S rDNA loci in a tetraploid cotton (*Gossypium hirsutum* L.) and its putative diploid ancestors. Chromosoma 105: 55–61.

Harrison GE, Heslop-Harrison JS (1995) Centromeric repetitive DNA sequences in the genus *Brassica*. Theor Appl Genet 90: 157–165.

Hasterok R, Maluszynska J (2000a) Cytogenetic analysis of diploid *Brassica* species. Acta Biol Cracoviensia, sr Bot 42: 145–153.

Hasterok R, Maluszynska J (2000b) Nucleolar dominance does not occur in root tip cells of allotetraploid *Brassica* species. Genome 43: 574–579.

Hasterok R, Jenkins G, Langdon T, Jones RN, Maluszynska J (2001) Ribosomal DNA is an effective marker of *Brassica* chromosomes. Theor Appl Genet 103: 486–490.

Hasterok R, Ksiazczyk T, Wolny E, Maluszynska J (2005a) FISH and GISH analysis of *Brassica* genomes. Acta Biol Cracoviensia, sr Bot 47: 185–192.

Hasterok R, Wolny E, Kulak S, Zdziechiewicz A, Maluszynska J, Heneen WK (2005b) Molecular cytogenetic analysis of *Brassica rapa-Brassica oleracea* var. *alboglabra* monosomic addition lines. Theor Appl Genet 111: 196–205.

Hasterok R, Wolny E, Hosiawa M, Kowalczyk M, Kulak-Ksiazczyk S, Ksiazczyk T, Heneen WK, Maluszynska J (2006) Comparative analysis of rDNA distribution in chromosomes of various species of Brassicaceae. Ann Bot 97: 205–216.

Heneen WK, Brismar K (2001) Maternal and embryonal control of seed colour by different *Brassica alboglabra* chromosomes. Plant Breed 120: 325–329.

Heneen WK, Jørgensen RB (2001) Cytology, RAPD, and seed colour of progeny plants from *Brassica rapa-alboglabra* aneuploids and development of monosomic addition lines. Genome 44: 1007–1021.

Heneen WK, Chen BY, Cheng BF, Jonsson A, Simonsen V, Jørgensen RB, Davik J (1995) Characterization of the A and C genomes of *Brassica campestris* and *B. alboglabra*. Hereditas 123: 251–267.

Henry Y, Bedhomme M, Blanc G (2006) History, protohistory and prehistory of the *Arabidopsis thaliana* chromosome complement. Trends Plant Sci 11: 267–273.

Heslop-Harrison JS (2000) Comparative genome organization in plants: from sequence and markers to chromatin and chromosomes. Plant Cell 12: 617–636.

Howell EC, Barker GC, Jones GH, Kearsey M J, King GJ, Kop EP, Ryder CD, Teakle GR, Vicente JG, Armstrong SJ (2002) Integration of the cytogenetic and genetic linkage maps of *Brassica oleracea*. Genetics 161: 1225–1234.

Howell EC, Armstrong SJ, Barker GC, Jones GH, King GJ, Ryder CD, Kearsey MJ (2005) Physical organization of the major duplication on *Brassica oleracea* chromosome O6 revealed through fluorescence in situ hybridization with Arabidopsis and *Brassica* BAC probes. Genome 48: 1093–1103.

Howell EC, Kearsey M J, Jones GH, King GJ, Armstrong SJ (2008) A and C genome distinction and chromosome identification in *Brassica napus* by sequential FISH and GISH. Genetics 180: 1849–1857.

Hu J, Quiros CF (1991) Molecular and cytological evidence of deletions in alien chromosomes for two monosomic addition lines of *Brassica campestris-oleracea*. Theor Appl Genet 81: 221–226.

Hua Y-W, Liu M, Li Z-Y (2006) Parental genome separation and elimination of cells and chromosomes revealed by AFLP and GISH analyses in a *Brassica carinata* × *Orychophragmus violaceus* cross. Ann Bot 97: 993–998.

Hussein MM, Abobakr MA (1976) Secondary associations in Brassica oleracea L. Egyp J Genet Cytol 5: 174–183.

Inomata N (1993) Crossability and cytology of hybrid progenies in the cross between *Brassica campestris* and three wild relatives of *B. oleracea*, *B. bourgeaui*, *B. cretica*, and *B. montana*. Euphytica 69: 7–17.

Jackson SA, Cheng Z, Wang ML, Goodman HM, Jiang J (2000) Comparative fluorescence *in situ* hybridization mapping of a 431-kb *Arabidopsis thaliana* bacterial artifcial chromosome contig reveals the role of chromosomal duplications in the expansion of the *Brassica rapa* genome. Genetics 156: 833–838.

Jaillon O, Aury JM, Noel B, Policriti A, Clepet C, Casagrande A, Choisne N, Aubourg S, Vitulo N, Jubin C et al. (2007) The grapevine genome sequence suggests ancestral hexaploidization in major angiosperm phyla. Nature 449: 463–467.

Jenczewski E, Eber F, Grimaud A, Huet S, Lucas MO, Monod H, Chèvre AM (2003) PrBn, a major gene controlling homeologous pairing in oilseed rape (*Brassica napus*) haploids. Genetics 164: 645–653.

Johnston JS, Pepper AE, Hall AE, Chen ZJ, Hodnett G, Drabek J, Lopez R, Price HJ (2005) Evolution of genome size in Brassicaceae. Ann Bot (Lond) 95: 229–235.

Jørgensen RB, Chen BY, Cheng BF, Heneen WK, Simonsen V (1996) Random amplified polymorphic DNA markers of the *Brassica alboglabra* chromosome of a *B. campestris-alboglabra* addition line. Chrom Res 4: 111–114.

Kamisugi Y, Nakayama S, O'Neil CM, Mathias RJ, Trick M, Fukui K (1998) Visualization of the *Brassica* self-incompatibility S-locus on identified oilseed rape chromosomes. Plant Mol Biol 38: 1081–1087.

Kaneko Y, Matsuzawa Y, Sarashima M (1987) Breeding of the chromosome addition lines of radish with single kale chromosome. Jpn J Breed 37: 438–452.

Kaneko, Y, Namai H, Matsuzawa Y, Sarashima M (1991) Maintenance and stability of the chromosome addition lines of radish with single kale chromosome. Jpn J Breed 41: 623–639.

Kaneko Y, Bang SW, Matsuzawa Y (2000) Early-bolting trait and RAPD markers in the specific monosomic addition line of radish carrying the e-chromosome of *Brassica oleracea*. Plant Breed 119: 137–140.

Kaneko Y, Yano H, Bang SW, Matsuzawa Y (2001) Production and characterization of *Raphanus sativus-Brassica rapa* monosomic chromosome addition lines. Plant Breed 120: 163–168.

Kaneko Y, Nagasawa N, Bang SW, Matsuzawa Y (2002) Homoeological relationships of *Brassica oleracea*. Plant Breed 121: 171–173.

Kaneko Y, Yano H, Bang SW, Matsuzawa Y (2003) Genetic stability and maintenance of *Raphanus sativus* lines with an added *Brassica rapa* chromosome. Plant Breed 122: 239–243.

Karpechenko GD (1922) The number of chromosomes and the genetic correlation of cultivated Cruciferae. Bull Appl Bot Genet Plant Breed 13: 3–14.

Kianian SF, Quiros CF (1992a) Genetic analysis of major multigene families in *Brassica oleracea* and related species. Genome 35: 516–527.

Kianian SF, Quiros CF (1992b) Generation of a *Brassica oleracea* composite RFLP map: linkage arrangements among various populations and evolutionary implications. Theor Appl Genet 84: 544–554.

Koch MA, Al-Shehbaz IA (2008) Molecluar systematics and evolution of "wild" crucifers (Brassicaceae or Cruciferae). In: SK Gupta (ed) Biology and Breeding of Crucifers. Taylor and Francis Group, London, UK, pp 1–19.

Koch MA, Kiefer M (2005) Genome evolution among cruciferous plants: a lecture from the comparison of the genetic maps of three diploid species—*Capsella rubella, Arabidopsis lyrata* subsp. *petraea,* and *A. thaliana*. Am J Bot 92: 761–767.

Koch M, Bishop J, Mitchell-Olds T (1999) Molecular systematics and evolution of *Arabidopsis* and *Arabis*. Plant Biol 1: 529–537.

Koch MA, Haubold B, Mitchell-Olds T (2000) Comparative evolutionary analysis of chalcone synthase and alcohol dehydrogenase loci in *Arabidopsis*, *Arabis*, and related genera (Brassicaceae). Mol Biol Evol 17: 1483–1498.

Koch MA, Haubold B, Mitchell-Olds T (2001) Molecular systematics of the Brassicaceae: evidence from coding plastidic *matK* and nuclear *Chs* sequences. Am J Bot 88: 534–544.

Koo D-H, Plaha P, Lim YP, Hur Y, Bang J-W (2004) A high-resolution karyotype of *Brassica rapa* ssp. *pekinensis* revealed by pachytene analysis and multicolor fluorescence *in situ* hybridization. Theor Appl Genet 109: 1346–1352.

Kuittinen H, de Haan AA, Vogl C, Oikarinen S, Leppala J, Koch M, Mitchell-Olds T, Langley CH, Savolainen O (2004) Comparing the linkage maps of the close relatives *Arabidopsis lyrata* and *A. thaliana*. Genetics 168: 1575–1584.

Kulak S, Hasterok R, Maluszynska J (2002) Karyotyping of *Brassica* amphidiploids using 5S and 25S rDNA as chromosome markers. Hereditas 136: 144–150.

Lagercrantz U (1998) Comparative mapping between *Arabidopsis thaliana* and *Brassica nigra* indicates that *Brassica* genomes have evolved through extensive genome replication accompanied by chromosome fusions and frequent rearrangements. Genetics 150: 1217–1228.

Lagercrantz U, Lydiate DJ (1996) Comparative genome mapping in *Brassica*. Genetics 144: 1903–1910.

Lawrence RJ, Earley K, Pontes O, Silva M, Chen ZJ, Neves N, Viegas W, Pikaard CS (2004) A concerted DNA methylation/histone methylation switch regulates rRNA gene dosage control and nucleolar dominance. Mol Cells 13: 599–609.

Lee KH, Namai H (1992) Stabilization of new types of diploids (2n=22, 24) through selfing of aneuploids (2n=21, 22) derived from crossing of sesquidiploids (2n=29, AAC) and *Brassica campestris* (2n=20, AA). Euphytica 60: 1–13.

Lee KH, Namai H (1994) Cytogenetic and morphological characteristics of new types of diploids (2n=22, 24, 40) derived from consecutive selfing of aneuploids in brassica crops. Euphytica 72: 15–22.

Leflon M, Eber F, Letanneur J-C, Chelysheva L, Coriton O, Huteau V, Ryder CD, Barker G, Jenczewski E, Chèvre A-M (2006) Pairing and recombination at meiosis of *Brassica rapa* (AA) × *Brassica napus* (AACC) hybrids. Theor Appl Genet 113: 1467–1480.

Leino M, Thyselius S, Landgren M, Glimelius K (2004) *Arabidopsis thaliana* chromosome III restores fertility in a cytoplasmic male-sterile *Brassica napus* line with *A. thaliana* mitochondrial DNA. Theor Appl Genet 109: 272–279.

Levan A, Frekga K, Sandberg A (1964) Nomenclature for centromeric position in chromosomes. Hereditas 52: 201–220.

Li MT, Li ZY, Zhang CY, Qian W, Meng JL (2005) Reproduction and cytogenetic characterization of interspecific hybrids derived from crosses between *Brassica carinata* and *B. rapa*. Theor Appl Genet 110: 1284–1289.

Li Z-Y, Ge X-H (2007) Unique chromosome behaviour and genetic control in *Brassica* × *Orychophragmus* wide hybrids: a review. Plant Cell Rep 26: 701–710.

Lim K-B, de Jong H, Yang T-J, Park J-Y, Kwon S-J, Kim JS, Lim M-H, Kim JA, Jin M, Jin Y-M, et al. (2005) Characterization of rDNAs and tandem repeats in the heterochromatin of *Brassica rapa*. Mol Cells 19: 436–444.

Lim KB, Yang TJ, Hwang YJ, Kim JS, Park JY, Kwon SJ, Kim JA, Choi BS, Lim MH, Jin M, et al. (2007) Characterization of the centromere and peri-centromere retrotransposons in *Brassica rapa* and their distribution in related *Brassica* species. Plant J 49: 173–183.

Lim YP, Plaha P, Choi SR, Uhm T, Hong CP, Bang JW, Hur YK (2006) Toward unraveling the structure of *Brassica rapa* genome. Physiol Plant 126: 585–591.

Lippman Z, Gendrel AV, Black M, Vaughn MW, Dedhia N, McCombie WR, Lavine K, Mittal V, May B, Kasschau KD, et al. (2004) Role of transposable elements in heterochromatin and epigenetic control. Nature 430: 471–476.

Lysak MA, Berr A, Pecinka A, Schmidt R, McBreen K, Schubert I (2006) Mechanisms of chromosome number reduction in *Arabidopsis thaliana* and related Brassicaceae species. Proc Natl Acad Sci USA 103: 5224–5229.

Lysak MA, Lexer C (2006) Towards the era of comparative evolutionary genomics in Brassicaceae. Plant Syst Evol 259: 175–198.

Lysak MA, Fransz PF, Ali HBM, Schubert I (2001) Chromosome painting in *Arabidopsis thaliana*. Plant J 28: 689–697.

Lysak MA, Pecinka A, Schubert I (2003) Recent progress in chromosome painting of Arabidopsis and related species. Chrom Res 11: 195–204.

Lysak MA, Koch MA, Pecinka A, Schubert I (2005) Chromosome triplication found across the tribe Brassiceae. Genome Res 15: 516–525.

Lysak MA, Cheung K, Kitschke M, Bureš P (2007) Ancestral chromosomal blocks are triplicated in Brassiceae species with varying chromosome number and genome size. Plant Physiol 145: 402–410.

Lysak MA, Koch MA, Beaulieu JM, Meister A, Leitch IJ (2009) The dynamic ups and downs of genome size evolution in Brassicaceae. Mol Biol Evol 26: 85–98.

Maere S, De BS, Raes J, Casneuf T, Van MM, Kuiper M, Van de PY (2005) Modeling gene and genome duplications in eukaryotes. Proc Natl Acad Sci USA 102: 5454–5459.

Maluszynska J, Heslop-Harrison JS (1993) Physical mapping of rDNA loci in *Brassica* species. Genome 36: 774–781.

Maluszynska J, Hasterok R (2005) Identification of individual chromosomes and parental genomes in *Brassica juncea* using GISH and FISH. Cytogenet Genome Res 109: 310–314.

Mandáková T, Lysak MA (2008) Chromosomal phylogeny and karyotype evolution in x=7 crucifer species (Brassicaceae). Plant Cell 20: 2559–2570.

Marhold K, Lihova J (2006) Polyploidy, hybridization and reticulate evolution: lessons from the Brassicaceae. Plant Syst Evol 259: 143–174.

Mayer K, Murphy G, Tarchini R, Wambutt R, Volckaert G, Pohl T, Dusterhoft A, Stiekema W, Entian KD, Terryn N, et al. (2001) Conservation of microstructure between a sequenced region of the genome of rice and multiple segments of the genome of *Arabidopsis thaliana*. Genome Res 11: 1167–1174.

McClintock B (1934) The relationship of a particular chromosomal element to the development of the nucleoli in *Zea mays*. Cell and Tissue Research 21: 294–328.

McCombie WR, de la BM, Habermann K, Parnell LD, Dedhia N, Gnoj L, Schutz K, Huang E, Spiegel L, Yordan C, et al. (2000) The complete sequence of a heterochromatic island from a higher eukaryote. Cell 100: 377–386.

McGrath JM, Quiros CF (1990) Generation of alien chromosome addition lines from synthetic *Brassica napus*: morphology, cytology, fertility, and chromosome transmission. Genome 33: 374–383.

McGrath JM, Quiros CF, Harada JJ, Landry BS (1990) Identification of *Brassica oleracea* monosomic alien chromosome addition lines with molecular markers reveals extensive gene duplication. Mol Gen Genet 223: 198–204.

Ming R, Hou S, Feng Y, Yu Q, Dionne-Laporte A, Saw JH, Senin P, Wang W, Ly BV, Lewis KL, et al. (2008) The draft genome of the transgenic tropical fruit tree papaya (*Carica papaya* Linnaeus). Nature 452: 991–996.

Mishima M, Ohmido N, Fukui K, Yahara T (2002) Trends in site-number change of rDNA loci during polyploid evolution in *Sanguisorba* (Rosaceae). Chromosoma 110: 550–558.

Mizushima U (1980) Genome analysis in *Brassica* and allied genera. In: S Tsunoda, K Hinata , C Gomez-Campo (eds) *Brassica* Crops and Wild Allies. Japanese Scientific Society Press, Tokyo, Japan, pp 89–105.

Morinaga T (1928) Preliminary note on interspecific hybridization in *Brassica*. Proc Imp Acad Tokyo 4: 620–622.

Morinaga T (1933) Interspecific hybridisation in *Brassica*: V. The cytology of F1 hybrids of *B. carinata* and *B. alboglabra*. Jpn J Bot 6: 467–475.

Morinaga T (1934) Interspecific hybridisation in *Brassica*: VI. The cytology of F1 hybrids of *B. juncea* and *B. nigra*. Cytologia 6: 62–67.

Nagpal R, Raina SN, Sodhi YS, Mukhopadhyay A, Arumugam N, Pradhan AK, Pental D (1996) Transfer of *Brassica tournefortii* (TT) genes to allotetraploid oilseed *Brassica* species (*B. juncea* AABB, *B. napus* AACC, *B. carinata* BBCC): homoeologous pairing is more pronounced in the three-genome hybrids (TACC, TBAA, TCAA, TCBB) as compared to allodiploids (TA, TB, TC). Theor Appl Genet 92: 566–571.

Nakayama S, Fukui K (1997) Quantitative chromosome mapping of small plant chromosomes by improved imaging on CHIAS II. Genes Genet Syst 72: 35–40.

Nicolas S, Le Mignon G, Eber F, Coriton O, Monod H, Clouet V, Huteau V, Lostanlen A, Delourme R, Chalhoub B, et al. (2007) Homeologous recombination plays a major role in chromosome rearrangements that occur during meiosis of *Brassica napus* haploids. Genetics 175: 487–503.

Nicolas SD, Leflon M, Liu Z., Eber F, Chelysheva L, Coriton O, Chèvre AM, Jenczewski E (2008) Chromosome 'speed dating' during meiosis of polyploid *Brassica* hybrids and haploids. Cytogenet Genome Res 120: 331–338.

Nishibayasahi S (1992) Banding in mitotic chromosomes of *Brassica campestris* var. *pekinensis* with a trypsin–Giemsa method. Genome 35: 899–901.

O'Neill CM, Bancroft I (2000) Comparative physical mapping of segments of the genome of *Brassica oleracea* var. *alboglabra* that are homoeologous to sequenced regions of chromosomes 4 and 5 of *Arabidopsis thaliana*. Plant J. 23: 233–243.

Olin-Fatih M (1994) A new method for differential staining of *Brassica* metaphase chromosomes, and karyotypes of *B. campestris*, *B. oleracea*, and *B. napus*. Hereditas 120: 253–259.

Olin-Fatih M (1996) The morphology, cytology, and C-banded karyotypes of *Brassica campestris*, *B. oleracea*, and *B. napus* plants regenerated from protoplasts. Theor Appl Genet 93: 420

Olin-Fatih M, Heneen WK (1992) C-banded karyotypes of *Brassica campestris*, *B. oleracea*, and *B. napus*. Genome 35: 583–589.

Paterson AH, Bowers JE, Bruggmann R, Dubchak I, Grimwood J, Gundlach H, Haberer G, Hellsten U, Mitros T, Poliakov A, et al. (2009) The *Sorghum bicolor* genome and the diversification of grasses. Nature 457: 551–556.

Paterson AH, Bowers JE, Burow MD, Draye X, Elsik CG, Jiang CX, Katsar CS, Lan TH, Lin YR, Ming R, et al. (2000) Comparative genomics of plant chromosomes. Plant Cell 12: 1523–1540.

Parkin IAP, Gulden SM, Sharpe AG, Lukens L, Trick M, Osborn TC, Lydiate DJ (2005) Segmental structure of the *Brassica napus* genome based on comparative analysis with *Arabidopsis thaliana*. Genetics 171: 765–781.

Pederson T, Politz JC (2000) The nucleolus and the four ribonucleoproteins of translation. J Cell Biol 148: 1091–1095.

Prakash S, Bhat SR, Quiros CF, Kirti PB, Chopra VL (2009) *Brassica* and its close allies: Cytogenetics and evolution. Plant Breed Rev 31: 21–187.

Preuss SB, Costa-Nunes P, Tucker S, Pontes O, Lawrence RJ, Mosher R, Kasschau KD, Carrington JC, Baulcombe DC, Viegas W, et al. (2008) Multimegabase silencing in nucleolar dominance involves siRNA-directed DNA methylation and specific methylcytosine-binding proteins. Mol Cells 32: 673–684.

Quiros CF (1999) Genome structure and mapping. In: C Gomez-Campo (ed) Biology of *Brassica* Coenospecies. Elsevier Science, Amsterdam, The Netherlands, pp 217–231.

Quiros CF, Ochoa O, Kianian SF, Douches D (1987) Analysis of the *Brassica oleracea* genome by the generation of *B. campestris-oleracea* chromosome addition lines: characterization by isozymes and rDNA genes. Theor Appl Genet 74: 758–766.

Rana D, Boogaart T, O'Neill CM, Hynes L, Bent E, Macpherson L, Park JY, Lim YP, Bancroft I (2004) Conservation of the microstructure of genome segments in *Brassica napus* and its diploid relatives. Plant J 40: 725–733.

Richards EJ, Elgin SC (2002) Epigenetic codes for heterochromatin formation and silencing: rounding up the usual suspects. Cell 108: 489–500.

Röbbelen G (1960) Beiträge zur Analyse des *Brassica*-Genoms. Chromosoma (Berlin) 11: 205–228.

Robert-Fortel I, Junera HR, Geraud G, Hernandez-Verdun D (1993) Three-dimensional organization of the ribosomal genes and Ag-NOR proteins during interphase and mitosis in PtK1 cells studied by confocal microscopy. Chromosoma 102: 146–157.

Roussel P, Andre C, Comai L, Hernandez-Verdun D (1996) The rDNA transcription machinery is assembled during mitosis in active NORs and absent in inactive NORs. J Cell Biol 133: 235–246.

Schelfhout CJ, Snowdon R, Cowling WA, Wroth JM (2004) A PCR based B-genome-specific marker in Brassica species. Theor Appl Genet 109: 917–921.

Schrader O, Budahn H, Ahne R (2000) Detection of 5S and 25 rRNA genes in *Sinapis alba, Raphanus sativus* and *Brassica napus* by double fluorescence *in situ* hybridization. Theor Appl Genet 100: 665–669.

Schranz ME, Mitchell-Olds T (2006) Independent ancient polyploidy events in the sister families Brassicaceae and Cleomaceae. Plant Cell 18: 1152–1165.

Schranz ME, Lysak MA, Mitchell-Olds T (2006) The ABC's of comparative genomics in the Brassicaceae: building blocks of crucifer genomes. Trends Plant Sci 11: 535–542.

Schubert I (2007) Chromosome evolution. Curr Opin Plant Biol 10: 109–115.

Sebastian RL, Howell EC, King GJ, Marshall DF, Kearsey MJ (2000) An integrated AFLP and RFLP *Brassica oleracea* linkage map from two morphologically distinct doubled-haploid mapping populations. Theor Appl Genet 100: 75–81.

Shishido R, Sano Y, Fukui K (2000) Ribosomal DNAs: an exception to the conservation of gene order in rice genomes. Mol Gen Genet 263: 586–591.

Sikka SM (1940) Cytogenetics of *Brassica* hybrids and species. J Genet 40: 441–509.

Simillion C, Vandepoele K, Van Montagu MC, Zabeau M, Van de PY (2002) The hidden duplication past of *Arabidopsis thaliana*. Proc Natl Acad Sci USA 99: 13627–13632.

Skarzhinskaya M, Fahleson J, Glimelius K, Mouras A (1998) Genome organization of *Brassica napus* and *Lesquerella fendleri* and analysis of their somatic hybrids using genomic *in situ* hybridization. Genome 41: 691–701.

Snowdon RJ (2007) Cytogenetics and genome analysis in *Brassica* crops. Chrom Res 15: 85–95.

Snowdon RJ, Köhler W, Köhler A (1997a) Chromosomal localization and characterization of rDNA loci in the *Brassica* A and C genomes. Genome 40: 582–587.

Snowdon RJ, Köhler W, Friedt W, Köhler A (1997b) Genomic *in situ* hybridization in *Brassica* amphidiploids and interspecific hybrids. Theor Appl Genet 95: 1320–1324.

Snowdon RJ, Friedt W, Kőhler A, Köhler W (2000a) Molecular cytogenetic localization and characterization of 5S and 25S rDNA loci for chromosome identification in oilseed rape (*Brassica napus* L.). Ann Bot 86: 201–204.

Snowdon RJ, Winter H, Diestel A, Sacristán MD (2000b) Development and characterisation of *Brassica napus-Sinapis arvensis* addition lines exhibiting resistance to *Leptosphaeria maculans*. Theor Appl Genet 101: 1080–1014.

Snowdon RJ, Friedrich T, Friedt W, Kőhler W (2002) Identifying the chromosomes of the A- and C-genome diploid *Brassica* species *B. rapa* (syn. *campestris*) and *B. oleracea* in their amphidiploid *B. napus*. Theor Appl Genet 104: 533–538.

Song K, Lu P, Tang K, Osborn TC (1995) Rapid genome change in synthetic polyploids of *Brassica* and its implications for polyploid evolution. Proc Natl Acad Sci USA 92: 7719–7723.

Struss D, Quiros CF, Röbbelen G (1992) Mapping of molecular markers on Brassica B-genome chromosomes added to *Brassica napus*. Plant Breed 108: 320–323.

Struss D, Bellin U, Röbbelen G (1991) Development of B-chromosome addition lines of *B. napus* using different interspecific *Brassica* hybrids. Plant Breed 106: 209–214.

Takamine N (1916) Über die ruhenden und die präsynaptischen. Phasen der Reduktionsteilung. Bot Mag (Tokyo) 30: 293–303.

Tang H, Wang X, Bowers JE, Ming R, Alam M, Paterson AH (2008) Unraveling ancient hexaploidy through multiply-aligned angiosperm gene maps. Genome Res 18: 1944–1954.

Thomas BC, Pedersen B, Freeling M (2006) Following tetraploidy in an Arabidopsis ancestor, genes were removed preferentially from one homeolog leaving clusters enriched in dose-sensitive genes. Genome Res 16: 934–946.

Thompson KF (1956) Production of haploid plants of marrow stem kale. Nature (Lond) 178: 748.

Town CD, Cheung F, Maiti R, Crabtree J, Haas BJ, Wortman JR, Hine EE, Althoff R, Arbogast TS, Tallon LJ, et al. (2006) Comparative genomics of Brassica oleracea and Arabidopsis thaliana reveal gene loss, fragmentation, and dispersal after polyploidy. Plant Cell 18: 1348–1359.

Truco MJ, Quiros CF (1994) Structure and organization of the B genome based on a linkage map in *Brassica nigra*. Theor Appl Genet 89: 590–598.

Truco MJ, Hu J, Sadowski J, Quiros CF (1996) Inter- and intragenomic homology of the *Brassica* genomes: implications for their origin and evolution. Theor Appl Genet 93: 1225–1233.

U N (1935) Genomic analysis of *Brassica* with special reference to the experimental formation of *B. napus* and peculiar mode of fertilization. Jpn J Bot 7: 389–452.

Vision TJ, Brown DG, Tanksley SD (2000) The origins of genomic duplications in Arabidopsis. Science 290: 2114–2117.

Volkov RA, Borisjuk NV, Panchuk II, Schweizer D, Hemleben V (1999) Elimination and rearrangement of parental rDNA in the allotetraploid *Nicotiana tabacum*. Mol Biol Evol 16: 311–320.

Wang XH, Luo P (1987) Studies on the karyotypes and C-banding patterns of Chinese kale (*Brassica alboglabra*) and cabbage (*B. oleracea* var. *capitata*). Acta Bot Sin 29: 149–155.

Wang XH, Luo P, Shu JJ (1989) Giemsa N-banding pattern in cabbage and Chinese kale. Euphytica 41: 17–21.

Wang YP, Snowdon RJ, Rudloff E, Wehling P, Friedt W, Sonntag K (2004) Cytogenetic characterization and *fae1* gene variation in progenies from asymmetric somatic hybrids between *Brassica napus* and *Crambe abyssinica*. Genome 47: 724–731.

Wang YP, Sonntag K, Rudloff E, Chen JM (2005) Intergeneric somatic hybridization between *Brassica napus* L. and *Sinapis alba* L. J Integr Plant Biol 47: 84–91.

Warwick SI, Sauder C (2005) Phylogeny of tribe Brassiceae (Brassicaceae) based on chloroplast restriction site polymorphisms and nuclear ribosomal internal transcribed spacer (ITS) and chloroplast trnL intron sequences. Can J Bot 83: 467–483.

Warwick SI, Al-Shehbaz IA (2006) Brassicaceae: Chromosome number index and database on CD-Rom. Plant Syst Evol 259: 237–248.

Warwick SI, Francis A, LaFleche J (2000) Guide to wild germplasm. *Brassica* and allied crops (tribe Brassiceae, Brassicaceae). 2nd edn. Agriculture and Agri-Food Canada: *http://www.brassica.info/info/publications/guide-wild-germplasm.php*

Wei W-H, Zhang S-F, Wang L-J, Chen B, Wu X-M, Song Y-C (2007) Karyotyping of *Brassica oleracea* L. based on C_0t-1 and ribosomal DNAs. Bot Stud 48: 255–261.

Wendel JF (2000) Genome evolution in polyploids. Plant Mol Biol 42: 225–249.

Wikström N, Savolainen V, Chase MW (2001) Evolution of angiosperms: calibrating the family tree. Proc Roy Soc Lond B 268: 2211–2220.

Yang T-J, Kim J-S, Lim K-B, Kwon S-J, Kim J-A, Jin M, Young JP, Lim M-H, Kim H-I, Hyung Kim S, et al. (2005) The Korea *Brassica* Genome Project: A glimpse of the *Brassica* genome based on comparative genome analysis with Arabidopsis. Compar Funct Genom 6: 138–146.

Yang T-J, Kim JS, Kwon S-J, Lim K-B, Choi B-S, Kim J-A, Jin M, Park JY, Lim M-H, Kim H-I, et al. (2006) Sequence-level analysis of the diploidization process in the triplicated *FLOWERING LOCUS C* region of *Brassica rapa*. Plant Cell 18: 1339–1347.

Yogeeswaran K, Frary A, York TL, Amenta A, Lesser AH, Nasrallah JB, Tanksley SD, Nasrallah ME (2005) Comparative genome analyses of *Arabidopsis* spp.: inferring chromosomal rearrangement events in the evolutionary history of *A. thaliana*. Genome Res 15: 505–515.

Zhao ZG, Hu TT, Ge XH, Du XZ, Ding L, Li ZY (2008) Production and characterization of intergeneric somatic hybrids between Brassica napus and Orychophragmus violaceus and their backcrossing progenies. Plant Cell Rep 27: 1611–1621.

Ziolkowski PA, Sadowski J (2002) FISH-mapping of rDNAs and Arabidopsis BACs on pachytene complements of selected Brassicas. Genome 45: 189–197.

Ziolkowski PA, Blanc G, Sadowski J (2003) Structural divergence of chromosomal segments that arose from successive duplication events in the Arabidopsis genome. Nucl Acids Res 31: 1339–1350.

Ziolkowski PA, Kaczmarek M, Babula D, Sadowski J (2006) Genome evolution in Arabidopsis/*Brassica*: conservation and divergence of ancient rearranged segments and their breakpoints. Plant J 47: 63–74.

7

Brassica rapa Genome Sequencing Project: Strategies and Current Status

*Jeong-Hwan Mun,[1] Tae-Jin Yang,[2] Soo-Jin Kwon,[1] and Beom-Seok Park[1]**

ABSTRACT

Brassica rapa is considered an ideal candidate to act as a reference species for *Brassica* genomic studies. Among the three basic *Brassica* species, *B. rapa* (AA genome) has the smallest genome (529 Mbp), compared to *B. nigra* (BB genome, 632 Mbp) and *B. oleracea* (CC genome, 696 Mbp). There is also a large collection of available cultivars of *B. rapa*, as well as a broad array of *B. rapa* genomic resources available. Under international consensus, various genomic studies on *B. rapa* have been conducted, including the construction of a physical map based on 22.5X genome coverage, end sequencing of 146,000 BACs, sequencing of >150,000 expressed sequence tags, and successful phase 2 shotgun sequencing of 589 euchromatic region-tiling BACs based on comparative positioning with the *Arabidopsis* genome. These sequenced BACs mapped onto the *B. rapa* genome provide beginning points for genome sequencing of each chromosome. Applying this strategy, all of the 10 chromosomes of *B. rapa* have been assigned to the sequencing centers in seven countries, Korea, the UK, China, India, Canada, Australia, and Japan. The two longest chromosomes, chromosome A3 and A9, have been sequenced except for several gaps, by NAAS in Korea. The other chromosomes will be sequenced in the other six countries, over the next few years. The imminent *B. rapa* genome sequence offers novel insights into the

[1]Genomics Division, Department of Agricultural Biotechnology, National Academy of Agricultural Science, RDA, 150 Suin-ro, Gwonseon-gu, Suwon 441-707, Korea.
[2]Department of Plant Science, Plant Genomics and Breeding Institute, and Research Institute for Agriculture and Life Sciences, College of Agriculture and Life Sciences, Seoul National University, Seoul, 151-921, Korea.
*Corresponding author: *pbeom@korea.go.kr*

organization and evolution of the *Brassica* genome. In parallel, the transfer of knowledge from *B. rapa* to other *Brassica* crops would be expected.

Keywords: *Brassica rapa*, genome, BAC, BAC-end sequence, expressed sequence tag, genetic map, physical map, chromosome sequencing

7.1 Introduction

The *Brassicaceae* family includes approximately 3,700 species in 350 genera. The species have diverse characteristics, many of which are of agronomic importance as vegetables, condiments, fodder, and oil crops (Beilstein et al. 2006). Economically, *Brassica* species contribute to approximately 10% of the world's vegetable crop produce and approximately 12% of the worldwide edible oil supplies (Economic Research Service USDA 2008). The tribe Brassiceae, which is one of 13–19 tribes in the Brassicaceae, consists of ~ 240 species and contains most crop species of *Brassica*. Species of particular importance are *B. napus* and *B. juncea* as sources of canola oil, *B. rapa* and *B. oleracea* as vegetable cole crops, and *B. nigra* as a source of the mustard condiment. In addition to the crop species, many of the wild species in the tribe Brassiceae have potential as new crops, sources of condiments, industrial oil, and other diverse products and/or host systems for molecular farming. Wild relatives possess a number of useful agronomic traits, including nuclear and cytoplasmic male sterility, resistance to disease, insect, and nematode pests, tolerance of cold, salt, and drought stresses. For this reason, an understanding of the genetic potential of Brassiceae wild relatives is critical for the establishment of long-term breeding programs of these crops.

Brassica crops are characterized by diverse morphologies with enlarged vegetative and floral meristems. The morphological diversity among the *Brassica* species appears to be linked to the genomic changes associated with polyploidy (Lukens et al. 2004). Comparative genetic mapping has revealed collinear chromosome segments in the Brassicaceae family (Schmidt et al. 2001), and conserved linkage arrangements between *Arabidopsis* and *Brassica* diverged from a common ancestor approximately 14.5–20.4 million years ago (Mya) (Bowers et al. 2003). The genomes of *Brassica* species contain triplicated homoelogous counterparts of the corresponding segments of *Arabidopsis* genome, due to triplication of the entire genome that occurred approximately 13–17 Mya (O'Neill and Bancroft 2000; Town et al. 2006; Yang et al. 2006). Furthermore, an additional natural allopolyploidization event, which happened during the last 10,000 years and resulted in a change of the chromosome numbers and genome size, played a role in the diversification of *Brassica* crops (Fig. 7-1) (U 1935; Rana et al. 2004; Johnston et al. 2005).

Of the six widely cultivated *Brassica* species, *B. rapa* (AA, $2n = 20$), *B. nigra* (BB, $2n = 16$), and *B. oleracea* (CC, $2n = 18$) are the monogenomic diploids. The interspecific breeding between these three diploid species resulted in the creation of three new species of allotetraploid hybrids, namely *B. juncea* (AABB, $2n = 36$), *B. napus* (AACC, $2n = 38$), and *B. carinata* (BBCC, $2n = 34$). Thus, investigation of the *Brassica* genome provides substantial opportunities to study the divergence of gene function and genome evolution associated with polyploidy, extensive duplication, and hybridization.

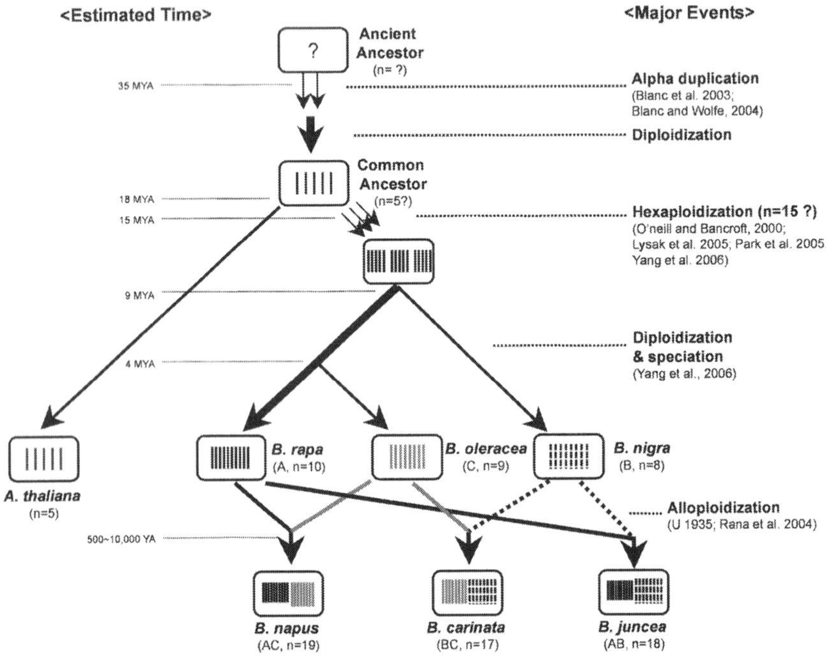

Figure 7-1 Schematic representation of the evolutionary process and genetic relationships between *Arabidopsis* and *Brassica* species with the chromosome numbers, based on literature reviewed and our research. Estimated time of major events represented as million years ago (Mya) on the left side; major events of genome divergence were summarized on the right side.

The close phylogenetic relationship between the *Brassica* species and the model plant, *Arabidopsis thaliana*, implies that knowledge transfer from the results of studies on *Arabidopsis* for *Brassica* crop improvement could be straightforward. The complex genome organization of the *Brassica* species as a result of multiple rounds of polyploidy and genome hybridization, however, renders the identification of orthologous relationships of genes between the genomes highly difficult. Genome triplication, subsequent extensive interspersed gene-loss or gene-gain events, and large scale chromosomal rearrangements including segmental duplications or deletions in the *Brassica* lineage all complicate the orthologous relationships of the loci between the two genomes (Town et al. 2006; Yang et al. 2006). For this reason, the genomes of several *Brassica* crop species have been characterized in detail over the past few years.

B. rapa is native to Europe and East Asia, where the existence of a large native *B. rapa* population has provided an important resource for the breeding program. There are wide morphological variations in *B. rapa*, including the leafy type (Chinese cabbage and pak choi), turnip type (vegetable turnip), and oil type (yellow sarson). Several *B. rapa* species are of regional agricultural importance, either as vegetable or oil crops. As a consequence of its native distribution and agronomic usage, *B. rapa* has great potential for use as a model for the study of both basic and applied aspects of plant biology. In particular, *B. rapa* ssp. *pekinensis* (Chinese cabbage), one of the most widely cultivated annual vegetable crops in Northeast Asia, exhibits characteristics that are useful for the study of genome characteristics such as diploidy and small genome size (529 Mbp). In response to the need for a simple genetic system with favorable genetic attributes for research among *Brassica* species, *B. rapa* has played a central role as a model species representing the *Brassica* "A" genome, and is the focus of multinational genome projects. Genomic studies on *B. rapa* ssp. *pekinensis* cv. *Chiifu* were initiated in 2003 when a bacterial artificial chromosome (BAC) library was constructed, and a collection of BAC-end sequences was initiated. Shortly afterward, a cytogenetic study, the generation of molecular markers, the construction of high-density genetic and physical maps, and eventually large-scale genome sequencing based on the clone-by-clone strategy (*http://www.brassica.info*) was initiated. The information and genomic resources produced during this course of investigations was useful for understanding the genetic system of *B. rapa*. Moreover, these resources will be beneficial for *Brassica* crop breeding, because they enable comparative genomic studies and subsequent transfer of knowledge from *B. rapa* to other *Brassica* crop species. In this chapter, we summarize the current state of information, the genomic resources pertaining to the *B. rapa* genome, and the progress on *B. rapa* genome sequencing, as conducted mainly by the National Academy of Agricultural Science (NAAS), RDA, Korea.

7.2 Characteristics of the *B. rapa* Genome

7.2.1 *Chromosome Structure*

Cytogenetic analyses have showed that the *B. rapa* genome is organized into relatively small, compact chromosomes, with genes concentrated in the euchromatic space, unlike centromeric repeat sequences and rDNAs in the heterochromatin which are arranged as tandem arrays (Lim et al. 2005, 2007). The individual chromosome size ranges from 2.1 µm to 4.5 µm, with a total chromosome length of 32.6 µm. An alternative cytogenetic map based on a pachytene DAPI (4′,6-diamidino-2-phenylindole dihydrochloride) and fluorescent in situ hybridization (FISH) karyogram was also constructed. The mean lengths of 10 pachytene chromosomes were determined by DAPI analysis to range from 23.7 µm to 51.3 µm, with a total chromosome length of 385.3 µm. Thus, chromosomes in the meiotic prophase stage are 12 times longer than those in the mitotic metaphase, and display a well-differentiated pattern of brightly fluorescing heterochromatin segments (Koo et al. 2004).

The relative positions of 33 sequence tagged site (STS) markers and five repetitive sequences on the metaphase chromosomes were examined by FISH, and the genetic and cytogenetic correspondences between the linkage groups and chromosomes were determined (Lim et al. 2005; Kim et al. 2006). As of 2008, more than 100 gene-containing BAC clones have been analyzed by FISH, and the distribution of five major centromeric and pericentromeric repeats have been determined. By the cytogenetic method, all chromosomes can be identified based on their lengths, centromere positions, heterochromatin patterns, and positions of various repeat sequences. Centromeric satellites were estimated to encompass approximately 30% of the total chromosomes in the mitotic metaphase, particularly in the core centromeric blocks of all the chromosomes. Moreover, in the pachytene FISH, the total length of the pericentromeric heterochromatic regions was estimated to be 38.2 µm, which is approximately 10% of the total chromosome length (Koo et al. 2004). Thus, overall, heterochromatin comprises approximately 40% of the total *B. rapa* genome. All gene-containing BACs are localized to the euchromatin. The relationship between the cytogenetic pachytene FISH and the sequence contig distance at several locations in the euchromatin has been estimated to range from 400 to 500 kbp per µm. The FISH karyotype has created a rational basis for integrating the molecular, genetic, and cytogenetic maps of *B. rapa*. More importantly, it provides a basis for intelligently targeting the gene space of *B. rapa* and for strategic clone-by-clone genome sequencing that is likely to facilitate the discovery of most genes.

7.2.2 Repetitive Sequences and Heterochromatin

The rapidly evolving centromere structure consists of highly repetitive sequences, such as tandem satellite repeats and centromere-specific retrotransposons. The centromeric repeats characterized in plant genomes often extend over several millions of nucleotides, with 150–180 motifs such as the pAL1 satellite in *Arabidopsis* (Copenhaver et al. 1999), CentO in rice (Zhang et al. 2004), CentC in maize (Ananiev et al. 1998), and MtRs in *Medicago truncatula* (Kulikova et al. 2004). The composition of the repetitive sequences in the *B. rapa* genome has been surveyed by a similarity search of 10,204 BAC-end sequences, with a previously reported tandem repeat that contains a *Hind*III site at both ends (Harrison and Heslop-Harrison 1995). Approximately 30% of the repetitive sequences showed a high similarity, and two kinds of 176 bp tandem repeats were classified with 82% sequence similarity to each other (Lim et al. 2005). From the FISH data, these 176 bp tandem repeats were localized on the centromeric regions of specific chromosomes, and named as centromeric tandem repeats of *Brassica* 1 & 2 (CentBr1 & CentBr2). The CentBr1 occupies centromeres on eight chromosomes (A1, A2, A4, and A6–A10) and CentBr2 resides on only 2 chromosomes (A3 and A5). This observation may reflect the predominance of CentBr1 in the *B. rapa* genome. The CentBr repeats were also present in other *Brassica* species studied, particularly in the *B. oleracea* whole-genome shotgun sequences, indicating that these repeats are the major components of the centromeric sequences of *Brassica* genome. Furthermore, sequence analysis of the heterochromatin-specific BAC clones identified additional repeat classes, including centromere-specific Ty1-copia-like retrotransposon (CRB), 238 bp-long degenerate tandem repeat (TR238) arrays, rDNAs, and pericentromere-specific Ty3-gypsy-like retrotransposons (PCRBr) (Lim et al. 2007). CRB was one of the major components of all centromeres in the three diploid *Brassica* species and their three reciprocal allotetraploid hybrids; however, TR238 and PCRBr were A-genome specific. Characterization of these specific centromeric or pericentromeric repeat elements may be important in identifying the heterochromatin/euchromatin borders. In addition to the centromeric or pericentromeric repeats searches, the BAC-end sequence search also identified many *B. rapa*-specific sequences (~ 50% of BES) that had no similarity with any sequence of *Arabidopsis*. This finding suggested that the *B. rapa* genome expansion appears to be a result of the amplification of *B. rapa*-specific sequences, many of which probably form heterochromatic blocks of transposons or tandem repeats (Lim et al. 2005; Yang et al. 2006).

We also identified a unique group of non-autonomous LTR retrotransposons, namely terminal-repeat retrotransposons in miniature (TRIM). TRIM elements are characterized by terminal repeats (TR), ranging from 100 to 250 bp in length, encompassing an internal domain of ~ 300 bp and creating 5 bp target site duplications. The internal sequence begins with a complement of the primer-binding site of tRNA-methionine and ends with typical polypurine tract motifs. From 96 Mbp BAC-end sequences of *B. rapa*, 4 distinct lineages of TRIMs (Br1–Br4), with lengths ranging from 364 bp to 1,311 bp, were identified (Yang et al. 2007). The estimated copy number of Br TRIMs was more than six times greater in the *Brassica* species than in *Arabidopsis*, suggesting that various TRIM elements were inserted into the *Brassica* genome after divergence from the *Arabidopsis* lineage. Although the distribution of the Br TRIMs in the *B. rapa* genome is unknown, many of them appear to be located in the euchromatic region, based on the distribution of *Arabidopsis* TRIM elements in the euchromatic gene space and the occurrence of the chimeric feature of TRIMs in *B. rapa* expressed sequence tags (ESTs). The abundant TRIMs in the euchromatin of the *B. rapa* genome are expected to play an important role not only in the reconstruction of the host genome but also in the modification of the gene features by insertion of promoter or terminator sequences residing inside the elements. This modification may act as the driving force for gaining a new function, even among the duplicated genes in the *Brassica* genome.

The euchromatic distribution and higher insertion polymorphisms of the Br TRIMs have potential as molecular markers to distinguish the various *Brassica* crop species. Thus, Kwon et al. (2007) developed a transposon-display system using the unique sequences of Br1 and Br2 TRIMs. The TRIM display system successfully accessed the genetic diversity in the Brassicaceae family, and effectively identified 16 commercial F_1 hybrids of *B. rapa* and other Brassica crops (Kwon et al. 2007).

7.2.3 Triplicated Nature of the B. rapa Genome

One of the important challenges in plant genomics is to redirect the knowledge gained from the study of model genomes into the important biological and agronomical questions among crop species, based on whole-genome sequencing of model genomes. Comparative genomic studies are well established in several plant families such as cereals and legumes (Paterson et al. 2003; Choi et al. 2004; Cannon et al. 2006). Because *Brassica* species are closely related to *A. thaliana*, comparative genomic studies are aimed to elucidate the *Brassica* genome organization and evolution for comparison with the *A. thaliana* genome. One of the profound findings from the comparative analysis was the recognition of the triplicated nature of the *Brassica* genome, indicating whole-genome triplication after the split from the *A. thaliana* lineage (Lysak et al. 2005; Town et al. 2006; Yang et al. 2006).

The triplicated nature of the *B. rapa* genome was revealed by comparative genetic and physical mapping and sequence level synteny comparison. The foundation of the comparative mapping has been a restriction fragment length polymorphism (RFLP) marker-based genetic map of cv. *Jangwon*. Kim et al. (2006) used 545 genetic markers based on *B. rapa* ESTs and *Arabidopsis* genes. Among the 520 RFLP markers, a majority of the markers detected > 1 locus, and only 62 EST markers showed a single locus anchor (Kim et al. 2006). Interchromosomal comparison of the linkage groups identified many homologous blocks in the genome. Comparative physical mapping between *A. thaliana* and *B. rapa* further revealed triplicated homologous blocks, including *FLOWERING LOCUS C* (*FLC*) regions in the *B. rapa* genome with collinearity of genes in the blocks (Park et al. 2005). These findings collectively suggested that the diploid *B. rapa* genome has been derived from a hexaploid ancestor. Sequence level analysis of the triplicated *FLC* regions of *B. rapa* revealed the diploidization process in the triplicated genome. Yang et al. (2006) compared four paralogous bacterial artificial chromosome (BAC) clones covering the *FLC* loci and the homologous 124 kbp segment of *Arabidopsis* chromosome 5. They estimated the time elapsed since the divergence of the paralogous and homologous lineages, and reported that the three paralogous subgenomes of *B. rapa* triplicated around 13–17 MYA, soon after the *Arabidopsis* and *Brassica* divergence occurred around 17–18 Mya (this result was summarized in Fig. 7-1). In addition, the *B. rapa* genome underwent segmental duplication around 0.8 Mya. The *B. rapa* genome segments showed extensive interspersed gene loss and insertion of specific transposons such as TRIMs relative to the *Arabidopsis* genome segment. The diploidization process was estimated to reduce 40% of the triplicated genome length, meaning that the *B. rapa* genome would contain only approximately a 1.7-fold higher number of genes than *Arabidopsis*.

The duplicated genes retained in the *B. rapa* genome may have a selective advantage owing to the gain of new functions (neofunctionalization) or partitioning of the original function (subfunctionalization) between the two duplicates. Genome triplication and diploidization in the *Brassica* lineage would affect the functional diversification of redundant genes, because these events could permit mutations in the ancestor genome loci that are normally under tight selective constraints; the resulting subfunctionalization or neofunctionalization of the duplicated genes might lead to phenotypic diversification of *Brassica*. A study on the functional diversification of triplicated genes would provide an insight into the role of polyploidization in the *Brassica* crop genomes.

7.3 Genomic Resources for *B. rapa*

7.3.1 BAC Libraries and BAC-end Sequences

Various genomic resources are indispensable for genomic study of any crop species. The genomic resources available for the *B. rapa* Genome Sequencing Project (BrGSP) are summarized in Table 7-1. A successful structural genomic study of the *B. rapa* genome relies on the quality and availability of detailed large-insert genomic libraries. As of 2008, five large-insert BAC libraries of *B. rapa* ssp. *pekinensis* cv. *Chiifu* were publicly available, providing approximately 53-fold genome coverage overall. These libraries were constructed using the restriction enzymes *Eco*RI, *Bam*HI, *Hind*III, and *Sau*3AI.

Table 7-1 Summary of the genomic resources available for the *B. rapa* Genome Sequencing Project.

Resources	Material	Characteristics	Number[a]	Reference[b]
BAC library	5 libraries		234,544 clones (53.4X)	KBGP
KBrB	cv. *Chiifu/Bam*HI	avr. insert size 120 kbp	55,296 clones (12.5X)	
KBrE	cv. *Chiifu/Eco*RI	avr. insert size 139 kbp	23,040 clones (6X)	
KBrH	cv. *Chiifu/Hind*III	avr. insert size 120 kbp	56,448 clones (12.8X)	
KBrS1	cv. *Chiifu/Sau*3AI	avr. insert size 100 kbp	55,296 clones (10.5X)	
KBrS2	cv. *Chiifu/Sau*3AI	avr. insert size 132 kbp	46,464 clones (11.6X)	
BAC-end sequence	146,688 clones	both end	260,637 reads (203 Mbp)	KBGP, NCBI
KBrB	55,296 clones	single-pass sequence	97,912 reads (75 Mbp)	KBGP, NCBI
KBrE	23,040 clones		43,168 reads (33 Mbp)	KBGP
KBrH	50,688 clones		88,951 reads (73 Mbp)	KBGP, NCBI
KBrS1	6,144 clones		8,117 reads (5 Mbp)	KBGP, NCBI
KBrS2	11,520 clones		22,489 reads (17 Mbp)	KBGP
Genetic map	3 reference maps			
JWF3p	cv. *Jangwon*	$F_{2:3}$ population 134 lines	1,287 cM, 545 markers	Kim et al. 2006
CK	cv. *Chiifu* × cv. *Kenshin*	DH F_2 population 78 lines	1,182 cM, 556 markers	Choi et al. 2007
VCS	cv. *VC1* × cv. *SR5*	DH F_2 population 91 lines	837 cM, 355 markers	KBGP
Physical map				
Build 2	67,468 BAC fingerprints	HICF map	1,428 contigs (717 Mbp)	Mun et al. 2008
Expressed sequence tag	33 cDNA libraries of cv. *Chiifu*	cDNA single-pass sequence	152,253 ESTs (91 Mbp)	KBGP
Microarray	2 NimbleGen chips			
KBGP-24K	24,000 unigenes	60mer, 6 probes/gene		Lee et al. 2008
KBGP-50K	32,000 unigenes, 17,000 CDS	60mer, 7 probes/gene		KBGP
BAC sequence	KBrB, KBrE, KBrH, KBrS clones	BAC shotgun sequence	Ongoing[c]	KBGP, NCBI

[a]Total sequence length or genome coverage is represented in parenthesis. Genome coverage was estimated based on the haploid genome equivalent of *B. rapa* as 529 Mbp.
[b]KBGP, Korea *B. rapa* Genome Project (*http://www.brassica-rapa.org*). NCBI, the National Center for Biotechnology Information.
[c]As of December 2008, 886 BAC sequences have been deposited in NCBI.

Using these BAC libraries, a total of 260,637 BAC-end sequences (BES) have been generated from 146,688 BAC clones (~ 203 Mbp), as a collaborative outcome of the multinational BrGSP community. Analysis of BES, combined with BAC sequence surveys, enabled the outlining of the features of the whole-genome structure of *B. rapa*. BLAST search of BES identified that up to 25% of BES were estimated to contain centromeric or pericentromeric repetitive sequences. An additional 15% of BES were matched with transposons and other repeat sequences. Based on this data and FISH analysis (see Section 7.2.1), the heterochromatic region was postulated to occupy > 40% of the *B. rapa* genome, while the euchromatic gene space was postulated to constitute < 60% of the *B. rapa* genome (Yang et al. 2005; Lim et al. 2007). Comparison of the BES with the *B. rapa* ESTs and *Arabidopsis* CDS led to the recognition of approximately 11% of the BES as protein-coding genes. Assuming that the average CDS size of the BAC survey sequences ranges from 1.1 to 1.3 kbp, the estimated number of genes in the whole genome would be at least 45,000 to 53,000, which is roughly similar to the estimation of Yang et al. (2006).

Comparison of the BES with the *Arabidopsis* genome was an efficient application of the BES dataset to select the seed BAC clones for *Brassica* genome sequencing (Yang et al. 2005). In silico comparative sequence matching of 91,000 *B. rapa* BES to the *A. thaliana* chromosome sequences identified approximately 50% of the BES showing significant sequence similarity along with overall collinearity to counterpart *Arabidopsis* chromosomal regions. Based on the comparative genetic map and microcollinearity between the two genomes, the BAC clones mapped onto the *Arabidopsis* genome can be chosen as seed BAC clones, even before a complete physical map is established. Practically, approximately 600 seed BAC clones were selected and sequenced by this method. The details of BES matching and seed BAC clone selection are explained in the section on genome sequencing (see Section 7.4.2). Almost all the BAC-end sequenced clones were also fingerprinted, and the BES data and physical contig information was combined in the BAC extension program of genome sequencing to select the BAC clones from a seed point to be sequenced. This tool uses the BLAST algorithm to match one or more input sequences with any sequence associated with the fingerprinted clones, similar to the Blast Some Sequence (BSS) tool in the FingerPrinted Contig (FPC) program.

7.3.2 Genetic Map

A genetic map is the basic starting point for any genomic study or sequencing. Genetic maps of *B. rapa* have been reported using various mapping populations. Most of the maps were constructed using hybridization-based markers. Recently, three high-density genetic maps of *B. rapa* ssp. *pekinensis* have been reported as references to guide the ongoing genome sequencing. One of them is an intercultivar map based on a cv. *Chiifu* × cv. *Kenshin* cross (Choi et al. 2007). The CK map used 78 doubled haploid lines, providing direct and accurate genetic information for *Chiifu* genome sequencing. For this reason, the BrGSP community chose the CK map as the reference map for the ongoing genome sequencing project. The CK map consists of a total of 556 markers over a length of 1,182 cM; however, the number of anchored BACs in the sequencing pipeline is limited. The second map is based on an intracultivar $F_{2:3}$ linkage map based on the cultivar *Jangwon* cross (Kim et al. 2006), and was first developed using 520 RFLP markers and 25 PCR-based markers over a length of 1,287 cM. Although the Jangwon map does not represent perfect information for *Chiifu* sequencing, it plays a crucial role in determining the positioning of the sequenced BAC clones. Most of the RFLP markers of the Jangwon map have been replaced by PCR-based sequence tagged site (STS) markers, including gene-targeting or simple sequence repeat (SSR) markers designed on the BAC sequences. As of 2008, a total of 513 sequenced BACs have been anchored on the Jangwon map, which contains 905 genetic loci over a length of 1,311 cM. A third genetic map has been established from a cross between cv. *VC1* and cv. *SR5* (*http://www.brassica-rapa.org*). The VCS map gives the positions of 225 sequenced BACs that are not included in the Jangwon map. In addition, the VCS map provides enhanced genetic resolution in regions where suppression of recombination or distorted segregation was shown in the previous maps. Integration of these three maps is underway using a set of common anchored markers, and an integrated map will soon be available to the sequencing community.

The availability of molecular markers, high-density genetic maps, and the genome sequence has significantly accelerated the identification and isolation of genes that are important for crop quality improvement and development, for example, the genes pertaining to disease resistance, flowering time, or biosynthesis of valuable phytochemicals. In addition, high levels of microsynteny between *B. rapa* and other *Brassica* species allows for the possibility of using syntenic positional cloning as a basis for isolating orthologous genes of interest in other crops of the Brassicaceae family.

7.3.3 Physical Map

The availability of a genome-wide, sequence-ready physical map is one of the crucial components for a successful clone-by-clone strategy. A physical map not only makes it possible to determine clones for genome sequencing with comprehensive coverage and reduced sequencing redundancy, but also enables one to simplify the sequence assembly by arranging the sequence contigs in order. Thus far, the utility of physical maps has been reported by major genome sequencing projects on the human (International Human Genome Sequencing Consortium 2001), *A. thaliana* (Marra et al. 1999), rice (Chen et al. 2002), and *M. truncatula* (Mun et al. 2006). These physical maps were constructed using a combination of techniques, including restriction enzyme-digested BAC fragment fingerprinting on agarose gels and assembly of the fingerprints using the FPC software package (Soderlund et al. 2000). The agarose method has been successful, but has limited throughput because of the need for manual band calling. Alternatively, fluorescence-labeled fingerprinting methods have been developed to make larger and more accurate contigs, with increased throughput using an automatic capillary sequencer (Gregory et al. 1997; Ding et al. 2001; Luo et al. 2003; Xu et al. 2004). Fluorescence-labeled capillary electrophoresis methods include the 3-enzyme method and the high-information content fingerprinting (HICF) methods, which use type IIS restriction enzymes or SNaPshot labeling techniques, respectively. As compared to the agarose method, the automatic workflow and higher resolution of these methods facilitates better physical map construction, both in terms of throughput and the quality of fingerprinting. The first genome-wide plant HICF physical map was constructed for maize (Nelson et al. 2005).

We constructed a genome-wide BAC-based physical map of *B. rapa* by the SNaPshot method (Mun et al. 2008). To create a robust sequence-ready physical map, a total of 99,456 BAC clones from the three independent BAC libraries (~ 22.5 × coverage) were fingerprinted by digestion with combinations of five restriction enzymes (*Eco*RI, *Bam*HI, *Xba*I, *Xho*I, and *Hae*III), followed by SNaPshot reagent labeling of four colors at the 3′-ends of the restriction fragments, then sizing on the ABI 3730xl capillary sequencer. Of the fingerprints, a total of 93,689 clones (94.2%) were successfully fingerprinted to be used for contig assembly. From the initial dataset, 26,221 BAC clones containing heterochromatic repetitive sequences were removed from the contig assembly, which significantly enriches the euchromatic contigs in the resulting build. The physical contig map was assembled using 67,468 high-quality, heterochromatic repeat-free BAC fingerprints from the initial dataset. These BAC clones represent 15.2× coverage of the *B. rapa* genome; they were condensed into 1,417 contigs, and the resulting contigs were manually edited to validate reliability. With the results of the contig evaluation, manual editing of the initial contig build yielded 1,428 contigs, with an average length of 512 kbp spanning 717 Mbp, 1.3× coverage of the genome. An unsatisfactory aspect of this assembly is its large number of Q clones, as the Q clones in this assembly corresponded to 15% of the clones. However, 3 specific deep contigs contributed to ~ 48.3% of all the Q clones in the build. Thus, when these deep contigs of the initial build were excluded due to false-positive overlaps, the Q clones in the remaining contigs correspond to 7.7% of all clones. This ratio is similar to the ratios reported in catfish (7.3%) (Quiniou et al. 2007) and maize (11%) HICF maps (Nelson et al. 2005).

The contigs produced in the course of the fingerprinting work were tagged with 315 anchored genetic markers. Practically, the important aspect of this was the integration of a physical map into a genetic map, enabling the positioning of 242 gene-rich contigs to specific locations on 10 chromosomes and providing seeds for the genome sequencing effort. The number of contigs associated with genetic loci is ~ 160.7 Mbp, or 30% of the total genome. The total coverage of the physical contigs suggests that most contigs do not have sufficient overlaps, and the gaps between the contigs need to be filled by additional fingerprinting. To improve the map, additional fingerprinting of approximately 30,000 clones of 2 new BAC libraries (KBrE and KBrS2) was performed. This data is being merged into the current build to continue the refinement of the physical map. In 2009, updating of the physical map was kept open for the *B. rapa* genome sequencing consortium. In parallel, linkage analysis of SSRs and single nucleotide polymorphisms (SNPs) in the seed BACs and BESs will be carried out to provide more anchoring points for the physical map. The details of these BACs and the associated markers can be found on the Korea *B. rapa* Genome Project (KBGP) website of NAAS (*http://www.brassica-rapa.org*). This information plays an important role in identifying the BAC clones for sequencing.

7.3.4 Expressed Sequence Tags and Transcriptome Analysis

To support genome annotation and functional study, we have sequenced a total of 152,253 *B. rapa* ESTs. These ESTs are obtained from 33 cDNA libraries representing a variety of organs and development stages and have been deposited in public databases. These ESTs were clustered into 39,095 unique sequences (unigenes), including 16,898 tentative consensus sequences. The current collection of *B. rapa* ESTs consistently represents the whole *B. rapa* genome, with the *Arabidopsis* genome as a reference. In silico mapping of the 39,095 unigenes on the *Arabidopsis* genome identified 85% of the unigenes, covering 75% of the overall *Arabidopsis* counterpart coding sequences. It was found that the remaining 15% of the *B. rapa* unigenes were not homologous with any gene in *A. thaliana*, rather, this 15% represented novel *B. rapa*-specific genes. Gene ontology analysis of the ESTs did not show any significant overestimation of specific categories in the *B. rapa* genome, as compared to the *Arabidopsis* genome. Analysis of the EST collections identified 21,409 full-length cDNA sequences. It is anticipated that these sequences will contribute to the formation of a *B. rapa*-specific set of sequences, not only for gene prediction programs for *B. rapa* and other *Brassica* species, but also to evaluate structure and alternative splicing of the predicted gene models.

The corresponding cDNA clones have been used to construct microarrays for expression profiling during development, under various biotic- and abiotic-stress conditions. Two microarrays, namely KBGP-24K and KBGP-50K, were developed using the NimbleGen platform. Both microarrays included the six (KBGP-24K) or seven (KBGP-50K) 60-nucleotide-long probes per gene. The 24K chip covered approximately 24,000 unigenes clustered by 127,144 ESTs from 20 cDNA libraries, whereas the 50K chip doubled the gene contents by including an additional 8,500 unigenes, plus 17,500 genes predicted from the seed BAC sequences in the genome sequencing pipeline. These microarrays examined the changes in the genome-wide gene expression of *B. rapa*, in response to transcriptional changes. Using the KBGP-24K chip, genome-wide transcriptome analysis was conducted in response to three abiotic stresses that significantly affect the productivity of *Brassica* crops: salt, cold, and drought (Lee et al. 2008). This analysis successfully identified stress-related genes along with novel transcription factor genes, suggesting the existence of a *B. rapa*-specific signaling pathway that works together with the common stress-response pathway under abiotic stress conditions. Information on the EST, unigene, and microarray experiments can be retrieved from the *B. rapa* EST and Microarray Database (BrEMD) website (*http://www.brassica-rapa.org/BrEMD*).

7.3.5 Functional Genomics and Information Resources

The online database (*http://www.brassica-rapa.org*) provides easy access to genomics information such as nucleotide sequences of the BAC clones with tentative annotation of the predicted genes, genetic maps along with the molecular markers, a physical map that can be viewed on WebFPC, cDNA sequences, and various transcriptome profiling data obtained by KBGP-24K microarray analysis. The BAC and cDNA libraries and clones can also be ordered through this site. The Korea *Brassica* Genome Resource Bank (KBGRB, *http://www.brassica-resource.org*) is a material resource database of *B. rapa*, and is supported by the Korea National Plant Research Resource Center. Resources can be obtained through this database, including the seeds of various *B. rapa* wild accessions and cultivars, recombinant inbred (RI) lines, mapping populations, and BAC and cDNA libraries.

A large number of T-DNA insertion mutant lines are being generated as resources for reverse genetics, and a targeting induced local lesions in genomes (TILLING) system initiated by the Multinational *Brassica* Genome Project (MBGP) TILLING Consortium has been developed. TILLING is a high-throughput, reverse genetic approach that combines the efficiency of ethylmethane sulfonate (EMS) with high-throughput genotyping methods to detect SNPs. The MBGP TILLING Consortium aims to share the EMS mutant populations developed from the *Chiifu* (Korea), rapid-cycling (USA), *R-O-18* (UK), and *L58* (China) cultivars for functional genomic studies. The large quantity of genetic information provided by the BrGSP will make the TILLING approach efficient and feasible.

7.4 Progress and Current Status of *B. rapa* Genome Sequencing

7.4.1 Sequencing of Euchromatic Regions Based on the Clone-by-Clone Strategy

Despite the importance of *Brassica* crops in plant biology and world agriculture, none of the *Brassica* species have been fully sequenced. Considering the importance of the *Brassica* species and the value of *B. rapa* as a model *Brassica*, sequencing most of its gene spaces will be very beneficial. Clone-by-clone sequencing and whole-genome shotgun (WGS) sequencing are both well-established genome sequencing approaches, and can be used to investigate the whole-genome structure of the species. We selected the clone-by-clone genome sequencing strategy because of its accuracy for the complex triplicated nature of the *Brassica* genome. The clone-by-clone strategy typically involves a "minimum tiling path" of large-insert (~ 100 kbp) clones, such as BACs of a known order, which is determined using a combination of genetic, physical, and/or cytogenetic mapping. Cytogenetic studies based on extensive fluorescence in situ hybridization (FISH) analysis of both metaphase and pachytene chromosomes have provided detailed insight into the organization of heterochromatic and euchromatic regions. These cytogenetic studies demonstrated that the genome of *B. rapa* is organized into distinct repetitive sequence-rich pericentromeric heterochromatin and gene-rich euchromatin. Sequencing of several BAC clones has confirmed that the gene density in *B. rapa* is relatively high on the order of one gene per 3–4 kbp (Yang et al. 2006). Each of the gene-rich BAC clones examined so far by FISH (> 100 BACs) was found to be localized to the visible euchromatic region of the genome. Concurrently, a whole-genome shotgun pilot sequencing of *B. oleracea* with 0.44-fold genome coverage generated transposable element-enriched sequences (Zhang and Wessler 2004; Ayele et al. 2005). Taken together, these data strongly point to a favorable genome organization where the overwhelming majority of the *B. rapa* euchromatic space can be sequenced in a highly efficient manner by the clone-by-clone strategy.

7.4.2 Initial Approach for Selection of Seed BACs Based on Comparative Genomics

The clone-by-clone sequencing starts from the defined seed points and builds outward. The fingerprint-based physical map, combined with BES and genetic anchoring data, provides a basis for selecting seed BAC clones and for creating a draft tiling path. Alternatively, comparative approaches using an already sequenced closely related model genome, such as the comparative tiling method, can be used as a backbone for in silico clone validation of seed BACs, even before the availability of a physical map. Indepth comparative sequence analyses of several sequenced *B. rapa* BAC clones revealed overall collinearity with a homologous region of the *Arabidopsis* counterpart. Thus, if we allocate *B. rapa* BAC clones on the *Arabidopsis* euchromatin using BES matches and scatter them onto the *B. rapa* chromosomes, selection of gene-rich BAC clones will be straightforward even without the information from the physical map (Fig. 7-2). Based on this idea, Yang et al. (2005) compared the BES of 46,000 *B. rapa* BAC clones with *Arabidopsis* sequences using BLASTZ. They mapped the BAC clones on the five *Arabidopsis* chromosomes by in silico matching based on unique, significant ($< E^{-6}$), and directional matches of the pair ends sequences of each BAC with a complement match within the 30–500 kbp interval (Yang et al. 2005). Approximately 50% of the BES showed significant sequence similarity with unique *A. thaliana* sequences, and a total of 4,317 BAC clones were mapped onto the *Arabidopsis* chromosomes by significant matching with both ends within the 30-500 kbp interval. These *B. rapa* BAC clones spanned 93 Mbp of the *A. thaliana* sequences, representing ~ 78% of the total *Arabidopsis* genome. BAC-FISH and STS mapping using BES of selected BAC clones positioned on the counterpart *Arabidopsis* chromosomes showed the real euchromatic locations of the BAC clones scattered on the *B. rapa* chromosomes. A single *B. rapa* BAC clone was calculated to span an average of 147 kbp of the *A. thaliana* counterpart sequence. Theoretically, 500 contiguous BAC clones will cover around 80 Mbp of the euchromatic regions of the *Arabidopsis* genome, assuming the average insert size of a BAC clone is 120 kbp. Therefore, if minimally overlapping BAC clones mapped onto the *Arabidopsis* genomes were selected and scattered onto the *B. rapa* chromosomes, they could provide a seed point for bidirectional outward genome sequencing. As of July, 2008, 589 minimally tiled *B. rapa* BAC clones spanning 75 Mbp of the *A. thaliana* genome were sequenced in phase 3 (finished sequences) or phase 2 (sequences that are fully oriented and ordered, but contain some small sequence gaps and low-quality regions). Most of them were distributed onto the 10 *B. rapa* chromosomes by STS mapping, FISH analysis, and physical contig information (Fig. 7-2). The 589 sequenced BAC clones were provided to BrGSP as seed BACs and used as starting points for chromosome sequencing (data available on the KBGP website). Integration of seed BACs with the physical map provides "gene-rich" contigs spanning ~ 160 Mbp (see Section 7.3.3). These "gene-rich" contigs will enable us to select clones to extend the existing sequence contigs.

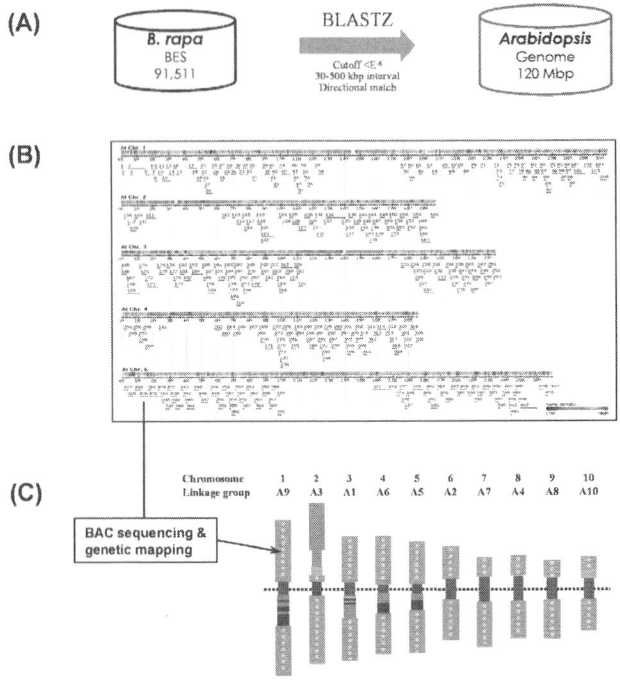

Figure 7-2 Schematic representation of the comparative tiling sequencing method to select seed BAC clones from the euchromatic regions of *B. rapa,* using in silico matching of BESs with known *Arabidopsis* sequences. (A) Based on the knowledge of the sequence level collinearity between *Arabidopsis* and *Brassica,* BAC-end sequences of *B. rapa* ssp. *pekinensis* cv. *Chiifu* were BLAST analyzed against *Arabidopsis* genome sequence and the corresponding BACs were selected based on the pair-end best hits within 30–500 kbp interval. (B) Selection of minimally overlapping 589 BAC clones based on *in silico* comparative allocation on the *Arabidopsis* genome. A total of 4,317 *B. rapa* BAC clones with an average insert size of 120 kbp were mapped in silico onto the euchromatic regions of the *Arabidopsis* genome. On in silico comparative allocation, 589 minimally overlapping BAC clones were mapped onto the comparative positions in the corresponding *Arabidopsis* genome sequence, then were selected and sequenced as phase 2 and scattered on the euchromatic regions of the 10 *B. rapa* chromosomes as seed BACs by a combination of genetic and cytogenetic mapping. Mixed-color bar represents each chromosome of *Arabidopsis* and blue bars depict *B. rapa* BAC contigs tiled on the counterparts of *Arabidopsis* genome. (C) Based on the sequence of each BAC clone, we developed STS markers and anchored the clones on the genetic maps. Overall, 513 sequenced BAC clones were mapped onto the 10 linkage groups and provided to each sequencing center of the BrGSP for completion of the chromosome sequencing. Detailed numbers of BAC clones on each chromosome are denoted in Table 7.3.

Color image of this figure appears in the color plate section at the end of the book.

7.4.3 Features of Protein Coding Genes

The predicted structures of the potential protein coding genes in the seed BAC clones, deposited in the High Throughput Genome Sequence (HTGS) database of the National Center for Biotechnology Information (NCBI), have been predicted ab initio using FGENESH (*www.softberry.com*) based on a *B. rapa* matrix (Table 7-2). Although the structural features of the protein coding genes in *B. rapa* are similar to those of the *B. oleracea* genes reported previously (Town et al. 2006), the former show a smaller average gene length (1.6 kbp) as compared to *Arabidopsis* (2.2 kbp). This difference appears to amount to almost one less exon per gene (4.7 and 5.8 exons per gene in *B. rapa* and *A. thaliana*, respectively) along with a shorter exon (225 bp in *B. rapa* and 268 bp in *A. thaliana*, respectively) and intron length (141 bp in *B. rapa* and 165 bp in *A. thaliana*, respectively) in *B. rapa*. The gene density in the sequenced BAC clones of the *B. rapa* genome (one gene per 4.2 kbp) is higher than that in the *Arabidopsis* genome (one gene per 4.5 kbp), indicating the compact organization of the euchromatic region of *B. rapa*. A similarity search of the protein coding genes of *B. rapa* against public databases indicated that approximately 18% of the predicted genes showed no significant similarity with any of the genes reported. This result is roughly consistent with the results of previous studies conducted using EST analysis (see Section 7.3.4) and synteny analysis of *FLC* regions (Yang et al. 2006). NAAS, the leading sequencing center for BrGSP, is developing an automated genome annotation pipeline that combines gene prediction and homology data, such as *B. rapa* ESTs and full-length cDNAs, to produce the best working model. High-quality sequences and first-pass annotation will be released by the center to the user community in a timely fashion.

Table 7-2 Comparison of overall composition of annotated protein coding genes in the *B. rapa* BAC sequences and *A. thaliana* genome.

Feature	*B. rapa*	*A. thaliana*[a]
Average gene size (kbp)	1.6	2.2
Number of exons per gene	4.7	5.8
Intron size (bp)	141	165
Exon size (bp)	225	268
Average gene density (kbp/gene)	4.2	4.5
Overall G/C content (%)	35.2	35.9
Exons	46.3	44.1
Introns	32.6	32.6
Intergenic regions	31.3	32.9

[a]Statistics for *A. thaliana* is based on the annotation of TAIR7 from the *Arabidopsis* Information Resource website (*http://www.arabidopsis.org/portals/genAnnotation/genome_snapshot.jsp*).

7.4.4 Sequencing Status and Progress

The BrGSP has decided to sequence the euchromatic arms of all 10 chromosomes, with sequencing responsibilities distributed as represented in Fig. 7-3. The project aims to initially produce a "phase 2" sequence with accessible trace files by shotgun, so that the researchers who require finished sequences from a specific region can finish them. BAC clones are sequenced by shotgun and standard assembly methods, then gene annotation is conducted by a combination of a semiautomatic method and manual editing. STS markers (including SSR or SNP markers generated from the BAC sequences) will localize BAC clones onto the reference genetic map.

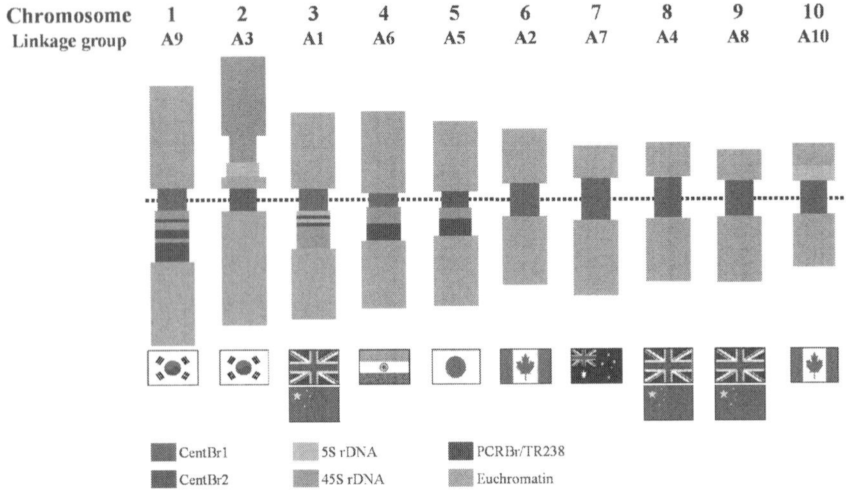

Figure 7-3 Idiogram of *B. rapa* chromosomes represented by euchromatic arms and heterochromatic repetitive sequence blocks. Each chromosome was assigned to a sequencing center in one of the seven countries of the BrGSP consortium.

Color image of this figure appears in the color plate section at the end of the book.

Although the details are different, each sequencing center has adopted similar BAC sequencing strategies. Large-insert BAC clones are isolated by a semi-automated "cleared lysate" procedure and are tested for clone fidelity. Shotgun libraries are produced from both small and large inserts generated by hydrodynamic shear, followed by size selection and cloning. The resulting size-fractionated, sheared fragments are blunt-end cloned after end-filling and agarose gel sizing. The libraries are sequenced, up to a total of 7- to 8-fold sequence coverage. The sequencing reactions are optimized to maximize the read lengths on the ABI 3730xl sequencers. Generally, 700-bp high-quality sequences are obtained by capillary running and read after vector trimming. Base-calling and sequence assembly are performed using the Phred/Phrap/Consed software package, and the new assemblies are released to the HTGS database of NCBI. Sequencing technology is a dynamic field, and it is possible that opportunities that will enable the centers to sequence more efficiently may arise during the course of the project. Advances in sequencing technology using next-generation sequencing machines could lower the sequencing costs. In fact, NAAS tested a combination of GS-FLX and ABI 3730xl sequencing machines that could reduce the sequencing costs by almost 20%; however, the lower resolution of the poly-mononucleotide tracks in the pyrosequencing data required additional validation.

As of December 2008, a total of 107 Mbp from 886 BAC clones (401 phase 3, 445 phase 2, and 40 phase 1 clones) have been sequenced and deposited in the HTGS database of NCBI by 3 sequencing centers: Korea (681 BACs), UK/China (200 BACs), and Australia (five BACs). The other participating sequencing center will soon increase its capacity to facilitate sequencing. Besides the reported BAC clones, NAAS sequenced an additional 884 BAC clones as a result of sequencing of the chromosomes A3 and A9 (Table 7-3). Using 670 BAC clones, the sequence scaffolds for the two chromosomes were constructed. In the case of A3, 381 minimum tiled BAC sequences generated seven sequence scaffolds spanning 34.9 Mbp; however, in the case of A9, 289 minimally tiled BAC clones generated 15 scaffolds comprising 28.5 Mbp (Fig. 7-4). Based on the estimated size of the *B. rapa* euchromatic region, the assemblies thus far constitute 90% and 85% of the entire euchromatic region of chromosome A3 and A9, respectively (our unpublished data). Additional effort is ongoing, in an attempt to fill the gaps between the scaffolds. The details of the sequences will be released soon with annotation.

Table 7-3 Current status of *B. rapa* BAC sequencing by NAAS, as of March 2009.

Chromosome	A1	A2	A3	A4	A5	A6	A7	A8	A9	A10	Unanchored	Total
Anchored seed BACs	56	62	94	14	53	58	50	47	58	34		526
Extended BACs	1	5	441	6	1	14	7	7	291	12		785
Total	57	67	535	20	54	72	57	54	349	46	543	1,854

7.5 Conclusion

B. rapa will provide an excellent reference genome sequence, as well as valuable information for understanding the genetic systems of the *Brassica* crop species. The most significant beneficiaries of *B. rapa* sequences will be *Brassica* crop researchers, from breeders to plant biologists. A large number of *B. rapa* EST collections, full-length cDNA sets, and the genome sequences will enable the identification and isolation of the genes of interest for agriculture. High-density genetic maps, STS markers, and associated sequence information have accelerated genome mapping and map-based

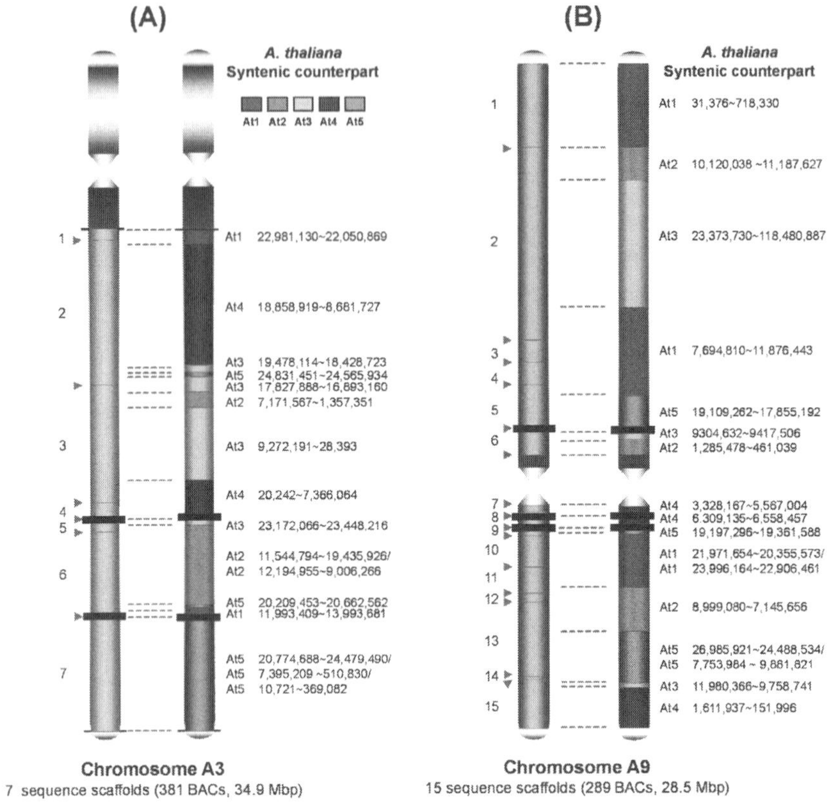

Figure 7-4 Current sequencing status of the two longest chromosomes, A3 and A9. BAC clones were sequenced as phase 2 and assembled into sequence scaffolds. (A) A total of 381 minimally tiled BAC clones generated seven sequence scaffolds spanning 34.9 Mbp for chromosome A3. (B) 15 scaffolds spanning 28.5 Mbp were assembled from 289 minimum tiled BAC clones for chromosome A9. Syntenic counterpart regions of the *A. thaliana* genome identified by sequence comparison were also represented in the colored boxes, along with position information. Red arrow heads indicate major gaps between scaffolds.

Color image of this figure appears in the color plate section at the end of the book.

cloning. The comparative genomics approach for *B. rapa* will also benefit genomic investigation of closely related *Brassica* crops, including *B. oleracea* and *B. napus*. Quantitative trait loci (QTL) or association mapping of valuable phytochemical-related genes in *B. napus* utilizes the available *B. rapa* genome sequence data and resources (Meng, Huazhung, pers. comm.). In addition, genome sequencing of other *Brassica* crops, particularly the construction of sequence assemblies and scaffolds of *B. napus*, will be improved owing to the presence of information on the *B. rapa* genome. Technically, newly emerging next generation sequencing techniques using GS FLX Titanium (Roche), Genome Analyzer (Illumina), and SOLiD (ABI) systems can be applied to other *Brassica* genomic studies with the *B. rapa* genome as a reference and eventually contribute to accelerate the genomics progress, biological study, and molecular breeding in every *Brassica* crops.

Acknowledgements

We thank the many participants of the Korea *Brassica rapa* Genome Project. This work was supported by the National Academy of Agricultural Science (05-1-12-2-1, PJ006759, and PJ006691) and by the BioGreen 21 Program (20050301034438), Rural Development Administration, Korea.

References

Ananiev EV, Phillips RL, Rines HW (1998) Chromosome-specific molecular organization of maize (*Zea mays* L.) centromeric regions. Proc Natl Acad Sci USA 95: 13073–13078.

Ayele M, Haas BJ, Kumar N, Wu H, Xiao Y, Van Aken S, Utterback TR, Wortman JR, White OR, Town CD (2005) Whole genome shotgun sequencing of *Brassica oleracea* and its application to gene discovery and annotation in *Arabidopsis*. Genome Res 15: 487–495.

Beilstein MA, Al-Shehbaz IA, Kellogg EA (2006) Brassicaceae phylogeny and trichome evolution. Am J Bot 93: 607–619.

Bowers JE, Chapman BA, Rong J, Paterson AH (2003) Unravelling angiosperm genome evolution by phylogenetic analysis of chromosomal duplication events. Nature 422: 433–438.

Cannon SB, Sterck L, Rombauts S, Sato S, Cheung F, Gouzy J, Wang X, Mudge J, Vasdewani J, Schiex T, Spannagl M, Monaghan E, Nicholson C, Humphray SJ, Schoof H, Mayer KF, Rogers J, Quétier F, Oldroyd GE, Debellé F, Cook DR, Retzel EF, Roe BA, Town CD, Tabata S, Van de Peer Y, Young ND (2006) Legume genome evolution viewed through the *Medicago truncatula* and *Lotus japonicus* genomes. Proc Natl Acad Sci USA 103: 14959–14964.

Chen M, Presting G, Barbazuk WB, Goicoechea JL, Blackmon B, Fang G, Kim H, Frisch D, Yu Y, Sun S, Higingbottom S, Phimphilai J, Phimphilai D, Thurmond S, Gaudette B, Li P, Liu J, Hatfield J, Main D, Farrar K, Henderson C, Barnett L, Costa R, Williams B, Walser S, Atkins M, Hall C, Budiman MA, Tomkins JP, Luo M, Bancroft I, Salse J, Regad F, Mohapatra T, Singh NK, Tyagi AK, Soderlund C, Dean RA, Wing RA (2002) An integrated physical and genetic map of the rice genome. Plant Cell 14: 537–545.

Choi HK, Mun J-H, Kim DJ, Zhu H, Baek JM, Mudge J, Roe B, Ellis N, Doyle J, Kiss GB, Young ND, Cook DR (2004) Estimating genome conservation between crop and model legume species. Proc Natl Acad Sci USA 101: 15289–15294.

Choi SR, Teakle GR, Plaha P, Kim JH, Allender CJ, Beynon E, Piao ZY, Soengas P, Han TH, King GJ, Barker GC, Hand P, Lydiate DJ, Batley J, Edwards D, Koo DH, Bang JW, Park BS, Lim YP (2007) The reference genetic linkage map for the multinational *Brassica rapa* genome sequencing project. Theor Appl Genet 115: 777–792.

Copenhaver GP, Nickel K, Kuromori T, Benito MI, Kaul S, Lin X, Bevan M, Murphy G, Harris B, Parnell LD, McCombie WR, Martienssen RA, Marra M, Preuss D (1999) Genetic definition and sequence analysis of *Arabidopsis* centromeres. Science 286: 2468–2474.

Ding Y, Johnson MD, Chen WQ, Wong D, Chen YJ, Benson SC, Lam JY, Kim YM, Shizuya H (2001) Five-color-based high-information-content fingerprinting of bacterial artificial chromosome clones using type IIS restriction endonucleases. Genomics 74: 142–154.

Economic Research Service USDA. (2008) Vegetables and melons outlook: *http://www.ers.usda.gov/Publications/VGS/Tables/World.pdf*

Gregory SG, Howell GR, Bentley DR (1997) Genome mapping by fluorescent fingerprinting. Genome Res 7: 1162–1168.

Harrison GE, Heslop-Harrison JS (1995) Centromeric repetitive DNA sequences in the genus *Brassica*. Theor Appl Genet 90: 157–165.

International Human Genome Sequencing Consortium (2001) A physical map of the human genome. Nature 409: 934–941.

Johnston JS, Pepper AE, Hall AE, Chen ZJ, Hodnett G, Drabek J, Lopez R, Price HJ (2005) Evolution of genome size in Brassicaceae. Ann Bot 95: 229–235.

Kim JS, Chung TY, King GJ, Jin M, Yang TJ, Jin YM, Kim HI, Park BS (2006) A sequence-tagged linkage map of *Brassica rapa*. Genetics 174: 29–39.

Koo DH, Plaha P, Lim YP, Hur Y, Bang JW (2004) A high-resolution karyotype of *Brassica rapa* ssp. *pekinensis* revealed by pachytene analysis and multicolor fluorescence in situ hybridization. Theor Appl Genet 109: 1346–1352.

Kulikova O, Geurts R, Lamine M, Kim DJ, Cook DR, Leunissen J, de Jong H, Roe BA, Bisseling T(2004) Satellite repeats in the functional centromere and pericentromeric heterochromatin of *Medicago truncatula*. Chromosoma 113: 276–283.

Kwon SJ, Kim DH, Lim MH, Long Y, Meng JL, Lim KB, Kim JA, Kim JS, Jin M, Kim HI, Ahn SN, Wessler SR, Yang TJ, Park BS (2007) Terminal repeat retrotransposon in miniature (TRIM) as DNA markers in *Brassica* relatives. Mol Genet Genom 278: 361–370.

Lee SC, Lim MH, Kim JA, Lee SI, Kim JS, Jin M, Kwon SJ, Mun JH, Kim YK, Kim HU, Hur Y,Park BS (2008) Transcriptome analysis in *Brassica rapa* under the abiotic stresses using *Brassica* 24K oligo microarray. Mol Cells 26: 595–606.

Lim KB, de Jong H, Yang TJ, Park JY, Kwon SJ, Kim JS, Lim MH, Kim JA, Jin M, Jin YM, Kim SH, Lim YP, Bang JW, Kim HI, Park BS (2005) Characterization of rDNAs and tandem repeats in the heterochromatin of *Brassica rapa*. Mol Cells 19: 436–444.

Lim KB, Yang TJ, Hwang YJ, Kim JS, Park JY, Kwon SJ, Kim J, Choi BS, Lim MH, Jin M, Kim HI, de Jong H, Bancroft I, Lim YP, Park BS (2007) Characterization of the centromere and peri-centromere retrotransposons in *Brassica rapa* and their distribution in related *Brassica* species. Plant J 49: 173–183.

Lukens LN, Quijada PA, Udall J, Pires JC, Schranz ME, Osborn TC (2004) Genome redundancy and plasticity within ancient and recent *Brassica* crop species. Biol J Linn Soc Lond 82: 665–674.

Luo MC, Thomas C, You FM, Hsiao J, Ouyang S, Buell CR, Malandro M, McGuire PE, Anderson OD, Dvorak J (2003) High-throughput fingerprinting of bacterial artificial chromosomes using the SNaPshot labeling kit and sizing of restriction fragments by capillary electrophoresis. Genomics 82: 378–389.

Lysak MA, Koch MA, Pecinka A, Schubert I (2005) Chromosome triplication found across the tribe *Brassiceae*. Genome Res 15: 516–525.

Marra M, Kucaba T, Sekhon M, Hillier L, Martienssen R, Chinwalla A, Crockett J, Fedele J, Grover H, Gund C, McCombie WR, McDonald K, McPherson J, Mudd N, Parnell L, Schein J, Seim R, Shelby P, Waterston R, Wilson R (1999) A map for sequence analysis of the *Arabidopsis thaliana* genome. Nat Genet 22: 265–270.

Mun J-H, Kim DJ, Choi HK, Gish J, Debelle F, Mudge J, Denny R, Endre G, Saurat O, Dudez AM, Kiss GB, Roe B, Young ND, Cook DR (2006) Distribution of microsatellites in the genome of *Medicago truncatula*: A resource of genetic markers that integrate genetic and physical maps. Genetics 172: 2541–2555.

Mun J-H, Kwon SJ, Yang TJ, Kim HS, Choi BS, Baek S, Kim JS, Jin M, Kim JA, Lim MH, Lee SI, Kim HI, Kim H, Lim YP, Park BS (2008) The first generation of a BAC-based physical map of *Brassica rapa*. BMC Genom 9: 280.

Nelson WM, Bharti AK, Butler E, Wei F, Fuks G, Kim H, Wing RA, Messing J, Soderlund C (2005) Whole-genome validation of high-information-content fingerprinting. Plant Physiol 139: 27–38.

O'Neill CM, Bancroft I (2000) Comparative physical mapping of segments of the genome of *Brassica oleracea* var. *alboglabra* that are homoelogous to sequenced regions of chromosomes 4 and 5 of *Arabidopsis thaliana*. Plant J 23: 233–243.

Park JY, Koo DH, Hong CP, Lee SJ, Jeon JW, Lee SH, Yun PY, Park BS, Kim HR, Bang JW, Plaha P, Bancroft I, Lim YP(2005) Physical mapping and microsynteny of *Brassica rapa* ssp. *pekinensis* genome corresponding to a 222 kbp gene-rich region of *Arabidopsis* chromosome 4 and partially duplicated on chromosome 5. Mol Genet Genom 274: 579–588.

Paterson AH, Bowers JE, Peterson DG, Estill JC, Chapman BA (2003) Structure and evolution of cereal genomes. Curr Opin Genet Dev 13: 644–650.

Quiniou SMA, Waldbieser GC, Duke MV (2007) A first generation BAC-based physical map of the channel catfish. BMC Genom 8: 40.

Rana D, van den Boogaart T, O'Neill CM, Hynes L, Bent E, Macpherson L, Park JY, Lim YP, Bancroft I (2004) Conservation of the microstructure of genome segments in Brassica napus and its diploid relatives. Plant J 40: 725–733

Schmidt R, Acarkan A, Boivin K (2001) Comparative structural genomics in the Brassicaceae family. Plant Physiol Biochem 39: 253–262.

Soderlund C, Humphray S, Dunham I, French L (2000) Contigs built with fingerprints, markers, and FPC V4.7. Genome Res 11: 934–941.

Town CD, Cheung F, Maiti R, Crabtree J, Haas BJ, Wortman JR, Hine EE, Althoff R, Arbogast TS, Tallon LJ, Vigouroux M, Trick M, Bancroft I (2006) Comparative genomics of *Brassica oleracea* and *Arabidopsis thaliana* reveal gene loss, fragmentation, and dispersal after polyploidy. Plant Cell 18: 1348–1359.

U N (1935) Genome analysis in *Brassica* with special reference to the experimental formation of *B. napus* and peculiar mode of fertilization. Jpn J Bot 7: 389–452.

Xu Z, Sun S, Covaleda L, Ding K, Zhang A, Wu C, Scheuring C, Zhang HB (2004) Genome physical mapping with large-insert bacterial clones by fingerprint analysis: methodologies, source clone genome coverage, and contig map quality. Genomics 84: 941–951.

Yang TJ, Kim JS, Lim KB, Kwon SJ, Kim JA, Jin M, Park JY, Lim MH, Kim HI, Kim SH, Lim YP, Park BS (2005) The Korea *Brassica* Genome Projects: a glimpse of the *Brassica* genome based on comparative genome analysis with *Arabidopsis*. Compar Funct Genom 6: 138–146.

Yang TJ, Kim JS, Kwon SJ, Lim KB, Choi BS, Kim JA, Jin M, Park JY, Lim MH, Kim HI, Lim YP, Kang JJ, Hong JH, Kim CB, Bhak J, Bancroft I, Park BS (2006) Sequence-level analysis of the diploidization process in the triplicated *FLOWERING LOCUS C* region of *Brassica rapa*. Plant Cell 18: 1339–1347.

Yang TJ, Kwon SJ, Choi BS, Kim JS, Jin M, Lim KB, Park JY, Kim JA, Lim MH, Kim HI, Lee HJ, Lim YP, Paterson AH, Park BS (2007) Characterization of terminal-repeat retrotransposon in miniature (TRIM) in *Brassica* relatives. Theor Appl Genet 114: 627–636.

Zhang X, Wessler SR (2004) Genome-wide comparative analysis of the transposable elements in the related species *Arabidopsis thaliana* and *Brassica oleracea*. Proc Natl Acad Sci USA 101: 5589–5594.

Zhang Y, Huang Y, Zhang L, Li Y, Lu T, Lu Y, Feng Q, Zhao Q, Cheng Z, Xue Y, Wing RA, Han B (2004) Structural features of the rice chromosome 4 centromere. Nucl Acid Res 32: 2023–2030.

8

Exploring the Paradoxes of the Brassica Genome Architecture

Isobel A. P Parkin,[1,*] and *Stephen J. Robinson*[1]

ABSTRACT

The natural history of Brassica species is exemplified by the diverse morphological and biochemical phenotypes displayed among those cultivated as vegetables. Unique genome structures underlie this variation as the result of fluid genome dynamics consisting of repeated rounds of duplication, chromosome rearrangement events and subsequent diploidisation. However, this rapid *Brassiceae* genome evolution has involved limited expansion of repetitive elements or transposition events characteristic of the evolutionary history of many plant tribes including the *Triticeae*. This has resulted in a remarkably high level of conservation among the *Brassicaceae* genomes. This chapter reviews the current knowledge of genome structure for *Brassiceae* species at the chromosomal karyotype and at the physical sequence level. Such analyses have provided both tantalising insights into polyploid genome evolution and the identification of apparent contradictory forces that have played a role in shaping and maintaining these important plant genomes.

Keywords: genome evolution, polyploidy, transposon expansion, comparative mapping, genome architecture, paleobotany

8.1 Introduction

Grown worldwide vegetable *Brassica* are an important and highly diversified group of crops that are largely represented by *Brassica oleracea* and *B. rapa* species. In the western hemisphere, *B. oleracea* is the predominant species encompassing a number of valuable crops harvested for both their

[1]Agriculture and Agri-Food Canada, Saskatoon Research Centre, 107 Science Place, Saskatoon, SK, S7N 0X2, Canada.
*Corresponding author: *isobel.parkin@agr.gc.ca*

vegetative and floral structures including cabbage, broccoli, cauliflower, Brussels sprouts, etc. In Asia, *B. rapa* is the more commonly cultivated species owing to the great cultural and economic importance of Chinese cabbage. Generally, of less economic relevance are the *B. rapa* and *B. napus* cultivars that provide the root vegetable turnip and turnip greens and the vegetables mustards of *B. juncea* species.

Among the Brassicacaeae there is a wide level of morphological variation found both between and within individual species including, inflorescence architecture, the size of roots and stems, patterning in leaf development and apical bud structures. This is most dramatically observed in the *B. oleracea* subspecies cauliflower (*botrytis*), broccoli (*italica*), cabbage (*capitata*), kale (*medullosa*), kohlrabi (*gongylodes*) and Brussels sprout (*gemmifera*) (Fig. 8-1). The natural history of the cultivated *Brassica* is interesting, as the genomes

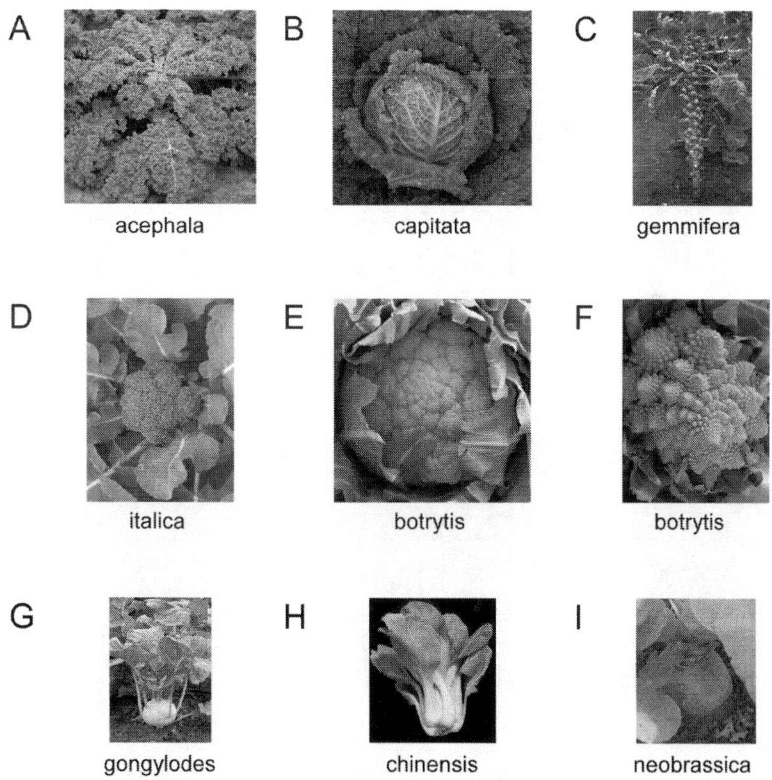

Figure 8-1 Morphological variation among the Brassica vegetable species. A-G are *Brassica oleracea*, H is *B. rapa* and I is *B. napus*. Variation in head size and compactness is observed among Kale (A), Cabbage (B) and Brussels sprouts (C). Variation in inflorescence architecture is observed among Broccoli (D), Cauliflower (E) and Romanesco broccoli (F). Variation in stem and root size is observed in, Kohlrabi (G), Bok Choi (H) and Swede/Rutabaga (I).

have undergone multiple rounds of polyploidisation and subsequent chromosomal rearrangement over a relatively short evolutionary period. The morphological variation so pronounced among *Brassica* species could result from changes in genome structure and gene regulation associated with the effects of polyploidisation. Studies focusing on these species offer the opportunity to investigate the impacts of polyploidy and the rapid evolution of phenotypic characteristics that has been observed (Lukens et al. 2004). Such research is facilitated by the close phylogenetic relationship of *Brassica* species to the plant model *Arabidopsis thaliana*, from which they are estimated to have diverged from a common ancestor approximately 15-24 Mya (Yang et al. 1999 Fig. 8-2). The extensive genetics and genomics resources currently available for *A. thaliana* can and have been exploited in the study of the *Brassica* species.

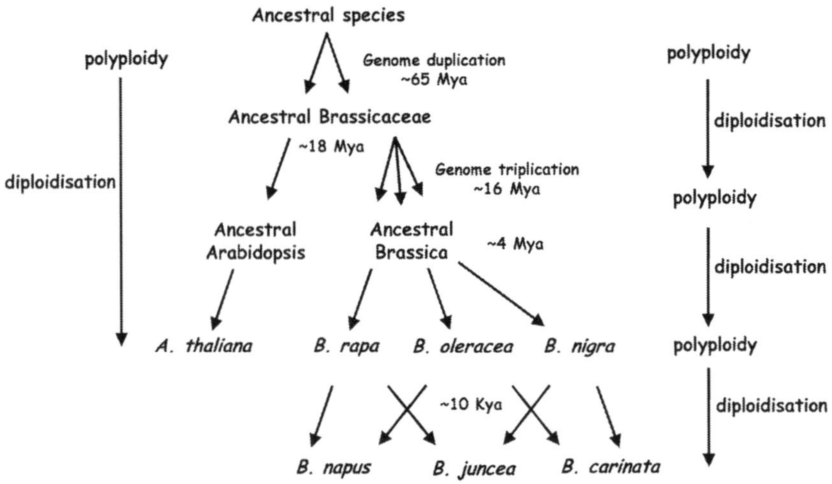

Figure 8-2 Schematic detailing the significant events in the evolution of the modern day Brassica genomes, adapted from Rana et al (2004).

In this chapter what is known of the genomic structure of the vegetable *Brassica* species will be discussed, at both the macro-level of chromosome organisation and at the micro-level of the nucleotide sequence. Much of the work described has been advanced through the availability of the well annotated genome sequence of *A. thaliana*, which currently provides the optimal foundation for comparative genetics and genomics studies within the Brassicaceae (Arabidopsis Genome Initiative 2000). The landscape of Brassica research will change rapidly in the next few years with the imminent release of the first sequenced Brassica genome, that of Chinese cabbage (*B. rapa*) (Hong et al. 2008). The importance of the current knowledge of

Brassica genome organisation for interpreting and exploiting the promise of many sequenced Brassica genomes will be discussed.

8.2 Genome Dynamics of the Brassicaceae

8.2.1 Genome Content

In the 1930s the relationship between six *Brassica* crop species was proposed and later validated determining the eponymous U's triangle (U 1935) (Fig. 8-3). The common vegetable *Brassiceae* species are members of U's triangle; *B. oleracea* ($2n = 18$, CC), *B. rapa* ($2n = 20$, AA) and *B. nigra* ($2n = 16$, BB) form the apices of the triangle and these species have fused in each pair wise combination to generate the larger and generally less diverse allopolyploid species *B. napus* ($2n = 38$, AACC), *B. juncea* ($2n = 36$, AABB) and *B. carinata* ($2n = 34$, BBCC). The Brassica genomes have undergone numerous rounds of chromosome doubling, nuclear hybridisations, and rearrangements, yet the genomes fall in the lower range of nuclear DNA content compared to other plant families (Johnston et al. 2005; Lysak et al. 2009). Among all reported Brassicaceae species none have been identified with a DNA C-value above 2.43 pg, in comparison to for example the cereal species of the Poaceae family, which range as high as 26.0 pg for a

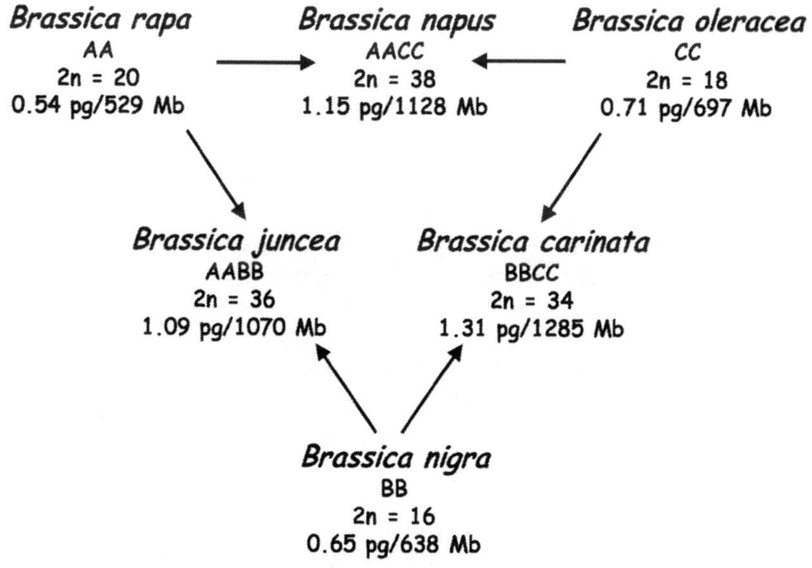

Figure 8-3 U's triangle describing the relationship among the six Brassica crop species (U 1935). The genome sizes are shown as 1C DNA content in pg and estimated nucleotide content in Megabase pairs (Mb) as presented in Johnston et al. (2005).

Triticale genome (Johnston et al. 2005; Lysak et al. 2009; *http://data.kew.org/cvalues/*). The low range of genome sizes, yet the acknowledged dynamic nature of genome evolution within the family offer an intriguing paradox. Some insight into the possible processes that maintain the low genome sizes has been gained from studies in *Arabidopsis thaliana*, which has one of the smallest known 1C DNA contents of 0.16 pg (Bennett et al. 2003) yet has evolved through paleopolyploidy (Blanc et al. 2003; Bowers et al. 2003; Henry et al. 2006). Although polyploid events and replication of repetitive sequences (see below) could lead to genome expansion, it appears that mechanisms to limit genome size are operating in the Brassicaceae family. For example, chromosomal rearrangements, perhaps as the result of non-reciprocal or unequal translocations, have been correlated with DNA loss in the evolution of *Arabidopsis* related species with reduced chromosome numbers (Yogeeswaran et al. 2005; Lysak et al. 2006). Interestingly the regulation of genome size does not appear to be influenced by selection (Lysak et al. 2009), thus the mechanisms and particularly the driving forces determining genome size remain to be completely elucidated.

8.2.2 Insights Offered by Comparative Genome Analysis

8.2.2.1 Comparisons Across the Brassica Species

Early cytological studies of chromosome pairing among *Brassica* species indicated the close relationship between the different species, particularly between the A genome of *B. rapa* and the C genome of *B. oleracea* (Attia and Röbbelen 1986a, 1986b). In resynthesised haploid *B. napus* lines where modern day *B. rapa* and *B. oleracea* genotypes were fused, homoeologous chromosome pairing was frequently recognised through the observation of multivalent chromosome formations. Yet, it was not until the application of molecular markers that it became apparent that such events were not the result of the forced intimacy, since homoeologous pairing between the A and C genomes occurred in natural *B. napus* lines at low but detectable levels (0.03–0.06 homoeologous recombination events per gamete) (Sharpe et al. 1995).

Molecular markers revolutionised the study of the genome organisation in plants, and in particular the application of restriction fragment length polymorphism (RFLP) analysis indicated the dawn of comparative mapping studies. Such research has allowed the complex genome architecture of numerous plant species to be elucidated, visualising underlying duplication events and chromosome rearrangements, and establishing evolutionary relationships (Schmidt 2002). RFLP markers are particularly informative for such analyses, since each locus resolved represents a duplicate copy of the molecular probe used to identify the genetic loci, immediately establishing common evolutionary elements (see Chapter 4).

Numerous genetic linkage maps, including RFLP maps, have been developed for *Brassica* vegetable species: *B. oleracea* (Slocum et al. 1990; Kianian and Quiros 1992; Bohuon et al. 1996; Lan et al. 2000; Babula et al. 2003), *B. rapa* (Kim et al. 2006; Choi et al. 2007; Suwabe et al. 2008), *B. juncea* (Axelsson et al. 2000; Mahmood et al. 2003; Panjabi et al. 2008) and *B. napus* (Ferreira et al. 1994; Parkin et al. 1995; Sharpe et al. 1995; Lombard and Delourme 2001; Piquemal et al. 2005; Qiu et al. 2006). The limitation for comparative analysis among these species has been the restricted use of common marker sets; however, in the late 1990s comparisons between the three Brassica diploid genomes of U's triangle became possible (Lagercrantz and Lydiate 1996; Parkin et al. 2003). These analyses were instrumental in developing our current understanding of chromosome architecture among the *Brassica* species, uncovering high levels of chromosome duplication and paradoxically considerable conservation of genomic segments across the genomes. The A and C genomes were found to be highly conserved with entire genetic linkage groups showing conservation of not only marker content but marker order (Fig. 8-4). The two genomes were separated by 16 detectable major chromosomal rearrangements, a number of which were associated with centromeric regions, suggesting that centromere fusion and fission are common forces leading to chromosome number deviations in *Brassica* species (Parkin et al. 2003). Comparisons of genetic linkage maps between all three *Brassica* diploid crop species indicate a larger number of rearrangements that separate the B genome from the A and C and none are shared with those that differentiate the A and C genomes (Lagercrantz and

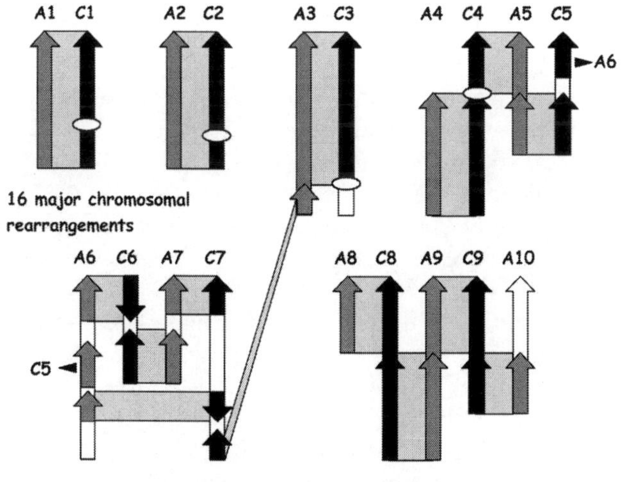

Figure 8-4 Cartoon representation of the relationship between the homoeologous regions of the Brassica A (grey) and C (black) genomes as defined in Parkin et al. (2003). The filled white circles show a subset of centromeres that have been positioned in the *B. oleracea* genome (A. Kelly and D. Lydiate, unpublished data).

Lydiate 1996; Panjabi et al. 2008). Strong conservation of extensive genome segments was also observed across all three genomes; despite this homology inherent difficulties have been found when attempting to transfer traits from the B genome into either of its related genomes (Struss et al. 1991; Chèvre et al. 1997) and homoeologous pairing was unexpectedly absent between the A and B genomes of resynthesised *B. juncea* (Axelsson et al. 2000). These data suggest that conservation of genome architecture may be restricted to the gross chromosomal structure and variation at the DNA sequence level is sufficient to prohibit pairing, this appears to be supported by the available expressed sequence tag (EST) data for B genome species (I. Parkin and A. Sharpe, unpubl. data).

The comparative genome analyses uncovered striking levels of genome duplication within all three Brassica genomes (Lagercrantz and Lydiate 1995; Parkin et al. 2003; Panjabi et al. 2008). These analyses noted that most maintained segments were arranged in three copies with varying levels of conservation, suggesting the *Brassica* species have evolved from a hexaploid ancestor likely formed from the fusion of three genomes of similar DNA content to the diminutive *A. thaliana*. The remnants of this ancestor have been further elucidated through comparisons with the plant model *A. thaliana* that have not only facilitated the resolution of evolutionary processes in this curious plant family, but also the exploitation of the substantial information and resources developed by the Arabidopsis community for Brassica crop improvement (Paterson et al. 2001).

8.2.2.2 Exploiting the A. thaliana Genome to Resolve Brassica Genome Architecture

Arabidopsis thaliana was adopted as a model for plant research largely due to its short generation time, small genome size and ease of transformation (Arabidopsis Genome Initiative 2000). Fortuitously for Brassica researchers although separated from the *Brassiceae* tribe approximately 15–24 Mya (Yang et al. 1999; Bowers et al. 2003) *A. thaliana* has proved an excellent resource for facilitating Brassica research, conspicuously in the resolution of genome and gene structure.

Initial comparative analyses aligned targeted regions between the genetic linkage maps of *Brassica* species and *A. thaliana* (Osborn et al. 1997; Cavell et al. 1998; Parkin et al. 2002), identifying islands of gene content and gene order that were maintained to varying degrees across the evolutionary divide. The publication of the complete genome sequence for *A. thaliana* (Arabidopsis Genome Initiative 2000) marked a radical change in the resolution of comparative mapping within the Brassicaceae. The origin of genetic loci in *Brassica* species could be accurately positioned in the sequenced *A. thaliana* genome based on the strong sequence identity observed between the species, 87% on average in genic regions (Lukens et al. 2003; Parkin et al. 2005). The most comprehensive analysis found that not only were the previously documented duplicated genomic segments preserved in the *A. thaliana* genome, but that the structure of the Brassica A and C genomes could be defined by approximately 21 of these regions (Parkin et al. 2005). These genomic segments ranged in physical length, from 2-9 Mb (the latter constituting almost 10% of the sequenced model genome) in the genome of *A. thaliana*, defining in some instances what appeared to be entire chromosome arms in the *Brassica* species, suggesting mechanisms are acting to maintain the structure of particular regions. Of note was the apparent conservation of chromosome landmarks such as centromeres, allowing the inference of as yet unmapped centromeric regions, both functional and archaic. These chromosome sites were also generally associated with breakpoints between conserved blocks and often correlated with numerous small segmental regions, which points to the fragile nature of such regions and their potential as rearrangement hotspots (Moore et al. 1997; Hall et al. 2006). Additional analyses have led to the resolution of at least 24 Brassicaceae blocks that can be replicated and rearranged to form the genomes of all members of the extended family (Schranz et al. 2006; Mandakova and Lysak 2008; Panjabi et al. 2008) (Fig. 8-5).

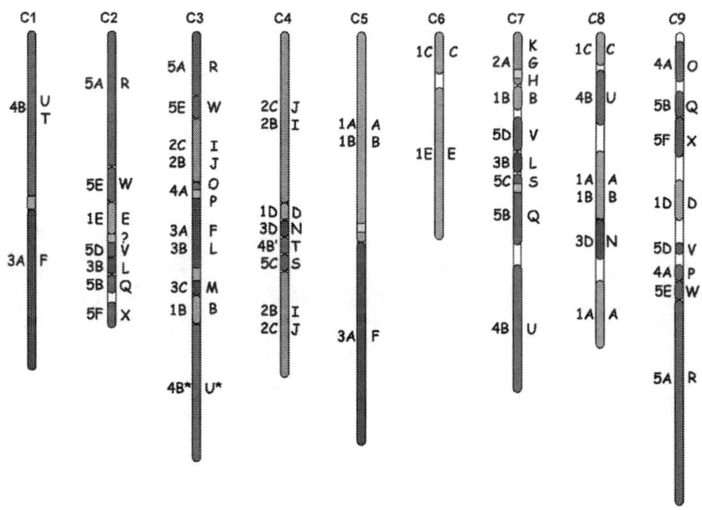

Figure 8-5 The *Brassica oleracea* genome (C1-C9) generated from the conserved ancestral genomic blocks. The blocks are coloured according to the corresponding homologous region of *A. thaliana*: chromosome 1—light blue; chromosome 2—orange; chromosome 3—dark blue; chromosome 4—green; and chromosome 5—red. The names of the blocks are shown to the left of each Brassica linkage group according to Parkin et al. (2005) and their equivalent regions as described by Schranz et al. (2006) are shown to the right.

Color image of this figure appears in the color plate section at the end of the book.

The identification of the conserved blocks confirmed the hexaploid nature of the *Brassiceae* ancestor, since invariably each of the blocks was found in three copies in each of the Brassica diploid genomes (Lysak et al. 2005; Parkin et al. 2005; Panjabi et al. 2008). This was also true for regions that had been replicated in the *A. thaliana* genome, itself an ancient polyploid (Blanc et al. 2003; Bowers et al. 2003; Henry et al. 2006), suggesting the duplications in *A. thaliana* pre-dated the triplication event that apparently differentiates the *Brassiceae* tribe. However, some of the conserved blocks appear to have undergone further segmental duplication and potentially some have been lost subsequent to the triplication event.

Polyploidy is an important and prevalent phenomenon in plant evolution since an estimated 70% of flowering plants are suggested to have polyploidy in their lineages (Masterson 1994). The putative benefit of this evolutionary strategy lies in the duplication of gene content that facilitates the diversification of phenotype to provide selective advantage over the diploid relatives. The resolution of newly formed polyploids necessitates diploidisation of the nuclear genome, ensuring normal chromosome association, stable inheritance of genetic factors, restoration of fertility and ultimately the establishment of robust new species (Doyle et al. 2008). The mechanisms leading to diploidisation of these novel genomes are not completely resolved; however, two pathways are thought to predominate, massive chromosomal restructuring that suppresses pairing between non-homologous chromosomes or the rapid resumption of homologous pairing controlled by heritable genetic factors (Jenczewski and Alix 2004). It appears in the history of the *Brassiceae* tribe that both mechanisms have played a role. In the *Brassica* diploids significant restructuring occurred after the triplication of the ancestral genome, making resolution of the ancestral karyotype problematic. However, there have been no observable chromosomal rearrangements since the recent generation of the *Brassica* allopolyploids (Parkin et al. 1995; Axelsson et al. 2000) and it has been proposed that they have inherited a genetic control of homologous pairing analogous to the *Ph1* gene of wheat (Sears 1976).

Comparative mapping at the chromosomal level has provided insights into the level and extent of genome conservation and rearrangement within the Brassicaceae, these analyses have also suggested that each of the 24 genomic segments was not completely preserved in all three duplicated copies. It would be anticipated that the resultant level of gene redundancy might require deletion of genes copies or potentially allow gene divergence to stimulate sub- and neo-functionalisation of duplicates further contributing to the plasticity of the polyploid genomes (Lynch and Conery 2000; Lynch and Force 2000). Some of these questions are beginning to be answered with fine scale analyses, comparing the Brassicaceae genomes at the physical level and more recently at the nucleotide level.

8.3 Dissecting the Remnants of the Hexaploid Ancestor: Lessons from Studies of Microsynteny among Brassicaceae Genomes

In the absence of a complete Brassica genome sequence an improved knowledge of genome organisation has been drawn from targeted comparative analyses at the physical level between Brassica genomes and regions of *A. thaliana* containing genes of interest.

The study of genome architecture in polyploid species is complicated by the inherent duplication found, and in comparisons between *Brassica* species and their close relative *A. thaliana*, the underlying paleopolyploid nature of the model genome must be considered. Initial studies physically defined regions of the Brassica genomes by identifying overlapping sets of large insert clones or bacterial artificial clones (BACs) that were homologous to a target segment of the *A. thaliana* genome (O'Neill and Bancroft 2000; Rana et al. 2004; Park et al. 2005). Notably these first studies elucidated the structure of *B. oleracea*, *B. rapa* and *B. napus* segments homologous to a 222 Kb gene-rich region of *A. thaliana*, which was itself segmentally duplicated in the model. The design of these analyses provided further corroboration of the triplication event that defines the genomes of the Brassiceae genomes and the fact that six regions were identified in both the A and C genomes compared to the duplicated region of *A. thaliana* placed the triplication event subsequent to the ancient duplications found in the *A. thaliana* lineage. It was apparent that the replication of the ancestral Brassica genome was followed by numerous changes in the microstructure with evidence of significant interspersed gene loss across the three genomic copies. Interestingly the microstructure of *B. napus* was found to closely mirror that of the diploid genomes indicating that maintenance of the contributing A and C genome within the newly formed allopolyploid nucleus was tenaciously controlled (Rana et al. 2004). The interspersed gene loss observed in comparison to *A. thaliana* is a common phenomenon in plant genome evolution and was originally speculated to be the result of unequal crossing over events (Bancroft 2001).

Sequencing of the six paralagous regions of the C genome identified in O'Neill and Bancroft (2000) has provided an illuminating glimpse into Brassica genome architecture and the processes driving change at the microstructure level in the *Brassiceae* (Town et al. 2006). In *B. oleracea*, 66% of the genes that would be predicted based on triplication of the annotated *A. thaliana* gene identifiers were found in the collinear regions (Fig. 8-6). Invariably no evidence remained of the inferred missing genes, although in 10 instances gene fragments were identified in collinear regions, these data suggested that breaks in collinearity were largely the result of small interspersed deletion events, the mechanism for such a prevalent evolutionary device in the *Brassiceae* remains to be elucidated. The presence

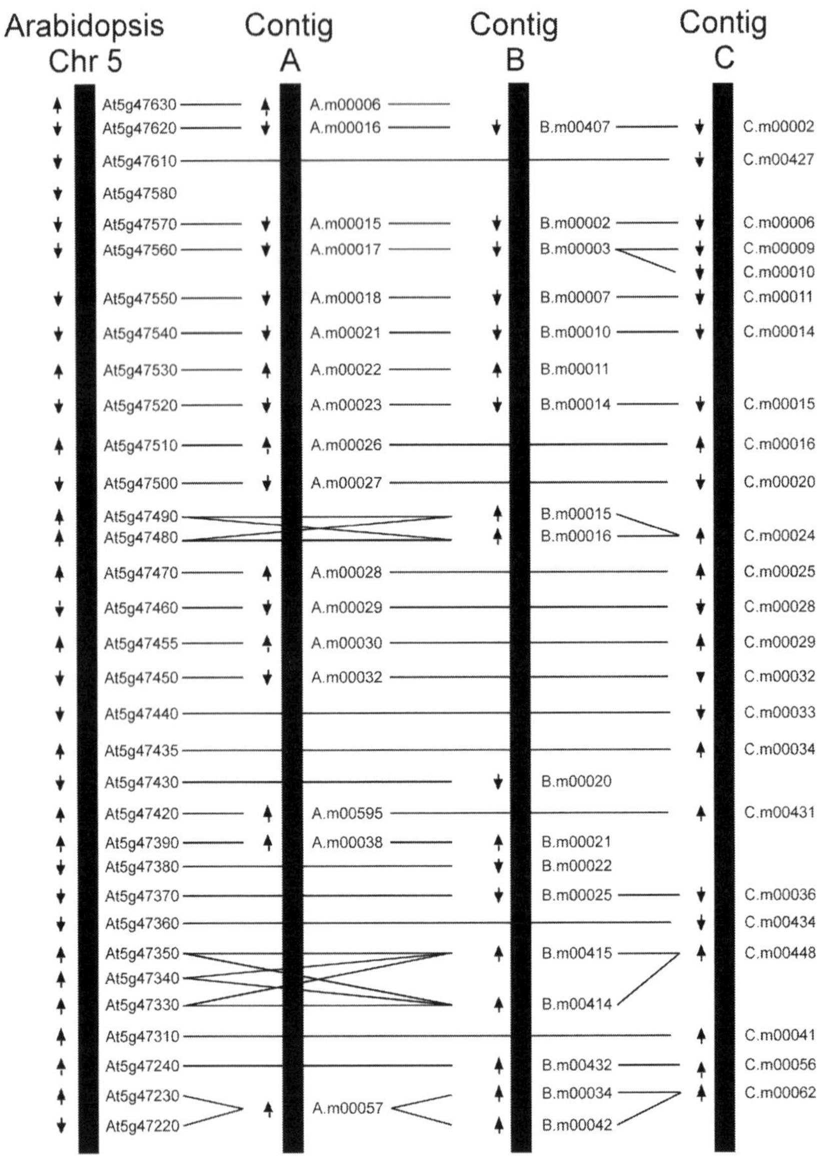

Figure 8-6 Microsynteny between *A. thaliana* and three paralogous regions of the *B. oleracea* genome, adapted from Town et al. (2006). The figure shows interspersed gene deletion across the duplicate copies in *B. oleracea* and the compression/expansion of tandemly repeated genes. Arrows indicate the presence and orientation of a gene model. The horizontal lines indicate homology between gene models.

of 13 genes in *A. thaliana* not found in the sequenced Brassica genomic regions identified transposition of genes as an additional force driving changes to microstructure. Single gene transposition is becoming accepted as a significant phenomenon in plant lineage evolution and recent data suggests this has occurred at relatively high levels during the history of *A. thaliana* (Freeling et al. 2008). Large scale genome sequencing of plants has uncovered high levels of tandemly repeated genes, in particular 17% of *A. thaliana* genes were found arranged in tandem arrays (Arabidopsis Genome Initiative 2000). The study of disease resistance loci has suggested such arrays are dynamic in nature and evolve through homologous recombination and unequal crossing over (Michelmore and Meyers 1998). Town et al. (2006) identified not only the expansion and reduction of arrays in *B. oleracea* compared to *A. thaliana* (Fig. 8-6) but also determined that some arrays have become fixed over evolutionary time, perhaps as a function of their organisation, inverted versus parallel duplication, or potentially as a consequence of the selective advantage of the additive gene expression for those particular genes.

A similar analysis of paralagous regions of the *B. rapa* genome homologous to the *A. thaliana* genomic segments containing the *FLOWERING LOCUS C (FLC)* noted similar levels of collinear gene loss and transposition events (Yang et al. 2006) (see Chapter 7 for additional details). The authors also noted that some genes were modified through the insertion of small repetitive elements, which has been suggested as a mechanism for attaining novel protein functions (Bennetzen 2000). The significance of the changes in the microstructure of the Brassica genomes has yet to be aligned with gene function, although preliminary data studying gene expression levels across paralagous regions of *B. oleracea* have suggested that the multiple gene copies lead to differentiation of transcript abundance across tissue types (F. Iniguez-Luy, H. Wu, B. Underwood, M.L. Federico, J.S. Kim, J.C. Pires, R.M. Amasino, T.C. Osborn, C.D. Town, unpublished).

Mechanisms such as single gene transposition along with transposon-mediated genomic rearrangements such as duplications, excision/insertion events or retro-transposition are probably responsible for some of the genome specific differences observed between the A and C Brassica genomes. Data describing these differences is slowly being accumulated and it appears that the C genome has undergone genome expansion compared to the A genome that may partially explain the observed difference in genome size between the two *Brassica* species (Fig. 8-7). Figure 8-7 demonstrates the insertion of Brassica specific sequences in the C genome compared to both the Brassica A and the model *A. thaliana* genome (Bancroft et al. 2009). These sequences comprise transposable elements and transcribed regions reminiscent of the previously defined *B. oleracea* self-incompatibility locus (Fujimoto et al. 2006) suggesting transposition of a larger genomic segment or intra-chromosomal duplication.

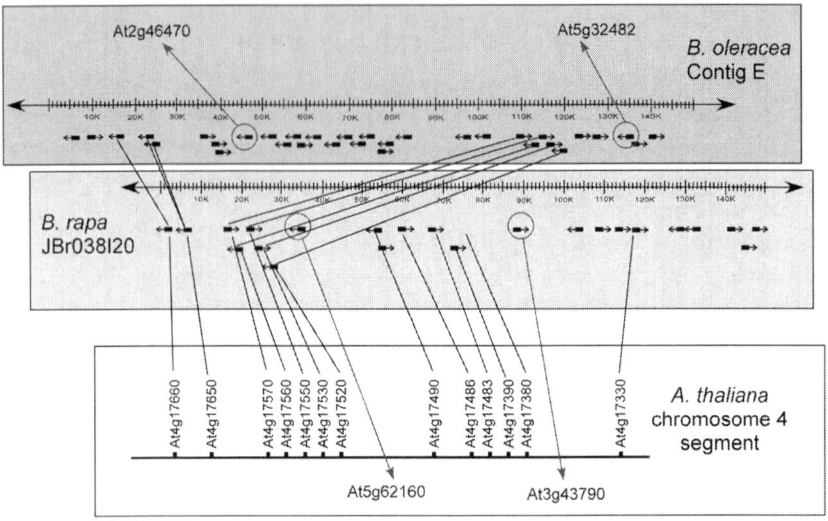

Figure 8-7 Comparative sequence analysis of a homologous region between *B. rapa*, *B. oleracea* and the model *A. thaliana*, based on data presented by Bancroft et al. (2009). This region shows expansion of the *B. oleracea* genome with Brassica specific sequences and the prevalence of transposition events in disrupting microsynteny across the genomes (circled gene models).

The detailed analyses of paralagous Brassica regions has provided an estimate of gene content in the A and C genomes that is not dissimilar to that found in *A. thaliana*, specifically when comparing homologous regions: one gene per 4.8 Kbp in *B. oleracea* compared to 4.5 Kbp in *A. thaliana* (Town et al. 2006) and one gene per 3.7 Kbp and 3.4 Kbp in *B. rapa* and *A. thaliana*, respectively (Yang et al. 2006). The average gene length and structure are also remarkably conserved among the three species. Although the available data can only be considered an appetiser for the whole genome it is apparent that when the level of observed gene duplication is taken into account, the increased size of Brassica genomes compared to *A. thaliana* can not be explained alone by genome duplication events. The available global sequence data generated for both *B. rapa* (Yang et al. 2005) and *B. oleracea* (Zhang and Wessler 2004) provide some insights into the expansion process. Analyses of the two data sets indicate that while both genomes contain a similar level of protein coding genes, *B. oleracea* contains a higher level of characterised transposable elements (20%) than its close relative *B. rapa* (12%). The role of transposable elements in genome evolution is a rapidly developing area of research particularly with the current exponential increase in available genome sequence data for plant species.

8.4 Contribution of Low Complexity Sequences to the Brassiceae Genomes

In the plant kingdom, repetitive elements are thought to be largely responsible for the extreme variation in genome size observed among angiosperm species, this is exemplified by the rapid increase in the nuclear content of maize through the accumulation of retrotransposons that contribute 50–60% of the genome (Bennetzen et al. 1998). In most eukaryotic organisms repetitive sequences are not evenly distributed across the genome but are found to cluster in heterochromatic regions, with telomeres and centromeres containing large numbers of diverse repeat sequences that contribute to the maintenance of chromosome stability and normal segregation (Henikoff et al. 2001). In *Brassica* species due to difficulties in resolving the karyotype, it is only recently that the positions of the centromeric regions have been accurately identified and only in the case of *B. napus* has this information been linked to the genetic maps, which has confirmed the conservation of centromere positions across *Brassiceae* species (Lim et al. 2007; Pouilly et al. 2008).

8.4.1 Pericentromeric and Centromeric Repeats

The centromere is an essential structure that ensures the faithful segregation of chromosomes at cell division, and consists largely of highly repetitive DNA sequences, specifically tandem satellite repeats (Jiang et al. 2003). The role of these tandem repeats has yet to be fully elucidated but they are believed to recruit DNA binding proteins during chromosome condensation facilitating the replication and segregation of the chromosomes (Csink and Henikoff 1998). Centromeric repeats have been characterized in a number of plant genomes and are notable for their extreme diversity with little sequence homology found across plant families, indicating they are rapidly evolving, which contrasts starkly with their fundamental role. The centromeres of *B. rapa* and *B. oleracea* are composed of 176-bp satellite repeat sequences (CentBr), which are not homologous with the centromeric repeats of either the model *A. thaliana* or the more closely related species, *B. nigra* (Harrison and Heslop-Harrison 1995; Lim et al. 2005; Lim et al. 2007). Further, the CentBr repeats in the *B. rapa* genome have been shown to represent two classes which are chromosome-specific, one being found on eight and the second on only two chromosomes (Lim et al. 2007) (see Chapter 7 for additional details). Sequencing of peri-centromeric regions is resolving the organisation of these complex structures and has identified a number of transposable elements among the satellite repeats that have been conserved across the Brassiceae species (Lim et al. 2007).

8.4.2 Transposable Elements

Transposable elements have undergone a renaissance from selfish junk elements to active facilitators of plant genome evolution; through gene capture, exon shuffling, gene transposition and the generation of novel polymorphism (Bennetzen 2005; Morgante 2006). A diverse range of transposable elements have been described in plants and are distinguished based on their mode of replication, either facilitating movement via an RNA intermediate (Class I) or via a direct DNA excision-insertion mechanism (Class II). Genome sequence analysis has identified that 20% and 12% of the *B. oleracea* and *B. rapa* genomes are contributed by transposable elements, respectively (Zhang and Wessler 2004; Yang et al. 2005).

The majority of transposable elements found in the vegetable Brassica genomes were classified as Class I retrotransposons, comprising 70-89% of the total identified elements. The accurate classification of transposable elements has only been completed for *B. oleracea*, which allowed comparisons with *A. thaliana* elements and provided insights into the evolution of these DNA molecules across the Brassicaceae genomes (Zhang and Wessler 2004). The *B. rapa* genome sequencing project is providing insights into the prevalence of these elements in the A genome and the analyses thus far is described in Chapter 7. The class I transposable elements are subdivided in two, those possessing long terminal repeats (LTRs) and those without LTRs that terminate with a polyadenaline tract, which are further sub-divided into long interspersed nuclear elements (LINEs) or short interspersed nuclear elements (SINEs).

In *B. oleracea*, the LTR-elements predominate the Class I type and are composed of families of *copia*-like and *gypsy*-like retrotransposons, in each case where related lineages were found in both the model and crop genome they were more prevalent in *B. oleracea*. All the *copia*-like elements found in *A. thaliana* were represented in *B. oleracea*; however, one family of elements was specific to the Brassica C genome. The *gypsy*-like elements appeared more evolutionary disparate with less than half the identified families being common to the two species and the remainder having emerged after their divergence from a common ancestor. LINEs are recognized as the most ancient class of retrotransposons in plants which was reflected in the high level of conservation observed between *B. oleracea* and *A. thaliana*. This was in stark comparison to SINEs where a novel class of elements was identified in *B. oleracea*, *BoS*, that was distinguished as the most abundant class of SINEs in the Brassicaceae and was found to be composed of an incredibly diverse range of related elements a proportion of which had apparently evolved through chimerism of more ancient elements (Zhang and Wessler 2005).

Class II transposons were much less abundant than their class I counterparts, approximately 6% of the *B. oleracea* genomic content was comprised of class II elements. Interestingly, two families of CACTA-like elements in *B. oleracea* appear to have undergone exponential expansion (*BoC1* and *BoC2*), contributing one third of the total class II elements. The expansion of class II elements in plant genomes is a relatively new observation since their preference for insertion in gene-rich regions would suggest increased density would be limited. It has yet to be determined if the expansion of class II elements in *B. oleracea* is specific to the C genome; however preliminary data have identified a smaller proportion of DNA-type transposable elements in available *B. rapa* sequence data (Hong et al. 2008).

The compact nature of the *A. thaliana* genome is reflected in the low abundance of transposable elements (~ 10% of the total genome), which is in contrast to that found in the Brassica lineages. The comparative analyses of such elements between *B. oleracea* and *A. thaliana* established the age of each element lineage, the level of replication in the two related genomes and determined their potential mode of evolution (Zhang and Wessler 2004). The difference between the genomes could result from a concerted loss or the prevention of expansion of transposable elements by an unknown control mechanism in the *A. thaliana* lineage. The latter hypothesis appears more likely since nearly all subfamilies of transposons have been found in both the Brassica and *A. thaliana* lineages and there is no single transposon family that has undergone significant expansion in the *A. thaliana* genome. In comparison to other plant families, all the Brassicaceae species studied have maintained remarkably low levels of transposable elements. In maize four families of LTR elements alone contribute 32% of the genomic content, in contrast to *B. oleracea* although there was a multiplication of both class I and class II elements there has been no amplification of specific retrotransposon families observed, instead many lower abundance families are maintained. Due to the paleoploid nature of the Brassica genomes this result would not be anticipated, since the genomic redundancy should provide an environment for transposons to flourish. These data suggest that mechanisms are regulating the transposition potential within the Brassicaceae genomes and as expression data accumulates it will be possible to identify active transposons from those no longer able to replicate, which may shed light on the genetic processes at play.

8.5 Conclusions

A plant's genome architecture partly determines the plasticity of a species and in affect controls its ability to respond to diverse environmental pressures. Polyploid species appear to have a selective advantage since their inherent genetic redundancy allows for rapid diversification of gene content and gene expression. Indeed the Brassica vegetable crops have evolved from a common hexaploid Brassicaceae progenitor to provide a diverse array of species, which vary in morphological, biochemical and geographical profiles. The release of the genome sequence for *B. rapa* is anticipated for 2010 and should begin to answer some of the remaining questions regarding Brassica genome structure. This information could finally provide a comprehensive picture of the level and distribution of genome duplication and potentially facilitate insights into the evolutionary mechanisms that have driven the generation of this valuable and diverse group of species. The identification of the estimated 50,000 genes in each of the Brassica diploid genomes will allow the role of duplicate gene copies in defining the overwhelming phenotypic diversity to be fully described. The escalating ability to capture sequence data accurately and economically should lead to the genomes of a number of Brassiceae and related species being released in the next few years, predicting an exciting and revelatory period for Brassica and polyploid genome research.

References

Arabidopsis Genome Initiative (2000) Analysis of the genome sequence of the flowering plant *Arabidopsis thaliana*. Nature 408: 796–815.

Attia T, Röbbelen G (1986a) Cytogenetic relationship within cultivated Brassica analyzed in amphihaploids from the three diploid ancestors. Genome 28: 323–329.

Attia T, Röbbelen G (1986b) Meiotic pairing in haploids and amphidiploids of spontaneous versus synthetic origin in rape, *Brassica napus* L. Genome 28: 330–334.

Axelsson T, Bowman CM, Sharpe AG, Lydiate DJ, Lagercrantz U (2000) Amphidiploid *Brassica juncea* contains conserved progenitor genomes. Genome 43: 679–688.

Babula D, Kaczmarek M, Barakat A, Delseny M, Quiros CF, Sadowski J (2003) Chromosomal mapping of *Brassica oleracea* based on ESTs from *Arabidopsis thaliana*: complexity of the comparative map. Mol Genet Genom 268: 656–665.

Bancroft I (2001) Duplicate and diverge: the evolution of plant genome microstructure. Trends Genet 17: 89–93.

Bancroft I, Trick M, Drou N, Fraser F, Soumpourou E, Cheung F, Park J-Y, Kwon S-J, Lim Y-P, Scott R, Town C, King G, Rawlings C, Barker G, Selby J, Long Y, Zhang C, Li R, Meng J (2009) The UK-China sequencing project: Exploiting the *B. rapa* genome sequence for oilseed rape improvement. In: Plant, Anim Micr Genome XVII Conf, San Diego, CA, USA.

Bennett MD, Leitch IJ, Price HJ, Johnston JS (2003) Comparisons with *Caenorhabditis* (approximately 100 Mb) and Drosophila (approximately 175 Mb) using flow cytometry show genome size in *Arabidopsis* to be approximately 157 Mb and thus approximately 25% larger than the *Arabidopsis* genome initiative estimate of approximately 125 Mb. Ann Bot (Lond) 91: 547–557.

Bennetzen JL (2000) Transposable element contributions to plant gene and genome evolution. Plant Mol Biol 42: 251–269.

Bennetzen JL (2005) Transposable elements, gene creation and genome rearrangement in flowering plants. Curr Opin Genet Dev 15: 621–627.

Bennetzen JL, SanMiguel P, Chen M, Tikhonov A, Francki M, Avramova Z (1998) Grass genomes. Proc Natl Acad Sci USA 95: 1975–1978.

Blanc G, Hokamp K, Wolfe KH (2003) A recent polyploidy superimposed on older large-scale duplications in the Arabidopsis genome. Genome Res 13: 137–144.

Bohuon EJR, Keith DJ, Parkin IAP, Sharpe AG, Lydiate DJ (1996) Alignment of the conserved C genomes of *Brassica oleracea* and *Brassica napus*. Theor Appl Genet 93: 833–839.

Bowers JE, Chapman BA, Rong J, Paterson AH (2003) Unravelling angiosperm genome evolution by phylogenetic analysis of chromosomal duplication events. Nature 422: 433–438.

Cavell AC, Lydiate DJ, Parkin IAP, Dean C, Trick M (1998) Collinearity between a 30–centimorgan segment of *Arabidopsis thaliana* chromosome 4 and duplicated regions within the *Brassica napus* genome. Genome 41: 62–69.

Chèvre AM, Barret P, Eber F, Dupuy P, Brun H, Tanguy X, Renard M (1997) Selection of stable *Brassica napus-B. juncea* recombinant lines resistant to blackleg (Leptosphaeria maculans). 1. Identification of molecular markers, chromosomal and genomic origin of the introgression. Theor Appl Genet 95: 1104–1111.

Choi S, Teakle G, Plaha P, Kim J, Allender C, Beynon E, Piao Z, Soengas P, Han T, King G, Barker G, Hand P, Lydiate D, Batley J, Edwards D, Koo D, Bang J, Park B-S, Lim Y (2007) The reference genetic linkage map for the multinational *Brassica rapa* genome sequencing project. Theor Appl Genet 115: 777–792.

Csink AK, Henikoff S (1998) Something from nothing: the evolution and utility of satellite repeats. Trends Genet 14: 200–204.

Doyle JJ, Flagel LE, Paterson AH, Rapp RA, Soltis DE, Soltis PS, Wendel JF (2008) Evolutionary genetics of genome merger and doubling in plants. Annu Rev Genet 42: 443–461.

Ferreira ME, Williams PH, Osborn TC (1994) RFLP mapping of *Brassica napus* using doubled haploid lines. Theor Appl Genet 89: 615–621.

Freeling M, Lyons E, Pedersen B, Alam M, Ming R, Lisch D (2008) Many or most genes in Arabidopsis transposed after the origin of the order Brassicales. Genome Res 18: 1924–1937.

Fujimoto R, Okazaki K, Fukai E, Kusaba M, Nishio T (2006) Comparison of the genome structure of the self-incompatibility (S) locus in interspecific pairs of S haplotypes. genetics 173: 1157–1167.

Hall AE, Kettler GC, Preuss D (2006) Dynamic evolution at pericentromeres. Genome Res 16: 355–364.

Harrison GE, Heslop-Harrison JS (1995) Centromeric repetitive DNA sequences in the genus *Brassica*. Theor Appl Genet 90: 157–165.

Henikoff S, Ahmad K, Malik HS (2001) The centromere paradox: Stable inheritance with rapidly evolving DNA. Science 293: 1098–1102.

Henry Y, Bedhomme M, Blanc G (2006) History, protohistory and prehistory of the *Arabidopsis thaliana* chromosome complement. Trends Plant Sci 11: 267–273.

Hong CP, Kwon SJ, Kim JS, Yang TJ, Park BS, Lim YP (2008) Progress in understanding and sequencing the genome of *Brassica rapa*. Int J Plant Genomics 2008: 582–837.

Jenczewski E, Alix K (2004) From diploids to allopolyploids: The emergence of efficient pairing control genes in plants. Crit Rev Plant Sci 23: 21–45.

Jiang J, Birchler JA, Parrott WA, Dawe RK (2003) A molecular view of plant centromeres. Trends Plant Sci 8: 570–575.

Johnston JS, Pepper AE, Hall AE, Chen ZJ, Hodnett G, Drabek J, Lopez R, Price HJ (2005) Evolution of genome size in Brassicaceae. Ann Bot 95: 229–235.

Kianian SF, Quiros CF (1992) Generation of a *Brassica oleracea* composite RFLP map: linkage arrangements among various populations and evolutionary implications. Theor Appl Genet 84: 544–554.

Kim JS, Chung TY, King GJ, Jin M, Yang T-J, Jin Y-M, Kim H-I, Park B-S (2006) A Sequence-Tagged Linkage Map of *Brassica rapa*. Genetics 174: 29–39.

Lagercrantz U, Lydiate DJ (1995) RFLP mapping in *Brassica nigra* indicates differing recombination rates in male and female meioses. Genome 38: 255–264.

Lagercrantz U, Lydiate DJ (1996) Comparative genome mapping in Brassica. Genetics 144: 1903–1910.
Lan TH, DelMonte TA, Reischmann KP, Hyman J, Kowalski SP, McFerson J, Kresovich S, Paterson AH (2000) An EST—enriched comparative map of *Brassica oleracea* and *Arabidopsis thaliana*. Genome Res 10: 776–788.
Lim KB, de Jong H, Yang TJ, Park JY, Kwon SJ, Kim JS, Lim MH, Kim JA, Jin M, Jin YM, Kim SH, Lim YP, Bang JW, Kim HI, Park BS (2005) Characterization of rDNAs and tandem repeats in the heterochromatin of *Brassica rapa*. Mol Cells 19: 436–444.
Lim KB, Yang TJ, Hwang YJ, Kim JS, Park JY, Kwon SJ, Kim JA, Choi BS, Lim MH, Jin M, Kim HI, de Jong H, Bancroft I, Lim YP, Park BS (2007) Characterization of the centromere and peri-centromere retrotransposons in *Brassica rapa* and their distribution in related *Brassica* species. Plant J 49: 173–183.
Lombard V, Delourme R (2001) A consensus linkage map for rapeseed (*Brassica napus* L.): construction and integration of three individual maps from DH populations. Theor Appl Genet 103: 491–507.
Lukens L, Zou F, Lydiate D, Parkin I, Osborn T (2003) Comparison of a *Brassica oleracea* genetic map with the genome of *Arabidopsis thaliana*. Genetics 164: 359–372.
Lukens LN, Quijada PA, Udall J, Pires JC, Schranz ME, Osborn TC (2004) Genome redundancy and plasticity within ancient and recent *Brassica* crop species. Biol J Linn Soc 82: 665–674.
Lynch M, Conery JS (2000) The evolutionary fate and consequences of duplicate genes. Science 290: 1151–1155.
Lynch M, Force A (2000) The probability of duplicate gene preservation by subfunctionalization. Genetics 154: 459–473.
Lysak MA, Koch MA, Pecinka A, Schubert I (2005) Chromosome triplication found across the tribe *Brassiceae*. Genome Res 15: 516–525.
Lysak MA, Berr A, Pecinka A, Schmidt R, McBreen K, Schubert I (2006) Mechanisms of chromosome number reduction in *Arabidopsis thaliana* and related *Brassicaceae* species. Proc Natl Acad Sci USA 103: 5224–5229.
Lysak MA, Koch MA, Beaulieu JM, Meister A, Leitch IJ (2009) The Dynamic Ups and Downs of Genome Size Evolution in *Brassicaceae*. Mol Biol Evol 26: 85–98.
Mahmood T, Ekuere U, Yeh F, Good AG, Stringam GR (2003) RFLP linkage analysis and mapping genes controlling the fatty acid profile of *Brassica juncea* using reciprocal DH populations. Theor Appl Genet 107: 283–290.
Mandakova T, Lysak MA (2008) Chromosomal phylogeny and karyotype evolution in x=7 crucifer species (*Brassicaceae*). Plant Cell 20: 2559–2570.
Masterson J (1994) Stomatal size in fossil plants: Evidence for polyploidy in majority of angiosperms. Science 264: 421–424.
Michelmore RW, Meyers BC (1998) Clusters of resistance genes in plants evolve by divergent selection and a birth-and-death process. Genome Res 8: 1113–1130.
Moore G, Roberts M, Aragon-Alcaide L, Foote T (1997) Centromeric sites and cereal chromosome evolution. Chromosoma 105: 321–323.
Morgante M (2006) Plant genome organisation and diversity: the year of the junk! Curr Opin Biotechnol 17: 168–173.
O'Neill CM, Bancroft I (2000) Comparative physical mapping of segments of the genome of *Brassica oleracea* var. *alboglabra* that are homoeologous to sequenced regions of chromosomes 4 and 5 of *Arabidopsis thaliana*. Plant J 23: 233–243.
Osborn TC, Kole C, Parkin IA, Sharpe AG, Kuiper M, Lydiate DJ, Trick M (1997) Comparison of flowering time genes in *Brassica rapa, B. napus* and *Arabidopsis thaliana*. Genetics 146: 1123–1129.
Panjabi P, Jagannath A, Bisht N, Padmaja KL, Sharma S, Gupta V, Pradhan A, Pental D (2008) Comparative mapping of *Brassica juncea* and *Arabidopsis thaliana* using Intron Polymorphism (IP) markers: homoeologous relationships, diversification and evolution of the A, B and C Brassica genomes. BMC Genom 9: 113.

Park JY, Koo DH, Hong CP, Lee SJ, Jeon JW, Lee SH, Yun PY, Park BS, Kim HR, Bang JW, Plaha P, Bancroft I, Lim YP (2005) Physical mapping and microsynteny of *Brassica rapa* ssp. *pekinensis* genome corresponding to a 222 kbp gene-rich region of *Arabidopsis* chromosome 4 and partially duplicated on chromosome 5. Mol Genet Genom 274: 579–588.

Parkin IAP, Sharpe AG, Keith DJ, Lydiate DJ (1995) Identification of the A and C genomes of amphidiploid *Brassica napus* (oilseed rape). Genome 38: 1122–1131.

Parkin IAP, Lydiate DJ, Trick M (2002) Assessing the level of collinearity between *Arabidopsis thaliana* and *Brassica napus* for *A. thaliana* chromosome 5. Genome 45: 356–366.

Parkin IAP, Sharpe AG, Lydiate DJ (2003) Patterns of genome duplication within the *Brassica napus* genome. Genome 46: 291–303.

Parkin IAP, Gulden SM, Sharpe AG, Lukens L, Trick M, Osborn TC, Lydiate DJ (2005) Segmental structure of the *Brassica napus* genome based on comparative analysis with *Arabidopsis thaliana*. Genetics 171: 765–781.

Paterson AH, Lan TH, Amasino R, Osborn TC, Quiros C (2001) Brassica genomics: a complement to, and early beneficiary of, the Arabidopsis sequence. Genome Biol 2: 3.

Piquemal J, Cinquin E, Couton F, Rondeau C, Seignoret E, doucet I, Perret D, Villeger MJ, Vincourt P, Blanchard P (2005) Construction of an oilseed rape (*Brassica napus* L.) genetic map with SSR markers. Theor Appl Genet 111: 1514–1523.

Pouilly N, Delourme R, Alix K, Jenczewski E (2008) Repetitive sequence-derived markers tag centromeres and telomeres and provide insights into chromosome evolution in *Brassica napus*. Chrom Res 16: 683–700.

Qiu D, Morgan C, Shi J, Long Y, Liu J, Li R, Zhuang X, Wang Y, Tan X, Dietrich E, Weihmann T, Everett C, Vanstraelen S, Beckett P, Fraser F, Trick M, Barnes S, Wilmer J, Schmidt R, Li J, Li D, Meng J, Bancroft I (2006) A comparative linkage map of oilseed rape and its use for QTL analysis of seed oil and erucic acid content. Theor Appl Genet 114: 67–80.

Rana D, van den Boogaart T, O'Neill CM, Hynes L, Bent E, Macpherson L, Park JY, Lim YP, Bancroft I (2004) Conservation of the microstructure of genome segments in *Brassica napus* and its diploid relatives. Plant J 40: 725–733.

Schmidt R (2002) Plant genome evolution: lessons from comparative genomics at the DNA level. Plant Mol Biol 48: 21–37.

Schranz ME, Lysak MA, Mitchell-Olds T (2006) The ABC's of comparative genomics in the *Brassicaceae*: building blocks of crucifer genomes. Trends Plant Sci 11: 535–542.

Sears ER (1976) Genetic control of chromosome pairing in wheat. Annu Rev Genet 10: 31–51.

Sharpe AG, Parkin IA, Keith DJ, Lydiate DJ (1995) Frequent nonreciprocal translocations in the amphidiploid genome of oilseed rape (*Brassica napus*). Genome 38: 1112–1121.

Slocum MK, Figdore SS, Kennard WC, Suzuki JY, Osborn TC (1990) Linkage arrangement of restriction fragment length polymorphism loci in *Brassica oleracea*. Theor Appl Genet 80: 57–64.

Struss D, Bellin U, Röbbelen G (1991) Development of B-genome chromosome addition lines of *B. napus* using different interspecific *Brassica* hybrids. Plant Breed 106: 209–214.

Suwabe K, Morgan C, Bancroft I (2008) Integration of Brassica A genome genetic linkage map between *Brassica napus* and *B. rapa*. Genome 51: 169–176.

Town CD, Cheung F, Maiti R, Crabtree J, Haas BJ, Wortman JR, Hine EE, Althoff R, Arbogast TS, Tallon LJ, Vigouroux M, Trick M, Bancroft I (2006) Comparative genomics of *Brassica oleracea* and *Arabidopsis thaliana* reveal gene loss, fragmentation, and dispersal after polyploidy. Plant Cell 18: 1348–1359.

U N (1935) Genomic analysis in Brassica with special reference to the experimental formation of *B. napus* and peculiar mode of fertilization. Jpn J Bot 7: 389–452.

Yang TJ, Kim JS, Lim KB, Kwon SJ, Kim JA, Jin M, Park JY, Lim MH, Kim HI, Kim SH, Lim YP, Park BS (2005) The Korea brassica genome project: a glimpse of the brassica genome based on comparative genome analysis with Arabidopsis. Compar Funct Genomics 6: 138–146.

Yang T-J, Kim JS, Kwon S-J, Lim K-B, Choi B-S, Kim J-A, Jin M, Park JY, Lim M-H, Kim H-I, Lim YP, Kang JJ, Hong J-H, Kim C-B, Bhak J, Bancroft I, Park B-S (2006) Sequence-level analysis of the diploidization process in the triplicated *FLOWERING LOCUS C* region of *Brassica rapa*. Plant Cell 18: 1339–1347.

Yang YW, Lai KN, Tai PY, Li WH (1999) Rates of nucleotide substitution in angiosperm mitochondrial DNA sequences and dates of divergence between Brassica and other angiosperm lineages. J Mol Evol 48: 597–604.

Yogeeswaran K, Frary A, York TL, Amenta A, Lesser AH, Nasrallah JB, Tanksley SD, Nasrallah ME (2005) Comparative genome analyses of *Arabidopsis* spp.: inferring chromosomal rearrangement events in the evolutionary history of *A. thaliana*. Genome Res 15: 505–515.

Zhang X, Wessler SR (2004) Genome-wide comparative analysis of the transposable elements in the related species *Arabidopsis thaliana* and *Brassica oleracea*. Proc Natl Acad Sci USA 101: 5589–5594.

Zhang X, Wessler SR (2005) BoS: a large and diverse family of short interspersed elements (SINEs) in *Brassica oleracea*. J Mol Evol 60: 677–687.

9

The Decoding of Gene Functions—Transcriptomics Tools Toward Molecular Physiology and Breeding

Agnieszka Ludwików,[1,*] *Krzysztof Krasowski,*[1] and *Leo Zeef*[2]

ABSTRACT

Brassica species are particularly important vegetable crops, that have been studied in many aspects of plant biology including seed growth, development, disease resistance and abiotic stress tolerance. In the past decade new genomic resources have been developed and subsequently used for a generation of tools to enable genome-wide screens in *Brassica*. Transcriptomic platforms have been established to facilitate the identification of genes underlying complex traits. Despite the substantial progress that has been made toward uncovering the network of gene functions in conditions investigated, much work still needs to be done to completely characterize associations between gene expression patterns and the genotypes in breeding populations. This chapter presents the current status of transcriptomic resources and efforts in determining the functions of genes in *Brassica* species.

Keywords: Brassica vegetables, genomics resources, microarrays, comparative genomics, stress tolerance

[1]Department of Biotechnology, Institute of Molecular Biology and Biotechnology, Faculty of Biology, Adam Mickiewicz University, Umultowska 89, 61-614 Poznań, Poland.
[2]School of Biological Sciences, University of Manchester, Manchester M139PT, UK.
*Corresponding author: *ludwika@amu.edu.pl*

9.1 Introduction

For the last decade plant functional genomics was developed based on model organisms such as *Arabidopsis thaliana* and more recently *Oryza sativa*. The knowledge gathered is now extensively used to study other plants including *Brassica* crop species. *Brassica* species have been studied in many aspects: seed growth, development, disease resistance and abiotic stress tolerance (van Poecke and Dicke 2002; Tuberosa and Salvi 2006; Broekgaarden et al. 2007, 2008). Vegetable Brassicas are recognized as a valuable source of vitamins, nutrients and anticancer compounds (Paterson et al. 2001). The large effort to bridge the gap between model plants and crop species functional genomics studies leveraged the development of many genomics resources for *Brassica*. Several tools were established to decipher gene functions (van Poecke and Dicke 2002; Tuberosa and Salvi 2006; Broekgaarden et al. 2007, 2008). These include expressed sequence tag (EST) collections, cDNA libraries, comprehensive mutant collections and DNA microarrays (Lim et al. 1996; Kwak et al. 1997). Special attention has been given to the generation of transcriptomic platforms, which are now released to the community and well supported. So far the *Brassica* transcriptome has been analyzed to address both agricultural and fundamental biological questions (Paterson et al. 2001; Kwun et al. 2004; Li et al. 2005; Soeda et al. 2005; Joosen et al. 2007; Kang et al. 2008; Lee et al. 2008; Xiang et al. 2008; Sarosh et al. 2009). The available microarray experiments uncovered genes with diverse functions that may act in regulation of embryogenesis and seed development (Lim et al. 1996; Soeda et al. 2005; Li et al. 2005; Kang et al. 2008; Xiang et al. 2008), stress response, and stress tolerance (Hammond et al. 2005; Vanderauwera et al. 2007; Zhou et al. 2007; Broekgaarden et al. 2008; Lee et al. 2008). Various analytical approaches were employed to facilitate the identification of new *Brassica* specific genes that respond to biotic and abiotic stimuli (Hammond et al. 2005; Broekgaarden et al. 2008; Lee et al. 2008). The available microarray studies revealed a large variety of functional assignments based on the conserved expression patterns of othologous genes. The microarray approach also has the potential to elucidate regulatory pathways controlling important traits, thereby increasing the effectiveness of plant breeding. This chapter presents the current status of transcriptomic resources and efforts in determining the functions of genes in *Brassica* species.

9.2 Functional Genomics Resources for *Brassica* species

9.2.1 Transcript Profiling Tools

To determine the relationships between gene structure and gene expression several techniques and platforms have been developed. These include methods such as cDNA library and EST production, serial analysis of gene expression (SAGE), massively parallel signature sequencing (MPSS), differential display PCR, suppression subtractive hybridization (SSH), real-time quantitative PCR, and microarray-based techniques. Broadly, transcript profiling techniques can be classified into two categories: sequencing-based and hybridization-based methods (Liu et al. 2007). With the exception of cDNA libraries, other techniques even if they are considered routine, are not frequently utilized for the estimation of transcript abundance in vegetable *Brassica* species. For example MPSS that leads to the faster discovery of novel single nucleotide polymorphisms (SNPs) and novel transcripts have not yet been used in vegetable *Brassica* research. The SAGE technique has been successfully applied to identify genes that are differentially expressed during seed development. However, among *Brassica* species this technique was applied just in the case of *B. napus* (GEO records: GSE14313; GSE7204; GPL8033; GPL4919). Suppression subtractive hybridization (SSH), one of the most popular methods for generating subtracted cDNA or genomic DNA libraries, was used by Cramer et al. (2006) to identify genes involved in *Alternaria brassicicola—Brassica oleracea* interaction.

The most popular approach to gene discovery in *Brassica* studies was the generation and screening of EST and cDNA libraries. This strategy provides global views of gene expression levels based on EST frequency. Currently, cDNA libraries have been used for identifying the genes associated with a physiological process or tissue specific expression (Lim et al. 1996; Kwak et al. 1997; Lim et al. 2000). For the functional genomics of *Brassica campestris*, a guard cell cDNA library has been constructed to gain an insight into the functions and development of this cell type (Kwak et al. 1997). The identified ESTs showed similarity to genes of different plant species including the Brassicaceae as well as non-plant genes. Several ESTs were also related to guard-cell functions such as regulation of stomatal movements. In other studies cDNA library-based transcript profiling was used with the aim to uncover the mechanism of the heading process in *B. rapa*. This study showed that majority of the identified ESTs share homology to the putative protein sequences of known biological functions and the highest score obtained for *A. thaliana* sequences (Rui-Juan et al. 2004). Furthermore, the partial sequencing of 1,216 randomly selected cDNA clones from *B. campestris* flower buds resulted in the identification of 904 cDNA clones (Lim et al. 1996). Nearly 50% of them had sequence homology to functionally defined genes. Using a similar methodology Lim et al. (2000) generated ESTs from different organs at various developmental stages of *Brassica rapa* L. ssp. *pekinensis*. This study can be applicable for the comprehension of the developmental processes of the *Brassica* plants and for the comparative analysis of *Arabidopsis* genome structure.

Efforts to generate ESTs are now being extended. Additional organ- or developmental stage-based cDNA libraries *B. oleracea* and *B. rapa* have been recently constructed. Continuously acquired data is expanding our knowledge about genome size and gene function in *Brassica*. Existing *B. napus* libraries cover a large amount of information regarding organ specific expression and development. Available cDNA libraries and ESTs were further used to facilitate the construction of transcriptomic microarrays (Brassica Genome Gateway). The microarray-based approach for gene expression profiling has shown enormous potential and is now the most widespread method (technique) applied in transcriptomes research. Microarray studies have been used in several *Brassica* species to better understand the changes in gene expression during growth and development, pathogenesis, abiotic stresses and many biological processes. These days, microarray-based gene expression monitoring is possible through high-density oligo or full-length cDNA *Brassica* arrays, *Arabidopsis* ATH1 GeneChip arrays and less frequently, custom low-density macroarrays utilizing *Arabidopsis* cDNA/EST as a probe. Unfortunately, the most recently developed BeadArrays (Kuhn et al. 2004) and high-throughput screening qPCR (e.g. TaqMan Gene Expression Assays) were not developed for *Brassica* research.

9.2.2 Microarray Resources

The development of a range of expression profiling tools for *Brassica* species have been achieved in the frame of different international networks. Five technology platforms have been developed including the Agilent eArray, the Combimatrix array, the NimbleGen array, and custom cDNA arrays spotted on nylon membrane (macroarray) or glass support. As oligo design supports sequence specificity, three out of four types of available *Brassica* microarrays have been designed up to 60-mer oligo probes (Table 9-1). The Agilent eArray, the Combimatrix and the NimbleGen arrays were designed based on available EST deposited in GenBank and large *Brassica* databases.

Table 9-1 *Brassica* microarray resources.

Technology platform	Probes	Probe length	Description
Agilent 2x 105K	91,854	60-mer	for use with all *Brassica* Species
Combimatrix 94K	~ 90,000	35-mer	for use with all *Brassica* Species
NimbleGen KGBP 24K	331,384	60-mer	*B. rapa* microarray
NimbleGen *B. rapa* 300K	360,312	60-mer	*B. rapa* microarray
Seed cDNA array	10,642	cDNA	for use with all *Brassica* Species
Custom cDNA arrays		cDNA	for use with all *Brassica* Species
Affymetrix ATH1		25-mer	*Arabidopsis* GeneChips arrays

95K microarray generated by the J. Craig Venter Institute (JCVI), The John Innes Centre (JIC) and Cogenics is a validated, one probe per gene, 91,854 60-mer oligonucleotide array synthesized in situ using Agilent Technologies' SurePrintTM technology in 2 × 105K format (an inkjet printing method). The updated description of the array such as EST assemblies, gene ontology distribution of ESTs associated with the array as well as estimations of species-specific genes are available on the *Brassica* Gateway website. Besides *Brassica* Unigenes, the array includes both negative controls (plant and human genes) and positive controls (insect housekeeping genes). The CombiMatrix 94K microarray system is the only one that offers ~ 90,000 *Brassica* probes on up to three times re-usable slides (Ma et al. 2004; Xiang et al. 2008).

Two arrays have been designed at the NimbleGen for functional genomics of *Brassica rapa*: 300K Microarray v2 and KGBP 24K Microarray with 47,548 and 24,963 unigenes, respectively. The 300K Microarray consists of 331,384 of 60-mer probes. Probes were synthesized in situ using a Digital Micromirror Device (DMD). An advantage of the NimbleGen platform is the relatively low cost of microarray preparation due to the absence of pre-existing masks as it occurs in Affymetrix GeneChips. Seven probes that cover 150 bp in the 3′ region of the gene represent each gene. The array includes genes from chloroplast, mitochondria and selection marker genes. Alternatively, KGBP 24K Microarray consists of 360,312 probes duplicated in 2 blocks. Six probes represent each gene and ~ 20% of genes with an unknown direction have 12 probes designed as follows: 6 probes in sense and 6 in an antisense direction. Sequences of probes and unigenes are available at BrGD (*http://www.brassica-rapa.org*). Other resources are also available. For example, SIPPE *Brassica napus* 8.1K cDNA Microarray (v1.0), NEFU *Brasicca rapa* 3k v1.0A as well as 10,642 *Brassica* Seed cDNA array were preliminary used for the identification of the genes involved in seed development processes or unique traits (Xiang et al. 2008). Additionally, for a comparative analysis ATH1 Genechip or other custom Arabidopsis arrays can be used. Originally, the *A. thaliana* ATH1-121501 Genechip array was used to study gene expression in *Brassica oleracea* (Hammond et al. 2005). Moreover, a 60k Brassica Affymetrix array, based on the JCVI assemblies, should be also available soon (*www.brassica.info*).

9.2.3 Microarray Analysis and Tools

One can be easily confused by the wide variety of microarray analysis tools currently available. However, if one looks beyond the specific software packages and algorithms, commonalities are apparent in the steps that need to be taken during microarray analysis (Allison et al. 2006). In this section we will present steps that should be performed and give some examples of software tools available. Since *Brassica* studies are likely to be performed on a variety of microarray platforms (Hammond et al. 2005; Katari et al. 2005; Soeda et al. 2005; Esmon et al. 2006; Hudson et al. 2007; Xiang et al. 2008) we have kept the information applicable to most microarray platforms.

9.2.3.1 Experimental Design

Knowing the possible sources of error is important. Since microarrays do not report absolute quantification but only comparative hybridization, this places great importance on the control to which the treated sample must be compared. Potential sources of difference between the treated sample and control, extraneous factors such as age differences of the subjects (when that is not the factor under investigation), or the manner and time of harvesting, processing RNA samples or running arrays, should be eliminated or at least brought to a minimum. To have confidence in differential expression of genes being measured by microarray, replication should be performed and this should be biological replication rather than technical, where the same RNA is placed on two or more arrays. Pooling is an acceptable way of improving statistical power without hugely increasing costs, however, replicates should still be performed. Pooling destroys information about the variability of RNA expression between individuals so verification of results should be done on unpooled samples (using quantitative RT-PCR for example).

9.2.3.2 Quality Control of Microarrays

Following the need to control extraneous factors that were emphasized in the previous paragraph, one seeks consistency between the arrays when it comes to quality analysis. There may be particular target metrics that the microarray manufacturer recommends, but in an experiment it is important to check that metrics like median intensity, distribution of intensities and background intensities are consistent. Small differences can be compensated for by normalization methods as we shall see in the next paragraph, but there is a limit to how effective these corrections can be. Checks should be done for technical artefacts and special defects on the array such as blobs and spot irregularities.

9.2.3.3 Image Processing and Normalization

For a given array type several methods may be available for estimation of the amount of RNA from fluorescent array images, while trying to minimize the extraneous variation that occurs owing to technical artefacts. The Bioconductor package, based on the R programming language, is an excellent open source and open development software project making these methods freely available (Gentleman et al. 2004). For example, when analyzing Nimblegen arrays, the "oligo" library can be used to perform the RMA method of image processing and normalization that is commonly used for analysis of Affymetrix arrays (Bolstad et al. 2003). Image processing and normalization will produce a spreadsheet of expression values for each gene. The data are commonly transformed into logarithmic scale (typically log base 2) for further analysis.

9.2.3.4 Quality Assessment of the Experiment

Performing a principal components analysis (PCA) is a valuable way of assessing the performance of the experiment. If replicates have been performed then grouping of replicates indicates that reliable differential expression has occurred between the conditions studied. If on the other hand a random distribution of the samples in PCA space is found then great caution needs to be exercized in the statistical analysis described in the next paragraph (when thousands of genes are analyzed at once there will always be some genes that appear differentially expressed in a statistically significant way due to random chance events). When more than one factor is being studied, for example drought and aphid resistance, a PCA can indicate the relative strengths of these effects on gene expression.

9.2.3.5 Inference

A statistical test is performed on a gene by gene basis to assess if differential expression has occurred. Using fold change alone is not acceptable. A parametric test such as the Student's t-test or ANOVA can be used on the data in logarithmic scale. Prior to this, removal of defective spots may be required, depending on the microarray platform used. Because of costs a typical microarray experiment contains a smaller number of replicates than is optimal. Variance shrinkage methods are useful in making use of the parallel-testing nature of microarrays to bolster statistics (Cui et al. 2005). The parallel-testing nature of microarrays leads to a different statistical problem when it comes to choosing a set of genes that show differential expression (referred to as a genelist). Selecting a genelist "with p values less than 0.05" leads to a multiple-testing error and so a false-discovery correction should be applied (Storey 2003).

9.2.3.6 Genelist Analysis

Once a set of statistically significant differentially expressed genes has been identified, the next challenge is to organize the list of genes. Analyzing the genes into classes, often based on Gene Ontology categories, is helpful in this regard. An intermediate step of comparative genomics may be required to make use of the information available in the *Arabidopsis* model species.

9.2.4 Genomic Projects and Databases

Numerous laboratories collaborate in several consortia to build resources that facilitate gene function discovery in *Brassica* crops. These resources include: a BBSRC Innovations in Crop Science Project headed by the AdVaB consortium, the Multinational *Brassica* Genome Project (MBGP), the Multinational *Brassica rapa* Genome Sequencing Project (BrGSP), and the Defra funded Oilseed Rape Genetic Improvement Network (OREGIN) (*Brassica.info*). In total, there are 10 *Brassica* Genome Projects: four for *B. napus* (EST project at Genoplante; Plasmid genome at the National Agricultural Research Center for the Hokkaido Region, Japan; Genome sequencing project at John Innes Centre and Beijing Genomics Institute), two for *B. rapa* (Genome sequencing at BrGSP and MBGP), two for *B. oleracea* (Genome sequencing at TIGR; physical map and comparative genomics at BBSRC) and single for *B. juncea* and *B. nigra* (both are part of MBGP). These organizations through their independent and collective activities develop *Brassica* resources, integrate technologies and build technical support for their practice usage.

Ongoing sequencing projects and other activities force the *Brassica* community to develop comprehensive databases and public repositories (Table 9-2). The biggest public repository database at NCBI, in cooperation with other *Brassica*-related resources, including ASTRA, BAGI, BGG, CSCL, and MBGB, have accumulated information for the *Brassica* species including 651 libraries, 19,343 nucleotide sequences, 850,713 ESTs, 970,602 GSSs, 2,066 probes, 10 genome projects, 46,855 UniGene and no SNP and dbGaP(Genotype and Phenotype data) records and many others. For vegetable *Brassica* species GeneBank deposits information about 144 cDNA libraries, 11,374 nucleotide sequences, and 249,090 ESTs records (Table 9-3).

As indicated by GeneBank, the sequencing of 118 *B. rapa* and 26 *B. oleracea* cDNA libraries generated 246,224 expressed sequence tags. Additional 2,866 ESTs are available for *B. carinata, B. juncea* and *B. nigra*. Several specific *B. rapa* cDNA libraries were also constructed, including whole plants library treated by light-chill or NaCl stress, calli in organogenesis stages on the tissue culture media library as well as cDNA library from leaves infected by the pathogen *Pectobacterium carotovora* subsp. *carotovora* (NCBI, BrGP). These libraries provide very useful information in relation to the biology of the stress response and regeneration of plants. Furthermore, ESTs from the *B. rapa* and *B. oleracea* libraries can be grouped into 20,114 UniGene

Table 9-2 Genomics resources for *Brassica* species.

Databases and public domains	Website URL
Brassica Genome Gateway	*http://brassica.bbsrc.ac.uk/*
Brassica Genomics	*http://www.brassicagenomics.ca/*
Brassica.info	*http://www.brassica.info/*
OREGIN	*http://www.oregin.info/*
WHRI *Brassica* Genomic Resources	*http://www2.warwick.ac.uk/fac/sci/whri/research/grc/brassica/*
The John Innes Genome Laboratory	*http://www.jicgenomelab.co.uk/libraries/brassica.html*
Brassica oleracea Genome Project	*http://www.tigr.org/tdb/e2k1/bog1/*
Brassica European Database	*http://documents.plant.wur.nl/cgn/pgr/brasedb/default.htm*
Brassica Working Group	*http://www.ecpgr.cgiar.org/Workgroups/brassica/brassica.htm*
AAFC *Brassica / Arabidopsis* Genomics Initiative (BAGI)	*http://brassica.agr.gc.ca/index_e.shtml*
Brassica rapa Genome Project	*http://www.brassica-rapa.org/BGP/NC_brgp.jsp*
Brassica rapa EST and Microarray database (BrEMD)	*http://www.brassica-rapa.org/BrEMD/*
National Center for Biotechnology Information (NCBI)	*http://www.ncbi.nlm.nih.gov/*

records (putative genes; Table 9-3). Having these estimations, we can speculate that the NCBI UniGene records cover less than 30% of the genes in the *B. rapa* genome (50,000 estimated transcripts) and 7% of *B. oleracea* genome (80,000 estimated transcripts), respectively. It is important to note that, the *Brassica rapa* Genome Project (BrGP) headed by two teams Korea *Brassica* Genomics (KBGP) and the Biogreen21 Program (BG21) reported a detailed analysis of 127,144 ESTs from the 26 different cDNA libraries of *Brassica rapa*. Following this information; the investigated EST sets generates 32,395 Unigene sequences that cover around 65% of the genes in *B. rapa* genome. The most current list of *Brassica* databases and related resources can be found on Brassica Genome Gateway webpage (*http://brassica.bbsrc.ac.uk/*) and Brassica.info (*http://www.brassica.info/*).

Table 9-3 *Brassica* resources.

Brassica species	ESTs	cDNA libraries	Nucleotide sequences	Unigene records
Brassica napus	656,145	542	11,871	27,143
Brassica rapa	194,741	140	4,242	14,606
Brassica oleracea	63,201	26	5,478	5,617
Brassica carinata	2,500	-	101	-
Brassica juncea	5,393	-	1,034	-
Brassica nigra	1,814	-	624	-
Total	923,794	708	23,350	47,366

Resources provided by the NCBI Entrez database; update for April 2010.

There was a significant acceleration of *Brassica* functional genomic research over the past few years. Unfortunately, it was not followed by the integration of public gene expression databases for *Brassica* species. The development of microarray platforms allowing specific in-species *Brassica* analyses as well as cross-species comparisons would advance the *Brassica* gene expression studies. Currently, the *Brassica rapa* EST and microarray database (BrEMD) store 10 microarray datasets and allows simple microarray data analysis as well as keyword search. There are only 18 different microarray experiments deposited in the existing repositories including the Gene Expression Omnibus database (GEO), EBI-ArrayExpress, ArrayExpress and NASCarrays. Some datasets are deposited in two databases. Fifteen microarray experiments are in GEO database: 10 experiments for *B. napus,* three for *B. rapa* and four for *B. oleracea.* ArrayExpress includes six microarray experiments: two for each: *B. napus, B. rapa* and *B. oleracea.* Additionally, NASCarrays has deposited three microarray experiments for *B. oleracea.* Beyond data analysis results this databases store information about data structure, protocols, and annotations in the MIAME compliant format.

In contrast to microarray data, comprehensive EST databases and online annotation tools have been well developed. The *Brassica* Expressed Sequence Tag Portal at Brassica Genomics (*http://www.brassicagenomics.ca/*) contains ESTs from four *Brassica* species (*B. napus, B. rapa, B. oleracea* and *B. carinata*), provides ESTs annotation and information about EST sequence homology. EST analysis tool at *Brassica* Genome Gateway (*http://brassica.bbsrc.ac.uk/*) allows both BLAST and EST ID search for Brassica 95k microarray v1 array. In the output it shows the homology search best hit (gene name, protein ID) and EST assembly components. Alternatively, *Arabidopsis* BLAST search at TAIR (www.arabidopsis.org) also provides an additional, advanced and well organized tool for gene annotation and metabolic pathway analyses in *Brassica.*

9.3 Comparative Genomics

The genomes of *Brassicaceae* are large and complex (Paterson et al. 2001; Paterson 2005; Xiang et al. 2008)(Paterson et al. 2001; Paterson, 2005; Xiang et al. 2008). Comparative analyses of chromosome segments have shown the conservation of gene order in *Arabidopsis* and *Brassica* (Paterson 2005; Ziolkowski et al. 2006). Given the high genomic collinearity, the comparative genomics approach gives an opportunity to transfer information from model species to crops. Comparative genomics improve our understanding in areas of phylogenomics (genomic data is used to construct more robust phylogenies), structural genomics (genome rearrangements: segmental duplications, deletions, insertions, transpositions, polyploidy), annotation and selection/evolution of traits (Paterson 2005; Chen and Ni 2006; Schranz et al. 2006; Ziolkowski et al. 2006; Schranz et al. 2007). In the past few years a comparison of *A. thaliana* relatives namely *B. oleracea* and *B. rapa* helped develop our knowledge about Brassicaceae.

9.3.1 *Arabidopsis thaliana—Brassica oleracea*

Comparative genomics is an effective method to annotate the sequenced genome by identifying genes, regulatory elements and improving gene structure (Colinas et al. 2002; Hammond et al. 2005; Katari et al. 2005; Esmon et al. 2006). A comparative approach has a great capacity to complement current annotation. An excellent example of the comparative genomics strategy is the transcriptome analysis of *Brassica oleracea*, by using an ATH1 Genechip array designed for *A. thaliana* (Hammond et al. 2005). Importantly, to improve the reliability of *Brassica* hybridization results the authors have identified probe-pairs that hybridized to the *B. oleracea* genomic DNA on the basis of the perfect-match (PM) probe signal intensity. The generated matrix was subsequently used for an analysis of *B. oleracea* response to a mineral nutrient stress. In another study Affymetrix Arabidopsis ATH1 were used for analyses of tropic responses (Esmon et al. 2006). The results suggested that the activation and accumulation of specific genes correlates with a lateral gradient of auxin in *Brassica* hypocotyls. *B. oleracea* anther-specific mRNAs were analyzed by using *Arabidopsis* cDNA macroarrays (Amagai et al. 2003). Similar *Arabidopsis* cDNA macroarrays have been used for a comparative analysis of drought stress response in *A. thaliana* and *Brassica oleracea* (Fig. 9-1) (Ludwików A., Babula-Skowronska, Szczepaniak M., Belter N., Dominiak E., Nowak W., Sadowski J., unpubl.). All these studies support the idea that *Arabidopsis* DNA microarrays and macroarrays with a high degree of confidence can be used to address gene expression profiling in *Brassica* species.

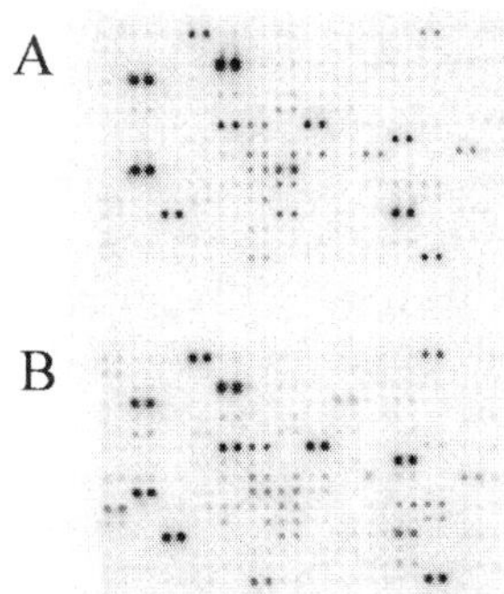

Figure 9-1 Macroarrays for the analysis of *Brassica oleracea* spp. *alboglabra* line A12DH gene expression in response to drought stress. The figure shows a representative macroarray hybridization result for control (A) and drought-treated (B) *B. oleracea* samples.

Due to the remarkable nucleotide sequence similarity between *B. oleracea* and *A. thaliana* (reported as high as 80%), full-length cDNA clones and EST resources for *A. thaliana* and other species of Brassicaceae are the most relevant for the precise annotation of new genes. To improve *A. thaliana* and *B. oleracea* annotation approximately 600,000 *Brassica oleracea* shotgun reads were sequenced by TIGR/CSHL/Washington University and analyzed by using the *Arabidopsis* genome and proteome databases (Katari et al. 2005). The authors focused their attention on the short arm of chromosome 4 of *A. thaliana* to test whether putative genes conserved between *Brassica* and *Arabidopsis* were expressed in comparison to non-conserved putative genes. The results revealed that putative genes tend to be expressed if they are identified in conserved regions. Additionally, the authors improved the *Arabidopsis* annotation by identifying new genes, demonstrating the expression of several hypothetical genes, and extending the *Arabidopsis* transcriptome by 850 genes. Moreover, incorrect gene structure was identified in approximately 21% of hypothetical genes, demonstrating the need of a re-annotation of this part of the *Arabidopsis* genome. Another comparative approach (Ayele et al. 2005) used the genome-wide alignment of Whole Genome Shotgun (WGS) sequences from *B. oleracea* to identify poorly annotated genes and genomic regions. The analysis resulted in the identification of conserved *Arabidopsis* genome sequences (CAGSs) that were subsequently merged into a potential functional unit. Moreover, several novel conserved units were identified that did exist apart from gene annotations. Overall, these studies demonstrate how information from *Brassica oleracea* sequences can be utilized to improve the annotation of the *Arabidopsis* genome.

Comparative genomics makes it possible to discover a large variety of functional elements by their conservation between *Arabidopsis* and *Brassica* (Ruvinsky and Ruvkun 2003; Doniger et al. 2005). Haberer et al. (2006) have used this strategy to study sequence conservation in orthologous promoter pairs. The employed methodology resulted in the identification of a nucleotide pattern in co-expression groups in *Arabidopsis* as well as putative orthologous upstream sequences in *Brassica*. Among many newly identified motifs, several known sequence patterns were revealed, which validated the undertaken approach. In another example, an *Arabidopsis* ATH1 was successfully used for the analysis of the flower (anther) development in the *Brassica oleracea Ms-cd1* mutant (Kang et al. 2008). In this particular approach, microarray profiling facilitated the identification of differential expression between male sterile *Ms-cd1* and wild type (male fertile) plants. Importantly, kinases, cell wall modification genes, and ion transport genes were suppressed by *Ms-cd1*. The authors concluded that *Ms-cd1* might affect an anther developmental pathway, which influences genes involved in signaling, cell wall modification, and ion transport in pollen grains.

9.3.2 *Arabidopsis thaliana—Brassica rapa*

Recent information obtained from the *A. thaliana* model has advanced our understanding of defense responses. However, there is still limited knowledge about the pathogen defense mechanism in related *Brassica* species. It is likely that related species may have unique pathogen responsive mechanisms and regulation of gene expression, which does not correspond to well known plant models. To investigate the above assumption the complexity of the gene expression of *A. thaliana* and *B. rapa* were examined in plants infected with fungal pathogen *Colletotrichum higginsianum* as well as after salicylic acid (SA), ethylene (ET), and methyl jasmonate (MeJA) treatment (Narusaka et al. 2006). Transcriptome analysis was performed using a cDNA microarray with 1,820 cDNA clones selected from the cDNA library of Chinese cabbage and 1,2K *Arabidopsis* cDNA microarray. In general, the constructed *B. rapa* microarray plate represented putative defense-related and regulatory genes. Most gene expression profiles between *B. rapa* and *Arabidopsis* were similar suggesting conservation of these genes function. However, genes that undergo different expression patterns between *B. rapa* and *Arabidopsis* were also identified revealing minor differences in *B. rapa* response to fungal pathogens in comparison to *A. thaliana*.

9.4 Application of Functional Genomics in Genomics-assisted Breeding

In the past decade, plant breeders have made a huge effort to create germplasm and plant cultivars having desirable traits. From these studies we have learnt that the success of this selection relies on the availability of wild alleles for the agronomic traits. As in many other branches of life science, plant breeding for crop improvement cannot exist without transcriptomics. By applying functional genomics approaches we are now capable of providing candidate genes that can be simultaneously used for understanding the biology of the plants and development of agronomically important traits (Ma et al. 2004; Tuberosa and Salvi 2006; Heffner et al. 2009; Sarosh et al. 2009). In general, the transcriptomics can be used for identification mechanisms that control the biological system. This initial knowledge serves as a basis for further molecular analyses, including the generation of transgenic plants and time-consuming studies of emerging phenotypes. It is important to note that despite such a huge potential of transcriptomics and other genomics methods there are still many bottlenecks. One of them is the variability of the design phenotype in a field. Simply, many transgenic plants do not copy its *in vitro* or chamber performance in the field. Nevertheless, it is highly desirable if the quality of the improved crop meets both yield and consumer satisfaction.

Despite all the undergoing breeding programs and the recent technological breakthroughs the overall contribution of genomics-assisted breeding to the release of new vegetable *Brassica* cultivars is almost none. So far we can point out just a few studies that have put new focus on breeding strategies for seed improvement and stress tolerance. These strategies utilize mainly classical techniques such as genetic mapping, association mapping, QTL mapping and traits combination (Tuberosa et al. 2002; Lou et al. 2007; Zhao et al. 2008; Heffner et al. 2009), and these will not be discussed here. Fortunately, genomics-based approaches for phenotype predictions utilizing microarrays as a tool are also accumulating. DNA microarrays (oligonucleotide and cDNA micro- and macroarrays) have been utilized to identify differentially expressed genes in a few *Brassica* species, including *B. rapa* and *B. oleracea*. Reported studies have made an initial but important contribution to vegetable *Brassica* crop biology. Applying gene expression profiling is now possible to identify regulatory gene networks that can be useful for trait-specific screens of agronomically important alleles.

9.4.1 Seed Development and Quality

Seed development is an important stage for seed quality (Wszelaki et al. 2007). Thus the understanding of the mechanisms controlling seed development and germination is essential to achieve high harvest yields (Soeda et al. 2005). Recent transcriptome analyses have advanced our knowledge of seed processing by providing information on how multiple genes interact to control the effect of priming (Soeda et al. 2005; Hudson et al. 2007; Xiang et al. 2008). From these analyses, it is now possible to observe how priming improves seedling performance and how can it be used in the field.

Two groups have successfully applied microarrays to study the effect of priming. Hudson et al. (2007) used Affymetrix ATH1 array to determine the mechanism of priming on the germination efficiency of cauliflower seeds (cultivars Lintop and Maverick). Microarray analyses differentiated three distinct expression patterns that correlated with hydratation treatment. Additionally, several functional categories of genes were identified to have particular expression profiles in investigated conditions. The authors propose that the effect of the priming process on the induction of protein synthesis could be a factor that determines seed durability. Furthermore, among 833 statistically significant *Arabidopsis* homologs, 59 were classified to be involved in protein turnover, protein fate, and ribosome biogenesis. Moreover, the majority of the genes were identified as ribosomal proteins, or were related to global metabolism. Significantly, the validation of results using real-time quantitative PCR confirmed the expression of four out of five orthologous *Brassica* transcripts, indicating slightly lower reproducibility of microarray hybridization results.

A second group (Soeda et al. 2005) investigated priming and germination exploiting cDNA microarrays containing genes that were isolated from cotyledon or germinating *B. napus* seeds. The experiment design, in comparison to the previous study, was supposed to prevent information loss occurring in comparative approaches. Significantly, applying strict criteria for detection of hybridization signal this group selected 1,100 cDNAs corresponding to 850 distinct *B. napus* genes. In total, clustering the samples revealed five significant expression patterns that depend on preferential expression in dry, osmoprimed or germinating seeds. Although, some parts of the results were rather confirmatory, a vast amount of data provided additional information of the gene expression levels in relation to the seed germination, osmopriming, and stress tolerance. Notably, the authors showed that the expression of stress-related genes predominantly occur during seed maturation and slow drying. Most of these genes were known to encode LEA proteins or ABA-inducible dehydrins (RAB18, Em6). In addition, putative signaling pathways involved in the regulation of these genes have also been revealed based on an analysis of data deposited in Genevestigator. Overall, these data suggested that the regulation of osmopriming is complex and to some extend overlaps with germination pathway. Therefore, identification of putative regulators is important for the understanding of germination programs and seedling survival in stress conditions. It is important to note that an insight gained from this study provided new gene candidates for seed quality.

Another useful approach to gene function study was applied by Xiang et al. (2008). To test the quality of the seed cDNA microarray they analyzed different stages of *Brassica* seed development and different types of tissues. The authors identified differentially regulated genes between embryo and leaf that are consistent with other findings in the field. Further functional analyses of these genes should provide a rich source of information on the biology and technology of the seeds.

9.4.2 Tolerance to Abiotic Stresses

Plants are continuously exposed to abiotic stresses and have evolved mechanisms to reduce their effects. Various abiotic stresses limit vegetable *Brassica* production and the most important include water stress (Lee et al. 2008; Ashraf et al. 2001; Ludwików unpubl.) (Lee et al. 2008; Ashraf et al. 2001; Ludwikow unpublished), nutrient deficit (poor degradated soil) (Hammond et al. 2005), high temperature (Björkman and Pearson 1998), wounding and pathogen attack (Park et al. 2005; Broekgaarden et al. 2007; Bruinsma et al. 2007; Sarosh et al. 2009) (Broekgaarden et al. 2007;

Broekgaarden et al. 2008; Bruinsma et al. 2007; Park et al. 2005; Sarosh et al. 2009). However, with the exception of biotic stresses, which will be discussed later, there are only two studies reporting the profile of *Brassica* plants in abiotic stress conditions (Hammond et al. 2005; Zhou et al. 2007; Lee et al. 2008). Gene expression profiling are novel approaches for the determination of genes and pathways involved in stress response and tolerance. Therefore, we are still far from detailed knowledge of signal transduction cascades activated under abiotic stress response in *Brassica* plants. Nevertheless, based on these initial reports, we summarized the progress and the current status of the gene profiling in the studies of *Brassicca* stress response. We also discussed the contribution of these studies to *Brassica* breeding.

Particular emphasis has been made to study *B. oleracea* transcriptional response to phosphorus stress (Hammond et al. 2005). The reported study has been designed to improve our knowledge about the regulation of gene expression in mineral nutrient stress and to investigate the consequences of P starvation. The analysis utilizing the Affymetrix ATH1 Genechip array revealed approximately 100 genes identified as significantly differentially regulated in the shoots in the response to phosphorus starvation. Only a few functional categories of genes were identified to have distinct expression profiles in P stress conditions. Among them several genes involved in the P turnover during P starvation were found to be up-regulated. Notably, many transcripts or functional groups identified for P starvation have been previously described in *A. thaliana*. Nevertheless, many identified genes can serve as good candidates for genetic manipulations improving nutrient efficiency in plants.

Another important study, developed to identify defensive responses against UV light in turnip (Zhou et al. 2007), indicated distinct genes regulated by UV-B and UV-A treatments while similar sets of genes were regulated by blue and UV-A light treatment. Several functional categories were identified to have particular expression profiles in stress conditions. As expected, ethylene- and oxidative stress-related genes showed higher expression levels upon UV-B exposure, while lower expression were observed for tissue or cell growth-related genes. In addition, what was the most important, a group of genes related to anthocyanin biosynthesis were found up-regulated in response to UV-A and this correlated with higher anthocyanin level in treated plants. Notably, this study has revealed a putative anthocyanin biosynthesis pathway mediated by unknown UV-A-specific photoreceptor. Anthocyanin pigment content has a critical role in the color quality of many fruits and vegetables, simply by making them attractive to the consumer. Thus, these results have great potential for practical applications in prevention and quality control in crop production.

As for many other crop species, the advances made in genomic analysis were limited by the extremely slow development of genomic resources including the construction of *Brassica* microarray. The most recent transcriptomic experiment using *Brassica* 24K Oligo Microarray (KBGP-24K) was designed to identify cold, salt and drought stress-responsive genes in *B. rapa* (Lee et al. 2008). By a cross-species comparison the authors selected many genes specifically responding to each factor: cold, salt, and drought, as well as a group of generally responsive genes regulated by all three forms of stress. Significantly, genes encoding several novel transcripts were identified and the differential expression of many known stress-induced genes was confirmed. This includes an ortholog of *Arabidopsis* CRT/DRE-binding factor gene (*BnCBF17*). Additionally, genes encoding putative transcription factors (members of AP2-EREBP, MYB, bHLH, bZIP, homeobox, and heat stress transcription factor families) responsible for abiotic stresses were also recognized. Taken together, the expression patterns for the majority of differentially expressed genes were rather similar to *Arabidopsis* homologues. The exception is *B. rapa SOS* genes whose expression level was not significant in this study. Nevertheless, there is a notable outcome of this research showing that such a direction is worth continuing. Identified candidate genes should be further evaluated for their phenotype in the field.

9.4.3 Transcriptomics to Improve Disease Resistance

Little is known about pathogen defense mechanisms in *Brassica* species. Although vegetable *Brassica* crop losses due to invading pathogens such as insect pests and fungi are significant, progress on identifying defense mechanisms that govern the response is extremely slow. It has been established that plants defend themselves against pathogens by activating both direct and indirect defense mechanisms (van Poecke and Dicke 2002; Ellis and Dodds 2003; Broekgaarden et al. 2007) (van Poecke and Dicke, 2002; Ellis and Dodds, 2003; Broekgaarden et al. 2007). In addition, there are several major signal transduction pathways involved in plant defense including salicylic acid (SA), jasmonic acid (JA), ethylene, ABA and the reactive oxygen species pathways (Fig. 9-2). However, our understanding of how these components interact and how other components are involved is still limited. Studies in *Arabidopsis* have demonstrated that an SA-dependent response is deployed against biotrophic pathogens, whereas ethylene or JA responses are important for the induction of resistance to insect and arthropod herbivores, some necrotrophic fungal pathogens, and nonpathogenic fungi (Thaler et al. 2004; Toth and Birch 2005; Park et al. 2005; Toth et al. 2006; Narusaka et al. 2006). Available gene-expression profiling studies analyzing host-pathogen interactions in *Brassica* (Park et al. 2005;

Ferry et al. 2006; Tuberosa and Salvi 2006; Broekgaarden et al. 2007; Bruinsma et al. 2007) have not revealed additional members of defense signaling pathways, but signify an interaction between JA, ET and SA signaling as crucial components of defense important for the understanding of plant response to biotic stresses.

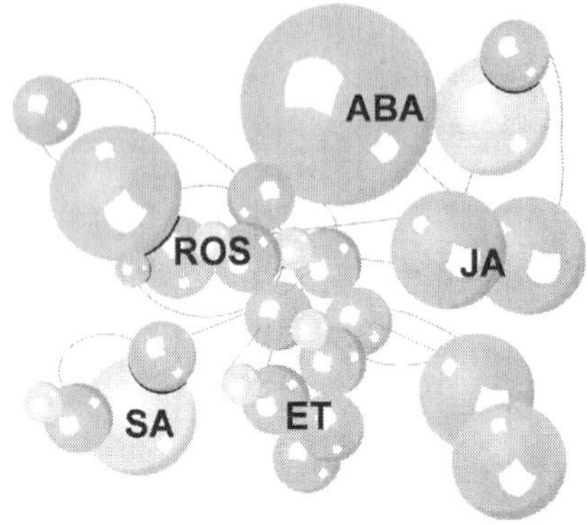

Figure 9-2 Signaling network in plant defense against pathogens.

To identify and characterize genes that are involved in inducible direct defense in *Brassica*, Broekgaarden et al. (2007) analyzed the gene expression profiles of two *B.oleracea* cultivars, Rivera and Christmas Drumhead, upon feeding by *Pieris rapae*. The two cultivars were shown to differ in transcriptional responses and the level of direct defense against pathogens. The analysis revealed a higher direct defense in Rivera compared to Christmas Drumhead cultivar. This correlated with differences in the transcriptional response including timing and cultivar-specific activation of some genes. However, what was most important, this study indicated that the jasmonate signaling network did not control a large portion of *P. rapae*-inducible genes. These results demonstrated that direct defense agains *P. rapae* was also influenced by other undefined factors.

Nevertheless, the involvement of JA in the regulation of direct and indirect defense has already been documented in many plant species (Zhu-Salzman et al. 2004; Liechti and Farmer 2006). JA is involved in the defense responses to oviposition and feeding damage (Hilker et al. 2005; Schroder et al. 2007; Walling 2008) (Schroder et al. 2007; Hilker et al. 2005; Walling, 2008). Several studies focused on the retardation of growth and leaf consumption of herbivores (Kingsolver and Woods 1997; Dunning et al. 2003; Williams et al. 2004) (Williams et al., 2004; Dunning et al., 2003; Kingsolver and Woods, 1997). In addition to these, Bruinsma et al. (2007) tested the effect of JA on the oviposition of *Pieris rapae* and *P. brassicae* herbivores in cabbage. Interestingly, they found that JA treatment reduces the acceptance of leaves for oviposition. However, JA-induced signaling changed the level of the insect's tolerance of leaves. Another study aimed at the characterization of the gene expression profiles in *B. rapa* in order to investigate the key lines of defense against *C. higginsianum* infection (Narusaka et al. 2006). The results confirmed that SA, JA and ET-dependent defenses are essential for host-pathogen interaction, however ET-dependent signaling pathway may be a key player in the early defense. Similarly, a cDNA microarray was utilized in the analysis of the defence response in canola upon treatment of the fungal pathogen *Sclerotinia sclerotiorum* (Yang et al. 2007). This analysis led to the identification of 300 transcripts that were responsive to this pathogen-challenge. In agreement to previous studies, this analysis also identified genes that were associated with JA biosynthesis and signaling. It is important to note that among identified genes an up-regulation of genes orthologous to AtERF2 and AtERF4 were reported, further confirming the involvement of ERFs in the defense pathways in related plant species. Additionally, the importance of the salicylate-dependent and ethylene-mediated pathways in defense responses to *Pseudomonas syringae* pv. *tomato* in Chinese cabbage was confirmed in the study of Park et al. (2005). Overall, nearly all of the presented results are in agreement with previous *Arabidopsis* studies. Again, further functional characterization of the revealed members of the defence signaling might be crucial for future rapid *Brassica* yield improvement.

9.5 Conclusions

Transcriptome analyses are powerful approaches for the decoding of gene functions. Due to a remarkable technical advance in transcriptomics new genomic resources have been developed for *Brassica*, an agriculturally important genus consisting of several vegetable crops. Many ESTs and cDNA libraries that have been generated were subsequently used to design *Brassica* DNA microarrays. As expected available microarray studies provide evidence for the effective application of microarrays to identify a large number of genes involved in seed growth, development, disease resistance, and abiotic stress tolerance. Despite the substantial progress that has been made toward uncovering the network of gene functions in investigated conditions, much work still needs to be done to completely characterize associations between gene expression patterns and the genotypes in breeding populations. Although functional genomics tools are very helpful in the detailed molecular description of agronomic traits, they alone cannot improve our understanding of biological systems shaping the phenotype. Therefore, much more attention should be paid now to the integration of high-throughput genomics with routine breeding strategies. The question remains: how will we take advantage of these opportunities in the future?

Acknowledgements

This study was supported by the Polish Committee for Scientific Research Grants PBZ-KBN-089/P06/2003, MNI 2P04A03329, PBZ-MNiI-2/1/2005, PZ-19/2006/2/JS, PBZ-MNiSW-2/3/2006/19/IGR/4 and 682/N-COST/2010/0. I. Makalowska is acknowledged for critical reading of the manuscript.

References

Allison DB, Cui X, Page GP, Sabripour M (2006) Microarray data analysis: from disarray to consolidation and consensus. Nat Rev Genet 7: 55–65.

Amagai M, Ariizumi T, Endo M, Hatakeyama K, Kuwata C, Shibata D, Toriyama K, Watanabe M (2003) Identification of anther-specific genes in a cruciferous model plant, *Arabidopsis thaliana*, by using a combination of *Arabidopsis* macroarray and mRNA derived from *Brassica oleracea*. Sexual Plant Reprod 15: 213–220.

Ashraf M, Nazir N, McNeilly T (2001) Comparative salt tolerance of amphidiploid and diploid *Brassica* species. Plant Sci 160: 683–689.

Ayele M, Haas BJ, Kumar N, Wu H, Xiao Y, Van AS, Utterback TR, Wortman JR, White OR, Town CD (2005) Whole genome shotgun sequencing of *Brassica oleracea* and its application to gene discovery and annotation in *Arabidopsis*. Genome Res 15: 487–495.

Björkman T, Pearson KJ (1998) High temperature arrest of inflorescence development in broccoli (*Brassica oleracea* var. *italica* L.). J Exp Bot 49: 101–106.

Bolstad BM, Irizarry RA, Astrand M, Speed TP (2003) A comparison of normalization methods for high density oligonucleotide array data based on variance and bias. Bioinformatics 19: 185–193.

Broekgaarden C, Poelman EH, Steenhuis G, Voorrips RE, Dicke M, Vosman B (2007) Genotypic variation in genome-wide transcription profiles induced by insect feeding: *Brassica oleracea-Pieris rapae* interactions. BMC Genom 8: 239.

Broekgaarden C, Poelman EH, Steenhuis G, Voorrips RE, Dicke M, Vosman B (2008) Responses of *Brassica oleracea* cultivars to infestation by the aphid *Brevicoryne brassicae*: an ecological and molecular approach. Plant Cell Environ 31: 1592–1605.

Bruinsma M, Van Dam NM, van Loon JJ, Dicke M (2007) Jasmonic acid-induced changes in *Brassica oleracea* affect oviposition preference of two specialist herbivores. J Chem Ecol 33: 655–668.

Chen ZJ, Ni Z (2006) Mechanisms of genomic rearrangements and gene expression changes in plant polyploids. Bioessays 28: 240–252.

Colinas J, Birnbaum K, Benfey PN (2002) Using cauliflower to find conserved non-coding regions in *Arabidopsis*. Plant Physiol 129: 451–454.

Cramer RA, Mauricio la Rota C, Cho Y, Thon M, Craven KD, Knudson DL, Mitchell TK, Lawrence CB (2006) Bioinformatic analysis of expressed sequence tags derived from a compatible *Alternaria brassicicola-Brassica oleracea* interaction. Molec Plant Pathol 7: 113–124.

Cui X, Hwang JT, Qiu J, Blades NJ, Churchill GA (2005) Improved statistical tests for differential gene expression by shrinking variance components estimates. Biostatistics 6: 59–75.

Doniger SW, Huh J, Fay JC (2005) Identification of functional transcription factor binding sites using closely related *Saccharomyces* species. Genome Res 15: 701–709.

Dunning CE, Redak RA, Paine TD (2003) Preference and performance of a generalist insect herbivore on *Quercus agrifolia* and *Quercus engelmannii* seedlings from a southern California oak woodland. For Ecol Manag 174: 593–603.

Ellis J, Dodds P (2003) Plant pathology: monitoring a pathogen-targeted host protein. Curr Biol 13: R400–R402.

Esmon CA, Tinsley AG, Ljung K, Sandberg G, Hearne LB, Liscum E (2006) A gradient of auxin and auxin-dependent transcription precedes tropic growth responses. Proc Natl Acad Sci USA 103: 236–241.

Ferry N, Edwards MG, Gatehouse J, Capell T, Christou P, Gatehouse AM (2006) Transgenic plants for insect pest control: a forward looking scientific perspective. Transgen Res 15: 13–19.

Gentleman R, Carey V, Bates D, Bolstad B, Dettling M, Dudoit S, Ellis B, Gautier L, Ge Y, Gentry J, Hornik K, Hothorn T, Huber W, Iacus S, Irizarry R, Leisch F, Li C, Maechler M, Rossini A, Sawitzki G, Smith C, Smyth G, Tierney L, Yang J, Zhang J (2004) Bioconductor: open software development for computational biology and bioinformatics. Genome Biol 5: R80.

Haberer G, Mader MT, Kosarev P, Spannagl M, Yang L, Mayer KF (2006) Large-scale cis-element detection by analysis of correlated expression and sequence conservation between *Arabidopsis* and *Brassica oleracea*. Plant Physiol 142: 1589–1602.

Hammond JP, Broadley MR, Craigon DJ, Higgins J, Emmerson ZF, Townsend HJ, White PJ, May ST (2005) Using genomic DNA-based probe-selection to improve the sensitivity of high-density oligonucleotide arrays when applied to heterologous species. Plant Meth 1: 10.

Heffner EL, Sorrells ME, Jannink JL (2009) Genomic selection for crop improvement. Crop Sci 49: 1–12.

Hilker M, Stein C, Schroder R, Varama M, Mumm R (2005) Insect egg deposition induces defence responses in *Pinus sylvestris*: characterisation of the elicitor. J Exp Biol 208: 1849–1854.

Hudson ME, Bruggink T, Chang SH, Yu W, Han B, Wang X, van der Toorn P, Zhu T (2007) Analysis of gene expression during *Brassica* seed germination using a cross-species microarray platform. Crop Sci 47: S-96.

Joosen R, Cordewener J, Supena ED, Vorst O, Lammers M, Maliepaard C, Zeilmaker T, Miki B, America T, Custers J, Boutilier K (2007) Combined transcriptome and proteome analysis identifies pathways and markers associated with the establishment of rapeseed microspore-derived embryo development. Plant Physiol 144: 155–172.

Kang J, Zhang G, Bonnema G, Fang Z, Wang X (2008) Global analysis of gene expression in flower buds of *Ms-cd1 Brassica oleracea* conferring male sterility by using an *Arabidopsis* microarray. Plant Mol Biol 66: 177–192.

Katari MS, Balija V, Wilson RK, Martienssen RA, McCombie WR (2005) Comparing low coverage random shotgun sequence data from *Brassica oleracea* and *Oryza sativa* genome sequence for their ability to add to the annotation of *Arabidopsis thaliana*. Genome Res 15: 496–504.

Kingsolver J, Woods HT (1997) Thermal sensitivity of growth and feeding in *Manduca sexta* caterpillars. Physiol Biochem Zool 70: 631–638.

Kuhn K, Baker SC, Chudin E, Lieu MH, Oeser S, Bennett H, Rigault P, Barker D, McDaniel TK, Chee MS (2004) A novel, high-performance random array platform for quantitative gene expression profiling. Genome Res 14: 2347–2356.

Kwak JM, Kim SA, Hong SW, Nam HG (1997) Evaluation of 515 expressed sequence tags obtained from guard cells of *Brassica campestris*. Planta 202: 9–17.

Kwun M, Choi Y, Yoon H, Park BS, Kang BJ, Chung YY (2004) Expression analysis of the pistil genes in controlling self-incompatibility of *Brassica campestris* by CO_2 gas using microarray. Mol Cells 18: 94–99.

Lee SC, Lim MH, Kim JA, Lee SI, Kim JS, Jin M, Kwon SJ, Mun JH, Kim YK, Kim HU, Hur Y, Park BS (2008) Transcriptome analysis in *Brassica rapa* under the abiotic stresses using *Brassica* 24K oligo microarray. Mol Cells 26: 595–605.

Li F, Wu X, Tsang E, Cutler AJ (2005) Transcriptional profiling of imbibed *Brassica napus* seed. Genomics 86: 718–730.

Liechti R, Farmer EE (2006) Jasmonate biochemical pathway. Sci Signal 2006: cm3.
Lim CO, Kim HY, Kim MG, Lee SI, Chung WS, Park SH, Hwang I, Cho MJ (1996) Expressed sequence tags of Chinese cabbage flower bud cDNA. Plant Physiol 111: 577–588.
Lim JY, Shin CS, Chung EJ, Kim JS, Kim HU, Oh SJ, Choi WB, Ryou CS, Kim JB, Kwon MS, Chung TY, Song SI, Kim JK, Nahm BH, Hwang YS, Eun MY, Lee JS, Cheong JJ, Choi YD (2000) Analysis of expressed sequence tags from *Brassica rapa* L. ssp. *pekinensis*. Mol Cells 10: 399–404.
Liu F, Jenssen TK, Trimarchi J, Punzo C, Cepko CL, Ohno-Machado L, Hovig E, Kuo WP (2007) Comparison of hybridization-based and sequencing-based gene expression technologies on biological replicates. BMC Genom 8: 153.
Lou P, Zhao J, Kim JS, Shen S, Del Carpio DP, Song X, Jin M, Vreugdenhil D, Wang X, Koornneef M, Bonnema G (2007) Quantitative trait loci for flowering time and morphological traits in multiple populations of *Brassica rapa*. J Exp Bot 58: 4005–4016.
Ma HM, Schulze S, Lee S, Yang M, Mirkov E, Irvine J, Moore P, Paterson A (2004) An EST survey of the sugarcane transcriptome. Theor Appl Genet 108: 851–863.
Narusaka M, Abe H, Kobayashi M, Kubo Y, Narusaka Y (2006) Comparative analysis of expression profiles of counterpart gene sets between *Brassica rapa* and *Arabidopsis thaliana* during fungal pathogen *Colletotrichum higginsianum* infection. Plant Biotechnol 23: 503–508.
Park YS, Jeon MH, Lee SH, Moon JS, Cha JS, Kim HY, Cho TJ (2005) Activation of defense responses in Chinese cabbage by a nonhost pathogen, *Pseudomonas syringae* pv. *tomato*. J Biochem Mol Biol 38: 748–754.
Paterson AH (2005) Polyploidy, evolutionary opportunity, and crop adaptation. Genetica 123: 191–196.
Paterson AH, Lan TH, Amasino R, Osborn TC, Quiros C (2001) *Brassica* genomics: a complement to, and early beneficiary of, the *Arabidopsis* sequence. Genome Biol 2: REVIEWS1011.
Rui-Juan G, Da-Peng D, Rong-Cai M, Ming-Qing C, Yue-Ming Y, Ya-Dong W, Shi-Jun R, Xin-Yu G (2004) EST analysis of the heading leaf of Chinese cabbage (*Brassica rapa* L. ssp. *pekinensis*) in the early phase of the heading stage. Chin J Agri Biotechnol 1: 119–124.
Ruvinsky I, Ruvkun G (2003) Functional tests of enhancer conservation between distantly related species. Development 130: 5133–5142.
Sarosh BR, Danielsson J, Meijer J (2009) Transcript profiling of oilseed rape (*Brassica napus*) primed for biocontrol differentiate genes involved in microbial interactions with beneficial *Bacillus amyloliquefaciens* from pathogenic *Botrytis cinerea*. Plant Mol Biol 70: 31–45.
Schranz ME, Lysak MA, Mitchell-Olds T (2006) The ABC's of comparative genomics in the Brassicaceae: building blocks of crucifer genomes. Trends Plant Sci 11: 535–542.
Schranz ME, Song BH, Windsor AJ, Mitchell-Olds T (2007) Comparative genomics in the Brassicaceae: a family-wide perspective. Curr Opin Plant Biol 10: 168–175.
Schroder R, Cristescu SM, Harren FJM, Hilker M (2007) Reduction of ethylene emission from Scots pine elicited by insect egg secretion. J Exp Bot 58: 1835–1842.
Soeda Y, Konings MC, Vorst O, van Houwelingen AM, Stoopen GM, Maliepaard CA, Kodde J, Bino RJ, Groot SP, van der Geest AH (2005) Gene expression programs during *Brassica oleracea* seed maturation, osmopriming, and germination are indicators of progression of the germination process and the stress tolerance level. Plant Physiol 137: 354–368.
Storey JD (2003) The positive false discovery rate: A Bayesian interpretation and the q-value. Annals of Statistics 31: 2013–2035.
Thaler JS, Owen B, Higgins VJ (2004) The role of the jasmonate response in plant susceptibility to diverse pathogens with a range of lifestyles. Plant Physiol 135: 530–538.
Toth IK, Pritchard L, Birch PR (2006) Comparative genomics reveals what makes an enterobacterial plant pathogen. Annu Rev Phytopathol 44: 305–336.
Tuberosa R, Salvi S (2006) Genomics-based approaches to improve drought tolerance of crops. Trends Plant Sci 11: 405–412.

Tuberosa R, Salvi S, Sanguineti MC, Landi P, Maccaferri M, Conti S (2002) Mapping QTLs regulating morpho-physiological traits and yield: case studies, shortcomings and perspectives in drought-stressed maize. Ann Bot (Lond) 89 Spec No: 941–963.

van Poecke RMP, Dicke M (2002) Induced parasitoid attraction by *Arabidopsis thaliana*: involvement of the octadecanoid and the salicylic acid pathway. J Exp Bot 53: 1793–1799.

Vanderauwera S, De BM, Van de SN, van de CB, Metzlaff M, Van BF (2007) Silencing of poly(ADP-ribose) polymerase in plants alters abiotic stress signal transduction. Proc Natl Acad Sci USA 104: 15150–15155.

Walling LL (2008) Avoiding effective defenses: strategies employed by phloem-feeding insects. Plant Physiol 146: 859–866.

Williams MM, Walsh DB, Boydston RA (2004) Integrating arthropod herbivory and reduced herbicide use for weed management. Weed Sci 52: 1018–1025.

Wszelaki AL, Doohan DJ, Alexandrou A (2007) Weed control and crop quality in cabbage (*Brassica oleracea* (capitata group)) and tomato (*Lycopersicon lycopersicum*) using a propane flamer. Crop Protect 26: 134–144.

Xiang D, Datla R, Li F, Cutler A, Malik MR, Krochko JE, Sharma N, Fobert P, Georges F, Selvaraj G, Tsang E, Klassen D, Koh C, Deneault JS, Nantel A, Nowak J, Keller W, Bekkaoui F (2008) Development of a *Brassica* seed cDNA microarray. Genome 51: 236–242.

Yang B, Srivastava S, Deyholos MK, Kav NNV (2007) Transcriptional profiling of canola (*Brassica napus* L.) responses to the fungal pathogen *Sclerotinia sclerotiorum*. Plant Sci 173: 156–171.

Yang KA, Lim CJ, Hong JK, Park CY, Cheong YH, Chung WS, Lee KO, Lee SY, Cho MJ, Lim CO (2006) Identification of cell wall genes modified by a permissive high temperature in Chinese cabbage. Plant Sci 171: 175–182.

Zhao J, Jamar DC, Lou P, Wang Y, Wu J, Wang X, Bonnema G, Koornneef M, Vreugdenhil D (2008) Quantitative trait loci analysis of phytate and phosphate concentrations in seeds and leaves of *Brassica rapa*. Plant Cell Environ 31: 887–900.

Zhou B, Li Y, Xu Z, Yan H, Homma S, Kawabata S (2007) Ultraviolet A-specific induction of anthocyanin biosynthesis in the swollen hypocotyls of turnip (*Brassica rapa*). J Exp Bot 58: 1771–1781.

Zhu-Salzman K, Salzman RA, Ahn JE, Koiwa H (2004) Transcriptional regulation of sorghum defense determinants against a phloem-feeding aphid. Plant Physiol 134: 420–431.

Ziolkowski PA, Kaczmarek M, Babula D, Sadowski J (2006) Genome evolution in *Arabidopsis/Brassica*: conservation and divergence of ancient rearranged segments and their breakpoints. Plant J 47: 63–74.

10

Proteomics and Metabolomics

Lucyna H. Misztal,* and *Izabela Ibragimow*

ABSTRACT

Proteome and metabolome are very complex and dynamic as they differ in their composition depending on many factors such as cell and tissue type, stage of development, and environmental conditions. Though proteome and metabolome deliver a large amount of data, there are not many reports describing whole proteome or metabolome. Not only are they complex, but there are also difficulties in reproducing their analyses. As plants of Brassicaceae family deliver food that is important for human nutrition and health, a large part of the conducted experiments concern the amount and composition of metabolites (especially glucosinolates) or proteins involved in anticancer activity. In recent years, there has been significant development of proteomic techniques, especially those related to mass spectrometry. In this chapter, we focus on the methods and techniques which have been most frequently used in research on proteome and metabolome in Brassica species.

Keywords: *Brassica* proteome, *Brassica* metabolome, proteome profiles, stress response, glucosinolates, anticancer activity

10.1 Introduction

The role of large scale approaches in transcriptome, proteome and metabolome analysis applied in gene expression studies is continuously growing. Numerous reports on selected individual transcripts, proteins, and metabolites could not resolve the problems connected with plant responses to, for example, environmental cues or gene manipulations performed by breeders and geneticists. Dynamic development of transcriptome profiling tools enabled deeper insight into many cell pathways and gene

---

Department of Biotechnology, Adam Mickiewicz University, Poznań, Poland.
*Corresponding author: *lmisztal@amu.edu.pl*

co-regulations over the past decade (see Chapter 9 for details). However, whole transcriptome analysis is just the first step in the study of gene expression because the correlation between changes in mRNA and protein content is not linear (Kersten et al. 2002; Millar 2004).

The large scale gene expression analyses, conducted so far, have concentrated mostly on model plants. Thus, a lot of information concerning the fundamental aspects of plant biology is available, although it needs to be emphasized that the course of some processes can be different in model plants and in crop plants. Vegetables such as cabbage, cauliflower, broccoli and others, belonging to the Brassicaceae family, are known to be important for human nutrition. Some unique metabolites produced in these plants have anticancer activity. In large scale analysis for better understanding of the network of connections between proteins and metabolites, it is of great importance to analyze simultaneously the proteins and metabolites, which would permit obtaining plants more resistant to stress and containing more compounds beneficial for human health.

On the other hand, analysis of non-model organisms is rather challenging because of the lack of coding sequence information, which makes protein identification more complex. Some strategies have been developed to circumvent these limitations.In the first strategy, peptide sequences are obtained in the mass spectrometry study. At present, several tools for bioinformatic peptide sequencing are available (for example NovoHMM and PepNovo algorithms). In the second, the data obtained are compared to the protein database of related organisms. This method allows identification of only those proteins that have at least one peptide with a sequence identical to the peptide from a protein in the database. In terms of the Brassicaceae species, such a comparative sequence analysis is facilitated by the fact that they are closely related to the model plant *Arabidopsis thaliana*, which is a member of the same family (van Wijk 2001; Grossman et al. 2007). This coding sequence homology enabled extensive comparative studies at the gene as well as protein levels within Brassicaceae species. For example, the antibodies specific for four *A. thaliana* superoxide dismutases (Kliebenstein et al. 1998) recognize their homologs from different species of *Brassica* genus (Fig. 10-1; Misztal et al. unpubl.).

Another field of large scale analyses, though poorly established as yet, is metabolite profiling. Metabolites are the end products of cellular processes. Their level and composition differ depending on the developmental level, tissue or in response to environmental changes. The analyses reported so far concerned the quantification of a few defined metabolites or identification of the overall metabolic changes. None of these analyses provides explanation of how metabolic networks work. The majority of recent studies concern the identification and quantification of multiple targets. Large scale analyses require reliable and high-throughput methods of isolation and analysis of proteins and metabolites.

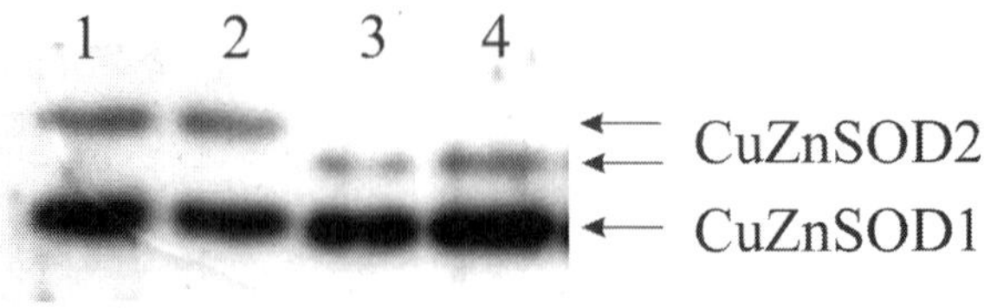

Figure 10-1 Specific identification of homologous variants of two superoxide dismutases Western blot detection of two superoxide dismutases in protein extract from *A. thaliana* and three members of Brassicaceae family based on antibodies raised for *A. thaliana* isozymes (L. Misztal et al. unpublished). 1-*A. thaliana*, 2-*B. oleracea* var. *alboglabra*, 3-*B. napus*, 4-*B. rapa* ssp. *pekinensis*.
Antibodies used were kindly provided by Prof. Daniel J. Kliebenstein.

10.2 Proteomics and Metabolomics Analysis

In this chapter, we focus on the presentation of the methods and techniques most frequently used in proteomic and metabolomic research. The development of transcriptomics, proteomics and metabolomics-oriented approaches, tools (including bioinformatics and statistics), technologies and equipment granted specialized "omics" platforms installation in dozens of research centers. They allow detection and throughout analysis of many molecules simultaneously.

In proteomics, two techniques play important roles: two dimensional electrophoresis (2-DE) for protein separation and mass spectrometry (MS) for protein sequencing and identification. Additionally, in proteome and metabolome studies mass spectrometry analyses are coupled with liquid chromatography (LC), which is known as the LC/MS technique allowing precise separation of different kinds of molecules. The general procedure for proteome analysis includes four steps: (1) sample preparation (extraction, digestion and enrichment), (2) separation of the proteins or the digested peptides, (3) identification of the proteins by mass spectrometry (top-down or bottom-up strategy) and (4) data mining for the information generated (functional classification or comparative analysis)(Fig. 10-2; Aebersold and Mann 2003).

10.2.1 Sample Preparation for Proteomics or Metabolomics Studies

For biologists the most important part is the correct sample preparation that can be followed by precise analysis, most commonly, by mass spectrometry. The choice of an efficient method for protein or metabolite extraction from cells or tissues and then enrichment of fractions of interest appears to be crucial in obtaining reproducible data. Proteomes and metabolomes are very dynamic and could very quickly change their composition because of a variety of cell and tissue types that are analyzed and the complexity of samples prepared.

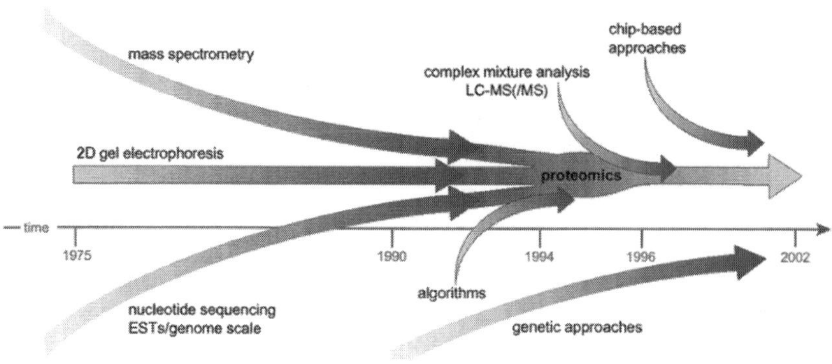

Figure 10-2 Time line of proteomics methods development. Reprinted from Patterson and Aebersold (2003).

For this reason, the proteomes or metabolomes are unique for each tissue, cell or organelle that unfortunately can change very quickly under the influence of different stimuli. This changeability is still a challenge in the precise preparation of samples for plant proteomic and metabolomic experiments. In contrast to DNA extraction techniques there are no standard procedures for protein extraction from different plant cells, tissues, or organelles; thus, specific extraction methods must be developed. Additionally, the presence of cell walls in plant material makes the protein extraction procedure more complicated in plant than in animal studies. This difference requires additional steps in plant extraction procedures (James 2004). The first step of all extraction methods for metabolomic and proteomic analyses must prevent sample degradation and rapidly stop all cellular enzymatic activities and metabolic processes that could change the metabolite or protein contents. The relatively high level of different secondary metabolites present in plant tissues could be problematic for accurate proteome analysis. This is important especially for successful high-quality 2-DE separation. Thus protein samples must be also sufficiently purified from salts, surfactants or other biomolecules (lipids, carbohydrates). Plant tissues usually contain a high concentration of proteases, therefore the extraction buffers used should also include inhibitors of proteases. For more information about sample preparation for proteomic and metabolomic analysis readers are directed to several reviews (Herbert et al. 2007; Hurkman and Tanaka 2007; Thelen 2007; Wang et al. 2008).

10.2.2 *Molecules Separation in Proteomics and Metabolomics*

Currently in proteomics, there are two core techniques for separation of complex protein samples: electrophoresis and chromatography. Among several electrophoretic methods the most popular are: denaturing polyacrylamide gel electrophoresis (SDS-PAGE) and two dimensional electrophoresis. Additionally, chromatographic techniques, especially liquid chromatography are also used extensively for protein purification and separation of peptides and small molecules (Table 10-1).

10.2.3 *Identification of Molecules (Mass Spectrometry, Nuclear Magnetic Resonance)*

The separation of hundreds of proteins by 2-DE was already possible in the late 1980s, but for proteomics researchers the lack of useful tools for rapid and robust identification of proteins of interest were still problematic. Before mass spectrometry, the most common approach to identify 2-DE separated proteins was the method called Edman degradation, based on enzymatic digestion of amino terminus of intact proteins. Released amino acids were identified by UV-absorbance spectroscopy. This technique was well established and automated but was also slow and had relatively poor sensitivity (Patterson and Aebershold 2003; Domon and Aebershold 2006). Currently, the most commonly used methods for molecules identification are mass spectrometry and nuclear magnetic resonance spectroscopy (NMR) (Table 10-2).

Mass spectrometry is a highly sensitive and accurate method for determination of molecular mass of different molecules.This method is especially important in proteomic approaches because it is unique and gives an opportunity to determinate the protein sequence from very complex protein or peptide mixture. The first report on mass spectrometry was published at the beginning of the last century by J. J. Thomson (1913). Since that time MS has become a well established technique for analysis of chemical structure of small molecules. MS is based on molecule ionization. For analysis of high-mass biological samples (peptides or proteins) researchers had to wait for decades because of the lack of suitable soft-ionization techniques for biological samples (which allows one to ionizate peptides without the breakdown of peptide bonds). Two soft desorption ionization methods for mass spectrometry of biological macromolecules were discovered in the late 1980s: matrix assisted laser desorption ionization (MALDI) and electrospray ionization (ESI) (Tanaka et al. 1987; Karas and Hillenkampf 1988; Fenn et al. 1989). For this discovery, John Fenn and Koichi Tanaka were awarded one-quarter of the Nobel Prize for chemistry in 2002.

Table 10-1 The techniques most commonly used for proteins, peptides or metabolites separation.

ELECTROPHORETIC METHODS			
Technique	**Principle of sample separation**	**Advantages/Disadvantages**	**Type of molecules**
One dimensional electrophoresis **(SDS-PAGE)**	• proteins are separated on denaturating (SDS) polyacrylamide gel according to molecular mass	• for protein separation only • relatively low resolution • separation acc. to molecular mass only	Proteins
Two dimensional electrophoresis, **(2-DE)**	• first dimension: isoelectrofocusing (IEF)—proteins are separated acc. to isoelectric point second dimension: SDS-PAGE	• the first method for high-throughput proteins separation • high resolution of protein separation	Proteins
2-D difference gel electrophoresis **(2D-DIGE)**	• protein samples are labeled with different fluorescent dyes and then separated by 2-DE (in the same gel) • gel is scanned by a laser scanner	• internal standard • useful for comparative proteomics • high resolution of protein separation • fluorescent dyes are more sensitive than silver nitrate staining	Proteins

CHROMATOGRAPHIC METHODES			
Technique	**Principle of sample separation**	**Advantages/Disadvantages**	**Type of molecules**
Column chromatography	• the sample is injected into mobile phase (liquid or gas). The mobile phase with the sample moves through stationary phase (immobilized porous substance) • individual components have different retention times (RT) due to the differences in their interactions with phases • in normal phase (NP) chromatography the mobile phase is less polar than the stationary phase In reverse phase (RP) system is in opposite	• molecules can be separated acc. to molecular weight, charge, hydrophobicity or affinity • liquid chromatography, (LC) or gas chromatography (GC) are most commonly used for proteomics and metabolomics	Proteins, peptides, other small molecules

High pressure liquid chromatography **(HPLC)**	• high pressure pumps are used to get a more resonate flow rate and enhance the separation efficiency	• separation efficiency depends mainly on the balance of the affinity of the sample components to the stationary phase and their solubility in mobile phase	
Reverse phase chromatography **(RP-HPLC)**	• the stationary phase is less polar than the mobile phase • molecules are separated on the basis of differences in hydrophobicity • the stationary phase usually prepared by surface-modification of silica beads (modifications cause the absorbance of the sample components)	• currently most widely used for proteomics and metabolomics • the eluate may be applied directly into a mass spectrometer by electrospray ionization (on-line) or collected for subsequent analysis (off-line) • good efficiency and resolution of separation	
Strong cation exchange **(SCX)**	• the stationary phase is usually made of surface modificated sulfonic acid group • two opposite charges attract each other • SCX is one of four types of ion exchange chromatography (the others are: weak cation exchange (WCX), strong anion exchange (SAX) and weak anion exchange (WAX))	• after SCX the mobile phase contains salts and eluate can't be directly applied to the mass spectrometer • good efficiency and resolution of separation	Proteins, peptides, other small molecules
Affinity chromatography **(AC)**	• technique based on specific interaction between separated analyte and specific functional group on stationery phase • the components of interest are retained while non interacting components pass through the column		Proteins, peptides

Table 10-2 The techniques most commonly used for proteins, peptides or metabolites identification.

MASS SPECTROMETRY

Technique	Principle of sample analysis	Advantages/Disadvantages	Type of molecules
Mass spectrometry **(MS)**	• first step: molecules are ionized (e.g., MALDI, ESI) • second step: ionized molecules are separated acc. to their mass-to-charge ratio (m/z) with different types of analyzers (e.g., TOF) • third step: ions detections • fourth step: *in silico* protein identification (with algorithms)	• highly sensitive and accurate method for the determination of molecular mass for different types of molecules	Protein, peptide, small molecules
Tandem mass spectrometry **(MS/MS)**	• peptide ion should be intact (peptide bonds) until fragmentation step • peptide are ionized with soft ionization techniques • two mass analyzers connected in tandem (e.g. TOF/TOF); the first mass analyzer characterizes the m/z of intact peptides; subsequently, selected peptide is fragmented and the other mass analyzer characterizes the m/z of ions	• for protein (peptide) sequencing • commonly used system is tandem mass spectrometry coupled which HPLC technique (HPLC-MS/MS) • determination of the structure of small molecules • high resolution and efficiency • requires low amount of a sample	Protein, peptide, small molecules

NUCLEAR MAGNETIC RESONANCE

Technique	Principle of sample analysis	Advantages/Disadvantages	Type of molecules
Nuclear magnetic resonance **(NMR)**	• isotopes with the spin and magnetic moment yields nuclear magnetic resonance (NMR) and can be observed in a strong magnetic field pulsed with electromagnetic radiation • the majority of biologically important elements occurs in isotopic forms. Some of these have integral (N^{14}) or fractional spin (C^{13}, H^1, N^{15}, F^{19}, P^{31}) and few have no spin (C^{12}, O^{16})	• provides detailed information about topology, dynamics and three-dimensional structure of metabolites or proteins	Protein, small molecules, metabolites

These ionization techniques together with improvement in sample protein preparation methods and the increasing number of full genome sequences (high quality protein databases) were crucial for high-throughput protein identification by mass spectrometry techniques. Currently, different kinds of spectrometers are used for measuring the masses of molecules (for review see Smith 2002; Domon and Aebersold 2006). Presently, the most commonly used is tandem mass spectrometry (MS/MS) coupled with high-pressure liquid chromatography (HPLC) technique. MS/MS analysis generates data on molecular mass and sequence of proteins. Due to MS development two major approaches are used for protein identification and characterization: the top-down and bottom-up approaches (also called shotgun proteomics or peptidomics). In the top-down approach the intact proteins are directly used for analysis and in the bottom-up approach the proteins are previously digested with an efficient protease (mostly trypsine) and peptides obtained are then used for analysis. All common mass spectrometers consist of three functional units: 1) ion source 2) mass analyzer and 3) ion detector and data acquisition unit (Eidhammer et al. 2007).

10.3 Current Status of Vegetable Brassica Proteomics

Obviously, most of the proteomics research within the Brassicaceae family concerns the model plant *A. thaliana* while there is not much information on the proteome of vegetable Brassicas. A few available publications describe large scale analyses on proteome comparison to identify genetic relationships within the Brassicaceae family (Marquez et al. 2001), pollen proteome (Murphy 2006), autopolyploidy relations in cabbage (Albertin et al. 2005) or pathogen attack response (Subramanian et al. 2005; Widarto et al. 2006). Besides general proteomic analysis, the majority of works concern individual proteins. Typically, analysis of a whole proteome focused on selected, well defined stress conditions, tissue, organ, organellum or developmental stage. Interestingly, it is not only the cost that limits wide scale comparative study. Complexity of the proteome system under study requires concentration on simple models. In addition, comparative studies of proteomes at different developmental stages or several time points during exposition to stress seem to be not sufficiently reproducible and challengeable at this stage of proteomics development. The proteome analysis can be limited for proteins of low abundance. These proteins can be masked in the presence of very abundant ones. Another problem concerns the proteins of such amino acid sequence that hampers their isolation or identification because of the problems with dissolving, digestion or ionization. Analysis of such proteins could be facilitated when a study concerns only one or a few proteins. Another advantage of analysis of a few proteins is the possibility of tracing the qualitative and quantitative differences in their expression in several developmental stages, tissues and stresses.

10.3.1 Structural and Developmental Proteome Profiles

Generally, a large part of proteomics-based research concerns the analysis of organellar proteome. In plants, proteome research covers most frequently chloroplast and mitochondrial proteomes. Pawlowski et al. (2005) described soluble and membrane bound fractions of cauliflower mitochondria with the use of two dimensional electrophoresis. The patterns obtained were evaluated with Imager Master 2-D Elite software. After silver nitrate staining, 561 protein spots were detected. The study of mitochondrial proteome conducted by Hochholdinger et al. (2004) showed that mitochondrial proteins encoded by the nuclear genome could differ even if nuclear genomes were identical. It suggests that mitochondrial genotypes can regulate the accumulation of such proteins. In the research conducted by Pawlowski et al. (2005) mitochondria were isolated from cauliflower buds by differential centrifugation in Percoll gradient. After fractionation into soluble, integral membrane and peripheral membrane, the mitochondrial proteins were resolved by 2-D electrophoresis. The patterns obtained were compared to the *A. thaliana* mitochondrial proteome. About 30 proteins showed similar pI and MW to the *A. thaliana* proteins, so they could be putative orthologs. They were for example pyruvate dehydrogenase, prohibitin or the elongation factor TU. Some of cauliflower proteins probably have different hydrophobicity and abundance than those of *A. thaliana* proteins, which makes their identification difficult. Some of the protein spots were extracted from gel and analyzed by ESI/MS, e.g., adenylate kinase A-B, ATP synthase, etc. were identified in this way. Differences observed between subfractions were qualitative (for example nine proteins from membrane fraction were not observed in soluble fraction) and quantitative (39 proteins being about 1.5-fold more intense in the membrane fraction; 62 proteins more intense in the soluble fraction).

Schmidt et al (2007) analyzed vacuoles isolated from cauliflower buds. In meristematic cells there are a great number of small vacuoles, while in the mature cells there is a large central vacuole. Thus far only a small number of vacuolar proteins have been identified. The authors isolated vacuoles from cauliflower buds, containing both meristematic and fully expanded cells. The isolated proteins were compared with *A thaliana* vacuolar proteome. The results allowed identification of 102 integral and 214 peripheral membrane proteins. The most prominent protein was phyrophosphatase. The authors also identified 57 proteins containing one or more membrane spanning domain. To verify the localization of the identified proteins, fusions of five of them with green fluorescent protein (GFP) were constructed. Four of these proteins (viz., putative cation transporter, nodulin -21, senescence related membrane protein and one protein of unknown function) were localized in the vacuolar membrane, whereas the fifth (white-brown ATP—binding cassette transporter) was localized in the plasma membrane. When small vacuoles from young cells fuse to form a single large vacuole they must increase their volume. It is related to the transport of inorganic ions into the vacuoles followed by water influx. Vacuoles in mature cells control cell homeostasis by regulating the concentration of metabolites and ions in cytoplasm by their uptake, storage and release. So, proteins participating in the transport of these compounds are very important. The proton pumps generate an electrochemical gradient driving solute uptake; antiporters contribute to the accumulation of uncharged solutes as sucrose. Among the antiporters of great interest are the Ca/proton ones as their expression can lead to salt tolerance. Another interesting group is that of cadmium/proton antiporters as they could be used in phytoremediation. The authors isolated tonoplast proteins from cauliflower buds, determined their purity by Western blot with specific antibodies and analyzed the ATPase activity. They found that an isolated fraction contained up to 1.3% plasma membranes, chloroplasts, mitochondria and other cellular structures. Five different extraction procedures were used: alkaline, saline, acetone, chloroform/methanol extractions and treatment with sodium phosphate. Proteins were fractionated by SDS-PAGE. Before MS, tryptic peptides were separated by reverse phase chromatography (RP-HPLC). Proteins were identified by SEQUEST algorithm search against the *A. thaliana* database on National Center for Biotechnology Information (NCBI). Using the methods described, the authors identified 102 cauliflower membrane proteins. Fifty three proteins were identified by a unique peptide hit. There were 28 proteins in saline, 36 in acetone, 13 in alkaline and 27 in phosphate extracted fractions. Of the 102 proteins, 56 proteins (55%) were isolated solely via one of the applied procedures. Fifty seven membrane proteins were not localized, and 36 could be annotated as non-vacuolar proteins. Four proteins localized in vacuolar membrane by GFP fusions have an unknown function. One of them (homologue of At1g16390) could be a putative cation transporter. This protein has 11 transmembrane domains.

One of the most frequently analyzed tissues and mechanisms are pollen, stigma and their interactions. It has been known for a long time that some proteins are very important in mediating self-incompatibility. Protein/glycoprotein constituent of pistil specifically inhibits self pollen growth. In *Brassica oleracea* stigma *S*-allele-specific glycoproteins have been detected. *S*-specificity resides probably in the protein part of the glycoprotein molecule. It has been also shown that one glycoprotein causes the inhibition of pollen applied to its surface. However, the manner of this interaction/inhibition is still unknown. The analysis by Roberts et al (1984) concerns pollen-stigma interactions and impact of movement of proteins. The authors observed that mature stigmas treated with cycloheximide lose their self-incompatibility. This means that for the control of self-pollination, synthesis of new proteins is necessary. On the other hand, tunicamycin application has no effect. Furthermore light microscope autoradiography has shown the movement of stigmatic proteins to the pollen but no penetration into the pollen was observed. The authors have hypothesized that the continuous cycling of glycoproteins is a kind of "second messenger", which inhibits pollen probably by preventing the normal rehydratation of its membrane. These glycoproteins are probably rapidly metabolized by pollen and therefore there is a necessity of their continuous synthesis, hence such an impact of cycloheximide.

The meaning of continuous protein synthesis was also noticed by Sarker et al. (1988). This study concerned pollen hydration and the phenomenon self-incompatibility, and was conducted by using metabolic inhibitors. The application of cycloheximide indicated that protein synthesis was necessary to regulate pollen hydration and self-incompatibility. Using tunicamycin, it was shown that the mechanism of self-incompatibility required some glycoprotein(s), which were not required for hydration control.

Extracellular protein coat of pollen was analyzed by Murphy (2006) in *A. thaliana, B. napus, B.oleracea* and *B. rapa* plants. The major component of the pollen coat is a specific group of proteins called pollenins, which comprise about 50–80% of all proteins. Murphy observed that there were many other proteins such as kinases, lipases, pectin esterase and caleosin.

Analyses of pollen grains were also conducted by Ruiter et al (1997). The main subjects of interest in this study were oleosins, which are glycine-rich proteins attached to the lipofilic part of pollen and absent in the remaining parts of the plant. Antibodies directed on the glycine-rich domain were prepared and used to detect proteins in extracts from sepals, carpels, pollen and stamens. This antibody detected several proteins in the pollen coat. The most abundant was 42 kDa protein. As the PCR analysis of genomic DNA and cDNA showed the existence of one and three different products respectively; the authors suggested that post-translational modifications were responsible for the several observed protein bands. In *Brassica napus*, oleosin-like proteins were always post-translationally modified. They had the N-terminal part removed. Protein bands after Western blot were observed in extracts from the pollen coat, pollen and whole stamens but not in the extracts from sepals, petals and carpels. In the fraction of the oil-body associated proteins isolated from pollen and stamen two oleosins were observed (42 and 34 kDa). After removal of the pollen coat by rinsing with cyclohexane no oleosins were detected, which confirms that oleosins are bound with the pollen coat. Additionally in the immature stamens 55 kDa protein was detected with the used antibody. The authors compared the deduced protein sequences with those from *B. alboglabra* and *B. napus* and observed little differences in the glycine-rich domain. These differences result from deletions/insertions of repeating units. The glycine-rich domain is different from the other glycine rich motifs. It is randomly coiled, which allows the protein to bind large amounts of water. This feature could explain the fact that oleosins remain soluble after the boiling of the probe.

Other objects of interest in proteomic research were the proteins of the plant vascular system. Analysis of proteins (SDS-PAGE followed by MS/MS analysis) from the xylem sap (Buthz et al. 2004) of broccoli, rapeseed, pumpkin and cucumber showed the existence of 10–20 proteins with masses from 10 to 100 kDa. Overall protein concentration was 500-times lower than that of the proteins isolated from phloem. The most abundant proteins were identified as peroxidases, chitinases and serine proteases. They were observed in all of the analyzed species. The other identified proteins were: glycine-rich protein, lectin-like xylem sap protein 30, aspartyl protease family protein, subtilisin-like protein, lipid transfer protein, etc. Altogether 30 different proteins were identified. The protein patterns of Brassica plants (broccoli, oilseed rape) were very similar, although not identical. Obvious differences were observed on a family level (between Brassicaceae and Cucurbitaceae) as well as on the species level (between cucumber and pumpkin). The major components of xylem exudates in all species were various peroxidases. The authors found four peroxidases in broccoli, five in oilseed rape, two in pumpkin and three in cucumber. Besides the proteins observed in all species, the authors also identified proteins specific to individual plant species. The lectin-like xylem sap protein 10 was found only in cucumber; aspartyl protease was identified only in *B. napus* xylem and serine carboxypeptidase was identified only in broccoli exudates.

10.3.2 Proteomics of Stress Response

The second important area of proteomic research is the analysis of plant responses to biotic and abiotic stresses such as insects (Subramanian et al. 2005) and other pathogens (Peck et al. 2001) or wounding (Kim et al. 2003). Kim and coworkers (2003) analyzed the response of Chinese cabbage (*B. rapa* ssp. *pekinensis*) to wound stress and salicylic acid (SA). Proteins were extracted from the control and treated leaves of Chinese cabbage and analyzed by 2-D electrophoresis. The protein spots differentiating the two analyzed pools were excised from the gel and identified. The authors showed that SA and wounding induced pathogenesis-related protein 1a (PR1a) both on mRNA and protein level. PR1a was induced both by SA and by wound stress. Quantitative differences between the control and treated plants were observed. Many other proteins were enriched after application of SA. Nine of them were excised from the gel and identified by

MALDI-TOF MS after trypsin digestion. The identified proteins included for example: phospholipase D, homolog of resistance protein RPS2, arginase, serine/threonine kinase BNK1 and L-ascorbate peroxidase. Wounding induces the appearance of six proteins, namely: heat shock cognate protein 70, polygalacturonase and peroxidase P7, which are known to play a role in the defense system. Other induced proteins were: lysine-ketoglutarate reductase/saccharopine dehydrogenase, S-locus glycoprotein 12, and efflux carrier whose function in the defense system is unclear.

Besides drought stress, substantial losses in vegetable Brassicas crops are caused by pathogens. *Leptosphaeria maculans* is a devastating fungal pathogen of Brassica crops. It causes the so-called "backleg". It is thus very important to understand the resistance mechanism of some plants to that pathogen. Subramanian et al. 2005 generated a cross between plants susceptible (*B. napus*) and resistant (*Brassica carinata*) to fungal pathogen (*Leptosphaeria maculans*) and analyzed the changes evoked in leaf proteomes. They used 2-D electrophoresis followed by tandem mass spectrometry. The results permitted identification of five proteins, unique to susceptible cultivar, (for example: fructose-bisphosphate aldolase, mutator-like trasposase, glutamine synthetase and triose-phosphate isomerase) and seven (carbonic anhydrase, thylakoid luminal 20 kDa protein, cyclin, B-type, sedoheptulose-bisphosphatase, nitrate reductase, hypothetical protein and superoxide dismutase) found only in a resistant genotype. They also observed 28 proteins differentially expressed in these genotypes in 48 hours after inoculation. The majority of the proteins identified in this study were antioxidant enzymes and enzymes involved in nitrogen metabolism, CO_2 fixation and photorespiration enzymes. Photosynthetic enzymes such as fructose biphosphate aldolase, triose phosphate isomerase, etc. were elevated in a resistant genotype after a pathogen attack. The proteins differentially regulated in these two genotypes were: malate dehydrogenase, putative cytochrome P450, rubisco large subunit, triose-phosphate, dehydroascorbate, hypothetical protein, unknown protein, fructose-bisphosphate aldolase, N-glyceraldehyde 2-phosphotransferase, sedoheptulose-bisphosphatase precursor, glutamine synthetase, ATP synthase (ß-subunit), peptidyl-prolyl cis-transisomerase, topoisomerase IV subunit-ß and peroxiredoxin type 2.

Liao et al (2008) studied the effects of di-N-butyl phthalate (DBP) —an endocrine disruptor on the growth of Chinese cabbage. Different concentrations (from 1 to 100 mg/L) of this compound were tested. In 2-DE analyses they noticed six protein spots differing in the control and the treated plants. They were: acyl-[acyl-carrier-protein] desaturase, root phototropism protein 3, ferredoxin-nitrite reductase, dihydroflavonol-4-reductase, aminoacyl-tRNA synthetase and ATP synthase (ß-subunit). The first three showed increased expression, the rest decreased in response to the stimulus investigated. The authors found that a closely related plant—Bok choy- responded differently.

Endocrine disruptors, and among them phthalate esters have been studied for years. They are widely used in industrial activities and exist commonly in the atmosphere, rain and soil. DBP is one of most popular phthalate ester and its half-life is 22 years in water solutions. Studies on a few species (*Raphanus, Browalia, Capsicum* and *Brassica campestris*) have shown that only DBP is absorbed and accumulated in plants, whereas plants do not take up the other phthalate esters. The biomass of the whole plant and chlorophyll content in leaves significantly decreased after DBP treatment. Changes in the protein content were analyzed by 2-D electrophoresis. Gels were stained with silver nitrate. Only six protein spots showed significant alterations. These proteins were identified by MALDI-TOF MS and peptide mass fingerprinting (PMF) analyses. Acyl-[acyl-carrier-protein] desaturase is involved in the biosynthesis of fatty acids (catalyzes the change of saturated to unsaturated acids), root phototropism protein is a signal transducer of phototropism response and ferredoxin nitrate reductase is involved in nitrate assimilation. Probably DPB could cause an increase in some physiological reactions. The level of the other enzymes (aminoacyl-tRNA synthetase, dihydroflavonol-4-reductase and ATP synthase) decreases in response to DBP, which can damage the cells and disturb the metabolism. The proteins, whose content decreases, are responsible for the biosynthesis of flavonoids, aminoacylation of tRNA and synthesis of ATP.

An interesting research concerning stress-specific proteome was conducted by Alvarez et al. (2009). They analyzed the root proteome of *Brassica juncea* in response to cadmium. Two methods were used: two dimensional difference gel electrophoresis (2D-DIGE) and isobaric tag for relative and absolute quantitation (iTRAQ). Each of the technologies used gave different sets of the proteins identified, however, these proteins belonged to similar functional categories. Proteins isolated from control and Cd-treated plants were labeled and separated on the same gel. The authors noticed an increase in the intensity of 26 and decrease in the intensity of 43 spots. In 69 spots with different intensity 102 proteins were identified. Up-regulated proteins were involved in sulfur assimilation, detoxification and redox homoeostasis; while down-regulated ones were involved in protein synthesis and processing. By the other method 130 differently expressed proteins were identified, among them 16 were also identified by the 2D-DIGE method. The results have shown that on large scale analyses it is not only important to repeat the analysis (the authors made three replicates for each method) but the choice of a proper method or a combination of a few methods is also important. Similar observations were made by Schmidt et al. (2007). They obtained different protein pools depending on the isolation procedure.

Equally important as the protein level change, is the analysis of post-translational modifications in some proteins. Such changes have been studied, e.g., by Peck et al. (2001) who compared phosphorylation changes between the control and elicitor treated plants. The authors analyzed the perception of microbial signal molecules in *A. thaliana* by using radioactive orthophosphate and 2-D electrophoresis. They observed a protein, which was phosphorylated within a few minutes after treatment. This protein was described as Atphos 43. They also found that the phosphorylation of similar proteins of rice and tomato was also induced quickly after a pathogen attack. The fact that rice and potato respond similarly to elicitors and the close relationship of *A. thaliana* with Brassicaceae allows one to suppose that the response of these plants could be similar.

10.3.3 Other Analyses

Proteomic as well as metabolomic research has also been used in other analyses; for example to establish genetic relationships between Brassica varieties (Marquez et al. 2001; Taylor et al. 2002). Marquez and coworkers compared the proteins of etiolated seedlings (aerial part) of cabbage, mustard, rapes, radishes and *A. thaliana* to establish the genetic relationships between them. Studies of the etiolated material allows analysis of a greater number of low abundant proteins, as the probe contains a lesser amount of ribulose bisphosphate carboxylase/oxygenase, which accounts for nearly 50% of green tissue proteins. The authors decided to analyze only about 750 spots well defined and reproducible, located in the central part of the gel. As not every sequence mutation changes and also the size and/or isoelectric point of the protein, only a fraction of the allelic variations were detected. There were 149 proteins common to the Brassica species and 49 common to all species analyzed. After analyses and construction of the phylogenetic tree the authors concluded that there are two lineages: one between *B. oleracea* and *B. rapa* and the other between *B. nigra* and *Raphanus* sp. (Marquez et al. 2001).

Seed proteins as a support in the discrimination of cultivars were also analyzed by Mukhlesur et al (2004). They analyzed 32 cultivars of the *B. rapa* from Bangladesh, Japan and China. The authors observed about 32 protein bands in each sample and 10 of them were polymorphic. Differences between the cultivars were quantitative rather than qualitative.

Another analysis of genetic relationships on the protein level was performed on the seed storage proteins (Sadia et al. 2009). The authors analyzed the seeds of *B. napus*, *B. juncea*, *B. carinata* and *B. rapa*. The choice to analyze seed proteins was made as they are not sensitive to changes in environmental conditions. Because of a small amount of proteins the analysis could be conducted on SDS-PAGE. Thirty cultivars were analyzed and about 29-31 protein bands were detected in one sample. Thirteen of them differed in intensity between cultivars and only two were absent in some samples. For example in *B. rapa* there was one protein band about 33 kDa, in the *B. napus* cultivars the authors observed the protein of about 31 kDa, in *B. carinata* both bands were visible, whereas in *B. juncea* none of them were present.

Another research group (Albertin et al. 2005) studied proteome changes in autoploidical genome duplication by 2-D electrophoresis of proteins isolated from haploid, diploid and tetraploid *B. oleracea*. The authors analyzed the homozygous lines of *B. oleracea*, *B. oleracea* var. *italica* and *B. oleracea* var. *alboglabra*. The proteomes of two organs, leaf and stem, were analyzed. Altogether 1,700 protein spots were observed, among them 558 were common to all the samples analyzed. Comparison of leaf and stem proteomes showed quantitative (about 24%) rather than qualitative (about 18%) differences. Proteins showing differences were mostly of stem origin, they are stem-specific or stem overexpressed, but overall the stem spots had lower abundance than leaf ones. No distinct qualitative variations were observed between the forms at different ploidy levels. Differences between the two genotypes examined were lower than that between organs. Analyses showed that ploidy level is not a major cause of changes in protein expression. A comparison of diploid and tetraploid proteomes allowed the identification of four different spots, whereas the analysis of different plants (biological replication/expected significantly by chance) showed 11 differences. The authors concluded that modification at the ploidy level did not induce changes in green tissue proteomes.

Besides the analysis of protein content and their post-translational modification, the third important branch of proteomic research is the analysis of changes in enzyme activity. Unfortunately, there is a very small number of research reported in this area. One of the works concerning enzyme activities is that by Saroop and coworkers. They analyzed soluble and bound peroxidase activities in the course of *B. juncea* seed development (Saroop et al. 2002). Three hydrogen donors were used: ferulic acid, caffeic acid and chlorogenic acid. The peroxidase activity rises after 20 days after anthesis. A reverse correlation was observed between peroxidase activity and water content. It suggests that this enzyme is involved in the termination of cell elongation. Changes in soluble peroxidase activity increased sharply between 20–30 days after anthesis, later the activity decreased. Levels of activity differed depending on the substrate. For ferulic acid minimal activity was observed and for caffeic and chlorogenic acid very similar results were obtained. The bound peroxidase activity was lower than that of the soluble one when the hydrogen donor was caffeic or chlorogenic acid. When ferulic acid was used, the activities were the same. The authors did not analyze how many isoenzymes were responsible for the activity of each fraction. Another article devoted to the catalytic properties of enzymes is the work of Tayefi-Nasrabadi (2008) concerning catalases from *B. oleracea* var. *gongylodes* (kohlrabi). Catalases are an important class of antioxidant enzymes participating in practically every stress defense. The author extracted protein from kohlrabi bulbs and performed kinetic analyses. He identified three catalase isoenzymes. These isoenzymes had a different optimal pH (4,5; 6,5 and 10) and the most efficient was the isoenzyme active at pH 6.5. The isoenzyme active at pH 4.5 was more sensitive to azide and more resistant to cyanide than the other ones. It also showed competitive inhibition. Three isolated isoforms also differed in the heat inactivation. The isoform active at pH 6.5 was inactive above 50°C; and the isoform active at pH 10 was inactive above 70°C. The differences in the effects of pH, temperature, substrate and inhibitors on the catalases activity suggest different roles of these isoenzymes.

Among the papers devoted to a few proteins, an interesting one is that by Köksal and Gülçin (2008). The authors purified and characterized peroxidase from cauliflower buds. Peroxidases are a class of enzymes, which catalyze the oxidation of many substrates in the presence of hydrogen peroxide. They are present in many organisms and are the key enzymes in control of plant growth and development. The enzyme described was isolated from crude protein extract by ammonium sulfate fractionation and subsequent ion exchange chromatography (CM-Sephadex A-50). The molecular mass of the isolated enzyme was determined by gel filtration chromatography as 44kDa. Electrophoresis in native polyacrylamide gel showed only one band—so there was only one isoenzyme. The optimum pH for this isoenzyme was 7.5 and the optimal temperature varied depending on the substrate used; it was 25°C for pyrogallol, 30°C for guaiacol and 2,2′-azino-bis(3-ethylbenzthiazoline-6-sulfonic acid (ABTS) substrates, 45°C for 4-methyl catechol, 50°C for catechol. Kinetic studies showed that this peroxidase had the highest affinity to pyrogallol. The authors suggested that this enzyme can be used for the synthesis of some aromatic compounds, removal of phenolic peroxides from water and industrial wastes. The literature devoted to proteomic research is scarce and concerns mainly other issues (different protein pools and stresses), so it is difficult to find any common denominator such as the same protein or type of stress. Quite a lot of articles are concerned with the differentiation of species and cultivars on the basis of protein profile. Unfortunately, the authors did not identify most of these proteins and only compared the profiles. From other analyses we can find out that most of the identified proteins belonged to the functional classes such as defense proteins (especially antioxidant enzymes) and metabolic enzymes. Surprisingly almost no proteins connected with signalization were identified. Only Alvarez et al. (2009) clearly stated that the level of some proteins from the signaling pathways differed between the control and the cadmium treated plants. Further analyses are needed to recognize the whole proteome of the vegetable Brassicas. Undoubtedly, the proteomic analyses of *A. thaliana* and *B. napus*, as well as recognition of the *B. campestris* genome will be a great help.

10.4 Current Status of Vegetable Brassica Metabolomics

Metabolomics has many advantages compared to transcriptomics and proteomics (Taylor et al. 2002). Firstly, a single change in the transcript or a protein can cause alterations in many metabolites or a substantial change of some metabolites. Such amplification of changes facilitates their detection. Secondly, metabolomic research does not need a full sequence of genome, as is needed in transcriptomics and proteomics research. Thirdly, metabolomics is more generic as genes and proteins differ between species and most metabolites are the same in all organisms. Four different approaches are used in metabolome analysis: target analysis, metabolite profiling, metabolomics and metabolic fingerprinting (Fiehn et al. 2001). Target analysis is the study of the substrate or direct product of one specific protein. Metabolite profiling (also called metabolic profiling) is the analysis of a pre-defined group of metabolites, for example members of particular metabolic pathways. Metabolomics concerns the identification and (sometimes) quantification of all metabolites from one cell, tissue, organ, etc. Metabolic fingerprinting (sometimes called metabonomics) is a comparison of some metabolite/ metabolites between genotypes/lines/probes. Each area of research requires different and strictly controlled methods of isolation and analysis of metabolites. Quite frequently proteomic research is accompanied with some metabolite analysis. Extracellular proteins from pollen coat were analyzed by Murphy (2006) in *A. thaliana, B. napus, B. oleracea* and *B. rapa*. Besides proteins, the second group of major components of pollen coat is lipids enriched in acylated compounds. They have a unusually high degree of saturation and different acyls. In coat protein there are mainly, campesterol, and campestdienol, while in lipids we usually observe acyls such as myristate, palmitate, and staearate. This lipidic layer protects male gamethophytes from dehydration. Metabolomic research is more diverse than the proteomic one. However, a large number of these analyses concern the response to various stresses. In relation to Brassicaceae, a large part of research is also connected with analysis of glucosinolate content.

10.4.1 Glucosinolates and Related Metabolic Compounds

The majority of metabolic analyses in Brassica vegetables are the metabolite profiling type, and the main subjects of interest are glucosinolates (GLS). All glucosinolates characterized (β-thioglucoside –N-hydroxysulfates) share a similar structure: β-D-glucopyranose with the thio group and chain derivative from amino acid (methionine, alanine, leucine, phenyloalanine or trypthophane) (Fig. 10-3). More than 120 chemically distinct glucosinolates have been identified in plants. The majority of GLS have been isolated from crucifer family members, and about 15 other plant families are known to contain glucosinolates, but not all are edible (Akaniaceae, Bataceae, Bretschneideraceae, Capparaceae, Caricaceae, Drypetes Euphorbiaceae, Gyrostemonaceae, Limnanthaceae, Moringaceae, Pentadiplantdraceae, Resedaceae, Salvodoraceae, Tovariaceae, Tropaeolaceae) (Rodman 1991a, b). For more detailed characterization of glucosinolates also see Chapter 1.

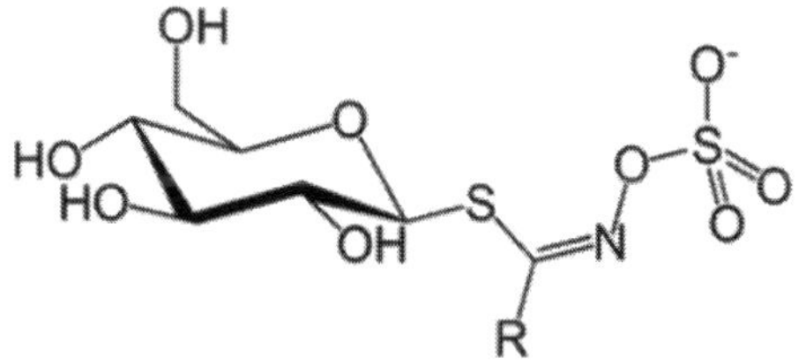

Figure 10-3 Structure of glucosinolate (*http://en.wikipedia.org/wiki/Glucosinolates*).

In literature we can find analyses of GLS content in different cultivars or developmental stages (Bayoumi et al. 2007; You et al. 2008), works on the impact of storage and preparation of vegetables on GLS digestibility (Rungampestry et al. 2007) and changes of these compounds in various environmental conditions (Liang et al. 2006; Widarto et al. 2006; Wurst et al. 2006; Jahangir et al. 2008).

Research conducted by Bayoumi et al (2007) concerned the metabolomic analysis of *B. rapa* (various cultivars). The most diverse cultivar metabolites were amino acids, carbohydrates, adenine, indole acetic acid, phenylpropanoids, flavonoids, and glucosinolates. The authors analyzed four cultivars of *B. rapa* (Raapstelen, Herfstraap, Witte Mei, and Oleifera) at two development stages (4 and 6 weeks old). The level of phenylpropanoids and glucosinolates was higher in older plants. As these compounds are involved in defense processes, it is suggested that there is need to increase the defense potential. In younger plants, the levels of amino acids and organic acids are higher. The lower level of these molecules in older plants can be caused by the fact that they are necessary for glucosinolates and phenylpropanoids synthesis.

Derivatives of GLS, generated by hydrolysis with myrosinase (Fig. 10-4), such as thiocyanates, isothiocyanates, nitriles, epithionitriles are involved in plant defense against various pathogens.

Isothiocyanates (ITC), such as sulforaphane (SNF) or its analogue erucin (Fig. 10-5) also have pharmaceutical meaning as anticancer agents.

On the other hand, some compounds resulting from the hydrolysis of glucosinolates can have harmful effects on human health. For example goitrin (Fig. 10-6) reduces the production of thyroid hormones and in some conditions it can form N-nitrosooxazolidone, a mutagen. For this reason You et al (2008) searching for the best source of sulforaphane (as a human diet complementation) also analyzed the goitrin content.

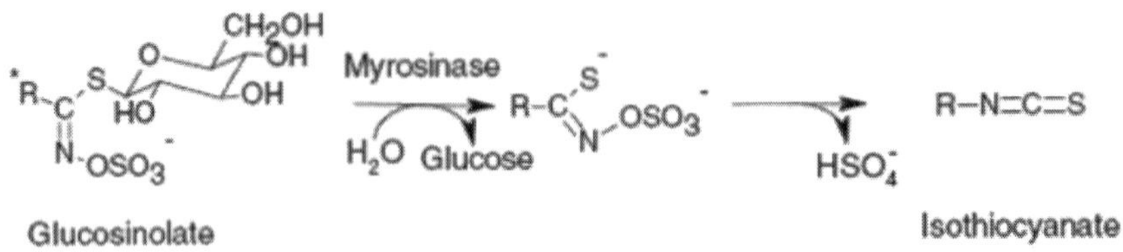

Figure 10-4 Hydrolysis of glucosinolates to its isothiocyanate by myrosinase (*http://lpi.oregonstate.edu/ss06/vegetables.html*).

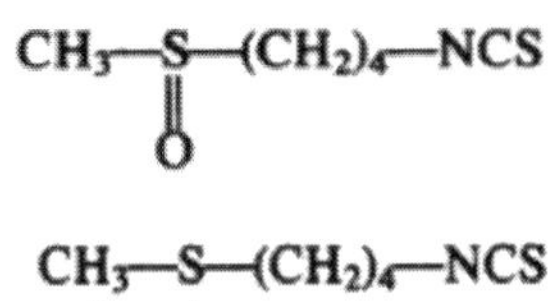

Figure 10-5 Structure of sulphoraphane (4-methylsulphinylbutyl isothiocyanate) and erucin (4-methylthiobutyl isothiocyanate). Reprinted from Zhang et al. (1992).

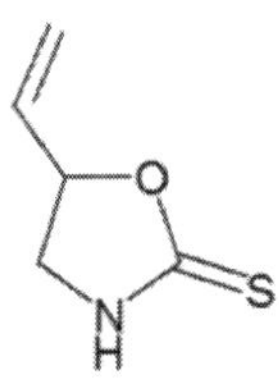

Figure 10-6 Structure of goitrin (*http://en.wikipedia.org/wiki/Goitrin*).

The HPLC analysis of seeds of 43 different *B. oleracea* varieties (You et al. 2008) showed that erucin was present mainly in cabbage (*B. oleracea* var. *capitata*) and kohlrabi (*B. olearacea* var. *gongylodes*) and sulforaphane in broccoli (*B. oleracea* var. *italica*). The content of isothiocyanates also depended on the climate and other environmental factors. The authors analyzed 43 cultivars belonging to 10 *B. oleracea* species. Among them only 21 cultivars contained detectable levels of sulforaphane and erucin. In broccoli, cabbage and kohlrabi there were significant differences in the isothiocyanates content between cultivars. When the seeds started to sprout the isothiocyanates level decreased sharply. Seeds of the *Brassica* species contain 10 times more isothiocyanates than the edible parts but they are toxic to humans and animals; so they can only be a source for the extraction of metabolites. The authors analyzed the level of sulforaphane, erucin and goitrin in seeds. They suggested using Brassica seeds as a diet complementary or as a source for the extraction of isothiocyanates. On the basis of the results obtained the most suitable source of food supplements were broccoli seeds. They contained a high level of sulforaphane and some cultivars (such as Green Dragon II) also had low levels of goitrin.

For Brassica (and other vegetables) some results are available on the impact of storage and cooking on metabolite composition. One of the works concerning glucosinolate level is by Rungampestry et al. (2007). Besides the environmental changes, the glucosinolate level is also affected by storage, processing and ingestion of vegetables. It is known that gucosinolates are hydrolyzed by plant myrosinases after tissue damage. In normal cells glucosinolates and myrosinase are spatially separated (placed in different compartments); after damage (for example by insects) myrosinase is released and glucosinolates are hydrolyzed to toxic compounds (such as isothiocyanates). Cooking can also change the glucosinolates level by impact of myrosinase activity. Besides the glucosinolate content, their anticancer activity also depends on the mode in which they are taken up by the organism. Eating raw Brassica vegetables causes the release of glucosinolates in the upper gastrointestinal tract; this releases the myrosinase present in plant tissue. Glucosinolates from cooked vegetables are released in the colon with the involvement of the colon microflora. Changes in the level of glucosinolates/isothiocyanates can also be caused by the introduction of transgenes. Zang et al. (2007) introduced three cDNAs coding cytochrome 450 familly enzymes from *A. thaliana* to Chinese cabbage. These are CYP79B2 and B3 from the biosynthetic pathway of indole-3-acetic acid (IAA) and indole- glucosinolate (IG) and CYP83B1 unique to IG biotynthesis. They analyzed the glucosinolate content of three independent lines transformed with each double and triple construct. They found that the lines containing CYP79B3 and CYP83B1 have the same level of IG as controls. Overexpression of CYP79B3 and CYP79B2 causes IAA overproduction that prevents the regeneration of callus. There were no indole glucosinolates in these lines which confirms the role of these enzymes in IG biosynthesis. The lines containing CYP79B3, CYP79B2 and CYP83B1 regenerated normally, and the obtained plants had elevated levels of glucobrassicin, 4-hydroxy glucobrassicin, and 4-methoxy glucobrassicin. Overexpression of CYP79B3 alone did not alter the IG level. HPLC analyses showed that the main differences in IG accumulation in transgenic plants concern glucobrassicin, while neoglucobrassicin (56% of all IG) was not affected. The authors conclude that it is possible to enhance the IG biosynthesis without affecting plant development by the simultaneous overexpression of three enzymes from this pathway. As mentioned above, glucosinolates are the main subjects of interest in Brassica metabolomics. Changes of their level are also analyzed in different stresses and developmental stages, which is described later in the text.

10.4.2 Metabolite Changes in Stress Conditions

Phospholipid catabolism in cauliflower treated with γ-irradiation has been studied by (Voisine et al. 1993). The authors observed the acceleration of membrane deterioration caused by this treatment, so they investigated the activity of lipolitic enzymes. The breakdown products obtained suggested that γ-irradiation changes the activity of the membrane bound phospholipase D, phosphatidic acid phosphatase and lypolitic acyl hydrolase. The enhancement of membrane deterioration is induced mainly by free radicals.

Recently, a report on plant interactions with below-ground biota was published (Wurst et al. 2006). The authors investigated changes in metabolite composition in *B. oleracea*. They analyzed the combined impact of nematode (*Meloidogyne incognita*) and earthworm (*Octolasion tyrtaeum*) and reported an increase in the N uptake and shoot biomass. As for glucosilonates, glucoiberin level was decreased in the presence of earthworms and the content of glucoraphanin was affected in the presence of both organisms, but the decrease in this content was more pronounced in the presence of each organism separately than when these two organisms were both present in the soil. The contents of the other glucosinolates (sinigrin, progoitrin, glucobrassicin, 4OH-glucobrassicin, MeOH-glucobrassicin, neoglucobrassicin) were not affected. Changes in the level of glucosinolates (especially glucoiberin) may affect the oviposition behavior of insects. The results obtained demonstrated that the presence of earthworms and/or nematodes in the soil can increase N availability and hence reduce the glucosinalate levels. On the other hand, changed glucosinolate levels can affect the above ground herbivore performance. The analyses performed by Liang et al. (2006) have shown that in *B. rapa* leaves treated with methyl jasmonate (MJ) changes in the contents of several metabolites were observed. They noticed a decrease in the content of sugar and amino acids, while the level of hydroxycinnamates and glucosinolates increased substantially. Treatment with MJ causes IAA and IG accumulation. It is known that MJ plays a role in a variety of biological processes including defense responses. The results obtained suggest that the components of key importance in the defense system of *B. rapa* leaves can be the metabolites such as hydroxycinnamates, indole derivatives and glucosinolates.

Industrial development has been increasing the contamination of soil with different metal ions. It is known that Brassica plants accumulate metals in large amounts; they are even used for the phytoremediation of contaminated soils (Jahangir et al. 2008). *B. rapa* plants were treated with increasing concentrations (50–500 mM) of three metal ions (C, Fe and Mn). After treatment metabolome changes were analyzed by NMR followed by chemometric methods. The authors observed changes in the contents of key metabolites, glucosinolates and hyroxycinnamic acids. Besides the metabolite changes after metal treatment, they also observed substantial differences in the components of roots and shoots. In roots they found large amounts of sugar, glucosinolates and some free amino acids, while in leaves there were larger amounts of phenylpropanoids. Regarding these differences metal impact was analyzed separately in roots and leaves. After treatment, the sugar levels decreased, which suggests that metal ions affect photosynthesis. A great number of changes were similar for all the metals considered, however the authors observed some differences. For example, when 100 mM Fe is used, substantial changes in the content of alanine, threonne, valine and glutamic acid differences are observed; while the addition of 100 mM Cu caused differences in serine concentration. Copper and iron have greater effect than manganese. As a result of the treatment with all the metals considered, the contents of amino acids and phenylpropanoids increased, which could be a detoxification response. It is known that amino acids and phenolics have metal-chelating properties. The changes mentioned above were observed for low and moderate metal concentrations. High concentrations (~ 500 mM) cause a decrease in the content of all the metabolites identified, such concentrations are toxic to plants. Plants tolerate metal to some extent, and with increasing metal concentrations they increase the level of some metabolites, such as amino acids and phenolics. However, when the metal concentration exceeds some critical point, the metal ions interfere with biosynthesis pathways causing a decrease in the contents of all metabolites and reduction of growth. The response of plants depends not only on the metal ion concentration but also on its type (Jahangir et al. 2008). A group of metabolites frequently found to change in response to various stresses are lipids. Nouairi et al (2006) examined the impact of cadmium on the membrane lipid content in *B. napus* and *B. juncea* plants. Lipid changes depended on cadmium concentration, but most of the differences were observed between species. Total lipid content decreased significantly in *B. napus*, while *B. juncea* showed a totally different reaction. Also the level of malondialdehyde (as an indicator of the peroxidation level of membranes) and that of nonproteinaceous thiol (as a chelationg agent) were measured. The impact of cadmium on lipid biosynthesis was analyzed by using of radiolabeled acetate. The authors observed lowered levels of lipids and their biosynthesis in *B. napus* leaves and increased levels of lipid peroxidation (6 times more of malondialdehyde in comparison to that in the control plants). There were mainly changes in plastids membranes (decreased level of glycolipids) with no or little changes in the extrachloroplastic lipids (slightly increased levels of phospholipids). In *B. juncea* lipid biosynthesis was stimulated by cadmium and there are no differences in the malondialdehyde content between the control and treated plants. Such stimulation of lipid biosynthesis can be explained by the fact that membranes are essential for the formation of vesicles involved in the compartmentalization of heavy metals. Changes in thiol concentration, which is important in chelating heavy metals, also support this fact. In oilseed rape there is no change in the thiol level, and in *B. juncea* the authors observed a significant increase (up to 20 times of the control level in leaves and up to 70 times in roots) in the content of thiol after cadmium treatment. This result explained the resistance of *B. juncea* to heavy metals and suggests that this plant could be a good phytoremediator.

Changes in *B. rapa* leaves metabolome after the attack of Lepidoptera (*Plutella xylostella* L. and *Spodoptera exigua* Hubner) were investigated by NMR (Widarto et al. 2006). The response induced by the herbivore attack is marked by production of a number of secondary metabolites. The metabolic changes can be specific depending on the attacking species or common to all treatments. After attack by insects, the plants at first initiate wound-response within a few minutes and then differences in signal transduction stimulate a specific response to a specific pathogen (Widarto et al. 2006). It was demonstrated earlier that herbivore attack causes changes in the glucosinolate level, some alkaloids and a wide variety of volatile organic compounds in plants. The study by Widarto et al. (2006) demonstrated that the metabolic changes caused by the 2nd and 4th instar larvae were different. In the leaves infested by the 2nd instars the increased levels of glucose, ferulic acid, and gluconapin were noted, whereas the leaves attacked by the 4th instars had increased levels of alanine and sinapoyl malate. In both tested species metabolome changes caused by young larvae were greater than those caused by older ones. It could be caused by the fact that younger larvae are not so mobile and heavily damage leaves in isolated places.

10.4.3 Other Analyses

As seen in the literature on proteins, the papers on metabolomics present great diversity of subjects. A metabolomics-based approach to plant genotype discrimination was proposed (Taylor et al. 2002). Two genotypes of *A. thaliana* (Col0 and C24) and their progeny that differ only in mitochondrial and chloroplast genotypes were analyzed. Two crosses were made differing in maternal inheritance. Only the polar phase isolated from leaves was analyzed. The authors used gas chromatography coupled with mass spectrometry (GC/MS) and identified 433 metabolites in the samples. Using the descriptive statistics they showed that the level of the key primary metabolites varies more than others. The information obtained was sufficient to discriminate between parental lines and each of the F_1 progeny. Both, metabolomic (Taylor et al. 2002) and proteomic analyses (Marquez et al. 2001; Mukhlesur et al. 2004; Sadia et al. 2009) are often used to characterize cultivars/germplasm collections. Metabolic research also includes the analysis of mineral content. Analyses conducted by Broadley and coworkers showed that mineral concentration differed significantly between varieties. The authors analyzed the content of Ca and Mg in 376 varieties of *B. oleracea* and its wild relatives. It was demonstrated that the shoot Ca and Mg content was heritable (up to 40%) and mapped quantitative trait loci (QTL) on chromosomes C2, C6, C7 and C8. The results proved that the shoot Ca and Mg content varies between analyzed lines and did not correlate with plant biomass (Broadley et al. 2008). The content of Ca varied from 1.7 to 3.3% and that of Mg from 0.35 to 0.8%. The highest level of these ions was observed in *B. oleracea* var. *capitata* and lowest in *B.oleracea* var. *gongylodes*. Although aflatoxins (AF) do not belong to the metabolites produced by plants, they can be present in their edible part, so it is important to analyze the conditions in which the aflatoxin content increases. Aflatoxins contamination of food is one of the most serious problems concerning human health; they can also cause liver cancer. One of the major factors increasing AF content in plants is drought stress (Guo et al. 2008). High temperature also increases AF content. Aflatoxins are produced mainly by *Aspergillus flavus*. Comparison of the susceptible and resistant cultivars has shown that susceptible lines induce the same proteins as resistant ones but at a lower level or much later after the pathogen attack. Differences affected pathogen induced proteins as well as constitutively expressed ones. Tobacco plants overexpressing glyoxalase I from *B. juncea* were more resistant to *Asperigillus*. This enzyme probably acts by the removal of methylgloxal, which is the AF inducing substrate.

10.4.4 Metabolite Changes in the Course of Development

Changes in antioxidant compounds during the seed sprouting of *B. oleracea* var. *costata* were analyzed (Sousa et al. 2007). Phenolic compounds such as the esters of sinapic acid and organic acids (conitic, citric, pyruvic, malic, shikimic, and fumaric) were identified. The authors observed a decrease in phenolic compound levels but no qualitative changes. Total organic acid content increased but some of the compounds (aconitic, citric and shikimic acid) occurred at a decreased level. Chromatographic analysis with diode array detector (HPLC/DAD) revealed the presence of 12 phenolic compounds in germinating sprouts. They included five sinapoyl glucosides, four disinapoyl glucosides, one trisinapoyl glucoside, sinapoyl choline, and one kaempferol glucoside acylated with sinapic acid. The total phenolic content showed significant quantitative changes. Its level decreased from

11.1 g/kg in 2-day-old sprouts to 1.6 mg/kg after 10 days. This decrease could be explained by using phenolics in cell wall biosynthesis and as antioxidants. Contrary to phenolic compounds, total organic acids content increased from 45.7 g/kg on day 2 to 66.9 g/kg on day 12. The main organic acids found at the time of germination are malic and citric acids, although the content of the latter decreased slightly from day 2 to 12. Despite the total organic acids increase, there are changes in the quantity of individual compounds at the time of sprouting, most of them in the first 4 days. As glucosinolates (including derivatives of their hydrolysis) have great impact on human health (especially as anti-cancer agents) the majority of the metabolic analyses based their estimation in different *Brassica* species and different environmental conditions.

10.5 Applications of Proteomics and Metabolomics Data (Clinical Aspects: Anti-Carcenogenic Effects—Human Proteome Research)

Cruciferous vegetables are commonly used in the human diet and contain many bioactive compounds, e.g. flavonoids, vitamin C, K, minerals such a selenium and glucosinolates (Moreno et al. 2006). The nutritional properties of Brassica crops and their use in human diet are described in Chapter 1. It has been shown that all of these biomolecules present in cruciferous vegetables may be protective natural compounds and should help prevent development of many cancers (Li et al. 2008; Boivin et al. 2009). The most intensely studied cruciferous biomolecules associated with cancer prevention have been glucosinolates. GLS are not bioactive, they must be first enzymatically hydrolyzed (by myrosinase) to isothiocyanates (Fenwick et al. 1983) (described previously in 10.3, Fig. 10-7). There are two most important and well-studied isothiocyanate members. The first one is sulforaphane, which was isolated and characterized, in the early 1990s mostly in broccoli, Brussels sprouts, cauliflower and kale (Zhang et al. 1992) and the second one is indole-3-carbinol (I3C) detected mainly in broccoli and cabbage (Moreno et al. 2006). Many studies conducted on animals (rodents) or human cells have shown that there are several mechanisms in which sulforaphane and isothiocyanates could act as cancer protective molecules. These include: 1) induction of Phase 2 detoxification enzymes, and 2) inhibition of carcinogen-activating Phase 1 enzymes, 3) induction of apoptosis, and 4) regulation of cell progression via cell cycle arrest, and 5) modulation of hormone metabolism, 6) also anti-inflammation. For a more detailed discussion of target genes and mechanisms of glucosinolates action the reader is directed to several extensive review articles (Bheemreddy and Jeffery 2006; Myzak and Dashwood 2006; Fimognari and Hreila. 2007; Clarke et al. 2008). Here we focus only on a few examples of glucosinolates' cancer protective mechanisms concerning mainly human cancer.

A

B

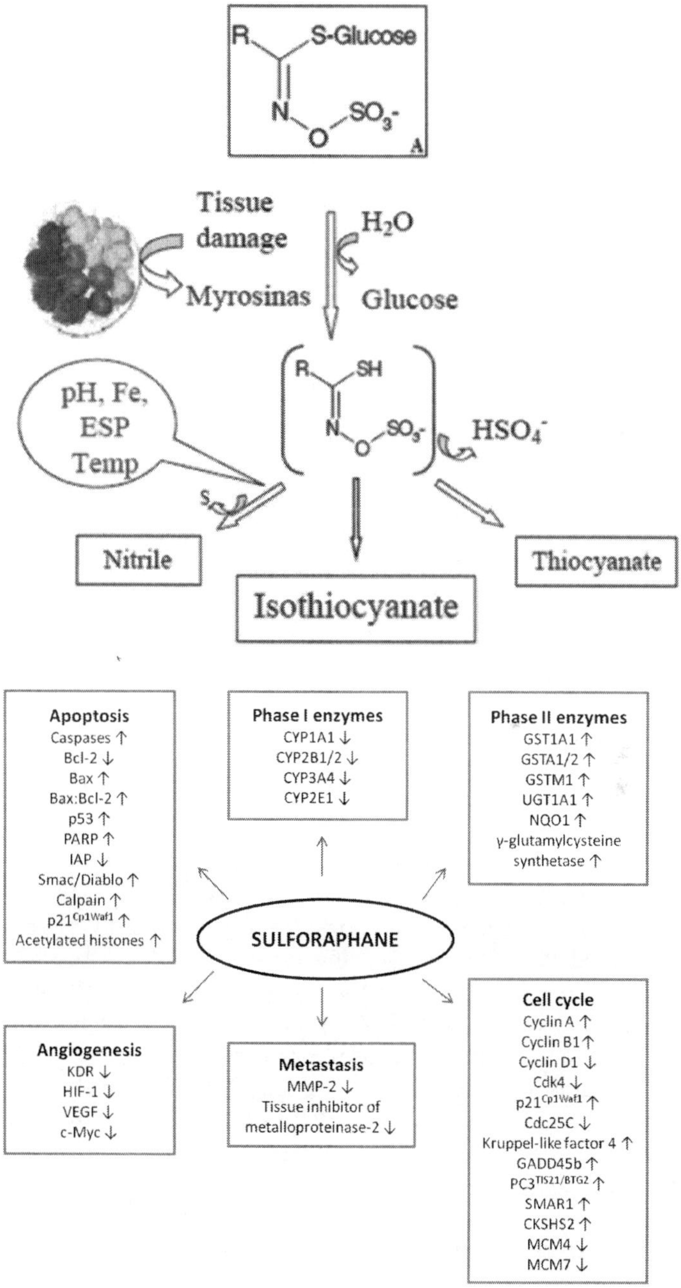
R
S-Glucose
N
O
SO_3^-
Tissue damage
Myrosinas
H_2O
Glucose
R
SH
N
O
SO_3^-
pH, Fe, ESP Temp
HSO_4^-
S
Nitrile
Isothiocyanate
Thiocyanate
Apoptosis
Caspases ↑
Bcl-2 ↓
Bax ↑
Bax:Bcl-2 ↑
p53 ↑
PARP ↑
IAP ↓
Smac/Diablo ↑
Calpain ↑
p21^{Cp1Waf1} ↑
Acetylated histones ↑
Phase I enzymes
CYP1A1 ↓
CYP2B1/2 ↓
CYP3A4 ↓
CYP2E1 ↓
Phase II enzymes
GST1A1 ↑
GSTA1/2 ↑
GSTM1 ↑
UGT1A1 ↑
NQO1 ↑
γ-glutamylcysteine synthetase ↑
SULFORAPHANE
Angiogenesis
KDR ↓
HIF-1 ↓
VEGF ↓
c-Myc ↓
Metastasis
MMP-2 ↓
Tissue inhibitor of metalloproteinase-2 ↓
Cell cycle
Cyclin A ↑
Cyclin B1 ↑
Cyclin D1 ↓
Cdk4 ↓
p21^{Cp1Waf1} ↑
Cdc25C ↓
Kruppel-like factor 4 ↑
GADD45b ↑
PC3$^{TIS21/BTG2}$ ↑
SMAR1 ↑
CKSHS2 ↑
MCM4 ↓
MCM7 ↓

C

Category	Gene	Change in activity
Phase I enzyme	CYP1A1	Decrease
	CYP2B1/2	Decrease
	CYP3A4	Decrease
	CYP4A10	Increase
	CYP4A14	Increase
	CYP39A1	Increase
Phase II enzyme	NQO-1	Increase
	GST	Increase
	γ-GCS	Increase
	UGT	Increase
Anti-oxidants	Thioredoxin reductase 1 and 3	Increase
	Glutathione peroxidase 3	Increase
Proteasome subunits	Alpha 1 and 3, beta 5, and 26S	Increase
Cell cycle and cell growth	Cyclins D1, E2 and T2	Increase
	CDK 4, 7 and 9	Increase
Stress response	Heat shock proteins 1, 1 alpha, 1 beta, 8 and 105	Increase
Transcription factors	CREB binding protein	Increase
	Aryl hydrocarbon receptor nuclear nuclear translocator-like	Increase

Figure 10-7 A. Schematic representation of glucosinolates hydrolysis. Reprinted from Bheemreddy and Jeffery (2006). B. Molecular targets of sulforaphane. Redrown according to Fimognari et al. (2007). C. Effects of sulforaphane on "blocking" mechanisms. Redrawn according to Clarke et al. (2008).

Jackson and Singletary (2004) have shown that sulforaphane can stop the growth of human breast cancer cells and identified the mechanisms of action of this compound. They discovered that SNF can stop cancer cells development by interfering with microtubule assembly, which are necessary for cell division and consequently inhibit the growth of tumor cells. These anti-proliferative effects of sulforaphane have been compared with cancer therapeutic drugs in the study reported by group led by Merlin Morris. The study has shown that higher concentration of SNF could stop cancer cell proliferation comparable to the effects of drugs, e.g., daunomycin or vinblastine (Tseng et al. 2004). Singh and coworkers have shown that sulforaphane can also inhibit the proliferation of the human PC-3 prostate cancer via inducing caspase-mediated apoptosis (Singh et al. 2004). Also Shan and colleagues have discovered that sulforaphane can induce apoptosis in the human bladder cancer cell (Shan et al. 2006). Furthermore, this compound could also inhibit the growth of several of the *Helicobacter pylori* strains, bacteria associated with gastric cancer development (Surh et al. 2001; Fahey et al. 2002).

Another representative of thiocyanates, indole-3-carbinol (I3C) in the human body is mostly converted into 3,3′-diindolylmethane (DIM), which is the bioactive dimer of I3C (Auborn et al. 2003). Several reports have shown that this compound also acts in multiple ways against the development of cancer. FirstlyI3C could negatively regulate estrogen hormone metabolism (Auborn et al. 2003). Estrogen hormones promote cancer cell growth and DIM may suppress it and consequently prevent the development of estrogen-enhanced cancer such as breast (Bradlow et al. 1991; Meng et al. 2000), cervical (Jin et al. 1999) or endometrial (Kojima et al. 1994) cancers. DIM can also inhibit the growth of prostate cancer cells via suppression of the proteins that speed up the growth of cancer cells (Smith et al. 2008). It was also shown that indole-3-carbinol inhibits the action of most prominent carcinogens derived from cigarettes called nitrosamines (e.g., NNK, nicotine-derived nitrosamine ketone, 4-(methylnitrosamino)-1-(3-pyridyl)-1-butanone). I3C may reduce the incidence of DNA damage caused by nitrosamines through changes or inhibition of its metabolism (Hecht et al. 1996; Taioli et al. 1997). Apart from the glucosinolates the most important biomolecules present on a high level in cruciferous vegetables are also carotenoids (lycopene), polyphenols, fibers and vitamins, especially vitamin C (Moreno et al. 2006), which all may act in cancer prevention as excellent antioxidants and additional stimulators of the immune system (Ferguson and Harris 1996; Yang et al. 1997; Sikora et al. 2008; Boivin et al. 2009).

Because of the rapid progress in high-throughput techniques over the past 20 years it was possible to demonstrate that the consumption of fruits and vegetables, especially cruciferous vegetables can reduce incidences of many types of cancer (Talalay and Zhang 1996; Hecht 1999). This progress facilitates better understanding of the role of various compounds in human health and helps to determine their proper amounts in the human diet (Talalay and Zhang 1996; Mithen et al. 2000).

Acknowledgements

This study was supported by the Polish Committee for Scientific Research Grants PZ-19/2006/2/JS and PBZ-MNiSW-2/3/2006/19/IGR/4.

References

Aebersold R, Mann M (2003) Mass spectrometry-based proteomics. Nature 422: 198–207.

Albertin W, Brabant P, Catrice O, Eber F, Jenczewski E, Chèvre A-M, Thiellement H (2005) Autopolyploidy in cabbage (*Brassica oleracea* L.) does not alter significantly the proteomes of green tissues. Proteomics 5: 2131–2139.

Alvarez S, Berla BM, Sheffield J, Cahoon RE, Jez JM, Hicks LM (2009) Comprehensive analysis of the *Brassica juncea* root proteome in response to cadmium exposure by complementary proteomic approaches. Proteomics 9: 2419–2431.

Auborn KJ, Fan S, Rosen EM, Goodwin L, Chandraskaren A, Williams DE, Chen D, Carter TH (2003) Indole-3-carbinol is a negative regulator of estrogen. J Nutr 133: 2470–2475.

Bayoumi I, Farid A, Kim HK, Choi YH, Verpoorte R (2007) Metabolic characterization of *Brassica rapa* leaves by NMR spectroscopy. J Agri Food Chem 55: 7936–7943.

Bheemreddy RM, Jeffery E (2006) Glucosinolates. In: D Heber, GL Blackburn, VLM Go, J Milner (eds) Nutritional Oncology. Elsevier Science & Technology, Burlington, San Diego, USA, pp 583–596.

Boivin D, Lamy S, Lord-Dufour S, Jackson J, Beaulieu E, Côté M, Moghrabi A, Barrette S, Gingras D, Béliveau R (2009) Antiproliferative and antioxidant activities of common vegetables: A comparative study. Food Chem 112: 374–380.

Bradlow HL, Michnovicz J, Telang NT, Osborne MP (1991) Effects of dietary indole-3-carbinol on estradiol metabolism and spontaneous mammary tumors in mice. Carcinogenesis 12: 1571–1574.

Broadley MR, Hammond JP, King GJ, Astley D, Bowen HC, Meacham MC, Mead A, Pink DAC, Teakle GR, Hayden RM, Spracklen WP, White PJ (2008) Shoot calcium and magnesium concentrations differ between subtaxa, are highly heritable, and associate with potentially pleiotropic loci in *Brassica oleracea*. Plant Physiol 146: 1707–1720.

Buthz A, Kolasa A, Arlt K, Walz Ch, Kehr J (2004) Xylem sap protein composition is conserved among different plant species. Planta 219: 610–618.

Clarke JD, Dashwood R, Hoet E (2008) Multi-targeted prevention of cancer by sulforaphane. Cancer Lett 269: 291–304.

Domon B, Aebersold R (2006) Mass spectrometry and protein analysis. Science 312: 212– 217.

Eidhammer I, Flikka K, Martens L, Mikalsen SO (2007) Computational Methods for Mass Spectrometry Proteomics. John Wiley & Sons, Chichester, West Sussex, England, pp 66–79

Fahey JW, Haristoy X, Dolan PM, Kensler TW, Scholtus I, Stephenson KK (2002) Sulforaphane inhibits extracellular, intracellular, and antibiotic-resistant strains of Helicobacter pylori and prevents benzo[a]pyrene-induced stomach tumors. Proc Natl Acad Sci USA 99: 7610–7615.

Fenn JB, Mann M, Meng CK, Wong SF, Whitehouse CM (1989) Electrospray ionization for mass spectrometry of large biomolecules. Science 246: 64–71.

Fenwick GR, Heaney RK, Mullin WJ (1983) Glucosinolates and their breakdown products in food and food plants. Crit Rev Food Sci Nutr 18: 123–201.

Ferguson LR, Harris PJ (1996) Studies on the role of specific dietary fibres in protection against colorectal cancer. Mutat Res 350: 173–184.

Fiehn O, Kloska S, Altmann T (2001) Integrated studies on plant biology using multiparallel techniques. Curr Opin Biotechnol 12: 82–86.

Fimognari C, Hreila P (2007) Suloraphane as a promising molecule for fighting cancer. Mutat Res 635: 90–104.

Grossman J, Fischer B, Baerenfaller K, Owiti J, Buhmann JM, Gruissem Q, Baginsky S (2007) Workflow to increase the detection rate of proteins from unsequenced organisms in high-throughput proteomics experiments. Proteomics 7: 4245–4254.

Guo B, Chen Z-Y, Lee RD, Scully BT (2008) Drought stress and preharvest aflatoxin contamination in agricultural commodity: Genetics, genomics and proteomics. J Integr Plant Biol 50: 1281–1291.

Hecht SS (1999) Chemoprevention of cancer by isothiocyanates, modifiers of carcinogen metabolism. J Nutr 129: 768S–774S.

Hecht SS, Trushin N, Rigotty J, Carmella SG, Borukhova A, Akerkar S, Rivenson A (1996) Complete inhibition of 4-(methylnitrosamino)-1-(3-pyridyl)-1-butanone induced rat lung tumorigenesis and favorable modification of biomarkers by phenethyl isothiocyanate. Cancer Epidemiol Biomark Prev 5: 645–652.

Herbert BR Righetti PG, Citterio A, Boschetti E (2007) Sample preparation and prefractionation techniques for electrophoresis-based proteomics. In: MR Wilkins, RD Appel, KL Williams,DF Hochstrasser (eds) Proteome Research. Concepts, Technology and Application. Springer-Verlag, Berlin Heidelberg, Germany, pp 14–40.

Hochholdinger F, Guo L, Schnable PS (2004) Cytoplasmic regulation of the accumulation of nuclear-encoded proteins in the mitochondrial proteome of maize. Plant J 37: 199–208.

Hurkman WJ, Tanaka CK (2007) High-resolution two-dimensional gel electrophoresis: A cornerstone of plant proteomics. In: J Samaj, J Thelen (eds) Plant Proteomics. Springer, Berlin, Germany, pp 14–28.

Jackson SJ, Singletary KW (2004) Sulforaphane inhibits human mcf-7 mammary cancer cell mitotic progression and tubulin polymerization. J Nutr 134: 2229–2236.

Jahangir M, Bayoumi I, Farid A, Choi YH, Verpoorte R (2008) Metal ion-inducing metabolite accumulation in *Brassica rapa*. J Plant Physiol 165: 1429–1437.

James P (2004) Automation of proteome analysis. In: DW Speicher (ed) Proteome Analysis. Interpreting the Genome. Elsevier, Philadelphia, USA, pp 305–326.

Jin L, Qi M, Chen DZ, Anderson A, Yang GY, Arbeit JM, Auborn KJ (1999) Indole-3-carbinol prevents cervical cancer in human papilloma virus type 16 (HPV16) transgenic mice. Cancer Res 59: 3991–3997.

Karas M, Hillenkamp F (1988) Laser desorption ionization of proteins with molecular masses exceeding 10,000 daltons. Ann Chem 60: 2299–301.

Kersten D, Bürkle L, Kuhn EJ, Giavalisco P, Konthur Z, Lueking A, Walter G, Eickhoff H, Schneider U (2002) Large-scale plant proteomics. Plant Mol Biol 48: 133–141.

Kim HN, Chal JS, Cho TJ, Kim HY (2003) Salicylic acid and wounding induce defense-related proteins in Chinese cabbage. Korean J Biol Sci 7: 313–219.

Kliebenstein DJ, Monde RA, Last RL (1998) Superoxide dismutase in Arabidopsis: an eclectic enzyme family with disparate regulation and protein localization. Plant Physiol 118: 637–650.

Kojima T, Tanaka T, Mori H (1994) Chemoprevention of spontaneous endometrial cancer in female Donryu rats by dietary indole-3-carbinol. Cancer Res 54: 1446–1449.

Köksal E, Gülçin I (2008) Purification and characterization of peroxidase from cauliflower (*Brassica oleracea* L. var. *botrytis*) buds. Prot Pept Lett 15: 320–326.

Li D, Wu K, Howie AF, Beckett GJ, Wang W, Bao Y (2008) Synergy between broccoli sprout extract and selenium in the upregulation of thioredoxin reductase in human hepatocytes. Food Chem 110: 193–198.

Liang Y-S, Choi YH, Kim HK, Linhorst HJM, Verpoorte R (2006) Metabolomic analysis of methyl jasmonate treated *Brassica rapa* leaves by 2-dimensional NMR spectroscopy. Phytochemistry 67:2503–2511.

Liao Ch-S, Yen J-H, Wang Y-S (2008) Growth inhibition in Chinese cabbage (*Brassica rapa* var. *chinensis*) growth exposed to di-*n*-butyl phthalate. J Hazard Matter 163: 625–631.

Marquez K, Sarazin B, Favre LCh, Zivy M, Thiellement H (2001) Comparative proteomics to establish genetic relationships in the Brassicaceae family. Proteomics 1: 1457–1462.

Meng Q, Goldberg ID, Rosen EM, Fan S (2000) Inhibitory effects of indole-3-carbinol on invasion and migration in human breast cancer cells. Breast Cancer Res Treat 63: 147–152.

Millar AH (2004) Location, location, location: surveying the intracellular real estate through proteomics in plants. Funct Plant Biol 31: 563–571.

Mithen RF, Dekker M, Verkerk R, Rabot S, Johnson IT (2000) The nutritional significance, biosynthesis and bioavailability of glucosinolates in human foods. J Sci Food Agri 80: 967–984.

Moreno DA, Carvajal M, López-Berenguer C, García-Viguera C (2006) Chemical and biological characterisation of nutraceutical compounds of broccoli. J Pharm Biomed Anal 41: 1508–22.

Mukhlesur RM, Hirata Y, Alam SE (2004) Genetic variation within *Brassica rapa* cultivars using SDS-PAGE for seed protein and isozyme analysis. J Biol Sci 4: 239–242.

Murphy DJ (2006) The extracellular pollen coat in members of the Brassicaceae: composition, biosynthesis, and functions in pollination. Protoplasma 228: 31–39.

Myzak MC, Dashwood RH (2006) Chemoprotection by sulforaphane: Keep one eye beyond Keap1. Cancer Lett 233: 208–218.

Nouairi I, Ammar WB, Youssef NB, Miled DDB, Ghorbal MH, Zarrouk M (2006) Variations in membrane lipid metabolism in *Brassica juncea* and *Brassica napus* leaves as a response to cadmium exposure. J Agron 5: 299–307.

Patterson SD, Aebersold RH (2003) Proteomics: the first decade and beyond. Nat Genet 33: 311–23.

Pawlowski T, Rurek M, Janicka S, Raczynska KD, Augustyniak H (2005) Preliminary analysis of the cauliflower mitochondrial proteome. Acta Physiol Plant 27: 275–281.

Peck SC, Nühse TS, Hess D, Iglesias A, Meins F, Boller T (2001) Directed proteomics identifies a plant-specific protein rapidly phosphorylated in response to bacterial and fungal elicitors. Plant Cell 13: 1467–1475.

Roberts IN, Harrod G, Dickinson HG (1984) Pollen-stigma interactions in *Brassica oleracea.* II. The fate of stigma surface proteins following pollination and their role in the selfincompatibility response. J Cell Sci 66: 255–264.

Rodman JE (1991a) A taxonomic analysis of glucosinolates containing plants, part 1: Phenetics. Syst Bot 16: 598–618.

Rodman JE (1991b) A taxonomic analysis of glucosinolates containing plants, part 2: Cladistics. Syst Bot 16: 619–629.

Ruiter RK, van Eldik J, van Herpen RMA, Schrauwen JAM, Wullems GJ (1997) Characterization of oleosins in the pollen coat of *Brassíca oleracea.* Plant Cell 9: 1621–1631.

Rungampestry V, Duncan AJ, Fuller Z, Ratcliffe B (2007) Effect of cooking brassica vegetables on the subsequent hydrolysis and metabolic fate of glucosinolates. Proc Nutr Soc 66: 69–81.

Sadia M, Malik SA, Rabbani MA, Pearce SR (2009) Electrophoretic characterization and the relationship between some *Brassica* Species. e-J Biol 5: 1–4.

Sarker RH, Ellenan CJ, Dickinson HG (1988) Control of pollen hydration in Brassica requires continued protein synthesis, and glycosylation is necessary for intraspecific incompatibility. Cell Biol 85: 4340–4344.

Saroop S, Chanda SV, Singh YD (2002) Changes in soluble and ionically bound peroxidase activities during *Brassica juncea* seed development. Bulg J Plant Physiol 28: 26–34.

Schmidt UG, Endler A, Schelbert S, Brunner A, Schnell M, Neuhaus HE, Mazars DM, Marty F, Baginsky S, Martinoia E (2007) Novel tonoplast transporters identified using a proteomic approach with vacuoles isolated from cauliflower buds. Plant Physiol 145: 216–229.

Shan Y, Sun C, Zhao X, Wu K, Cassidy A, Bao YP (2006) Effect of sulforaphane on cell growth, G0/G1 phase cell progression and apoptosis in human bladder cancer T24 cells. Int J Oncol 29: 883–888.

Sikora E, Cieślik E, Leszczyńska T, Filipiak-Florkiewicz A, Pisulewski PM (2008) The antioxidant activity of selected cruciferous vegetables subjected to aquathermal processing. Food Chem 107: 55–59.

Singh AV, Xiao D, Lew KL, Dhir R, Singh SV (2004) Sulforaphane induces caspase-mediated apoptosis in cultured PC-3 human prostate cancer cells and retards growth of PC-3 xenografts in vivo. Carcinogenesis 25: 83–90.

Smith RD (2002) Trends in mass spectrometry instrumentation for proteomics. Trends Biotechnol 20: 3–7.

Smith S, Sepkovic D, Bradlow HL, Auborn KJ (2008) 3,3′-Diindolylmethane and genistein decrease the adverse effects of estrogen in LNCaP and PC-3 prostate cancer cells. J Nutr 138: 2379–2385.

Sousa C, Lopes G, Pereira DM, Taveira M, Valentao P, Seabra RM, Pereira JA, Baptista P, Ferreres F, Andrade PB (2007) Screening of Antioxidant Compounds During Sprouting of *Brassica oleracea* L. var. *costata* DC. Combin Chem High Throughput Screen 10: 377–386.

Subramanian B, Bansal VK, Kav NNV (2005) Proteome-level investigation of *Brassica carinata*-derived resistance to *Leptosphaeria maculans*. J Agri Food Chem 53: 313–324.

Surh YJ, Chun KS, Cha HH, Han SS, Keum YS, Park KK (2001) Molecular mechanisms underlying chemopreventive activities of anti-inflammatory phytochemicals: Down-regulation of COX-2 and iNOS through suppression of NF-κB activation. Mutat Res 480–481: 243–268.

Taioli E, Garbers S, Bradlow HL, Carmella SG, Akerkar S, Hecht SS (1997) Effects of indole-3-carbinol on the metabolism of 4-(methylnitrosamino)-1-(3-pyridyl)-1-butanone in smokers. Cancer Epidem Biomark Prev 6: 517–522.

Talalay P, Zhang Y (1996) Chemoprotection against cancer by isothiocyanates and glucosinolates. Biochem Soc Trans 24: 806–810.

Tanaka K, Ido Y, Akita S, Yoshida Y, Yoshida T, Matsuda, H and Xiao-tian L (eds) (1987) Proc Second Japan-China Joint Symp on Mass Spectrometry, 185. Osaka, Japan, 15–18 September 1987.

Tayefi-Nasrabadi H (2008) Catalytic properties of three catalases from Kohlrabi (*Brassica oleracea gongylodes*). Afr J Biotechnol 7: 472–475.

Taylor J, King RD, Altmann T, Fiehn O (2002) Application of metabolomics to plant genotype discrimination using statistics and machine learning. Bioinformatics 18: S241–S248.

Thelen JJ (2007) Introduction to proteomics: a brief historical perspective on contemporary approaches. In: J Samaj, J Thelen (eds) Plant Proteomics. Springer Verlag, Berlin, Heidelberg, Germany, pp 1–12.

Thomson JJ (1913) Rays of positive electricity and their application to chemical analysis. Proc Roy Soc London, 89: 1–20.

Tseng E, Scott-Ramsay EA, Morris ME (2004) Dietary organic isothiocyanates are cytotoxic in human breast cancer MCF-7 and mammary epithelial MCF-12A cell lines. Exp Biol Med 229: 835–842.

Voisine R, Vézina L-P, Willemot C (1993) Modification of phospholipid catabolism in microsomal membranes of γ-lrradiated cauliflower (*Brassica oleracea* 1). Plant Physiol 102: 213–218

Wang W, Tai F, Chen S (2008) Optimizing protein extraction from plant tissues for enhanced proteomics analysis. J Separ Sci 31: 2032–2039.

Widarto HT, Van der Meijden E, Lefeber AM, Erkelens C, Kim HK, Choi YH, Verpoorte R (2006) Metabolomic differentiation of *Brassica rapa* following herbivory by different insect instars using two-dimensional nuclear magnetic resonance spectroscopy. J Chem Ecol 32: 2417–2428.

Wijk van KJ (2001) Challenges and prospects of plant proteomics. Plant Physiol 126: 501–508.

Wurst S, Langel R, Rodger S, Scheu S (2006) Effects of belowground biota on primary and secondary metabolites in *Brassica oleracea*. Chemoecology 16: 69–73.

Yang CS, Lee MJ, Chen L, Yang GY (1997) Polyphenols as inhibitors of carcinogenesis. Environ Health Perspect 105: 971–976.

You Y, Wu Y, Mao J, Zou L, Liu S (2008) Screening of Chinese *Brassica* species for anti-cancer sulforaphane and erucin. Afr J Biotechnol 7: 147–152.

Zang Y-X, Lim M-H, Park B-S, Hong S-B, Kim DH (2007) Metabolic engineering of indole glucosinolates in Chinese cabbage plants by expression of Arabidopsis CYP79B2, CYP79B3, and CYP83B1. Mol Cell 25: 231–241.

Zhang Y, Talalay P, Cho CG, Posner GH (1992) A major inducer of anticarcinogenic protective enzymes from broccoli: isolation and elucidation of structure. Proc Natl Acad Sci USA 89: 2399–2403.

11

Role of Bioinformatics as a Tool for Vegetable *Brassica* Species

Chris Duran,[1] Zoran Boskovic,[1] Jacqueline Batley,[2] and *David Edwards[1,*]*

ABSTRACT

With the rapidly increasing volumes and complexity of genetic and genomic data available for *Brassica* crops, there is an increasing need for bioinformatics tools to analyze and organize this data. This chapter reviews a selection of tools and databases available for *Brassica* bioinformatics, describing their purpose and utility for *Brassica* researchers. The majority of *Brassica* bioinformatics tools have been developed for the discovery and analysis of molecular genetic markers. These include autoSNPdb, SSRPrimer CMAP3D and Brassica.info. With the growth of gene and genome sequence data, as well as gene expression and phenotypic information, there will be an increasing requirement to integrate and merge this information using current and future bioinformatics applications.

Keywords: autoSNPdb, SRRPrimer, CMap, EnsEMBL, Brassica.info

11.1 Introduction

With the continued development of molecular techniques, the quantity and variety of data obtained by researchers is growing at an exponential rate. This increase in data production makes data analysis and integration a major challenge and increasingly sophisticated methods are required to draw meaningful conclusions from the analysis. It is becoming extremely

[1]Australian Centre for Plant Functional Genomics, School of Land, Crop and Food Sciences, Institute for Molecular Bioscience, University of Queensland, Brisbane, QLD 4072, Australia.
[2]School of Land, Crop and Food Sciences, ARC Centre of Excellence for Integrative Legume Research, University of Queensland, Brisbane, QLD 4072, Australia.
*Corresponding author: *dave.edwards@uq.edu.au*

difficult for humans to interpret the experimental results without the use of dedicated bioinformatics tools, which enable the management and analysis of large quantities of information.

The demand for bioinformatics tools has increased with increasing data and is leading to a virtual revolution in science, with an increasing focus on data analysis rather than data production. With the escalating complexity of bioinformatics analysis, there is the potential to use the wrong tool, or to misinterpret data. It is, therefore, important that researchers are aware of the capability and limitations of the tools being applied as well as limitations related to data quality. In this chapter we describe a range of bioinformatics tools and databases available to *Brassica* researchers that may be applied to assist molecular genetic analysis in these species leading to improved *Brassica* varieties.

Bioinformatics systems fall into several categories, including central data repositories such as the Protein Data Bank (PDB, *http://www.rcsb.org/pdb*), which contains information about experimentally-determined protein structures, nucleic acids and complex assemblies; the National Centre for Biotechnology Information (NCBI, *http://www.ncbi.nlm.nih.gov/*), a repository for DNA sequence information; and UniProt (*http://www.uniprot.org/*), a protein sequence and functional information resource. In addition to central repositories, there are several more specialized databases, such as AutoSNPdb (Duran et al. 2008) and CMap (Lim et al. 2007), as well as general data analysis tools, such as the molecular marker discovery tools, SNPServer (Savage et al. 2005), SSRPrimer (Robinson et al. 2004) and SSR Taxonomy Tree (Jewell et al. 2006); the sequence clustering tool CLUSTAL-W (Larkin et al. 2007); and the sequence comparison tool BLAST (Basic Local Alignment Search Tool; Altschul et al. 1990). There are also general *Brassica* databases such as brassica.info (*http://www.brassica.info/*) and the *Brassica* Genome gateway (*http://brassica.bbsrc.ac.uk/*); and comparative genomics tools such as: *Brassica*—Arabidopsis Comparative Genome Viewer (*http://brassica.agr.gc.ca/navigation/viewer_e.shtml*), the AtEnsembl viewer (*http://atensembl.arabidopsis.info*) and the ATIDB database resource (*http://atidb.org*).

This chapter reviews the available *Brassica* tools, detailing their purpose and utility for scientists and breeders.

11.2 Genetic Resources

Genetic studies examine genetic variation in relation to heritable traits. The majority of variation between individuals is found at the nucleotide level, and molecular genetic markers are powerful tools for the association of nucleotide variation with heritable traits. Two forms of sequence based markers, simple sequence repeats (SSRs), also known as microsatellites, and single nucleotide polymorphisms (SNPs) have become the major markers currently applied in modern genetic analysis. These are supplemented with anonymous marker systems such as amplified fragment length polymorphisms (AFLPs; Vos et al. 1995) and diversity array technology (DArT; Wenzel et al. 2004). Molecular genetic markers are used for genetic mapping and the identification of genes underlying heritable traits, which makes them excellent tools for crop improvement by marker-assisted selection.

11.2.1 Marker Discovery

The application of molecular genetic markers for crop improvement has traditionally been limited by the huge costs related to marker discovery and association with valuable agronomic traits. Early marker discovery methods were laboratory based, expensive and time consuming. However, with the increased application of high-throughput sequencing technology, there is a vast amount of DNA sequence data available in the public domain, allowing bioinformatics developers to apply computational methods for the identification and annotation of large numbers of markers rapidly and at low cost. Having access to large numbers of markers, such as SNPs and SSRs, at a high density in the genome provides a powerful resource for molecular geneticists. SNPs and SSRs are routinely used in animal and crop breeding programs (Gupta et al. 2001), for genetic diversity analysis, cultivar/breed identification, phylogenetic analysis, characterization of genetic resources and association of genetic loci with valuable traits (Rafalski 2002). Several tools have been developed for the discovery of SNPs and SSRs from DNA sequence data and a selection of these is listed below.

11.2.1.1 AutoSNPdb

AutoSNPdb combines SNP discovery software and sequence annotation with a relational database (Duran et al. 2008). The software implements the autoSNP SNP discovery pipeline (Barker et al. 2003; Batley et al. 2003) and a custom sequence annotation pipeline within a relational database to enable mining for SNP and indel polymorphisms associated with genes of known function or comparative physical map position. A non-redundant set of *Brassica* gene sequences were mined for SNPs, annotated by comparison with GenBank and UniRef90, and compared with the complete *Arabidopsis* genome sequence. A web interface enables searching and visualization of the data, including the display of sequence alignments and SNPs (Fig. 11-1). The tool allows users to identify SNPs between *Brassica* cultivars or within genes of predicted function. The database hosts details of 203,036 predicted *Brassica* SNPs in 169,721 unigene sequences and can be accessed at: *http://acpfg.imb.uq.edu.au/autosnpdb.php*.

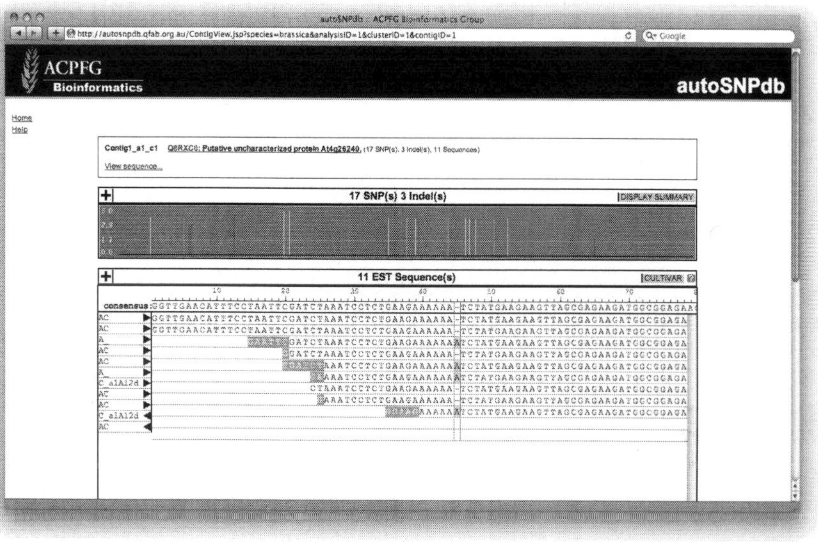

Figure 11-1 Web interface of autoSNPdb showing sequence alignment and SNPs.

11.2.1.2 SSRPrimer

SSRPrimer (Robinson et al. 2004; Jewell et al. 2006) provides an automated process to identify and design PCR primers for SSR loci by combining the SSR discovery tool Sputnik and the PCR primer design software Primer3. The scripts take multiple sequences in FASTA format as input and produce lists of SSRs and associated PCR primers in tabular format. This web-based tool is also available as a stand alone version for very large datasets. SSRPrimer has been successfully applied to a wide range of species including shrimp (Perez et al. 2005), citrus (Chen et al. 2006), mint (Lindqvist et al. 2006), strawberry (Keniry et al. 2006), *Brassica* (Burgess et al. 2006; Batley et al. 2007; Hopkins et al. 2007; Ling et al. 2007), *Sclerotinia* (Winton et al. 2007) and *Eragrostis curvula* (Cervigni et al. 2008). Analysis of public *Brassica* data identified 31,155 SSR molecular markers with associated PCR primers. The latest version of the SSRPrimer tool, SSRPrimerII is available at *http://acpfg.imb.uq.edu.au/SSRDiscovery.php*. Additionally, *Brassica* SSRPrimer data generated using SSRPRimer are available at *http://acpfg.imb.uq.edu.au/brassica.php*.

11.2.1.3 SSRPoly

SSRPoly was the first tool to be developed to identify polymorphic SSRs from DNA sequence data. The software consists of a set of Perl scripts and MySQL tables and can be implemented on UNIX, Linux and Windows platforms. The application of SSRPoly to *Brassica* EST sequences has led to the prediction of 1,528 polymorphic SSRs. The tool can be downloaded from *http://acpfg.imb.uq.edu.au/ssrpoly.php.*

11.2.2 Genetic Mapping

Insight into plant genome organization can be obtained by constructing genetic linkage maps using molecular genetic markers. Molecular markers offer the opportunity to identify the genetic locations of large numbers of regions that govern important agronomic traits, and the resulting molecular genetic maps provide a means to link heritable traits with underlying genome sequence variation. Genetic mapping places markers on linkage groups based on their segregation in a population. Sequence based markers, such as SSRs and SNPs, allow linking of heritable traits with regions of the genome providing a link to the underlying sequence variation associated with observed heritable phenotypic variation. There are several tools available for the generation of genetic maps and the mapping of heritable traits to these maps. These include MapMaker (Lander et al. 1987) and JoinMap (Stam 1993). In addition, there are repositories of *Brassica* genetic mapping data and associated viewers, these include CMAP (*http://acpfg.imb.uq.edu.au*) and the online *Brassica* repository brassica.info.

11.2.2.1 Brassica.info

Brassica.info is an online repository of *Brassica* information with links to external *Brassica* resources. Resources available from this site include details of genetic maps, markers, sequencing projects and transcriptome information. New data is constantly being collated and updated as it becomes available. This repository can be accessed at *http://www.brassica.info.*

11.2.2.2 Comparative Mapping

Comparative mapping based on the alignment of chromosomes using common molecular markers helps researchers translate information from one map to another and allows the transfer of knowledge from one genome to another related genome (Chao et al. 1989). Comparative mapping is of particular relevance to the breeding of the *Brassica* crops where conservation between the three *Brassica* genomes permits the transfer of knowledge between these diploids and also with the allopolyploid species.

Restriction fragment length polymorphism (RFLP), SSR and increasingly SNP markers are applied for comparative genetic mapping, since they are often transferable between related species. The linkage arrangement of markers can be compared between closely related species, if the same molecular markers are used for genetic mapping. This has been demonstrated in *Brassica,* where it has been shown that the linear order of genes is conserved over a large evolutionary timescale between the amphidiploid AB and AC genomes and the diploid progenitor genomes. Comparative genetic mapping may be extended to more divergent species. *Brassica* species are in the same family as *Arabidopsis thaliana,* and these genera diverged approximately 15–21 million years ago (Koch et al. 2000). DNA sequences of homologous genes are similar between the two taxa, and it is, therefore, possible to use markers from one species to map related loci in the other species. Several tools have been developed for comparative genetic mapping, the most common of which is CMap. In addition, the increasing availability of genome sequence data permits the linking of genetic maps with sequenced genomic regions of *Brassica* and related species. This is exemplified by the data within the AtEnsEMBL database. These tools will become more important as genome sequences become increasingly available.

11.2.2.3 CMAP Database

The *Brassica* CMAP database provides comparative genetic information for *Brassica* species. Marker information has been collated from various sources and is presented using a standard graphical interface (Fig. 11-2). The relative locations of markers on *Brassica* chromosomes are presented, and sequence based markers have also been mapped onto the *Arabidopsis* genome. The CMAP database can be accessed through *http://acpfg.imb.uq.edu.au/brassica.php.*

11.2.2.4 AtEnsembl

AtEnsembl is a database and genome viewer for *A. thaliana. Brassica* genomic information in the form of Unigenes and whole genome shotgun data from the *Brassica* C genome has been mapped onto the *Arabidopsis* viewer and can be seen in relation to *Arabidopsis* genomic information including gene expression data and genome insertion mutants. The mapping of *B. rapa* bacterial artificial chromosome (BAC) ends produced as part of the multinational *Brassica* genome sequencing project enables the rapid identification of Chinese cabbage BAC clones predicted to host genes that have been characterized in *Arabidopsis* (Love et al. 2005).

A

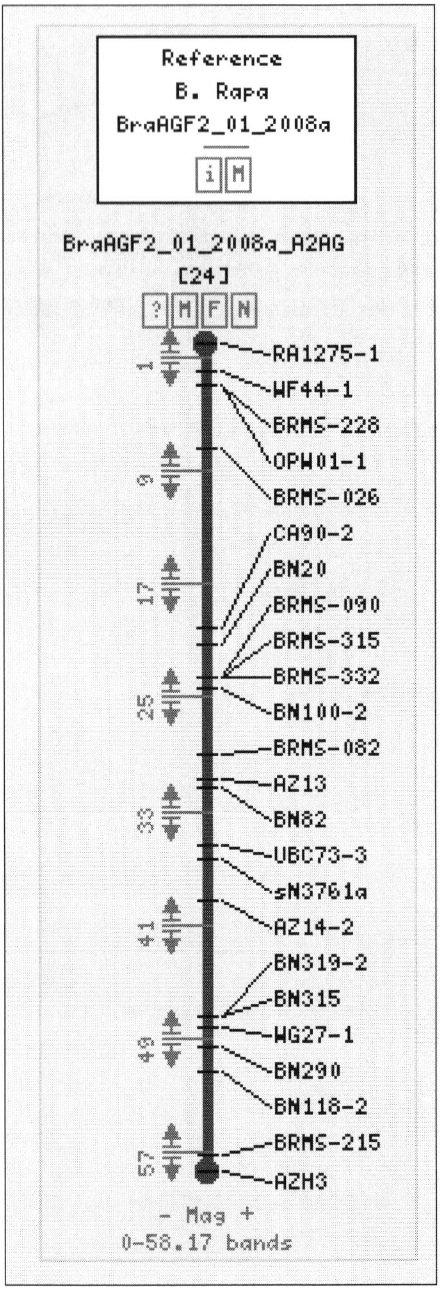
Reference
B. Rapa
BraAGF2_01_2008a
i M
BraAGF2_01_2008a_A2AG
[24]
? M F N
1
9
17
25
33
41
49
57
RA1275-1
WF44-1
BRMS-228
OPW01-1
BRMS-026
CA90-2
BN20
BRMS-090
BRMS-315
BRMS-332
BN100-2
BRMS-082
AZ13
BN82
UBC73-3
sN3761a
AZ14-2
BN319-2
BN315
WG27-1
BN290
BN118-2
BRMS-215
AZH3
- Mag +
0-58.17 bands

B

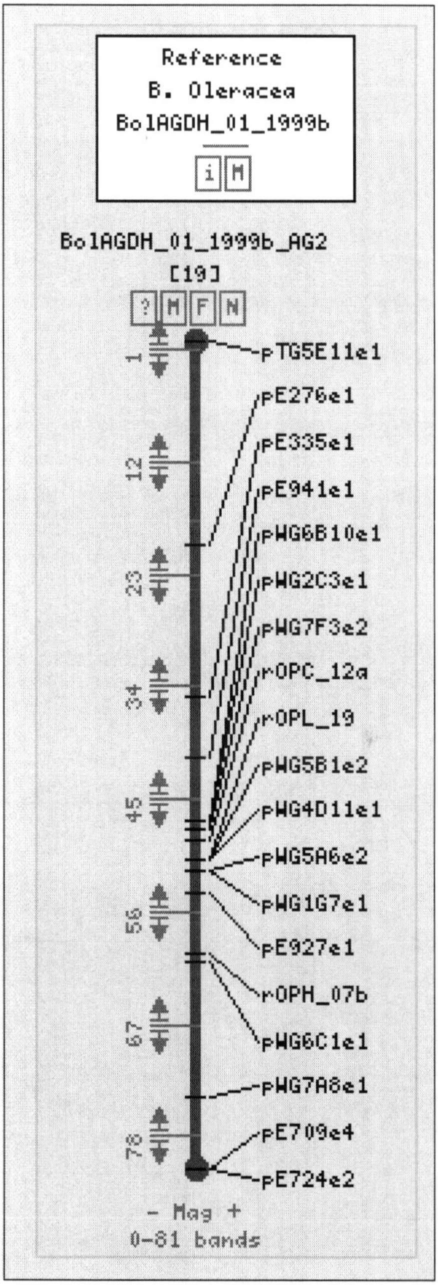

Figure 11-2 The standard graphical interface of the genetic map viewer CMAP showing mapping information for (a) *Brassica rapa* and (b) *Brassica oleracea.*

11.3 Genomic Resources

11.3.1 Brassica Genome Sequencing

The value of complete genome sequences for species is becoming increasingly recognized. Significant efforts are underway for the full genome sequencing of the important Asian vegetable Chinese cabbage (*B. rapa*). Currently there are two major *Brassica* sequencing projects in development as described below.

11.3.1.1 Brassica rapa Genome Sequencing

The Multinational *Brassica* Genome Sequencing Project aims to sequence the complete genome of Chinese cabbage using a BAC by BAC approach. Three BAC libraries have been produced and end sequences have been obtained for 110,000 BACs. Sequencing is underway for each of the 10 chromosomes and the sequences for several hundred BACs have been submitted to the public GenBank database, including sequences corresponding to the complete *Arabidopsis* genome. In addition to this project, several groups are undertaking whole genome shotgun sequencing of *B. rapa* using the Illumina GAII system. The Illumina sequencing is highly complementary to the BAC sequencing project and will assist in the final finishing and polishing of the completed genome sequence.

11.3.1.2 Brassica oleracea Genome Sequencing

In a collaboration between Cold Spring Harbor and TIGR, 0.5× coverage of the *B. oleracea* genome was produced using whole genome shotgun sequencing. These sequences have been mapped onto the *Arabidopsis* genome and the results of this comparison can be found in AtEnsEMBL.

11.3.2 Expressed Sequence Tags

Expressed sequence tags (ESTs) correspond to regions of expressed genes. ESTs can be used as a gene discovery tool and also to demonstrate which genes are expressed in a particular tissue at a specific time. Sequencing ESTs is a powerful method of gene discovery and gene sequence determination, and ESTs are also a valuable resource for SNP discovery. There are various resources available that provide information on *Brassica* ESTs (Table 11-1).

Table 11-1 Number of *Brassica* ESTs contained in different databases.

Database	**Number of *Brassica* ESTs**
dbEST at NCBI	~ 1 million
UniGene database at NCBI	~ 500,000
Cleansed EST database	~ 500,000
AutoSNPdb	131,266

11.3.2.1 dbEST

dbEST is an online repository of ESTs maintained at the NCBI. It hosts over one million *Brassica* ESTs and is the primary repository for all EST information. It can be accessed from *http://www.ncbi.nlm.nih.gov/dbEST/.*

11.3.2.2 UniGene Database at NCBI

A UniGene is derived from a group of ESTs that seem to come from the same gene. The Unigene database currently contains around 50,000 *Brassica* entries that can be accessed from *http://www.ncbi.nlm.nih.gov/sites/entrez?db=unigene.*

11.3.2.3 CleanEST Database

CleanEST (Lee and Shin 2008) represents a collation of EST libraries that are classified according to species and the sequencing center. It is called a cleansed EST database because it solves the problem of inconsistent terminology and the presence of contaminated sequences. The database currently contains about 500,000 *Brassica* entries and can be accessed at *http://cleanest.kobic.re.kr.*

11.3.2.4 AutoSNPdb Clusters

As well as SNP information, AutoSNPdb (Duran et al. 2008) also hosts information on 825,195 ESTs that assemble into169,721 unigenes consisting of 38,455 contigs and 131,266 singletons. .

11.3.3 Gene Expression Information

Along with knowing the gene content of *Brassica,* it is important to understand which genes are expressed in each of the different tissues at different times and under different environmental conditions. Gene expression microarrays are the favored tool for such high-throughput gene expression analysis. Several independent formats of gene expression microarray have been developed for *Brassica* species, including Agilent (*http://brassica.bbsrc.ac.uk/array_info.html*), Nimblegen (*http://www.intl-pag.org/16/abstracts/PAG16_P07a_708.html*), Combimatrix (*http://www.brassica-rapa.org/BGP/NC_DNA_chip.jsp*) and spotted cDNA arrays (*http://www.intl-pag.org/14/abstracts/PAG14_P740.html*). In addition to *Brassica* specific arrays, the *Arabidopsis* Affymetrix GeneChip has also been successfully applied for *Brassica* transcriptomics (*http://affymetrix.arabidopsis.info/xspecies/*). It is expected that with the increased availability of second generation sequencing technology this would be applied for transcriptomics in *Brassica*. However there are currently no published reports of this application.

11.4 Future Prospects

There are many tools and information resources available for genetic and genomic analysis in *Brassica*. However, the everchanging technologies will demand increasingly complex and powerful systems to translate the massive amount of data being generated to improve *Brassica* vegetable crops. The principle challenges and opportunities come from the application of new sequencing technologies. One may predict that in the near future, researchers will be able to sequence the genome of varieties or wild relatives to identify the underlying basis of phenotypic variation, and then use this information directly for crop improvement. This will require changes in the way we manage *Brassica* data, particularly phenotypic data, to ensure that robust associations are made between the emerging genome sequences and valuable agronomic traits.

References

Altschul SF, Gish W, Miller W, Myers EW, Lipman DJ (1990) Basic local alignment search tool. J Mol Biol 215: 403–310.

Barker G, Batley J, O' Sullivan H, Edwards KJ, Edwards D (2003) Redundancy based detection of sequence polymorphisms in expressed sequence tag data using autoSNP. Bioinformatics 19: 421–422.

Batley J, Barker G, O'Sullivan H, Edwards KJ, Edwards D (2003) Mining for single nucleotide polymorphisms and insertions/deletions in maize expressed sequence tag data. Plant Physiol 132: 84–91.

Batley J, Hopkins CJ, Cogan NOI, Hand M, Jewell E, Kaur J, Kaur S, Li X, Ling AE, Love C, Mountford H, Todorovic M, Vardy M, Walkiewicz M, Spangenberg GC, Edwards D (2007) Identification and characterization of simple sequence repeat markers from *Brassica napus* expressed sequences. Mol Ecol Notes 7: 886–889.

Burgess B, Mountford H, Hopkins CJ, Love C, Ling AE, Spangenberg GC, Edwards D, Batley J (2006) Identification and characterization of simple sequence repeat (SSR) markers derived in silico from *Brassica oleracea* genome shotgun sequences. Mol Ecol Notes 6: 1191–1194.

Cervigni GD, Paniego N, Díaz M, Selva JP, Zappacosta D, Zanazzi D, Landerreche I, Martelotto L, Felitti S, Pessino S, Spangenberg G, Echenique V (2008) Expressed sequence tag analysis and development of gene associated markers in a near-isogenic plant system of *Eragrostis curvula*. Plant Mol Biol 67: 1–10.

Chao S, Sharp PJ, Worland AJ, Warham EJ, Koebner RMD, Gale MD (1989) RFLP-based genetic maps of wheat homologous group-7 chromosomes. Theor Appl Genet 78: 495–504.

Chen C, Zhou P, Choi YA, Huang S, Gmitter FG Jr (2006) Mining and characterizing microsatellites from citrus ESTs. Theor Appl Genet 112(7): 1248–1257.

Duran C, Appleby N, Clark T, Wood D, Imelfort M, Batley J, Edwards D (2008) AutoSNPdb: an annotated single nucleotide polymorphism database for crop plants. Nucl Acids Res 37: 951–953.

Gupta PK, Roy JK, Prasad M (2001) Single nucleotide polymorphisms: A new paradigm for molecular marker technology and DNA polymorphism detection with emphasis on their use in plants. Curr Sci 80: 524–535.

Hopkins CJ, Cogan NOI, Hand M, Jewell E, Kaur J, Li X, Lim GAC, Ling AE, Love C, Mountford H, Todorovic M, Vardy M, Spangenberg GC, Edwards D, Batley J (2007) Sixteen new simple sequence repeat markers from *Brassica juncea* expressed sequences and their cross-species amplification. Mol Ecol Notes 7: 697–700.

Jewell E, Robinson A, Savage D, Erwin T, Love CG, Lim GAC, Li X, Batley J, Spangenberg GC, Edwards D (2006) SSRPrimer and SSR Taxonomy Tree: Biome SSR discovery. Nucl Acids Res 34: 656–659.

Keniry A, Hopkins CJ, Jewell E, Morrison B, Spangenberg GC, Edwards D, Batley J (2006) Identification and characterization of simple sequence repeat (SSR) markers from Fragaria x ananassa expressed sequences. Mol Ecol Notes 6: 319–322.

Koch M, Haubold B, Mitchell-Olds T (2000) Evidence for homology of flowering time genes VFR2 from Brassica rapa and FLC from *Arabidopsis thaliana*. Theor Appl Genet 102: 425–430.

Lander ES, Green P, Abrahamson J, Barlow A, Daly MJ, Lincoln SE, Newberg LA (1987) MAPMAKER: an interactive computer package for constructing primary genetic linkage maps of experimental and natural populations. Genomics 93: 398.

Larkin MA, Blackshields G, Brown NP, Chenna R, McGettigan PA, McWilliam H, Valentin F, Wallace IM, Wilm A, Lopez R, Thompson JD, Gibson TJ, Higgins DG (2007) Clustal W and Clustal X version 2.0. Bioinformatics 23: 2947–2948.

Lee B, Shin G (2008) CleanEST: a database of cleansed EST libraries. Nucl Acids Res 37: 686–689.

Lim GAC, Jewell EG, Li X, Erwin TA, Love C, Batley J, Spangenberg G, Edwards D (2007) A comparative map viewer integrating genetic maps for *Brassica* and *Arabidopsis*. BMC Plant Biol 7: 40.

Lindqvist C, Scheen AC, Yoo MJ, Grey P, Oppenheimer DG, Leebens-Mack JH, Soltis DE, Soltis PS, Albert VA (2006) An expressed sequence tag (EST) library from developing fruits of an Hawaiian endemic mint (*Stenogyne rugosa, Lamiaceae*): characterization and microsatellite markers. BMC Plant Biol 6: 16.

Ling AE, Kaur J, Burgess B, et al. (2007) Characterization of simple sequence repeat markers derived *in silico* from *Brassica rapa* bacterial artificial chromosome sequences and their application in *Brassica napus*. Mol Ecol Notes 7: 273–277.

Love CG, Robinson AJ, Lim GA, Hopkins CJ, Batley J, Barker G, Spangenberg GC, Edwards D (2005) Brassica ASTRA: an integrated database for *Brassica* genomic research. Nucl Acids Res 33: 656–659.

Pérez F, Ortiz J, Zhinaula M, Gonzabay C, Calderón J, Volckaert FA (2005) Development of EST-SSR markers by data mining in three species of shrimp: *Litopenaeus vannamei, Litopenaeus stylirostris,* and *Trachypenaeus birdy*. Mar Biotechnol 7: 554–569.

Rafalski JA (2002) Novel genetic mapping tools in plants: SNPs and LD-based approaches. Plant Sci 162: 329–333.

Robinson AJ, Love CG, Batley J, Barker G, Edwards D (2004) Simple sequence repeat marker loci discovery using SSR primer. Bioinformatics 20(9): 1475–1476.

Savage D, Batley J, Erwin T, Logan E, Love CG, Lim GAC, Mongin E, Barker G, Spangenberg GC, Edwards D (2005) SNPServer: a real time SNP discovery tool. Nucl Acids Res 13: 493–495.

Stam P (1993) Construction of integrated genetic-linkage maps by means of a new computer package—JoinMap. Plant J 3: 739–744.

Vos P, Hogers R, Bleeker M, Reijans M, van de Lee T, Hornes M, Frijters A, Pot J, Peleman J, Kuiper M, Zabeau M (1995) AFLP—A new technique for DNA-fingerprinting. Nucl Acids Res 23(21): 4407–4414.

Wenzl P, Carling J, Kudrna D, Jaccoud D, Huttner E, Kleinhofs A, Kilian A (2004) Diversity Arrays Technology (DArT) for whole-genome profiling of barley. Proc Natl Acad Sci USA 101: 9915–9920.

Winton LM, Krohn AL, Leiner RH (2007) Microsatellite markers for *Sclerotinia subarctica* nom. prov., a new vegetable pathogen of the High North. Mol Ecol Notes 7: 1077–1079.

Index

C

P

Q

R

Color Plate Section

Chapter 2

Figure 2-1 The shape of "Aikang qing"(left), "Shanghai qing"(middle) and "Siyue man" (The photo was presented by a Chinese researcher Tianhwa Fu).

Figure 2-2 The shape of "Seoul" Chinese cabbage.

Figure 2-4 A rose shaped head (A) and semi-folded head (B) of hybrid cultivars in Chinese cabbage.

Figure 2-5 Yellow inner leaf colored spring Chinese cabbage.

Figure 2-6 Winter cultivation area in South-coast province of Korea.

Figure 2-7 Procedure of inoculation method of clubroot in Chinese cabbage (A; Seed germination, B; Input 20 mL of inoculm, C; Trans planting 15 day-old seedlings using 50 pot plug tray and urethane sponge.

Figure 2-8 Hydroponic culture of clubroot inoculated Chinese cabbage and the symptoms. (A; 2 weeks after transplanting, B; 6 weeks after transplanting, C; Symptoms of clubroot, left; sensitive, middle; medium sensitive, right; resistance).

Chapter 6

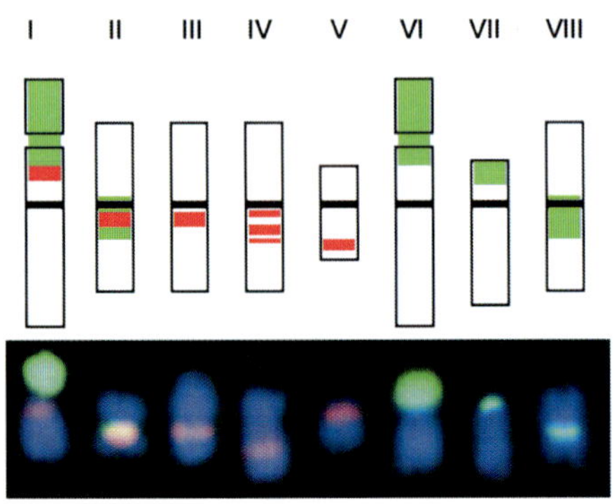

Species	Genome	A	A	B	C	A	B,C	B,C	A
B. nigra	BB	-	-	1	-	-	2	1	-
B. oleracea	CC	-	-	-	1	-	1	1	-
B. rapa	AA	1	1	-	-	1	-	-	3
B. carinata	BBCC	-	-	1	1	-	2	2	-
B. juncea	AABB	1	2	1	-	1	2	1	2
B. napus	AACC	1	3	-	1	1	1	1	1

Figure 6-5 Types of mitotic chromosomes with 5S rDNA (red) and 45S rDNA (green) and their occurrence within the genomes of six *Brassica* species of the U triangle. Chromosome types (Roman numerals) are numbered according to Hasterok et al. (2001). Modified from Hasterok et al. (2001, 2006) with permission.

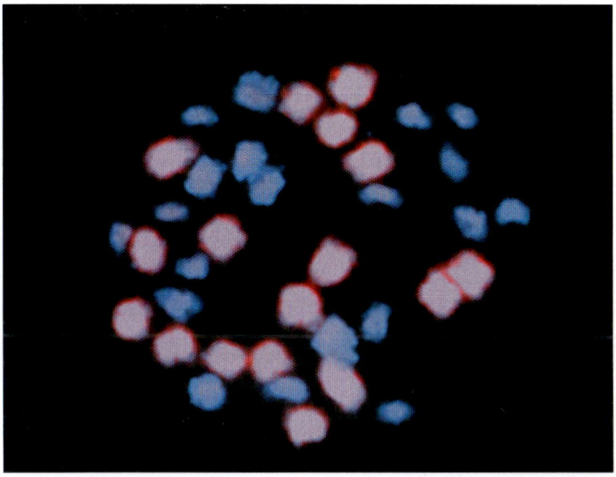

Figure 6-7 GISH in *Brassica napus*. GISH with labeled genomic DNA of *B. oleracea* (red) revealed 18 mitotic chromosomes corresponding to the C genome. Unlabeled DNA of *B. rapa* (A genome) and the intergenic spacer region of *B. oleracea* 45S rDNA were used as a blocking DNA. Chromosomes counterstained with DAPI. Courtesy of S. J. Armstrong (University of Birmingham).

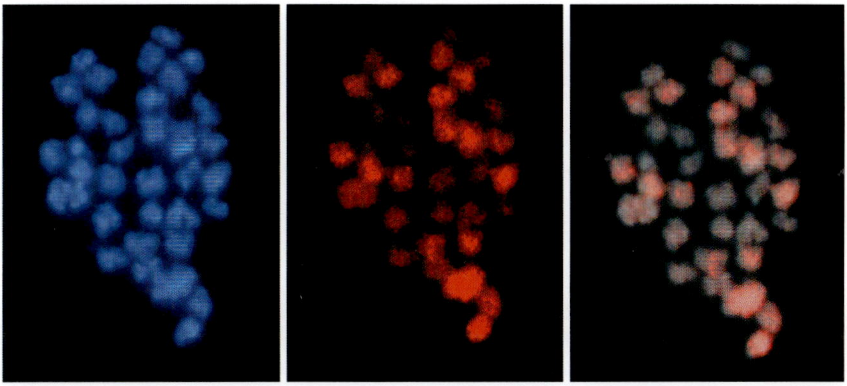

Figure 6-8 Fluorescence in situ hybridization of a CACTA transposon *Bot1* to metaphase chromosomes of *B. napus* cv. "Yudal". The *Bot1* probe (red) labels 18 chromosomes corresponding to the C genome; only weak hybridization to 20 chromosomes of the A genome is apparent. Chromosomes counterstained with DAPI and shown in blue (left) or gray (right). Adapted from Alix et al. (2008) with permission.

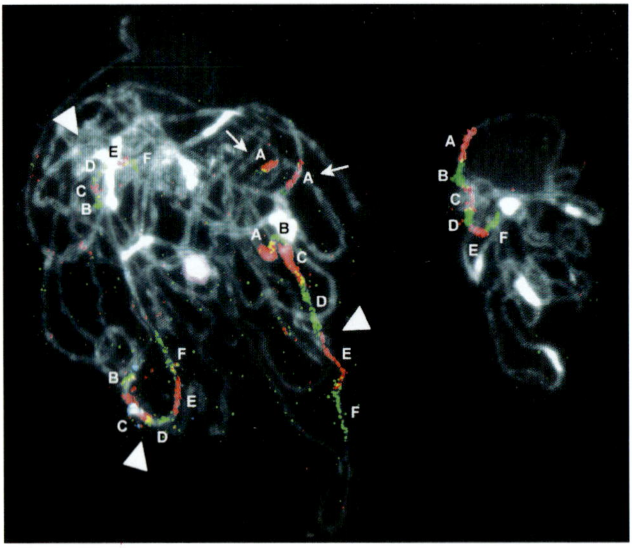

Figure 6-9 An example of comparative chromosome painting in *B. oleracea*. A 5.4 Mb long Arabidopsis BAC contig from chromosome 3 divided into six differentially labeled segments, was hybridized to pachytene chromosomes of *A. thaliana* (right) and *B. oleracea* (left). Whilst the BAC contig reveals a single chromosome region in *A. thaliana*, three copies of the contig were found in *B. oleracea* (arrowheads). Bioinformatic and genetic mapping studies suggest that the arrangement of A-F segments observed in *A. thaliana* has ancestral-like structure [(Henry et al. 2006, also data from *http://chibba.agtec.uga.edu/duplication/index/home* (Tang et al. 2008)]. This is observed in one of the three *B. oleracea* copies, as well. However, segment A was separated from segments B-F in the two other copies (dissociated A segments indicated by arrows). Modified from Ziolkowski et al. (2006).

Chapter 7

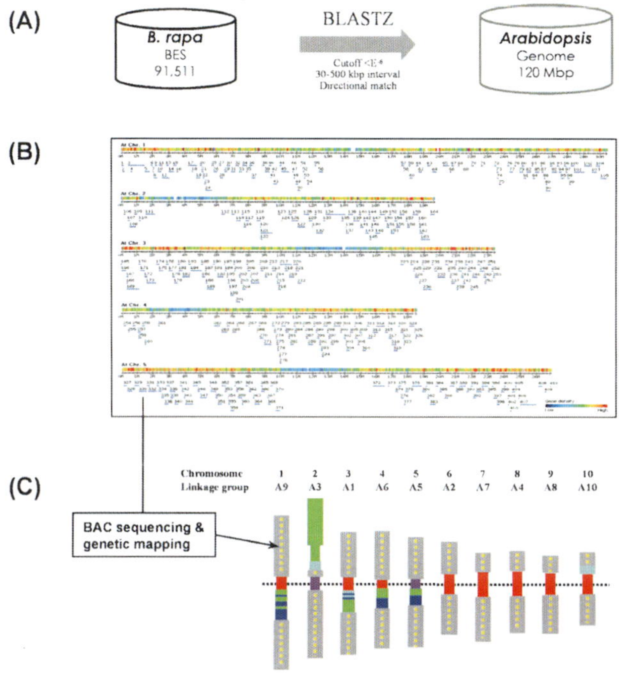

Figure 7-2 Schematic representation of the comparative tiling sequencing method to select seed BAC clones from the euchromatic regions of *B. rapa*, using in silico matching of BESs with known *Arabidopsis* sequences. (A) Based on the knowledge of the sequence level collinearity between *Arabidopsis* and *Brassica*, BAC-end sequences of *B. rapa* ssp. *pekinensis* cv. *Chiifu* were BLAST analyzed against *Arabidopsis* genome sequence and the corresponding BACs were selected based on the pair-end best hits within 30–500 kbp interval. (B) Selection of minimally overlapping 589 BAC clones based on *in silico* comparative allocation on the *Arabidopsis* genome. A total of 4,317 *B. rapa* BAC clones with an average insert size of 120 kbp were mapped in silico onto the euchromatic regions of the *Arabidopsis* genome. On in silico comparative allocation, 589 minimally overlapping BAC clones were mapped onto the comparative positions in the corresponding *Arabidopsis* genome sequence, then were selected and sequenced as phase 2 and scattered on the euchromatic regions of the 10 *B. rapa* chromosomes as seed BACs by a combination of genetic and cytogenetic mapping. Mixed-color bar represents each chromosome of *Arabidopsis* and blue bars depict *B. rapa* BAC contigs tiled on the counterparts of *Arabidopsis* genome. (C) Based on the sequence of each BAC clone, we developed STS markers and anchored the clones on the genetic maps. Overall, 513 sequenced BAC clones were mapped onto the 10 linkage groups and provided to each sequencing center of the BrGSP for completion of the chromosome sequencing. Detailed numbers of BAC clones on each chromosome are denoted in Table 7.3.

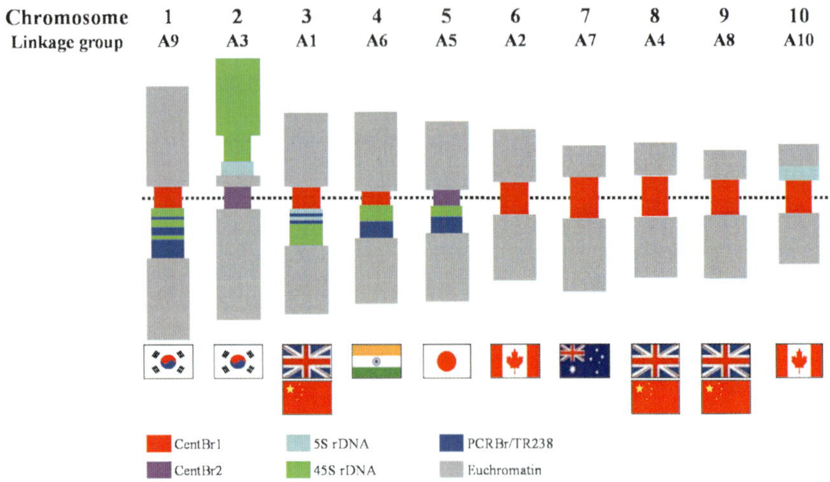

Figure 7-3 Idiogram of *B. rapa* chromosomes represented by euchromatic arms and heterochromatic repetitive sequence blocks. Each chromosome was assigned to a sequencing center in one of the seven countries of the BrGSP consortium.

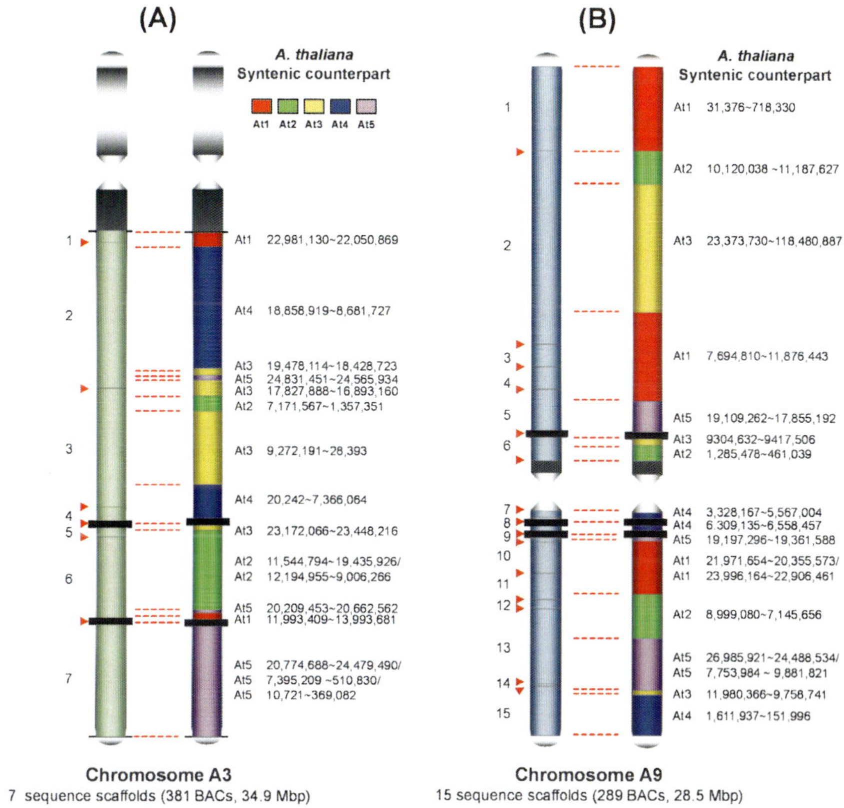

Figure 7-4 Current sequencing status of the two longest chromosomes, A3 and A9. BAC clones were sequenced as phase 2 and assembled into sequence scaffolds. (A) A total of 381 minimally tiled BAC clones generated seven sequence scaffolds spanning 34.9 Mbp for chromosome A3. (B) 15 scaffolds spanning 28.5 Mbp were assembled from 289 minimum tiled BAC clones for chromosome A9. Syntenic counterpart regions of the *A. thaliana* genome identified by sequence comparison were also represented in the colored boxes, along with position information. Red arrow heads indicate major gaps between scaffolds.

Chapter 8

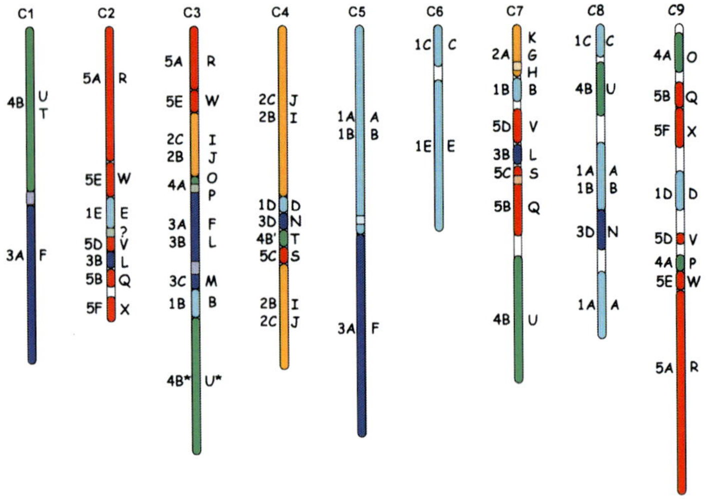

Figure 8-5 The *Brassica oleracea* genome (C1–C9) generated from the conserved ancestral genomic blocks. The blocks are coloured according to the corresponding homologous region of *A. thaliana*: chromosome 1—light blue; chromosome 2—orange; chromosome 3—dark blue; chromosome 4—green; and chromosome 5—red. The names of the blocks are shown to the left of each Brassica linkage group according to Parkin et al (2005) and their equivalent regions as described by Schranz et al (2006) are shown to the right.